FRAMINGHAM STATE COLLEGE
3 3014 00058 1241

D1191536

# EVOLUTION IN THE HIGHER BASIDIOMYCETES

# EVOLUTION IN THE HIGHER BASIDIOMYCETES

*An International Symposium*

*Ronald H. Petersen*

EDITOR

THE UNIVERSITY OF TENNESSEE PRESS

KNOXVILLE

Library of Congress Catalog Card Number 73-100410
Standard Book Number 87049-109-1

# PREFACE

Perhaps no higher praise can be given a scholar and his work than for him to be honored by eminent specialists of his own discipline. Such a tribute was paid to Lexemuel Ray Hesler in August, 1968, when, on the occasion of his eightieth birthyear, mycologists from Holland, Great Britain, France, Mexico, Canada, Czechoslovakia, and from many areas of the United States assembled on The University of Tennessee campus in Knoxville to participate in the symposium on which this volume is based. The meeting honoring Dr. Hesler proved to be highly appropriate, for the papers presented—and indeed the discussions—reflected a liveliness of intellect and enthusiasm for learning that have characterized the honoree throughout his long career.

Over the years both botanical research and university administration have been enriched by L. R. Hesler's contributions as teacher, research worker, administrator, writer, and, above all, avid student of fungi. Following undergraduate preparation at Wabash College, his interest in higher fungi was awakened in graduate school at Cornell University where he later took his first job as assistant professor of plant pathology. It was not until 1925, however, six years after moving to The University of Tennessee as Head of the Department of Botany, that his attention came to be focused more narrowly on the agarics. A hot, wet June and the curiosity of Knoxville residents combined to effect the change. Dr. Hesler's phone rang frequently that summer with questions, mainly from the ladies of Knoxville, inquiring about the identity and edibility of the mushrooms and other fungi springing up in their yards and vacant lots. As a result, the youthful botanist turned to the field and the laboratory to provide the answers, and these two sources—field and laboratory—have been his academic homes ever since.

There were times during his long and distinguished career at the University—first as Head of the Department of Botany and then for twenty-four years as Dean of the College of Liberal Arts—when his interest in the agarics had to be subordinated to administrative responsibilities. Even so, he managed throughout those years to continue his research and to present the results of his findings to the profession. He collaborated in writing several laboratory manuals and published more than 50 technical articles and reports. But the large segments of time necessary for the completion of book-length manuscripts were not available to him. Thus, his formal retirement in 1958 actually opened the way to fulfillment of a long-cherished dream. Since that time, productively busy in the field and in the laboratory of Hesler Biology Building, he has written or collaborated in writing six books, and another is on the way. His first was the popular *Mushrooms of the Great Smokies*, and that was followed by North American floral monographs on *Hygrophorus*, *Crepidotus*, *Pholiota* (all with Alexander H. Smith), *Gymnopilus*, and *Entoloma* for the Southeast.

The papers collected here provide a permanent record of the best thinking of some of the world's leading systematists of fungi. The material not only tells us "where we are" (or think we are) in several significant areas, but it should also stimulate the yield of new data and techniques and suggest new hypotheses to be tested. No attempt is made here to summarize or evaluate the papers presented, but perhaps the reader's attention should be directed to an underlying current (treated more fully in the *BioScience** report on the symposium) which became apparent as the symposium progressed and which emphasized that the meeting came at a time of change. Although participants made repeated references to the dubiousness of the Friesian scheme, especially in its apparent lack of phylogenetic cohesiveness, no other formalized system was proposed to take its place. In fact, much time and content were devoted to breaking from the old, with the admission that patterns were only now beginning to emerge to supplant it. But, for the first time, hitherto tangential character-types were accepted without question as equal to the time-tested (and stressed) morphological fruit body features.

Both tape-recorded and manuscript sources were utilized in preparing the final manuscript for this volume. The twenty-one papers and discussions were given in several languages, and the discussions

* Vol. 19, no. 1:73–75. 1969.

often continued for forty-five minutes or more, frequently with much enthusiasm. Thus, the printed rendition of these proceedings has been subject to the foibles of the spoken word, as well as to the imprecision of translation. The editor hopes, however, that in this work and in his further editing of the discussions for continuity and conciseness, he has succeeded at least modestly in keeping the comments and concepts of the participants faithful to the proceedings. Help with this task was given generously by many people: the authors, other participants, and colleagues in the Department of Botany at The University of Tennessee, especially Dr. Raymond Holton, who provided suggestions as well as released time for the program. To all of these, and to Mrs. Alida Sales, who typed the manuscript, I offer my sincere thanks. The assistance of the National Science Foundation is, of course, most gratefully acknowledged, for without its financial support the program could not have taken place.

R. H. P.

L. R. Hesler

*Participants:*

1. J. R. Raper
2. René Pomerleau
3. R. H. Petersen
4. M. A. Donk
5. Rolf Singer
6. Albert Pilát
7. D. P. Rogers
8. D. A. Reid
9. K. A. Harrison
10. A. H. Smith
11. Noël Arpin
12. O. K. Miller, Jr.
13. V. E. Tyler, Jr.
14. H. D. Thiers
15. Jacques Boidin
16. R. L. Gilbertson
17. M. K. Nobles
18. Edward Hacskaylo
19. M. P. Christiansen
20. Roger Heim (*inset*)

# CONTENTS

## ILLUSTRATIONS

# I

## INTRODUCTORY ADDRESS

M. A. DONK
*Senior Mycologist*
*Rijksherbarium, Leiden, Holland*

# PROGRESS IN THE STUDY OF THE CLASSIFICATION OF THE HIGHER BASIDIOMYCETES

❦

Not so very long ago, barely two and a half centuries, a few botanists set themselves to mapping out a new world, the Kingdom of Fungi. This required a considerable measure of daring, tenacity of purpose, and formidable underestimation of the difficulties ahead. Among the early explorers there was one of outstanding merit—Persoon—who sketched one of the first maps. It was his map that served as a basis for later elaboration until toward the end of the last century the task seemed almost finished, even in detail, at least as far as the Hymenomycetes were concerned. For this group the Friesian version had become a tradition, not in the least due to its having been accepted in Saccardo's "Sylloge."

This lecture aims at showing on broad lines how the process of transmutation of the Friesian classification of the Hymenomycetes into a more natural one is now operating.

## The Friesian Tradition versus Modern Trends: an Unfinished Struggle

It was inevitable that for a long time mycology should concentrate principally on cataloguing species, in the first place on those of the macromycetes—fungi with big fruit bodies. This preliminary work on the Hymenomycetes was done without the aid of the microscope, with the consequence that the most important features used in classifying these fungi were visible to the naked eye. It culminated in what we now call the Friesian system, with its subsequent modifications.

The principal characters emphasized by Persoon and retained by Fries were found in the hymenial configuration. In his final classification Fries (1874) stuck to the early directives and divided what is now called the Aphyllophorales and Agaricales into five families. The

Clavariaceae, with erect fruit body and amphigenous hymenium, was opposed to the others, all of which had unilateral hymenium, viz., Thelephoraceae, Hydnaceae, Polyporaceae, and Agaricaceae, with smooth, toothed, tubular, and lamellate hymenophore, respectively. It has become customary to supplement these by at least two further families, the Meruliaceae and the Cantharellaceae, with reticulately and radiately veined hymenium, respectively. Some other artificial families sprouting from the Friesian tradition are the Corticiaceae, Stereaceae, and Cyphellaceae; these were separated from the Thelephoraceae and are primarily based on the shape of the fruit body.

It is striking how few genera the Friesian scheme needed; an enormous number of species were stuffed into such genera as *Agaricus*, *Polyporus*, and so on. Fries and his followers were confirmed "lumpers" at the generic level. It is therefore not surprising that there developed the tendency to ease the situation by fragmenting genera. The most important "splitters" at this stage were Karsten, Schroeter, Quélet, and Murrill. Yet these authors were not really true innovators—they did little more than raise to generic rank groups that Fries had previously distinguished within his genera which relied primarily on outer characters. It is true that in many cases these authors added to the generic diagnostic character certain features of the spores and occasionally of the cystidia, usually microscopic characters of which Fries had hardly been aware. The revised definitions of *Lactarius* and *Russula* are examples. However, these genera had been recognized at an early date; it was the generic characters that were improved. Only rarely did this work lead to highly original natural delimitations, as was the case with *Rhodophyllus* (or *Entoloma s.l.*) by Quélet.

The results of a closer study of the basidia, in the first place by the Tulasne brothers, did not really shatter faith in the Friesian genera. The various basidial types they described appeared in the main to correspond rather well with already accepted genera. This line of research merely provided a basis for dividing the Hymenomycetes into an increased number of families, and eventually orders. Those now referred to the Heterobasidiae will be left out of the discussion.

Although trust in the various hymenial configurations as a suitable basis for the classification of the Hymenomycetes should have been badly shaken almost from the start by the increasing number of intermediate forms that became known, together with the great variation displayed by many species, Fries's scheme has been maintained until the present day. Deviation from it, or from some of its time-honored emendations, still arouses suspicion in the minds of some mycologists.

It is the lure of simplicity that keeps the Friesian tradition alive. The orderly mind is easily satisfied by it. The families form an ascendant series based on the hymenial configuration and its efforts to increase the sporulating surface; and at a lower level the species or genera again form ascending series based on the shape of the fruit body, often from thinly effused crusts to centrally stalked caps, from simple to compound, from primitive to advanced. All of this also satisfied the didactically minded evolutionist and the authors of textbooks.

## The Patouillardian Era

The first really important iconoclast who earnestly tried to free himself from the Friesian tradition was Patouillard. In his *Essai taxonomique* (1900) he gave a synthesis of a strongly modified classification of the Hymenomycetes. This work is now unanimously considered to be the pinnacle of his achievements. He was also a splitter, but his new genera, or emendations of genera, bear the stamp of originality; most of them have survived, although for nomenclatural reasons often under different names. Still more important for the classification of the Hymenomycetes were his drafts of some new groups above the rank of genus. His *Série des Igniaires* and his *Série des Phylactéries* were comprised of genera with strongly diverse hymenial configuration. They formed the bases of the modern families of Hymenochaetaceae and the strongly emended Thelephoraceae.

The advantages as well as the defects of Patouillard's system were clearly exposed when it was applied in Bourdot and Galzin's magnum opus (1928). The exceptionally high standard of that work made it indispensable to the students of the Heterobasidiae and the Aphyllophorales and inevitably it directed attention to Patouillard's system, which had scarcely become known and appreciated outside France. That Lloyd ridiculed Patouillard was undoubtedly one of the reasons why Patouillard's system was ignored by North American and many other mycologists for a long time. It has been only during the last few decades that his classification has inspired a later generation of mycologists, and this has led, unavoidably, to a rapid increase in the number of genera and families.

It is now generally agreed that the Friesian system was built up of life-forms or grades; it was mainly physiognomic, which really means that its taxa were not "natural." A close survey of present practice shows that there is still a strong tendency to cling stubbornly to an elaborated Friesian version, like the treatment presented by Killermann (1928), whereas at the same time "radicals" are proposing far-reaching changes which are often only too eagerly accepted, without sufficient critical examination. Perhaps nowhere among the fungi is this disjunction as pronounced as in the Aphyllophorales. For this group many authors keep to the traditional 7 families of Aphyllophorales, but I am inclined to accept 23 families on a more-or-less tentative basis, in the firm belief that more will follow (Donk, 1964). It should be borne in mind that the breaking up of some of the big Friesian conglomerates has not yet really begun. Still found among the 23 families are such chaotic masses as the Corticiaceae, Stereaceae, and the residues of the Polyporaceae, Hydnaceae, and Clavariaceae.

In table 1 the augmented set of families is contraposed to the traditional one in such a way that it can be seen from which of the traditional families the contents of each newly defined family are derived. By this method an arrangement according to the time-honored types of hymenial configuration from smooth to lamellar is arrived at again. It is notable that in several of the newly defined families the full scale of hymenial configuration is entirely or at least largely represented. This repeatedly recurring tendency should warn the phylogenetically minded mycologist that hymenophoral differentiation might well be easily attainable during evolution and for this reason could very well be of low taxonomic value.

As long as it remains impossible to offer an advanced new classification, the only solution for improving the still current elaboration of the Friesian classification is to compromise and introduce into it such innovations as appear acceptable to the informed specialist.

## Several Theoretical Considerations

There are many difficulties in store for the taxonomist who sets himself to improve upon the Friesian system of the Hymenomycetes.

TABLE 1. DISTRIBUTION OF HYMENIAL CONFIGURATION OVER THE FAMILIES OF APHYLLOPHORALES.

| Traditional families / Modern families | Clavaria-ceae | Thelepho-raceae | Hydnaceae | Polypora-ceae | Merulia-ceae & Cantha-rellaceae | Agarica-ceae |
|---|---|---|---|---|---|---|
| Auriscalpiaceae | | | √ | | | √ |
| Bankeraceae | | | √ | | | |
| Bondarzewiaceae *s. str.* | | | | √ | | |
| Cantharellaceae | | √ | | | √ | |
| Clavariaceae* | √ | (√)[1] | | | | |
| Clavulinaceae | √ | | | | | |
| Coniophoraceae | | √ | √ | √ | √ | |
| Corticiaceae* | (√)[2] | √ | √ | √ | √ | |
| Echinodontiaceae | | | √ | | | |
| Fistulinaceae | | | | √[3] | | |
| Ganodermataceae | | | | √ | | |
| Gomphaceae | √ | √ | √ | | √ | |
| Hericiaceae | √ | √ | √ | | | |
| Hydnaceae* | | | √ | | | |
| Hymenochaetaceae | √ | √ | √ | √ | | √ |
| Lachnocladiaceae | √ | √ | √ | | | |
| Polyporaceae* | | | | √ | | √ |
| Punctulariaceae | | √ | | | | |
| Schizophyllaceae | | √ | | | √ | (√)[4] |
| Sparassidaceae | √ | √[1] | | | | |
| Stereaceae* | | √ | | | | |
| Thelephoraceae | √ | √ | √ | √ | √ | √ |
| Tulasnellaceae | | √ | | | | |

* Artificial families.
[1] Effused conditions of fruit bodies reported.
[2] Clavaroid conditions of fruit bodies reported.
[3] Mutually free tubes.
[4] The so-called gills in one genus (*Schizophyllum*) are not comparable to normal gills.

Some of the problems I mention might appear to be far-fetched, but I firmly believe that awareness of them is fundamental to a correct readjustment in approach.

It has long been known that in the laboratory the agaric *Armillaria mellea* may be induced to form haploid mycelia on which basidia are produced (Kniep, 1911). This state is a far cry from the fruit body as we know it from nature. Assuming that this corticioid form also occurs occasionally in the field, the mycologist will be confronted with a complete negation of the agaric model; instead of dealing with a fruit body consisting of stalk, cap, gills, hymenium, and basidia, he is faced with basidia only, and haploid at that. He will have to search for some reliable characters in this corticioid state to be able to recognize it with certainty the next time. But this is not all; he will also have to adjust the specific description of *A. mellea* in such a way as to include it.

When Molliard (1909, 556) had succeeded in obtaining in culture scattered basidia on a hypochnoid mycelium of *Fistulina hepatica*, he may have been one of the first mycologists to raise the problem. "It seems to me," he wrote (translated from the French), "that this fact poses an important question in connection with the group of the Hypochnae. When very highly differentiated Basidiomycetes are capable of producing diffuse hymenia under certain conditions, then it is really permissible to ask whether among the forms classified among the Hypochnae there might not be at least a few that belong to genera which were defined by their aggregate state."

Another example: *Grifola* [*Polyporus*] *frondosus* is a big polypore with many caps springing from a common, thick stalk. It may produce a resupinate and basidiferous, or so-called *Poria*-, state in culture (Cartwright, 1940). Suppose that this species had first become known through this condition and had been described as a species of *Poria*; then the discovery of the completely developed fruit body in nature would have required the transfer of the fungus to a quite different genus.

These examples may seem to have been sought out by design, but have they? I believe that in an only slightly diluted form they are characteristic of the present-day taxonomy of the Aphyllophorales. For however this may be, variations of the fruit body within a species are sometimes enormous. This is well known. The series that in many a modern family can be arranged all the way from corticioid fruit bodies to centrally stalked caps may also be encountered within the limits of a single species. A case in point is *Merulius* [*Serpula*] *lacrymans*: its fruit body may be a thin, effused membrane, but it may also be effuso-reflexed, or sessile and thick, or even assume the shape of stalked caps. *Polyporus* [*Abortiporus*] *biennis* usually forms more-or-less centrally stalked big fruit bodies, but it also occurs in nature as a thin poria and in all the intermediate stages. *Sebacina incrustans*, a tremellaceous species, is often strictly effused, but it also commonly forms cristate outgrowths, while very occasionally clavarioid fruit bodies resembling *Clavaria* [*Clavulina*] *rugosa* have even been found.

The student who wishes to regard the genera of certain modern families of Aphyllophorales as forming phylogenetic lines of progressively ascending (or descending) series, thus as members of series of which the origin of the extremes are far removed from each other in time, certainly must not avoid familiarizing himself with the versatility of these and many other fungi. In respect to the shape of the fruit body, the whole gamut of a family may be realized within a single species.

The objection might be raised that in all these cases the structure of the fruit body has remained the same. This is often true, but not always, and I cannot resist drawing attention to the phenomenon of what may be called dual or twin fruit bodies. Some species are known to be capable of normally producing two different kinds of *basidiferous* fruit body, each of which, when found isolated, would be referred to a different taxon. One of these species tends to eliminate the traditional boundaries between the Corticiaceae and the Clavariaceae. Corner (1950) described this striking example for *Pterulicium*, a genus closely related to *Pterula* but distinguished from it by developing, besides the truly negatively geotropic *Pterula* fruit body, an effused patch covered with a normal fertile hymenium when the patch faces downward. This patch generally accompanies the clavarioid fruit body but sometimes may occur separately. Whereas the clavarioid fruit body has a context dimitic by skeletals, the corticioid is devoid of skeletals and has a monomitic context. If these two types of fruit body were not known to be expressions of the same species, one of them would undoubtedly be classed with *Pterula* among the Clavariaceae and the other with *Corticium* among the Corticiaceae.

The taxonomist must try to see through these disguises and establish the really fundamental characters. This is a seemingly hopeless task in this instance and in cases similar to the one referred to in connection with *A. mellea*. Nevertheless, he has already overcome difficulties of this kind; the recognition of the close relationship between the corticioid genus *Ramaricium* and the clavarioid genus *Ramaria* by Eriksson (1954) is an example.

It would be easy to go on in this vein and also to list quite a number of examples that show wide variation in the hymenial configuration, just as it would be easy to mention examples in which it is not only the shape of the fruit body that varies but also the hymenial configuration. The latter is realized in the emended genus *Sistotrema*, whose members were taken from various traditional families. In short, the taxonomic mycologist working with the Hymenomycetes must be prepared for these variations and must be daring enough to draw unexpected and far-reaching conclusions.

One last theoretical consideration in this connection. Apart from a very few exceptions, the Hymenomycetes discharge their spores forcibly from either a vertical hymenophore as it occurs in the clavarioid groups, or far more often from a hymenophore directed downward. Buller's *Researches on Fungi* (1909, 1922, 1924) have shown that the adjustments of the fruit body in the interests of this type of spore liberation are manifold and often highly complicated. He has also pointed out that the greater the increase in the spore-producing hymenial area of a fruit body, the more precise the adjustment of the hymenophore to the strictly horizontal position must be for the hymenophore to perform its task efficiently. This is not all; the fruit body must afford an opportunity for the falling spores to be carried away by air currents, which in terrestrial agarics means that the cap is raised above the ground by a stalk.

It follows that the context of a stalked or sessile fruit body should be capable of performing diverse tasks. During its growth it should respond to various morphogenetic forces in order to bring the hymenophore into a horizontal position. It should be firm enough to keep the hymenophore in a horizontal position during the whole period of spore discharge, which in polypores may be many years. The hymenophore itself should so develop and adjust itself during the formation of its spines, tubes, or gills that the spores will fall undisturbed. The modern methods of hyphal analysis have furnished numerous details about how many of these tasks are carried out, and from this information the taxonomist now derives a whole set of characters by which he defines his genera or even families. Isn't this building on thin ice, at least as far as taxa above the rank of genus are concerned? Are these truly "conservative" characters?

What would one expect to happen to the fruit body if a stalked fungus fruit body were to migrate from the ground to the side of a tree stem, or even to the underside of a log? One might expect that through adaptation the stalk would become reduced and that the cap would eventually become sessile on the upstanding tree stem, while this dimidiate cap would in turn become resupinate or even effused when the formation of the fruit body shifted to the underside of the log. I need not elaborate on the changes in construction of the context that could be expected in response to such a migration: a conceivable sequence is from a monomitic to a di- and trimitic context and finally, in the most reduced effused fruit bodies, again monomitic, in this last case, however, of a type different from the fleshy stalked fruit body at the other end of the series. In the case of the *Pterulicium* example already mentioned, the context goes from dimitic by skeletals in the clavarioid state to monomitic in the corticioid state. In short, one might expect in a natural series in which the fruit body became altered by migration from the soil to vegetable substratum that many valuable taxonomic characters would disappear and new ones would appear during this evolutionary process.

The next question to be raised concerns what one might expect to happen if a species were to lose its capacity for forcible spore liberation, or in other words turn from hymenomycetous to gasteromycetous. In certain respects the answer is not so difficult. The slightly curved spiculum from which the ballistospore is shot away and the characteristic features coupled with the asymmetrical ballistospore would be changed. The sterigmata or spicula would become straight, or might disappear completely (the spore becoming sessile), or else grow out to various lengths often on one and the same basidium. The ballistospore would lose its characteristic asymmetrical profile in side view, the apiculus would disappear, and the connection with the sterigma would become unadapted. How do we know this? By actual observation in series where this process of gasteromycetation has occurred. I wish to remind you of the many "bridges" between the agarics and the Gasteromycetes. The best known of these is found in the Russulales. Another example is to be found among the Tremellaceae, in which Rogers (1947) described *Xenolachne*, a genus that easily compares with *Exidiopsis* or *Sebacina*. Still another example is the auriculariaceous genus *Phleogena*, where the spores have become practically sessile, symmetrical dry spores. As to the other adaptations of the fruit body proper that served to support forcible spore liberation, I leave it to you to visualize the enormous changes that might result. Not only would the adaptations for forcible spore discharge become superfluous, but also new modes for spore dispersal would have to be developed. In this respect the Gasteromycetes are a great source of inspiration; in this group it looks as though "Nature tries again," to use an expression of Ingold.

## Taxonomy and Evolution

Before the manuscript of this lecture was trimmed to its prescribed length, it included a chapter on "adverse influences" that tend to bias the conclusions the taxonomist draws up from facts he has gathered empirically. Most of this discourse I cut out except for mentioning the stand of a growing number of taxonomists against what is considered the misuse of the theory of evolution. At present, more often than not, the biological student is instilled with the dictum that a natural system must reflect the evolution of a group as closely as possible. I would oppose this and suggest that the mycologist who tries to build up a natural system should divorce himself rigorously from any implanted evolutionism. Only after he has delivered his results may he call in the evolutionist to have a look at them. I do not wish to discourage the mycologist's fantasy and his willingness to work from an adaptable ideological pattern designed to arrange his results in a surveyable system, but he must not confuse this ideological pattern with a phylogenetic system, as is so often done. Much more is needed for a really well-founded phylogenetic system than the known fungi of today. For instance, what we know about fossil fungi, let us say of agarics, is so extremely little that it does not contribute anything to our knowledge. Thus, the very important element of time must be eliminated as a support in building up a phylogenetic system.

Moreover, we know only a portion of the fungi of today, and few Hymenomycetes from the tropics. According to an estimate by the

Commonwealth Mycological Institute, the registrar's office for fungi, it may be assumed that not more than about 100,000 species of fungi have been described, with about 150,000 still to follow. Today the yearly increase is only about 1,500, and of these a considerable number are poured into the workshop of the taxonomist concerned with Hymenomycetes. One newly detected missing link may kaleidoscopically change the pattern of a neatly construed classification.

Some mycologists believe that the Gasteromycetes are primitive and that the agarics have been derived from them. It is true that the taxonomist is able to build up quite a number of apparently natural series that cross the borderline and belong partly to the Gasteromycetes and partly to the agarics. By accepting the agarics as the more advanced, it is unavoidable also to agree that the mechanism for the forcible discharge of the basidiospores has been acquired repeatedly during the course of evolution, and that in all cases this has resulted in precisely the same mechanism. How much easier it is to suppose that agarics have repeatedly lost their capacity for forcibly discharging their spores and have thus passed into the domain of the Gasteromycetes. However, acceptance of such an ideological plan would not necessarily mean that we have hit on the true course of evolution of these groups.

## Dwindling Families

All the traditional families of the Hymenomycetes have somehow appeared to be artificial. This also applies to several of those that have been introduced more recently. Such taxa rapidly become smaller and smaller. What is currently called the Cyphellaceae is an instructive example (Donk, 1966) of something not uncommon in the systematics of mycology. Some mycologists retain a family in the traditional sense and consider it as good a family as any, whereas others are convinced that it is nothing but a handy bin from which part of the contents have already been taken out and disposed of by scattering them over various groups, although the bin is still needed for keeping what remains. We do not yet know what to do with this remainder, principally because the published accounts are inadequate and the species have not yet been scrutinized anew in the light of present-day taxonomy.

The Cyphellaceae was launched by Patouillard (1900) as a sub-tribe that was later raised to the rank of family. The type species is *Cyphella digitalis*. The species referred to this family may be described briefly as resembling minute cup-fungi, but they are provided with basidia instead of asci. Formerly everything that closely or remotely answered to this formula was likely to be transferred to the Cyphellaceae, even *Fistulina*, because the mutually free tubes of this genus were considered to be many cups united on a common stroma represented by the fruit body, of which the tubes form the hymenophore. Although Fries referred most of the Cyphellaceae he knew to the Thelephoraceae because of their smooth hymenium, he referred those with densely crowded tubes or cups, like *Solenia* and *Fistulina*, to the Polyporaceae.

Some of the genera that have been included in the Cyphellaceae conform only partly to the formula. The genus *Aleurodiscus* has a few species producing typically cup-shaped fruit bodies, closely related to the type of *Cyphella*, whereas in some other species the fruit bodies are more *Stereum*-like, and in still others, completely corticioid. In fact, most of this genus protrudes beyond the set limits of the family. Another series, of which certain members have been placed in the Cyphellaceae, is now often considered a single genus, viz., *Leptoglossum*; the other end of the practically uninterrupted series is formed by typical agarics. In still other cases, certain Cyphellaceae in some respects closely resemble certain cup-shaped agarics and although there is a gap in these series, the microscopic details speak forcibly for a close relationship. In this way many genera that first had to be defined were removed from the Cyphellaceae and incorporated among the agarics.

The cyphellaceous or discomycete-like habit is encountered in practically all groups of Hymenomycetes; thus the agarics, for instance, have *Resupinatus*; the Auriculariaceae, *Hirneola*; the Tremellaceae, *Exidia*; the Dacrymycetaceae, *Femsjonia*. Except for the agarics these genera were recognized as distinct from *Cyphella* because of their gelatinous context. If it had not been discovered before the introduction of the Cyphellaceae that they also had special types of basidia, they might also have been referred to this group. All this shows that the cup-shaped fruit body has originated along many convergent lines.

The numerous cyphellaceous genera that at the present are associated with the agarics are to be regarded as having "reduced" rather than "primitive" fruit bodies. The trouble with these supposedly reduced forms is that so little has been left of the fruit body for the deduction of their relationship: a hymenium, a thin layer of hyphae, and, in the most favorable cases, hairs on the exterior. One sometimes feels like a peddler going from genus to genus with his often limited notes of a cyphellaceous genus, to find response. In those cases of agarics where, for instance, the microscopic details of the surface of the cap have been well studied one will occasionally succeed, but contact cannot be established except when details at both sides have been carefully worked out.

The position of the type species of *Cyphella*, *C. digitalis*, is not altogether clear, although whatever the case it seems close to *Aleurodiscus*. I certainly would not be surprised if the family name were to be taken up again after the strongly reduced genus *Cyphella* has started to act as a crystallization center for a family very different from the one to which it has been applied in the past.

One of the main difficulties preceding an orderly re-evaluation of an artificial family or group is the unmixing of the cocktail. More often than not the constituent genera also appear to be artificial, which requires a lot of splitting up and reclassifying into smaller genera. To mention an example: the old genus *Clavaria* has not only been broken up into numerous parts, but quite a number of these have been transferred to other families like the Clavulinaceae, Gomphaceae, Hericiaceae, Hymenochaetaceae, Thelephoraceae, and so on.

## Some Modern Families

It is now about time to mention some of the more recently established families and show that they may consist of a motley mixture taken from no less than six Friesian grades based on the hymenial configuration. As a tribute to the genius of Patouillard it is only fair to mention first his *Série des Phylactéries* (1900). In its modern form it is called Thelephoraceae. Its present circumscription is widely removed from the traditional family of the same name, with which it shares only the strongly reduced, name-bringing genus *Thelephora*. In most of the genera the shape of the fruit body is typical for each genus, except in *Thelephora* itself, which covers clavarioid, stalked and pileate, effuso-reflexed, or effused fruit bodies.

It is perhaps useful to remark that so far nobody has been able to formulate a generic distinction between the strictly effused fruit bodies of *Thelephora* and those of *Tomentella*, all the species of which produce such fruit bodies. Also, mycologists are now fully

TABLE 2. THELEPHORACEAE (emended)

| Hymenial configuration / Fruit body | smooth | cantharelloid | toothed | tubulate | lamellate |
|---|---|---|---|---|---|
| clavarioid | *Thelephora pr.p.* | | | | |
| | *Scytinopogon* | | | | |
| effused | *Kneiffiella* | | | | |
| | *Tomentella* | | *Caldesiella* | | |
| | *Thelephora pr.p.* | | | | *Lenzitopsis* |
| reflexed to sessile | *Thelephora pr.p.* | | | | |
| stalked and pileate | *Thelephora pr.p.* | *Polyozellus* | *Hydnodon* ? | | |
| | | | *Hydnellum* | | |
| | | | *Sarcodon* | *Boletopsis* | |

aware that the distinction between *Tomentella* and *Caldesiella*, a corticioid and a hydnoid genus, is difficult to uphold. Questions like these attract attention more quickly if the related genera are grouped together instead of being scattered all over the Aphyllophorales.

The two principal stalked and hydnoid genera of the Thelephoraceae are *Sarcodon* and *Hydnellum*. Each of these included a group of species with a different type of spore; these lack some of the features of the true Thelephoraceae. The spores are colorless with an even outline and have their own kind of asperulation. The habit of the fruit body is so strikingly similar to that of the brown-spored species that many contemporary mycologists will not even go so far as to assign generic rank to the white-spored elements; at most they will admit twin pairs of genera: *Sarcodon–Bankera* and *Hydnellum–Phellodon*. To my mind, on the basis of the spore character, supported by some other unrelated features, it is preferable to segregate these white-spored groups into a distinct family, the Bankeraceae, and thus put it to the test (Donk, 1961).

The naturalness of the thus remodeled Thelephoraceae is not seriously questioned at present, and an increasing number of mycologists are accepting it, notwithstanding the fact that a really sharp family character is difficult to draw up. The principal method of

TABLE 3. GOMPHACEAE

| Hymenial configuration / Fruit body | smooth | cantharelloid | toothed | tubulate | lamellate |
|---|---|---|---|---|---|
| clavarioid | *Ramaria* | | | | |
| | *Lentaria* | | | | |
| | (*Kavinia*) | | | | |
| effused | *Ramaricium* | | *Kavinia* | | |
| stalked and pileate | | *Gomphus et al.* | *Beenakia* | | |

learning and appreciating the present conception is to look at the hymenium and the spores under the microscope for oneself. Recent investigations have lent additional support. For instance, the spore wall is acyanophilous and thelephoric acid is known among the Aphyllophorales only from various genera of this family.

The next example has been worked out during the past few years (Donk, 1961, 1964). The Gomphaceae derives its genera from at least four grades: Corticiaceae, Clavariaceae, Cantharellaceae, and Hydnaceae. Here is an excellent example of the thesis that the discovery of completely new genera is still in the air. The corticioid genus *Ramaricium* was discovered in Sweden by John Eriksson (1954), and the first species of *Beenakia*, a stalked and pileate hydnum, was described from Australia by Reid. *Beenakia* now consists of two species, the second having been found in Africa. Both were carefully studied by Maas Geesteranus (1963b, 1967). I am sure that if the microscopical analysis of *Beenakia* were to be put before a trained mycologist specialized in clavarias he would ascribe it to some species of *Ramaria*. Different as the fruit bodies in this family may look, a careful comparison of the analyses of the hyphal structure throughout the fruit body as well as of the hymenial elements in all three genera shows that there is an extraordinary similarity. From now on I shall refer to these microscopical analyses as the "inner pictures."

Another family recently introduced, this time by Maas Geesteranus (1963a), is the Auriscalpiaceae. In contradistinction to the Hericiaceae, to be discussed presently, it shows a dimitic hyphal structure with skeletal hyphae and noninflating generative hyphae. It agrees in several other respects as well, for instance in its system of oil-filled, thin-walled, nonseptate hyphae (which I now call gloeoplerous hyphae), darkening in sulphuric-benzaldehydes, and also amyloid, asperulate spores. The three genera had never before been connected with one another, which is not surprising considering their very different fruit bodies. *Gloiodon* produces seemingly dorsiventral fruit bodies with spines on the underside, but in reality they are formed by a diageotropic coralloid system of branches embedded in a dense tomentum and forming downward-directed teeth. *Auriscalpium* has stalked caps with a toothed hymenophore. The third genus is *Lentinellus*, which nearly all mycologists still regard as a typically agaric genus. However, if the inner pictures of these genera are compared, a very far-reaching similarity will speak for their mutual relationship.

## The Making of a Family

It is perhaps worthwhile to outline the growth of a family in the mycological workshop. The method I prefer is strictly empirical and in principle very simple. Pick out a small and homogeneous genus, then start looking for another that comes the closest to it; do the same with this second genus—if any—and so on. This method may be called taxonomy *par enchaînement*, or by chain formation. The example is the Hericiaceae in the narrow sense (Donk, 1964). Under the present circumstances it is of course unavoidable that quite a number of details be omitted and that for the sake of brevity the presentation be somewhat simplified.

*Hericium*, in its most restricted sense, is a genus of about half a dozen rather well-known species. Its inner picture, in combination with some chemical features, is quite characteristic, although not altogether unique. The hyphal structure is monomitic, with inflating generative hyphae in the flesh. In addition, there is a system of hyphae with oily contents, with thin walls and no septa; these are the gloeoplerous hyphae which end in the hymenium as gloeocystidia. The spores are rather small, more-or-less short-ovoid, with somewhat thickened, distinctly amyloid, and asperulate walls.

On critical examination of the fruit body doubts will soon arise about the still current incorporation of *Hericium* in the old family Hydnaceae. There is no toothed hymenophore: the teeth are in reality merely downward-directed ends of the branches of a coralloid fruit body. This is best seen in *Hericium clathroides* and *H. coralloides.* If positive geotropism could be replaced by a negative reaction, the result would be a typical, coralloidly branched clavaria. Nature itself has made this very experiment in all its gradations between and including the two extremes. As I pointed out several years ago (Donk, 1964, 1965), *Clavicorona pyxidata,* which was formerly included in *Clavaria* because of its strictly negatively geotropic fruit bodies, has an inner picture closely matching that of *Hericium.* Among the Clavariaceae *Clavicorona* can be recognized by its peculiar kind of branching; the branches resemble inverted cones which acquire this shape because their tips stop growing; the new branches are formed peripherally in small numbers along the rim of the flat top.

Some species now placed in *Clavicorona* do not branch, in which case the fruit body consists of a single inverted cone. These unbranched species were placed in various Friesian genera, as for instance *Craterellus.* This is not at all surprising if the external morphology of the fruit body was the determining factor. *Craterellus taxophilus* has inamyloid spores, while others now also placed in *Clavicorona* have dextrinoid instead of amyloid spores. This fading out of a feature usually rated as of high taxonomic importance has by now been repeatedly observed in many other genera as well. Just the same, it is annoying.

To return to *Hericium*: it is not even necessary to look elsewhere for striking examples of a reversal of the geotropic morphogenetic forces. A search among the published figures of the genus itself yields several clear-cut instances in which it can be seen that downward growth may set in at a late stage. Quite recently Leathers & Smith (1967) described two species in which the positively geotropic reactions were almost, or completely, reversed. The two authors placed these species in *Clavicorona,* but I suggest that they could better be accommodated in *Hericium,* judging from depicted fruit bodies.

The rather loose branching found in *H. coralloides* may by strong condensation and webbing lead to apparently tuberous fruit bodies, less easily recognizable as essentially coralloidly branched. However, careful inspection of a section leaves little doubt that this situation is still the case in *H. erinaceus*. The response of the growing, coralloid fruit body to the morphogenetic forces of geotropism may vary and even lead to apparently typically dorsiventral fruit bodies. This is true of *Creolophus*: a strongly condensed and webbed system of branches grows out first diageotropically to form a horizontal fruit body from which finally the so-called teeth of the underside only grow out to full length downward.

Thinking in terms of reduction by habit, I wondered how an extremely strongly reduced *Hericium* would look. This drew my attention to the genus *Mucronella*, which forms minute downward-directed, unbranched fruit bodies resembling a crowd of isolated teeth. At first, microscopical examination was not very promising. Although the structure was monomitic with inflating generative hyphae, no system of gloeoplerous hyphae was found and in the first species studied the spores appeared inamyloid and smooth; thus, in this case several features considered important for defining the genus *Hericium* were lacking. Then I came across North American material in which the spores proved to be distinctly amyloid. Shortly after I had pointed out this gap between *Hericium* and *Mucronella*, Domański (1965) found in the Bialowieża virgin forest in Poland the missing link, which he called *Dentipratulum*. Not only did it have a system of gloeoplerous hyphae ending as gloeocystidia in the hymenium, but it also had amyloid and eventually faintly asperulate spores.

It would be easy to draw up a diagram arranging these constituent elements of the Hericiaceae according to a chosen pattern and to connect them by lines. In this way a figure is arrived at that is often called a phylogenetic tree, but of course it would be too much of a good thing if this ideological pattern were really to coincide with the lines of evolution.

This process of building up taxa of higher rank by chain-formation could be continued for the Hericiaceae in various directions, but I shall leave it at this stage, except for a few closing remarks. It is not unlikely that in the near future a decision will have to be made as to whether or not the clavarioid Hericiaceae as a whole are closely connected with other groups possessing a system of gloeoplerous hyphae and amyloid, ornamented spores. Such groups are numerous:

for instance, the order of the Russulales, which covers agaric and gasteromycetous elements; the Echinodontiaceae; the Bondarzewiaceae; certain genera now included in the Lachnocladiaceae and Corticiaceae; and so on. One of these groups has already been mentioned, viz., the Auriscalpiaceae. It differs from the Hericiaceae mainly in the dimitic hyphal system with noninflating generative hyphae. All these taxa may eventually be considered as part-chains of one big system without necessarily losing their status as distinct families. They might appear to be members of one big order, but then certainly an order also containing quite a number of reduced species or genera in which the original, most prominent features chosen for its characterization have vanished.

## The Modern Arsenal

The steady improvement of the classification of the Aphyllophorales has meant that more and more diverse characters must be taken into consideration for a correct evaluation of similarity. Resemblances may point either to relationship or to its reverse, "convergence." Among the Aphyllophorales a coralloid fruit body was formerly considered to be a sure indication of taxonomic relationship. Now we approach this character with more caution. It is not accepted on its own as valid but only in conjunction with other correlated characters. What we really need in our pursuit of a better classification is a constantly increasing stock from which to draw. The search for new characters is now a significant aim of the taxonomist. One trouble is that the taxonomist who finds a given character useful in certain groups often tends to overemphasize its importance and is often too eager to apply it on too wide a scale. A striking example of this is the position of the nuclear division-spindles in the basidium as applied by Maire (1902) and van Overeem (1923). Another trouble is that each new character now added to the stock and considered of possible value requires an enormous amount of detailed observation. How many tests have to be made before we know approximately where the amyloid spore wall occurs throughout the Hymenomycetes!

Cytology entered into the picture when Van Tieghem and Juel drew attention to the fact that two main basidial types could be distinguished in connection with the direction of the nuclear division-spindles in the basidium. Maire overemphasized the taxonomic importance of the two basidial types. After a period of negligence, data published by Penancier (1961) and Boidin have shown that the recognition of these types may be very useful in delimiting genera and, in the opinion of Donk (1933, 1964), also of certain families. The emended Cantharellaceae, the Clavulinaceae, and the Hydnaceae in its most restricted form all have stichic basidia.

Considerable success has resulted from the introduction of chemical tests. For instance, the amyloidity of the spore wall has proved to be of importance in defining several families, for example, the Russulaceae, Hericiaceae, Auriscalpiaceae; eventually it may appear to be helpful in tracing the limits of an order embracing these and other families. Yet, amyloidity is often not an absolute character. It sometimes drops out even within the limits of small genera.

The use of cotton blue was introduced in the systematics of the Hymenomycetes by Nannfeldt & Eriksson (1954) when they found it to be a supporting feature in defining the Coniophoraceae, where the inner spore wall proved to be strongly cyanophilous. Quite recently this same feature appeared to underline the close affinity that I suspect exists between the Coniophoraceae and the Paxillaceae, a family of Agaricales. There are also strong indications that cyanophily should enter into the definition of the Gomphaceae in connection with the ornamentation and the outer layer of the spore wall. Recently Kotlaba & Pouzar (1964) paid renewed attention to the absorption of cotton blue by membranes of spores and also of hyphae. The test is now becoming more and more important in delimiting genera of polypores.

Among the Aphyllophorales it has been found that thelephoric acid occurs only in various genera of the emended Thelephoraceae, although it may occur in all. The so-called xanthochroic reaction of the Hymenochaetaceae deserves to be more systematically studied.

A new approach to a better understanding of the fruit body was inaugurated by Corner. He was the first person to study more than thin sections and he stressed that to be understood the more complex tissues must be teased out by fine needles under a dissecting microscope in order to disentangle the hyphae. This work requires considerable skill and is often time consuming, which may account for the relatively few carefully worked out analyses of more intricately built fruit bodies published so far. Even so, however, it is now believed that a proper understanding of the hyphal construction is essential for the elucidation of the taxonomy of the polypores and other groups of Aphyllophorales. More will be said about this in a discussion of the polypores. On the same occasion the so-called cultural characters, which have gained in importance during the last few decades, will also be mentioned. I am sure that other participants in this symposium will inform you about recent developments in this regard.

There is still much to be done before the systematics of the Hymenomycetes quiets down.

## References


Bourdot, H., and A. Galzin. 1927 (1928). . . . Hyménomycètes de France. Sceaux. 764 p.

Buller, A. H. R. 1909, 1922, 1924. Researches on fungi. 3 vols. London.

Cartwright, K. S. G. 1940. Note on a heart rot of oak tree caused by *Polyporus frondosus* Fr. Forestry 14: 38–41.

Corner, E. J. H. 1950. A monograph of *Clavaria* and allied genera. London. Ann. Bot. Mem. 1:1–740.

Domański, S. 1965. [Polish title.] Wood-inhabiting fungi in Bialowieża virgin forest in Poland. II. The mucronelloid fungus of the *Hericium*-group: *Dentipratulum bialoviesense*, gen. et sp. nov. Acta mycol. 1: 5–11.

Donk, M. A. 1933. Revision der niederländischen Homobasidiomycetae-Aphyllophoraceae II. Utrecht. 238 p.

———. 1961. Four new families of Hymenomycetes. Persoonia 1: 405–407.

———. 1964. A conspectus of the families of Aphyllophorales. Persoonia 3: 199–324.

———. 1965. Veelvoudige overeenkomsten bij Hymenomyceten. Versl. gewone Verg. Afd. Natuurk. K. Ned. Akad. Wet. 74: 24–32.

———. 1966. A reassessment of the Cyphellaceae. Acta bot. neerl. 15: 95–101.

Eriksson, J. 1954. *Ramaricium* n. gen., a corticioid member of the *Ramaria* group. Svensk bot. Tidskr. 48: 188–198.

Fries, E. M. 1874. Hymenomycetes europaei. . . . Upsaliae. 755 p.

Killermann, S. 1928. . . . Reihe Hymenomyceteae. Nat. Pfl. Fam., 2. Aufl., 6: 99–283.

KNIEP, H. 1911. Ueber das Auftreten von Basidien im einkernigen Mycel von *Armillaria mellea* Fl. Dan. Z. Bot. 3: 529–553.

KOTLABA, F., AND Z. POUZAR. 1964. Preliminary results of the staining of spores and other structures of Homobasidiomycetes in cotton blue and its importance for taxonomy. Feddes Rep. 69: 131–142.

LEATHERS, C. R., AND A. H. SMITH. 1967. Two new species of clavarioid fungi. Mycologia 59: 456–462.

MAAS GEESTERANUS, R. A. 1963a. Hyphal structures in Hydnums. II. Proc. K. Ned. Akad. Wet. (C) 66: 426–436.

———. 1963b. Hyphal structures in Hydnums. III. Proc. K. Ned. Akad. Wet. (C) 66: 437–446.

———. 1967. Quelques champignons hydnoides du Congo. Bull. Jard. bot. natn. Belg. 37: 77–107.

MAIRE, R. 1902. Recherches cytologiques et taxonomiques sur les Basidiomycètes. Bull. Soc. mycol. Fr. 18, Suppl. 209 p.

MOLLIARD, M. 1909. Sur une forme hypochnée du *Fistulina hepatica* Fr. Bull. Soc. bot. Fr. 56: 553–556.

NANNFELDT, J. A., AND J. ERIKSSON. 1954. On the hymenomycetous genus *Jaapia* Bres. and its taxonomic position. Svensk bot. Tidskr. 47: 177–189.

OVEREEM, C. VAN. 1923. Beiträge zur Pilzflora von Niederländisch-Indien. [5. Ueber javanische Clavariaceae]. Bull. Jard. bot. Buitenz. III 5: 254–280.

PATOUILLARD, N. 1900. Essai taxonomique sur les familles et les genres des Hyménomycètes. Lons-le-Saunier. 184 p.

PENANCIER, N. 1961. Recherches sur l'orientation des fuseaux mitotiques dans la baside des Aphyllophorales. Trav. Lab. "La Jasinia" 2: 57–71.

ROGERS, D. P. 1947. A new gymnocarpous Heterobasidiomycete with gastromycetous basidia. Mycologia 39: 556–564.

## DISCUSSION

GILBERTSON: In your discussion of the Auriscalpiaceae, you did not mention the genus *Gloeodontia*, described by Dr. Boidin. I wonder why you omitted this.

DONK: In this paper I took into consideration only the Hericiaceae in the strictest sense because I did not want to become involved on this occasion in the question of how this family is to be delimited. It is still difficult to agree about the real distinctions between the Auriscalpiaceae and the Hericiaceae. Within these complexes there are groups that can be arranged into chains that seem to go on and on; we don't know yet where they will end or connect with each other. Therefore, *Gloeodontia* was left out in order to avoid a too intricate definition of the Hericiaceae.

SMITH: I want to compliment Dr. Donk on giving a very good presentation of all of the problems that we will be discussing for the rest of the week. I would also like to advance an idea. I agree perfectly with what you said about the Friesian system's being artificial. There were many things included in the same group on the basis of the configuration of the hymenophore which on further study actually did not belong there. That was the first step in the development of the taxonomy of the higher fungi. It is apparent to me now after reading Dr. Donk's work and that of others involved in hyphal systems that we are entering a second stage, which is the elaboration of the hyphal system. I assume that we will eventually find as many parallelisms in the formation of different hyphal systems and different hyphal types in the sporocarp as there were parallelisms in the hymenophore. The third phase of mycology will come when we take all of the characters into account, including culture work, and synthesize a classification on that basis. We must keep in mind all the time that, from the standpoint of the fungus, we are considering the evolution of the basidiocarp as it relates to the survival of the species. The fungus, reacting with the environment in the course of evolution, does repeatedly develop many of these systems. I disagree with Dr. Donk's including *Lentinellus cochleatus* in the same group with *Auriscalpium vulgare*. I think this is just as far in error as putting some things in the Agaricaceae simply because they have gills.

DONK: In principle I said precisely the same as what Dr. Smith just expressed. Much of what we learn of the fruiting body of the Hymenomycetes makes sense only when it is seen as an instrument for effective dispersal of the spores; this also applies to the hyphal make-up. I have stated that exclusive emphasis on the results of hyphal analysis is building on thin ice.

# 2

# SUPPORTIVE CHARACTERS IN SYSTEMATICS

VARRO E. TYLER, JR.

*Dean of the School of Pharmacy and Pharmacal Sciences*
*Purdue University, Lafayette, Indiana*

# CHEMOTAXONOMY IN THE BASIDIOMYCETES

Before undertaking a specific discussion of chemotaxonomy in the Basidiomycetes, it seems desirable to devote a few introductory remarks to chemotaxonomy *per se*. The concept that plants which are related in form or structure are also related in their ability to biosynthesize and accumulate secondary constituents is a relatively ancient one, predating the presently utilized system of plant classification (Hegnauer, 1958). However, in its present useful form the science is quite new, dating back little more than a decade. This delayed development of a long-recognized concept may be attributed to the paucity of phytochemical data which existed until recent times. Sufficient knowledge of the chemical constituents of plants was not obtained until modern analytical methods, such as the various kinds of chromatographic procedures, were developed and applied to plant tissues. Thus, the present-day science of chemotaxonomy rests upon a sound but very limited knowledge of the constituents of plants, which in turn is based on modern analytical instrumentation and techniques.

Chemical taxonomy, or biochemical systematics, as some prefer to call it, is founded upon several fundamental principles, an understanding of which is mandatory if we are to recognize the science's utility and limitations. Most basic of these is the premise that the chemical constituents of a plant which are genetically controlled and which may be objectively described are useful criteria for helping determine the relationship of that plant to others. An extension of this principle is the so-called frequency rule, which states that compounds that are biosynthetically complex are more restricted in their occurrence and therefore of more taxonomic significance than the biosynthetically simple compounds. Finally, it must be recognized that convergence and divergence occur with respect to molecular characteristics, that is, chemical compounds, just as they do in the case of morphological features (Alston, Mabry & Turner, 1963; Alston & Turner, 1963; Erdtman, 1963). Much of the criticism which has been leveled against the utility of chemotaxonomy has resulted from a lack of understanding of these basic axioms. Consideration of suitable examples will help to clarify this point.

The key words in the first principle are *helping determine.* As Hänsel (1956) has pointed out (in translation), "To take without further evidence either the sporadically or ubiquitously occurring compounds as a guide to plant classification would be just as artificial as the attempts at classification based solely on leaf form or the number of stamens."

Critics of chemotaxonomy have often used the widespread distribution of nicotine as an example of the shortcomings of the method. Nicotine has been detected in more than a dozen genera of plants ranging from *Equisetum* to *Zinnia* (Hegnauer, 1963). To one versed in biochemistry and particularly biosynthetic pathways, this is not at all surprising. Nicotine is easily derived by the coupling of ubiquitously occurring nicotinic acid with an ornithine equivalent (Dawson, 1960), a reaction sequence (see fig. 1) so biosynthetically simple

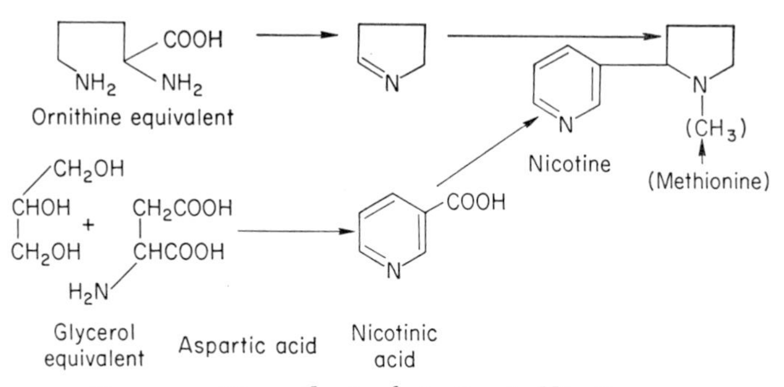


FIGURE 1. Biosynthesis of nicotine in *Nicotiana* sp.

that Hegnauer (1957) has postulated its occurrence in most higher plants. It is thus apparent that nicotine does not meet the first requirement of biosynthetic complexity which is mandatory for systematic significance.

One must also proceed cautiously in attributing relationship based on similarities of structures of secondary constituents. It is not evident from structures alone whether the compounds are homologous or merely analogous. Since homology presupposes identical precursors and identical pathways of formation, whereas analogy only considers the nature of the end-product, a knowledge of biosynthesis is obviously required before true chemotaxonomic significance can be determined.

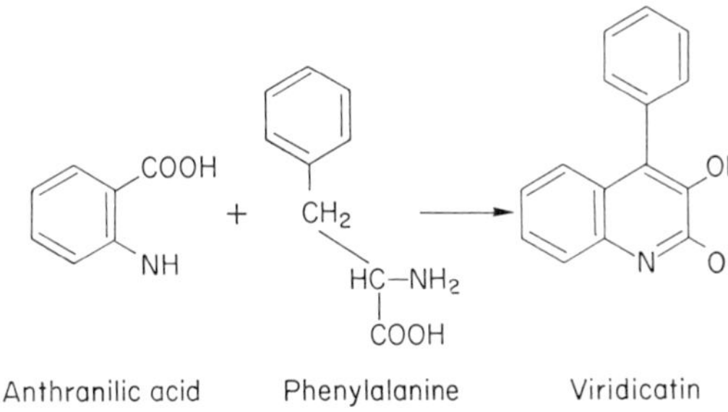

Anthranilic acid + Phenylalanine → Viridicatin

FIGURE 2. Biosynthesis of viridicatin.

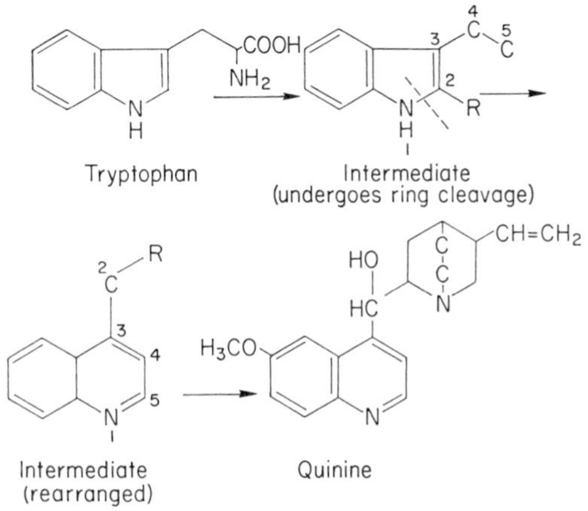

Tryptophan → Intermediate (undergoes ring cleavage) → Intermediate (rearranged) → Quinine

FIGURE 3. Biosynthesis of quinine.

An example of this situation is found among the quinoline alkaloids. Viridicatin, an alkaloid produced by *Penicillium viridicatum* Westling, has a basic structural similarity to quinine obtained from *Cinchona* species. However, the biosynthetic pathways of formation are different; the former is derived by condensation of anthranilic acid and phenylalanine, the latter by ring cleavage and rearrangement of tryptophan (see figs. 2 and 3). Thus, the two alkaloids are analogous, not homologous (Luckner, 1963).

The ergoline alkaloids furnish the best example of convergence of secondary plant constituents yet identified. For many years, these compounds were believed to be restricted in their occurrence to species of the genus *Claviceps*. It was somewhat surprising when derivatives of this fairly complex tetracyclic nucleus were found in other genera of fungi, specifically in *Aspergillus* and *Rhizopus*. Entirely unexpected was the discovery of such compounds in the seeds and vegetative parts of species of *Ipomea* and *Rivea* of the family Convolvulaceae. Since it has now been demonstrated that these alkaloids are synthesized from the same precursors in both fungi and higher plants (see fig. 4), their homologous character must be accepted

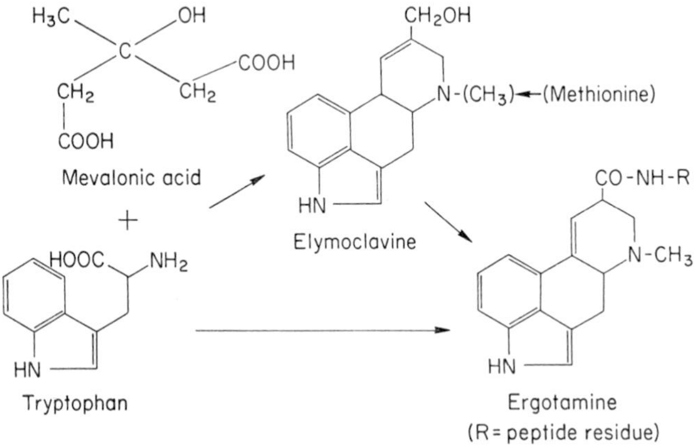

FIGURE 4. Biosynthesis of ergot alkaloids.

(Mothes, 1966). This is a convincing example of the phenomenon of convergence of chemical characteristics.

I hope that these words of introduction to the principles of chemotaxonomy have prepared you for the remarks which follow with specific reference to the Basidiomycetes. If they have pointed out just a few of the basic principles and the pitfalls of the chemotaxonomic method, they will have served their purpose.

## PIONEERING CHEMOTAXONOMIC STUDIES

One of the first of the modern mycologists to recognize the implications of chemical constituents on fungal taxonomy was Heim. As early as 1942, he discussed the relationship of mushroom pigments to systematics (Heim, 1942), and later he reviewed the importance of chemical criteria in the study of the relations of Macromycetes (Heim, 1958).

The first experimental work specifically directed toward obtaining chemotaxonomic information on the higher fungi was undertaken by Fries (1958). He prepared alcoholic extracts of the fruiting bodies of 95 species of Hymenomycetes and, after chromatographing these on paper strips, examined the chromatograms for fluorescent spots under ultraviolet light—so-called chromatographic fingerprint analysis. This approach, which does not even involve identification of the compounds in the extracts and is quite primitive, is nevertheless typical of the chemotaxonomic methodology of the period and does permit some tentative conclusions to be drawn. Fries reported his belief that the method could serve as an aid in differentiating between taxonomic units, especially in the genera *Amanita, Tricholoma,* and *Boletus.*

Bonnet (1959) reported the results of an extensive chromatographic investigation of the constituents of Basidiomycetes. Although the data collected were not analyzed specifically from the chemotaxonomic viewpoint, they nevertheless have considerable utility in this field, and the work may be considered a pioneering effort.

Beginning with the present decade, chemotaxonomic studies became more sophisticated as well as more numerous. Their results are best presented by organizing them with respect to the taxon examined. Consequently, the remainder of this presentation will consist of summaries of data obtained about specific groups of Basidiomycetes and attempts to interpret the systematic significance of these data.

## The Genus Amanita

Because of their poisonous nature, relatively large size, and general availability, carpophores of *Amanita* species have been subjected to rather extensive chemical studies. Examination of the amino acid patterns of seven species of *Amanita* and three of *Vaginata* revealed a striking similarity in all (Catalfomo & Tyler, 1961). Basic differences involved only three amino acids, phenylalanine, proline, and γ-aminobutyric acid, all three of which were detected with certainty only in species of *Vaginata* and in *Amanita calyptroderma* Atk.

& Ballen. Smith (1949) has pointed out that at times the annulus of the latter species is evanescent, making it difficult to distinguish from species of *Vaginata*. The inability to accumulate large quantities of proline in both these groups tends to support their close relationship, even though subsequent studies on large amounts of plant material have indicated that both proline and γ-aminobutyric acid are present as minor components in *A. muscaria* (Fr.) Hook. (Talbot & Vining, 1963). Still, because of their biochemical simplicity and widespread occurrence, free amino acids may be said to have comparatively little utility as taxonomic indicators.

Relatively complex cyclopeptides referred to collectively as amanita toxins (see fig. 5) have been isolated or detected in a number of species of the genus *Amanita* (Tyler *et al.*, 1966; Wieland, 1968). These include *A. bisporigera* Atk., *A. phalloides* (Fr.) Secr., *A. tenuifolia* Murr., *A. verna* (Fr.) Vitt. s. Boud., and *A. virosa* Secr. With the exception of *A. bisporigera*, all of these species are classified by Singer (1962) in section 6, *Euamanita*, of subgenus II, *Euamanita*. *A. bisporigera* is listed under section 5, *Amidellae*. On the basis of its high concentration of α- and β-amanitin, as well as its botanical affinity to *A. virosa* (Gilbert, 1940), it would seem that the species should properly be classified in section 6, *Euamanita*.

Amanitas which are misplaced, chemically speaking, as members of section 6, *Euamanita*, are species of the stirps *citrina*, specifically, *A. citrina* S. F. Gray, *A. porphyria* (Fr.) Secr., *A. aestivalis* Sing., and *A. brunnescens* Atk. and its varieties. The first two of these species are characterized by the presence of relatively large numbers of tryptamine derivatives, including bufotenine, bufotenine-*N*-oxide, serotonin, *N*-methylserotonin, *N*,*N*-dimethyltryptamine, and 5-methoxytryptamine (see fig. 6) (Tyler & Gröger, 1964b). If *A. tomentella* Kromb. is considered distinct from *A. porphyria*, it, too, falls in this category (Tyler, 1961). *A. aestivalis* and *A. brunnescens* are quite different chemically from the aforementioned species. Neither contains bufotenine or similar tryptamine derivatives (Tyler, 1961), and both are no doubt devoid of amanita toxins, although this has been proven conclusively only for *A. brunnescens* (Isaacs & Tyler, 1963).

These findings point up the extreme chemical diversity of section *Euamanita* of the subgenus *Euamanita*. As presently classified, this taxon contains species rich in peptide toxins (e.g., *A. phalloides*), species containing a large variety of tryptamine derivatives (e.g.,

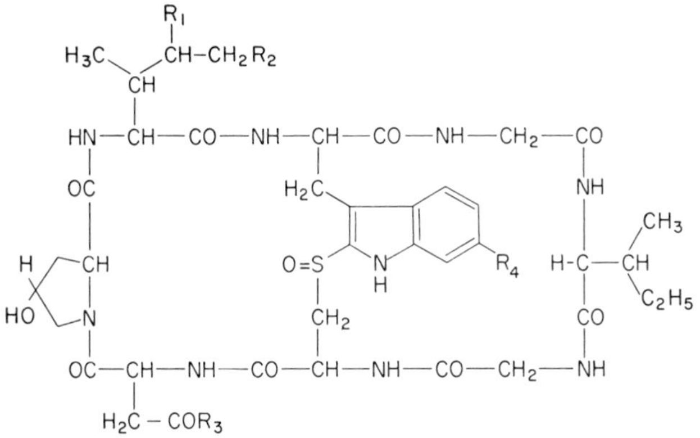

| | | $R_1$ | $R_2$ | $R_3$ | $R_4$ |
|---|---|---|---|---|---|
| a | α-Amanitin | OH | OH | $NH_2$ | OH |
| b | β-Amanitin | OH | OH | OH | OH |
| c | γ-Amanitin | H | OH | $NH_2$ | OH |
| d | Amanin | OH | OH | OH | H |
| e | Amanullin | H | H | $NH_2$ | OH |

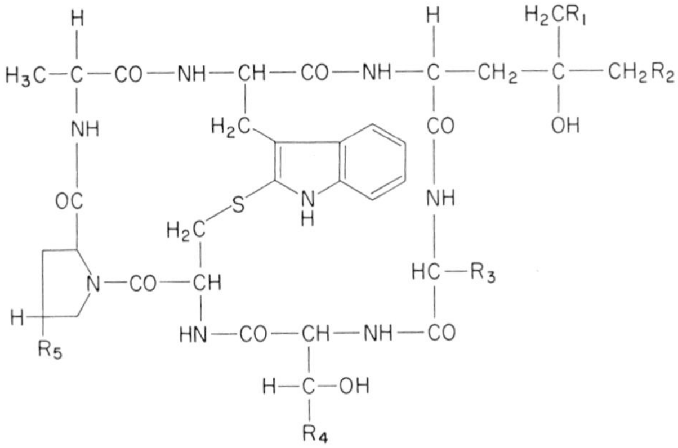

| | | $R_1$ | $R_2$ | $R_3$ | $R_4$ | $R_5$ |
|---|---|---|---|---|---|---|
| a | Phalloidin | OH | H | $CH_3$ | $CH_3$ | OH |
| b | Phalloin | H | H | $CH_3$ | $CH_3$ | OH |
| c | Phallisin | OH | OH | $CH_3$ | $CH_3$ | OH |
| d | Phallicidin | OH | H | $CH(CH_3)_2$ | $CO_2H$ | OH |
| e | Phallin B (tentatively) | H | H | $CH_2C_6H_5$ | $CH_3$ | H |

FIGURE 5. Structures of the amanita toxins.

*A. citrina*), and species containing neither of these types of secondary compounds (e.g., *A. brunnescens*). On the basis of chemical evidence a reinvestigation of the botanical classification of species in this section is certainly warranted.

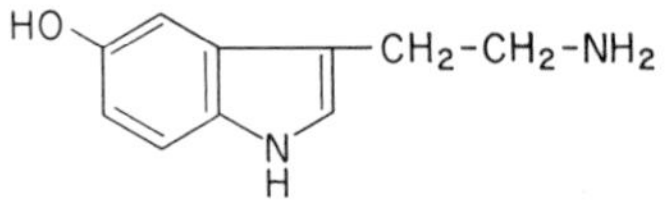

HO — $CH_2$-$CH_2$-$NH_2$ — N H

Serotonin

HO — $CH_2CH_2$-N-$(CH_3)_2$ — N H

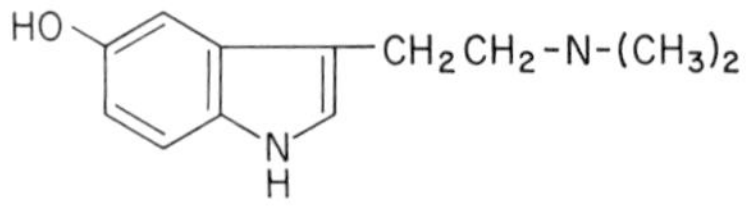

Bufotenine

FIGURE 6. Structures of 5-substituted tryptamine derivatives.

Before completing our consideration of the toxin-containing and related *Amanita* species, it is necessary to point out that the occurrence of these cyclopeptide toxins is not restricted to members of this genus. Studies have now shown that the poisonous character of species of the section *Naucoriopsis* of the genus *Galerina*, specifically, *G. autumnalis* (Peck) Smith & Sing., *G. marginata* (Fr.) Kühn., and *G. venenata* A. H. Smith, can be attributed to the occurrence in them of α- and β-amanitin. Since biosynthetic information on these compounds is lacking, we can only conclude that this is another example of evolutionary convergence of chemical characteristics.

During the past several years, the long-sought psychotropic principles of *A. muscaria* have been identified as isoxazole derivatives, ibotenic acid and its closely related and easily derived decarboxylation product muscimol (Eugster, 1967; Theobald *et al.*, 1968). An oxazole compound, muscazone, has also been detected, but its contribution to the central nervous system activity of the mushroom requires clarification (see fig. 7). A survey by Benedict, Tyler, and Brady (1966) of eighteen species of *Amanita* and two of *Vaginata* revealed the presence of ibotenic acid-muscimol only in recent specimens of *A. muscaria* and *A. pantherina* (Fr.) Secr. In spite of the occurrence of isoxazole derivatives in these two species, the compound has only scant systematic significance for section 2, *Amanita*, of the genus, since the closely related *A. gemmata* (Fr.) Gill. and *A. cothurnata* Atk. were found to be devoid of both (see table 1).

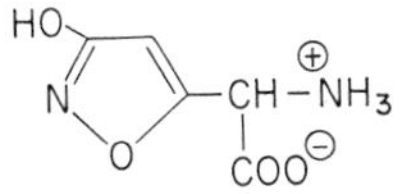

Ibotenic acid

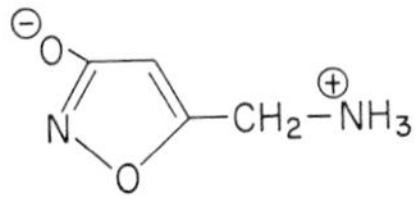

Muscimol

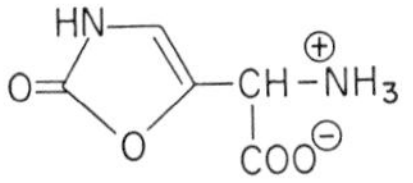

Muscazone

FIGURE 7. Structure of psychotropic principles isolated from *Amanita muscaria*.

TABLE 1. *Amanita* SPECIES DEVOID OF IBOTENIC ACID-MUSCIMOL

| | | |
|---|---|---|
| *A. agglutinata* | *A. cothurnata* | *A. silvicola* |
| *A. aspera* | *A. flavoconia* | *A. solitaria* |
| *A. bisporigera* | *A. gemmata* | *A. strobiliformis* |
| *A. calyptroderma* | *A. phalloides* | *A. verna* |
| *A. chlorinosma* | *A. porphyria* | *Vaginata fulva* |
| *A. citrina* | *A. rubescens* | *V. livida* |

The absence of ibotenic acid-muscimol from *A. gemmata* (*A. junquillea* Quél.) was used as the basis for a series of experiments designed to clarify a long-standing taxonomic problem. In the Puget Sound region of western Washington a species-complex exists whose members show all degrees of morphological intergradation (pileus color, size, stature, etc.) between *A. pantherina* and *A. gemmata*. A large number of specimens, representing both species as well as a number of intermediate forms, were collected. The transitional forms were divided into three groups, primarily on the basis of color of the mature pileus, and these, together with typical representatives of each species, were analyzed for ibotenic acid-muscimol. The results are summarized in table 2.

Carpophores of typical *A. pantherina* contained normal quantities of the two isoxazoles (average 0.42% dry weight); the typical *A. gemmata* specimens contained none. Specimens selected as being approximately intermediate between the two extremes contained small amounts (average 0.02%) of isoxazoles. In general, the quantity increased (average 0.35%) in specimens more closely related to *A. pantherina*, but in specimens resembling more nearly *A. gemmata*, both compounds were absent. These results conform to well-established chemotaxonomic principles derived from the study of higher plant hybrids (Alston & Turner, 1963) and lead to the conclusion that the intermediate forms represent actual hybrids between the two species. It is possible that hybrid forms of this sort account for Eugster's (1967) inability to detect ibotenic acid-muscimol in *A. pantherina* of Swiss origin, but this explanation requires experimental verification.

TABLE 2. OCCURRENCE OF IBOTENIC ACID-MUSCIMOL IN *Amanita pantherina-gemmata* INTERGRADES

Average of five samples (%)

| *A. pantherina* | Intergrade approaching *A. pantherina* | Intermediate form |
|---|---|---|
| I = 0.14 | I = 0.05 | I = 0.02 |
| M = 0.28 | M = 0.30 | M = trace |

| Intergrade approaching *A. gemmata* | *A. gemmata* |
|---|---|
| I = 0 | I = 0 |
| M = 0 | M = 0 |

I = Ibotenic acid, M = Muscimol

In marked contrast to the above findings are those obtained by examination of two color varieties of *A. muscaria*, i.e., var. *formosa* (Fr.) Sacc. with yellow pileus and var. *alba* (Peck) Coker with white pileus. The concentration of ibotenic acid-muscimol (0.17–0.18%) detected in both these varieties did not vary from that (0.18%) found in the typical var. *muscaria* with orange-red pileus. From this it was concluded that the difference in pileus color in this species is due merely to variation in the genes which control pigmentation and that this variation is not directly related to the genetic control of the biosynthetic pathway leading to isoxazole derivatives. In other words, in *A. muscaria* pigmentation and isoxazole accumulation occur independently of one another, and pileus color gives no indication of the concentration of ibotenic acid-muscimol and thus of the toxicity of the variety.

Examination of fourteen species of *Amanita* for the ability to accumulate urea has produced some interesting data (Tyler, Benedict, & Stuntz, 1965). With one exception, species classified in subgenus I, *Amanita*, were either urea-negative or failed to accumulate that compound in quantity. The exception is a member of the section *Vaginatae*, *A. vaginata* (Fr.) Vitt., which many authors have referred to other genera, such as *Vaginata* or *Amanitopsis*, because it lacks an annulus. However, a species which did not accumulate urea was also found in this section. Species of subgenus II, *Euamanita*, generally contained appreciable amounts of urea. The results point to the need for further investigation, particularly of the relationship of the members of the section *Vaginatae* both to other *Amanita* species and to each other.

## The Genus Inocybe

Both chemical and biological assays for muscarine have been developed and applied to thirty or more species of *Inocybe* with results that are in generally good agreement (Brown *et al.*, 1962; Malone *et al.*, 1961, 1962). Of the species tested biologically, *I. nigrescens* Atk. lacked muscarinic activity and that of *I. picrosma* Stuntz (0.005%) and *I. albodisca* Peck (0.003%) was marginal. *I. napipes* Lange (2.10–3.15%) and *I. mixtilis* (Britz.) Sacc. (1.33%) were the most potent, and a very high order of muscarinic activity was also present in *I. griseolilacina* Lange (0.84%), *I. lacera* (Fr.) Quél. (0.85–1.00%), and *I. decipientoides* Peck (0.78%). The remaining species showed moderate activity.

Other secondary metabolites detected in an investigation by Robbers, Brady, and Tyler (1964) of 39 *Inocybe* species included choline, ergothioneine, 5-hydroxytryptophan, imidazole acetic acid, and tryptamine. It was recognized that these latter compounds, together with several unidentified constituents, were of relatively minor taxonomic value, but their occurrence was still useful in differentiating certain species. These chemical characteristics were then used to construct a chemotaxonomic key, a portion of which is illustrated in figure 8.

1. Ergothioneine detected with Pauly's (orange spot at $R_F$ 0.14 BAcW and $R_F$ 0.41 BPyW) and with T. and R. (orange-yellow spot at $R_F$ 0.14 BAcW and $R_F$ 0.11 BMeW)........... 2
1. Ergothioneine not detected with Pauly's and T. and R..................................20
    2. 5-Hydroxytryptophan detected with PDAB (purple spot at $R_F$ 0.12 BAcW and $R_F$ 0.46 BPyW), with Pauly's (pink spot at $R_F$ 0.12 BAcW and $R_F$ 0.46 BPyW) and with Xanth. (blue spot at $R_F$ 0.12 BAcW and $R_F$ 0.46 BPyW)................*I.* 1838
    2. 5-Hydroxytryptophan not detected with PDAB, Pauly's and Xanth............. 3
3. Tyrosine detected with Xanth. (blue spot at $R_F$ 0.22 and $R_F$ 0.57 BPyW)............. 4
3. Tyrosine not detected with Xanth................................................14
    4. Yellow spot formed with Pauly's at $R_F$ 0.11 BAcW and $R_F$ 0.32 BPyW............ 5
    4. Yellow spot not formed with Pauly's..........................................12
5. Muscarine detected with T. and R. (orange spot at $R_F$ 0.47 BAcW and $R_F$ 0.48 BMeW).. 6
5. Muscarine not detected with T. and R............................................11
    6. Muscarine detected with Pauly's (grey-purple spot at $R_F$ 0.47 BAcW and $R_F$ 0.65 (BPyW)....................................................................... 7
    6. Muscarine not detected with Pauly's.......................................... 9
7. Asparagine detected with Nin. (blue spot at $R_F$ 0.06 BAcW and $R_F$ 0.38 PhW).....*I.* 4893
7. Asparagine not detected with Nin................................................ 8
    8. Purple spot formed with Nin. at $R_F$ 0.20 BAcW and at $R_F$ 0.80 PhW...*I. obscuroides*
    8. Purple spot not formed with Nin...................................*I. griseolilacina*
9. Yellow spot formed with T. and R. at $R_F$ 0.12 BAcW and at $R_F$ 0.11 BMeW.......... *I.* 2147 or *I.* 2149
9. Yellow spot not formed with T. and R............................................10
    10. Asparagine detected with Nin. (blue spot at $R_F$ 0.06 BAcW and $R_F$ 0.38 PhW).... *I.* 4895
    10. Asparagine not detected with Nin..........................................*I.* 4790
11. Red [illegible]urple spot form[illegible]d with PDAB [illegible]nd purple spot formed w[illegible]h Nin. at $R_F$ 0.50 BAcW ar[illegible] [illegible] 80 PhW. [illegible] *albodisc*[illegible]
11. [illegible] spot [illegible] with [illegible] nd [illegible] ot n[illegible] with [illegible] *I.* 10[illegible]

FIGURE 8. Portion of a chemotaxonomic key for *Inocybe* species.

Ergothioneine and muscarine were used as major characteristics because their occurrence seemed to delimit the genus into well-defined groups. 5-Hydroxytryptophan and tryptamine had limited occurrence and were used to characterize the individual species in which they were found. Choline and imidazole acetic acid, which occurred widely in *Inocybe* species, were of limited value in the preparation of the key. Primary metabolites, such as asparagine, glutamic acid, and ornithine, were used to differentiate between species only when no other character was available. Any unusual colored spot on the chromatogram that appeared in only one of the species was considered for use in separating that species in the key. It was reasoned that since such constituents did not have a wide distribution, they were possibly secondary metabolites which were peculiar to the particular species.

The occurrence of ergothioneine appeared to delimit certain stirpes in the genus *Inocybe sensu* Stuntz. It was detected in all the species investigated in the stirpes *laetior, subbrunnea, obscura, terrifera, geophylla, subdestricta, cinnamomea, oblectabilis,* and *umbrina* and in *I. kauffmanii* A. H. Smith in the stirps *kauffmanii.* From this representative sampling it may be concluded that ergothioneine can be used as a measure of biochemical relationship. However, there appeared to be no definite relationship between the morphological features which separate the subgenera in *Inocybe* and the occurrence of ergothioneine in the stirpes. For example, the stirps *geophylla* and the stirps *umbrina* are not considered to be closely related on the basis of morphological characteristics; nevertheless, both stirpes have species that contain ergothioneine. This presents the problem of how to interpret both morphological and biochemical characteristics. Both are genetically controlled, but there is no finite knowledge which permits greater systematic significance to be attributed to either chemical or morphological characteristics.

Examination of the chemotaxonomic classification at the species level revealed some interesting relationships. For example, the two closely related species *I. nigrescens* and *I. xanthomelas* Boursier & Kühn. are separated only on the basis of the detection of γ-aminobutyric acid, glutamic acid, ornithine, and phenylalanine, all compounds with a low level of systematic significance. On the other hand, some species which are very similar morphologically are quite dissimilar chemically. Examples are *I. kauffmanii* and *I. picrosma,* which differ chiefly in that the latter species has an acrid odor and an incarnate stain at the base of the stipe not found in the former. Neither muscarine, ergothioneine, tyrosine, nor valine could be detected in *I. picrosma,* but all these constituents were present in *I. kauffmanii.* In general, the comparison between the chemotaxonomic classification and the morphological classification of the *Inocybe* species is very encouraging, and it suggests the utility of secondary chemical constituents in the study of the systematics of this difficult genus.

## The Genus Panaeolus

All species of this genus which have been examined by Tyler and coworkers (Tyler & Gröger, 1964a; Tyler & Malone, 1960; Tyler & Smith, 1963) were found to accumulate a number of 5-hydroxytryptamine derivatives. In contrast to similar compounds detected in *Amanita* species, none of the tryptamines in *Panaeolus* was *N*-methylated. Serotonin (5-hydroxytryptamine) and 5-hydroxytryptophan have now been detected in the following species: *P. acuminatus* (Secr.) Quél. and its form *gracilis* Tyler & Smith, *P. campanulatus* (Fr.) Quél., *P. foenisecii* (Fr.) Kühn., *P. fontinalis* A. H. Smith, *P. semiovatus* (Fr.) Lundell, *P. sphinctrinus* (Fr.) Quél., *P. subbalteatus* (Berk. & Br.) Sacc., and *P. texensis* Tyler & Smith. 5-Hydroxyindole acetic acid was also detected in relatively fresh specimens of *P. campanulatus*, *P. foenisecii*, and *P. acuminatus*. The concentration of serotonin in the two former species was determined quantitatively (Weir & Tyler, 1963) and found to be 0.008% and 0.024% respectively (fresh-weight basis). The latter concentration is one of the highest reported for plant material.

Biosynthetically speaking, serotonin is a relatively simple compound that is widely distributed in both animals and higher plants. In the Basidiomycetes it has been detected only in the genus *Panaeolus* and in two species of *Amanita*, *citrina* and *porphyria*. However, in *Amanita* the compound forms a relatively minor component of a very complex mixture of tryptamines, most of which are *N*-methylated. Contrariwise, in *Panaeolus*, serotonin is the major tryptamine present and is unaccompanied by detectable amounts of *N*-methylated derivatives.

A curious anomaly exists in the literature with respect to the tryptamines found in one species of this genus, *Panaeolus sphinctrinus*. Heim and Hofmann (1958) reported the isolation of psilocybin (4-phosphoryl *N*, *N*-dimethyltryptamine) from cultivated carpophores of their strain RPI. No experimental details of the analysis and no description of the specimens were provided. Furthermore, other studies on this species by the same authors have failed to confirm these results (Heim, 1963). Since the report was not in agreement with the analytical data obtained from all other members of the species examined, an effort was made to obtain an authentic collection of *P. sphinctrinus*. Analysis (Tyler & Gröger, 1964a) of carpophores determined by Singer to be the same species as the Mexican material originally collected by Schultes, and originally but erroneously reported to be used as a hallucinogen by Mexican Indians (Singer, 1958), yielded results consistent with those obtained for other species of *Panaeolus*. Serotonin and 5-hydroxytryptophan were present; psilocybin could not be detected. On the basis of these findings, two conclusions may be drawn regarding the reports of hallucinogenic activity or psilocybin in *P. sphinctrinus.* They are based either on misidentifications of this difficult species or on the existence of different chemical races of it. The same conclusions must be drawn regarding reports attributing hallucinogenic activity to *P. subbalteatus* (Stein, Closs, & Gabel, 1959), a species also shown to contain serotonin and 5-hydroxytryptophan but to be devoid of psilocybin (Tyler & Smith, 1963).

Heim (1967) believes that the detection of psilocybin in a *Copelandia* species (Heim, Hofmann & Tscherter, 1966) supports the possible occurrence of that compound in *P. sphinctrinus.* This is based on the fact that *Copelandia* could be considered as a section of *Panaeolus.* Our own opinion, supported by chemical data, is that *Copelandia* is rightfully segregated from *Panaeolus,* not merely by differences in the occurrence and character of the cystidia but also by the tendency of the context to discolor bluish on exposure to the air. This color change which has been studied by Levine (1967) is a very useful biochemical marker, brought about by the dephosphorylation of psilocybin to psilocin and subsequent oxidation of that compound to form a fairly stable free radical with a deep blue color (see fig. 9).

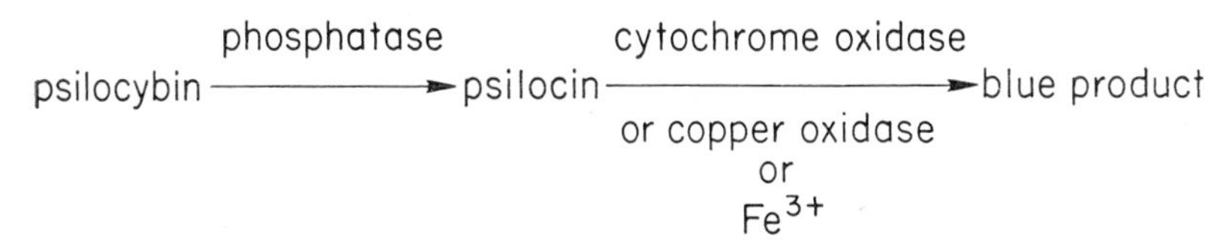


FIGURE 9. Reaction producing blue-colored product from psilocybin.

Since the color change is dependent upon the initial presence of psilocybin or psilocin, and since it is not characteristic of *Panaeolus* carpophores but is used by taxonomists to segregate members of *Copelandia* from this genus (Singer, 1962), it may be validly argued that psilocybin could not occur in a species of *Panaeolus* as normally defined.

The observation by Singer and Smith (1958) that one carpophore of *P. subbalteatus* out of several hundred had a cesious to bluish base and an earlier report that the cultivated sclerotia of this species have a striking bluish-green color require experimental clarification. Such coloration may be caused by pigments of a different chemical type than that resulting from the oxidation of psilocybin-psilocin, or in the latter case it may be attributed to difficulties in the identification of basidiomycete sclerotia. It is also conceivable that genetic changes induced by the artificial cultural conditions employed resulted in the formation of chemical races with unusual biosynthetic abilities. In any event, these observations cannot be utilized as a valid argument for the presence of psilocybin in normal carpophores of the genus *Panaeolus*.

## The Genera Psilocybe and Conocybe

The 4-substituted tryptamines, psilocybin and/or psilocin (see fig. 10), have been detected in a number of species of the section

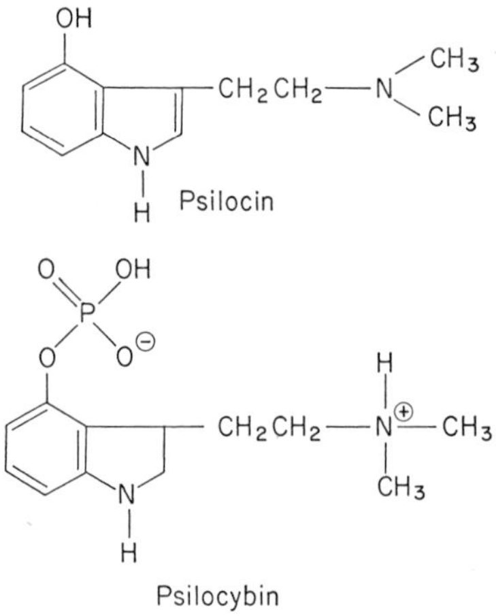

Figure 10. Structures of 4-substituted tryptamine derivatives.

*Caerulescentes* of the genus *Psilocybe*. Species in which the compounds have actually been detected by chemical means include: *P. aztecorum* Heim, *P. baeocystis* Sing. & Smith, *P. caerulescens* Murr., *P. cubensis* (Earle) Sing. (*Stropharia cubensis* Earle), *P. cyanescens* Wakefield, *P. fimetaria* (Orton) Watling, *P. mexicana* Heim, *P. mixaeensis* Heim, *P. pelliculosa* (A. H. Smith) Sing. & Smith, *P. quebecensis* Ola'h & Heim, *P. semilanceata* (Secr.) Kummer and its variety *caerulescens* Cooke, *P. semperviva* Heim & Cailleux, *P. wassonii* Heim, and *P. zapatecorum* Heim (Benedict *et al.*, 1962; Benedict, Tyler, & Watling, 1967; Heim & Hofmann, 1958; Hofmann, 1966; Hofmann, Heim, & Tscherter, 1963; Ola'h & Heim, 1967). Because of the relatively uncommon occurrence of tryptamine derivatives substituted in position 4, psilocybin and psilocin may be said to possess considerable chemotaxonomic value.

For this reason, the identification of psilocybin in *Conocybe cyanopus* (Atk.) Kühn. (Benedict *et al.*, 1962) and subsequently in *Conocybe smithii* Watling (Benedict, Tyler, & Watling, 1967) aroused considerable interest. In the first place, these findings tended to verify the earlier claim that a *Conocybe* species, *C. siligineoides* Heim, had been employed as a hallucinogen by the Mazatec Indians (Heim, 1963). Secondly, the occurrence must either be accepted as an indication that certain species of *Conocybe* and *Psilocybe* are more closely related than previously believed, or it is a remarkable example of the convergence of biochemical characteristics. At the present time, there are insufficient data available to permit a definitive judgment in this matter.

## The Genus Agaricus

Species of *Agaricus* are rich in free amino acids, several unusual derivatives of which seem to possess considerable potential for systematic utility. After Jadot, Casimir and Renard (1960) isolated *N*-(γ-L-glutamyl)-*p*-hydroxyaniline from *Agaricus hortensis* (Cooke) Pilát, Heinemann, and Casimir (1961) determined the occurrence of that compound in a number of species of the genus. Although the chromatographic procedures used by the latter investigators may not have yielded entirely unequivocal results, their findings are nevertheless of interest.

Of the 17 species of *Agaricus* examined, all contained the compound, at least in traces, and the only other specimen in which it was detected was the closely related *Lepiota rhacodes* (Vitt.) Quél. However, a large number of *Lepiota* species was not examined, and the authors speculate that others, especially those in which the context tends to redden when cut or bruised, may contain this compound.

Levenberg (1964) has isolated a similar compound from *A. bisporus* (J. Lange) Pilát, β-*N*-(γ-L(+)-glutamyl)-*p*-hydroxymethylphenylhydrazine (fig. 11), which has been given the trivial name agaritine. The compound was detected in 10 of 15 different species of *Agaricus* examined for its presence. The five species in which it could not be identified were all members of the *silvaticus* subgroup.

Several species appeared to contain closely related compounds which remain unidentified. In view of these results, the genus *Agaricus* should prove to be a very fertile field for chemotaxonomic investigation.

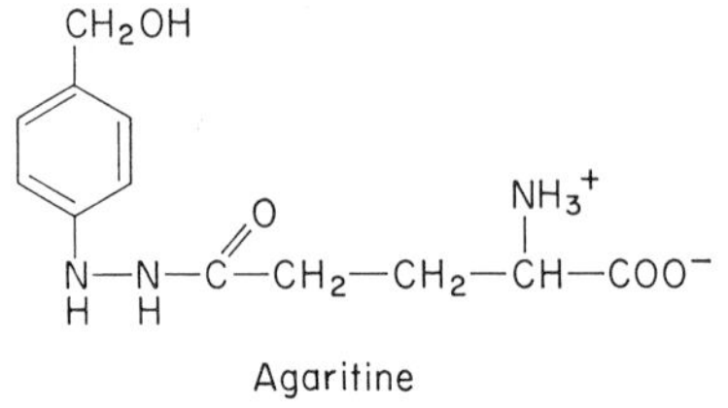

$CH_2OH$

$NH_3^+$

O

$\underset{H}{N}—\underset{H}{N}—C—CH_2—CH_2—CH—COO^-$

Agaritine

FIGURE 11. Structure of agaritine.

## THE GENUS CANTHARELLUS AND OTHER CAROTENOID-CONTAINING SPECIES

Carotenoids are of rather limited occurrence in Basidiomycetes. They are widely distributed in and may be considered typical of the genus *Cantharellus*, having been reported in *C. cibarius* Fr., *C. cinereus* Fr., *C. cinnabarinus* Schw., *C. friesii* Quél., *C. infundibuliformis* Fr., *C. lutescens* Fr., and *C. tubaeformis* Fr. (Fiasson & Arpin, 1967; Hegnauer, 1962). Fiasson and Arpin (1967) have discussed the interesting differences which occur in the composition of the carotenoids in some of these species. For example, in *C. tubaeformis*, partially saturated, aliphatic carotenoids of a "primitive" type predominate; a similar situation exists in *C. lutescens*. However, in other species, such as *C. cibarius*, *C. cinnabarinus*, and *C. friesii*, the carotenoids are of more "evolved" bicyclic, often oxygenated types. It is noteworthy that of all the chanterelles studied, only *C. tubaeformis* and *C. lutescens*, the two species with predominantly "primitive-type" carotenoids, also are the only two possessing hollow stipes.

Arpin (1967) has also demonstrated the presence of carotenoids in *Gerronema venustissimum* (Fr.) Sing. [=*Clitocybe venustissima* (Fr.) Sacc.], but this does not necessarily indicate any real affinity between this species and the genus *Cantharellus*. The principal pigment in *G. venustissimum* is $\gamma$-carotene which has been found only in traces in *Cantharellus* species and which occurs in a number of types of fungi other than Basidiomycetes. Incidentally, this species of *Gerronema* has been placed by some authorities in the genus *Hy-*

*grophoropsis*, but this seems particularly unsuitable since Arpin has shown that the pigment in the type species of this genus, *H. aurantiaca* (Fr.) Maire apud Martin-Sans, although of a similar color, is actually not carotenoid in nature.

Singer's (1962) placement of *Clitocybe venustissima* in the genus *Gerronema* and his subsequent transfer of *Omphalia chrysophylla* (Fr.) Kummer to the same genus, although originally made on morphological grounds, are strongly supported by chemical evidence. Arpin's detection of carotenoids in *G. venustissimum* has now been followed by the identification by Fiasson and Bouchez (1968) of carotenoids in *G. chrysophyllum* (Fr.) Sing. Although the composition of the carotenoid fractions of the two species is not identical, this is nevertheless a remarkable example of the chemical confirmation of relationships established on the basis of other criteria. Since *G. chrysophyllum* shows a number of similarities to chanterelles on both morphologic and chemical grounds, it does raise the question as to the validity of exclusion by some authors of the latter group from the Agaricales.

## The Genus Hydnellum

Sullivan, Brady, and Tyler (1967) investigated the occurrence of certain terphenylquinones (see fig. 12), including atromentin, aurantiacin, dihydroaurantiacin dibenzoate, and thelephoric acid in seven species and varieties of *Hydnellum*. Thelephoric acid was identified in all species and is probably a characteristic of the genus, even though its occurrence is by no means restricted to *Hydnellum* species (Shibata, Natori, & Udagawa, 1964). No correlation was found between the presence or absence of clamp connections, which Maas Geesteranus (1957) used as a significant feature in his classification of these species, and the distribution of terphenylquinones.

It was concluded that differentiation of the species based on their terphenylquinone content was possible but not practical, since useful morphologic criteria are more readily evaluated. However, in the case of *H. scrobiculatum* (Secr.) Karst., its two varieties *scrobiculatum* and *zonatum* (Fr.) Harr., which are difficult to distinguish by classical means, may be readily separated on the basis of the occurrence of aurantiacin in the latter variety and its absence from the former.

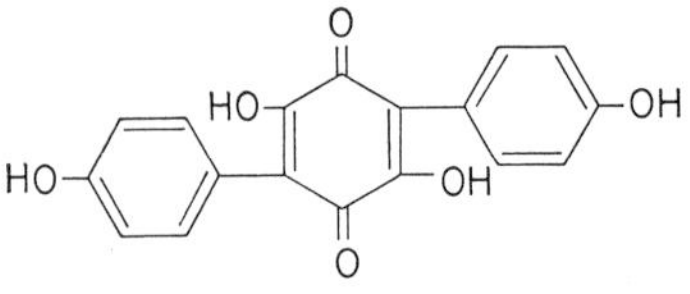

Atromentin

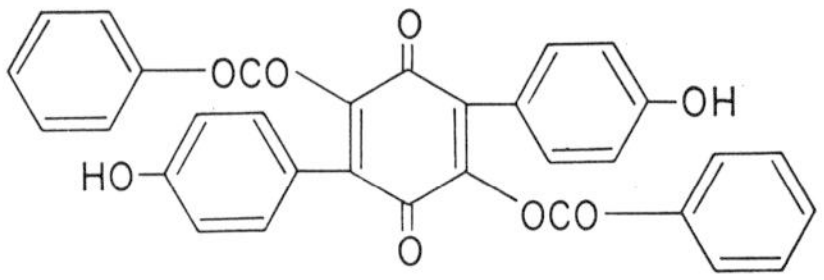

Aurantiacin

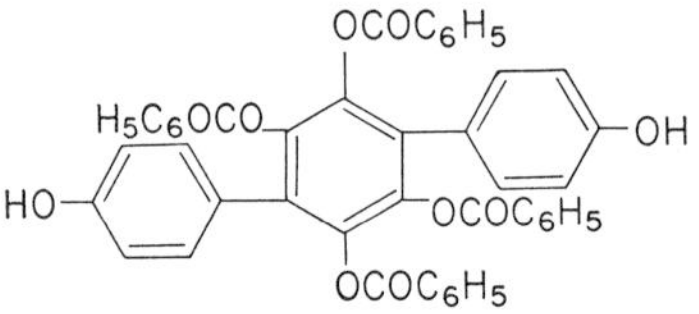

1,2,4,5-Tetrakis(benzoyl)-3,6-bis(*p*-hydroxyphenyl)benzene

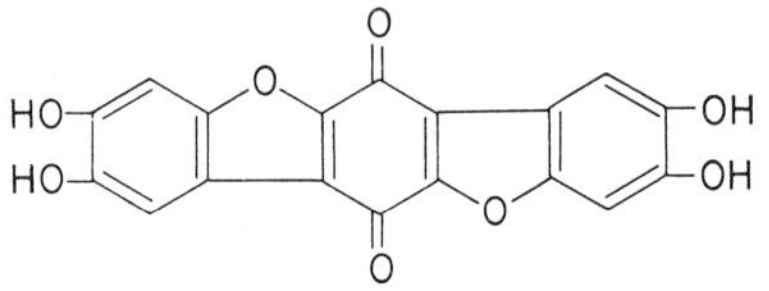

Thelephoric acid

FIGURE 12. Structures of terphenylquinones.

## THE GENUS TRICHOLOMA

A preliminary chemotaxonomic appraisal of the genus *Tricholoma* has been conducted by Benedict *et al.* (1964). Some interesting results were obtained based on the chromatographic patterns of compounds in the mushroom extracts which formed colored spots with Ehrlich's reagent (*p*-dimethylaminobenzaldehyde). The data obtained suggested that *Armillaria zelleri* Stuntz & Smith and *Tricholoma aurantium* (Fr.) Ricken are closely related to each other and to *Tricholoma saponaceum* (Fr.) Kummer. Consideration should be given to transferring the two former species to subgenus I, *Contextocutis*, of the genus *Tricholoma* where *T. saponaceum* is now positioned.

Urea, which is readily detectable by this method, was found to be absent from all *Tricholoma* species examined except *T. sclerotoideum* Morse. Singer (1962) has mentioned the possibility of transferring this species to *Clitocybe* based on its similarity to *Clitocybe inornata* (Fr.) Gill. This proposed transfer is supported by the chemical data which indicate that *Clitocybe* species ordinarily contain appreciable amounts of urea (Tyler, Benedict, & Stuntz, 1965).

*T. inamoenum* (Fr.) Quél. and *T. sulphureum* (Fr.) Kummer

showed identical chromatographic profiles of Ehrlich-positive spots. This raises the question as to whether they are actually different species or merely varieties of the same basic type.

## The Families Boletaceae and Strobilomycetaceae

Because of their relative biosynthetic simplicity and wide occurrence, the carbohydrates are generally considered to be of little value in systematics. The viewpoint has been confirmed by a recent study of the distribution of sugars and sugar alcohols in species of Boletaceae and Strobilomycetaceae (Benedict & Tyler, 1969). Examination of 18 species representing 7 different genera of these two families revealed a relatively uniform carbohydrate pattern in all.

Of the six sugars and sugar alcohols identified in the various species, three (glucose, trehalose, and mannitol) were uniformly present. A fourth, arabitol, was generally present but could not be detected in *Xerocomus chrysenteron* (St. Am.) Quél. or *Boletellus zelleri* (Murr.) Snell, Sing. & Dick. The distribution of fructose and heptulose (probably sedoheptulose) was much more limited; fructose was found only in *Boletus miniato-olivaceous* var. *sensibilis* Peck and two *Leccinum* species, heptulose in *Leccinum aurantiacum* (St. Am.) S. F. Gray and *Boletellus zelleri.*

Taxonomic conclusions which may be drawn from these data are meager. Some may prove of value at the species level, but a much greater number of species will have to be examined before this can be definitely established.

## The Subclass Gastromycetes

Demoulin (1967) has reported the results of a chromatographic analysis of extracts of 13 different species of Gastromycetes and has drawn a number of interesting taxonomic conclusions from them. The occurrence of homocystine and possibly other sulfur-containing amino acids or peptides was restricted to *Scleroderma* species. High concentrations of urea and the presence of an unidentified fluorescent compound characterized all species of the Lycoperdaceae examined. The distribution of γ-guanidobutyric acid was restricted to species of Lycoperdaceae and to *Geastrum triplex* Jungh. An alkaloid-like compound detected in the latter species may possess considerable taxonomic utility.

## The Chemotaxonomic Significance of Urea

In order to demonstrate the systematic utility of a single, easily identified, structurally simple yet biosynthetically complex compound, Tyler, Benedict, and Stuntz (1965) undertook a comprehensive investigation of the distribution of urea in 344 species of higher fungi, including 335 species of Basidiomycetes. On the basis of the analytical results, the species were grouped into three categories, (*a*) urea-negative, (*b*) urea-nonaccumulating, and (*c*) urea-accumulating species. The wide scope of the study provided a considerable number of results of taxonomic interest, of which only a few significant examples can be summarized here.

Seemingly conflicting results were obtained with the genus *Collybia*. Of the four species examined, two (*C. acervata* [Fr.] Kummer and *C. maculata* [Fr.] Kummer) were urea-negative and two (*C. confluens* [Fr.] Kummer and *C. dryophila* [Fr.] Kummer) were urea-accumulating species. However, as noted by Smith (1949), difficulty has always been encountered in distinguishing *Collybia* species from *Marasmius*, and intermediate forms were normally classified as *Collybia*. Since all *Marasmius* species examined were urea-positive, the data support transfer of *C. confluens* and *C. dryophila* to that genus. Transfer of the urea-negative *Marasmiellus albus-corticus* (Secr.) Sing. from *Marasmius* is also supported by these results (see table 3).

In the Strophariaceae, species of the genus *Stropharia* did not accumulate large amounts of urea, but that compound was consistently detected in moderate amounts except in two urea-negative species, *S. hornemanii* (Fr.) Lund. & Nannf. and *S. rugosoannulata* Farlow apud Murr. Both of these species had previously been classified by Singer (1949) in the genus *Naematoloma*, in which urea was either absent or barely detectable in all species examined. *Stropharia* is now separated from *Naematoloma* by the noncellular hypodermium and regular hymenophoral trama in adult specimens, but the urea accumulation patterns cast some doubt upon the absolute specificity of these characters and suggest that a reinvestigation of the problem might prove fruitful (see table 4).

The urea accumulation pattern detected in the genus *Rhodophyllus* was particularly interesting, and its interpretation proved challenging. Of five species examined, two were urea-negative, two contained intermediate concentrations, and one possessed a very high concentration of urea. Attempts to reconcile these results revealed that the two urea-negative species, *R. sericellus* (Fr.) Quél. and *R. serrulatus* (Fr.) Quél., are classified by Dennis, Orton, and Hora (1960) as species of *Leptonia*. The two species containing moderate

TABLE 3. UREA CONTENT OF *Collybia* AND *Marasmius* SPECIES

| Species | Urea content |
|---|---|
| *Collybia acervata* | − |
| *Collybia maculata* | − |
| *Collybia confluens* | + |
| *Collybia dryophila* | + |
| *Marasmius cohaerens* | + |
| *Marasmius oreades* | + |
| *Marasmius plicatulus* | + |
| *Marasmius scordonius* | + |
| *Marasmiellus albus-corticus* | − |

TABLE 4. UREA CONTENT OF *Stropharia* AND *Naematoloma* SPECIES

| Species | Urea content |
|---|---|
| *Stropharia aeruginosa* | + |
| *Stropharia ambigua* | + |
| *Stropharia coronilla* | + |
| *Stropharia semiglobata* | + |
| *Stropharia hornemannii* | − |
| *Stropharia rugosoannulata* | − |
| *Naematoloma capnoides* | ± |
| *Naematoloma dispersum* | ± |
| *Naematoloma fasciculare* | − |

concentrations, *R. nidorosus* (Fr.) Quél. and *R. sinuatus* (Fr.) Sing., are classified in the genus *Entoloma*, and the species rich in urea, *R. sericeus* (Fr.) Quél., in the genus *Nolanea*. These results are based on a limited sample and must not be accepted as conclusive. Nevertheless, they indicate that useful results would most probably be obtained by an extended study of the occurrence of urea in *Rhodo-*

*phyllus* species (see table 5). I am pleased to report that such a study is now in progress at the University of Washington.

It was hoped that studies of urea accumulation patterns might provide objective evidence to support existing hypotheses of the

TABLE 5. UREA CONTENT OF *Rhodophyllus* SPECIES

| Species | [Section-Singer] | Urea content |
|---|---|---|
| *Rhodophyllus sericellus* | Candidi | − |
| *Rhodophyllus nidorosus* | Romagnesia | + |
| *Rhodophyllus serrulatus* | Leptonia | − |
| *Rhodophyllus sericeus* | Sphaerospori | ++ |
| *Rhodophyllus sinuatus* | Entoloma | + |

| Species [Reclassified-Dennis *et al.*] | Urea content |
|---|---|
| *Leptonia sericella* | − |
| *Leptonia serrulata* | − |
| *Entoloma nidorosum* | + |
| *Entoloma sinuatum* | + |
| *Nolanea sericea* | ++ |

phylogenetic relationships of the higher fungi. These presently include: (*a*) derivation of Agaricales from Polyporales, (*b*) derivation of Agaricales from Gastromycetes, and (*c*) derivation of Agaricales from both Polyporales and Gastromycetes. Available evidence led to the conclusion that the capacity to accumulate large amounts of urea is absent in primitive forms of higher fungi, that it is acquired in relatively advanced forms, but that it disappears again as development continues.

Unfortunately, these findings prevented any definitive conclusion. Since we may expect the most primitive and the most advanced forms of Agaricales to be deficient in urea, and since the pertinent species of Polyporales and Gastromycetes which have been postulated as giving rise to or descending from the Agaricales are also devoid of urea, it is not possible to substantiate the course of phylogenetic development from this evidence. The distribution of urea in the Basidiomycetes neither supports nor denies the origin of the Agaricales from Gastromycetes or Polyporales.

## Future Applications of Chemotaxonomy

During the course of this presentation a number of specific problems have been mentioned, about which, it is felt, significant information could be obtained by application of the chemotaxonomic method. Most of these are situations where differences of opinion have arisen because sufficient data to permit definitive judgments are not available. Chemical characteristics will prove most useful in solving such problems, for such information may be viewed as additional evidence of an objective nature supporting the more subjective histological and morphological features of the fungi.

Eventually it may be possible to develop an entire system of chemical classification based, for example, on the sequence of nucleotides in the DNA molecules of various fungal species. However, until techniques suitable for such sophisticated analysis are perfected, information must be based on the ability of an organism to accumulate secondary metabolites which reflect in an indirect and imperfect manner the true genetic composition. We are still very much uninformed about the true relationship of such chemical data to biological form and structure. Until much additional information is obtained, chemotaxonomy will continue to be an imperfect science. However, to ignore the valuable contribution which this, as yet, unperfected discipline can make to systematics would be unwise. To paraphrase the words of Charles F. Kettering, systematic mycology "must have a certain amount of intelligent ignorance to get anywhere."

## References

Alston, R. E., T. J. Mabry, and B. L. Turner. 1963. Perspectives in chemotaxonomy. Science 142: 545–552.

Alston, R. E., and B. L. Turner. 1963. Biochemical Systematics. Prentice-Hall, Inc., Englewood Cliffs, N. J. 404 p.

Arpin, N. 1967. Recherches chimiotaxinomiques sur les champignons. Sur la présence de carotènes chez *Clitocybe venustissima* (Fries) Sacc. Compt. Rend. 262: 347–349.

Benedict, R. G., and V. E. Tyler. 1969. Occurrence of sugars and sugar alcohols in the Boletaceae. Herba Hungarica 7: 17–21.

Benedict, R. G., V. E. Tyler, Jr., and L. R. Brady. 1966. Chemotaxonomic significance of isoxazole derivatives in *Amanita* species. Lloydia 29: 333–342.

Benedict, R. G., V. E. Tyler, and R. Watling. 1967. Blueing in *Conocybe*, *Psilocybe*, and a *Stropharia* species and the detection of psilocybin. Lloydia 30: 150–157.

Benedict, R. G., *et al.* 1962. Occurrence of psilocybin and psilocin in certain *Conocybe* and *Psilocybe* species. Lloydia 25: 156–159.

Benedict, R. G., *et al.* 1964. Preliminary chemotaxonomic appraisal of certain *Tricholoma* species. Planta Med. 12: 100–106.

Bonnet, J.-L. 1959. Application de la chromatographie sur papier a l'étude de divers champignons (Basidiomycètes-Hyménomycètes). Bull. Soc. Mycol. France 75: 215–352.

Brown, J. K. *et al.* 1962. Paper chromatographic determination of muscarine in *Inocybe* species. J. Pharm. Sci. 51: 853–856.

Catalfomo, P., and V. E. Tyler, Jr. 1961. Investigation of the free amino acids and amanita toxins in *Amanita* species. J. Pharm. Sci. 50: 689–692.

Dawson, R. F. 1960. Biosynthesis of the *Nicotiana* alkaloids. Am. Scientist 48: 321–340.

Demoulin, V. 1967. Intérêt de certaines substances a fonctions basiques en chimiotaxinomie des Gastéromycètes. Bull. Soc. Mycol. France 83: 342–353.

Dennis, R. W. G., P. D. Orton, and F. B. Hora. 1960. New check list of British agarics and boleti. Trans. Brit. Mycol. Soc., Suppl. June: 1–225.

Erdtman, H. 1963. Some aspects of chemotaxonomy, pp. 89–125. *In* T. Swain, ed., Chemical plant taxonomy. Academic Press, New York.

Eugster, C. H. 1967. Über den Fliegenpilz. Naturforschenden Gesellschaft in Zurich, Switzerland. 39 p.

Fiasson, J.-L., and N. Arpin. 1967. Recherches chimiotaxinomiques sur les champignons. V.—Sur les caroténoïdes mineurs de *Cantharellus tubaeformis* Fr. Bull. Soc. Chim. Biol. 49: 537–542.

Fiasson, J.-L., and M.-P. Bouchez. 1968. Recherches chimiotaxinomiques sur les champignons. Les carotènes de *Omphalia chrysophylla* Fr. Compt. Rend. 266: 1379–1381.

Fries, N. 1958. Paper chromatography as a diagnostic aid in Hymenomycetes. Ann. Acad. Reg. Sci. Upsalien. 2: 5–16.

Gilbert, E. J. 1940. Amanitaceae. Iconographia Mycologia (Suppl. 1) 27: 1–425.

Hänsel, R. 1956. Pflanzenchemie und Pflanzenverwandschaft. Arch. Pharm. 289: 619–628.

Hegnauer, R. 1957. Botanisch-systematische Betrachtungen zur Biologie der Alkaloide. Abhandl. Deut. Akad. Wiss. Berlin, Kl. Chem. Geol. Biol. 1956(7): 10–17.

———. 1958. Chemotaxonomische Betrachtungen. VI. Phytochemie und Systematik: Eine Rück- und Vorausschau auf die Entwicklung einer Chemotaxonomie. Pharm. Acta Helv. 33: 287–305.

———. 1962. Chemotaxonomie der Pflanzen Bd. 1. Thallophyten, Bryophyten, Pteridophyten und Gymnospermen. Birkhäuser Verlag, Basel. 517 p.

———. 1963. The taxonomic significance of alkaloids, pp. 389–427. *In* T. Swain, ed., Chemical plant taxonomy. Academic Press, New York.

Heim, R. 1942. Les pigments des champignons dans leurs rapports avec la systématique. Bull. Soc. Chim. Biol. 24: 48–79.

———. 1958. Les critères d'ordre chimique dans l'étude des affinités chez les Macromycetes. Uppsala Univ. Årsskr. 6: 48–58.

———. 1963. Les champignons toxique et hallucinogènes. Éditions N. Boubée & Cie, Paris. 327 p.

———. 1967. Conclusions. II. Études d'ordre chimique, pp. 215–216. *In* R. Heim, ed., Nouvelles investigations sur les champignons hallucinogènes. Muséum National d'Historie Naturelle, Paris.

Heim, R., and A. Hofmann. 1958. La psilocybine et la psilocine chez les Psilocybes et Strophaires hallucinogènes, p. 262. *In* R. Heim and R. G. Wasson, eds., Les champignons hallucinogènes du Mexique. Muséum National d'Historie Naturelle, Paris.

Heim, R., A. Hofmann, and H. Tscherter. 1966. Sur une intoxication collective à syndrome psilocybien causée en France par un *Copelandia.* Compt. Rend. 262: 519–523.

Heinemann, P., and J. Casimir. 1961. Distribution des acides aminés libres dans le genre *Agaricus* Fr. sensu stricto (=*Psalliota*). Rev. Mycol. 26: 24–33.

Hofmann, A. 1966. Alcaloïdes indoliques isolés de plantes hallucinogènes et narcotiques du Mexique, pp. 223–241. *In* Phytochimie et plantes médicinales des terres du pacifique. Centre National de la Recherche Scientifique, Paris.

Hofmann, A., R. Heim, and H. Tscherter. 1963. Présence de la psilocybine dans une espèce européenne d'Agaric, le *Psilocybe semilanceata* Fr. Compt. Rend. 257: 10–12.

ISAACS, B. F., AND V. E. TYLER, JR. 1963. β-Amanitin in an amanita from Oregon. Mycologia 55: 124–127.

JADOT, J., J. CASIMIR, AND M. RENARD. 1960. Séparation et caractérisation du L (+)-γ-(*p*-hydroxy) anilide de l'acide glutamique à partir de *Agaricus hortensis.* Biochim. Biophys. Acta 43: 322–328.

LEVENBERG, B. 1964. Isolation and structure of agaritine, a γ-glutamyl-substituted arylhydrazine derivative from Agaricaceae. J. Biol. Chem. 239: 2267–2273.

LEVINE, W. G. 1967. Formation of blue oxidation product from psilocybin. Nature 215: 1292–1293.

LUCKNER, M. 1963. Über neue Arbeiten zur Biosynthese der Alkaloide. 3. Teil: Die Bildung von Verbindungen mit Chinolinringsystem. Pharmazie 18: 93–107.

MAAS GEESTERANUS, R. A. 1957. The stipitate hydnums of the Netherlands. II. Fungus 27: 50–70.

MALONE, M. H., *et al.* 1961. A bioassay for muscarine activity and its detection in certain *Inocybe*. Lloydia 24: 204–210.

———. 1962. Relative muscarinic potency of thirty *Inocybe* species. Lloydia 25: 231–237.

MOTHES, K. 1966. Biogenesis of alkaloids and the problem of chemotaxonomy. Lloydia 29: 156–171.

OLA'H, G.-M., AND R. HEIM. 1967. Une nouvelle espèce nord-américaine de *Psilocybe* hallucinogène: *Psilocybe quebecensis* G. Ola'h et R. Heim. Compt. Rend. 264: 1601–1604.

ROBBERS, J. E., L. R. BRADY, AND V. E. TYLER, JR. 1964. A chemical and chemotaxonomic evaluation of *Inocybe* species. Lloydia 27: 192–202.

SHIBATA, S., S. NATORI, AND S. UDAGAWA. 1964. List of fungal products. Charles C. Thomas, Springfield, Ill. 170 p.

SINGER, R. 1949. The Agaricales (mushrooms) in modern taxonomy. Lilloa 22: 1–832.

———. 1958. Mycological investigations on teonanácatl, the Mexican hallucinogenic mushroom. Part I. The history of teonanácatl field work and culture work. Mycologia 50: 239–261.

———. 1962. The Agaricales in modern taxonomy. 2nd ed. J. Cramer, Weinheim, Germany. 915 p.

SINGER, R., AND A. H. SMITH. 1958. Observations on agarics causing cerebral mycetisms. IV. About the identity of the weed panaeolus or poisonous panaeolus. Mycopathol. Mycol. Appl. 9: 280–284.

SMITH, A. H. 1949. Mushrooms in their natural habitats. Sawyer's Inc., Portland, Ore. 626 p.

STEIN, S. I., G. L. CLOSS, AND N. W. GABEL. 1959. Observations on psychoneurophysiologically significant mushrooms. Mycopathol. Mycol. Appl. 11: 205–216.

SULLIVAN, G., L. R. BRADY, AND V. E. TYLER, JR. 1967. Occurrence and distribution of terphenylquinones in *Hydnellum* species. Lloydia 30: 84–90.

TALBOT, G., AND L. C. VINING. 1963. Pigments and other extractives from carpophores of *Amanita muscaria.* Canad. J. Bot. 41: 639–647.

THEOBALD, W., *et al.* 1968. Pharmakologische und experimental-psychologische Untersuchungen mit 2 Inhaltsstoffen des Fliegenpilzes (*Amanita muscaria*). Arzneimittel-Forsch. 18: 311–315.

TYLER, V. E., JR. 1961. Indole derivatives in North American mushrooms. Lloydia 24: 71–74.

TYLER, V. E., JR., R. G. BENEDICT, AND D. E. STUNTZ. 1965. Chemotaxonomic significance of urea in the higher fungi. Lloydia 28: 342–353.

TYLER, V. E., JR., AND D. GRÖGER. 1964a. Occurrence of 5-hydroxytryptamine and 5-hydroxytryptophan in *Panaeolus sphinctrinus.* J. Pharm. Sci. 53: 462–463.

———. 1964b. Investigation of the alkaloids of *Amanita* species. II. *Amanita citrina* and *Amanita porphyria.* Planta Med. 12: 397–402.

TYLER, V. E., JR., AND M. H. MALONE. 1960. An investigation of the culture, constituents, and physiological activity of *Panaeolus campanulatus.* J. Am. Pharm. Assoc., Sci. Ed. 49: 23–27.

TYLER, V. E., JR., AND A. H. SMITH. 1963. Protoalkaloids of *Panaeolus* species. Abhandl. Deut. Akad. Wiss. Berlin, Kl. Chem. Geol. Biol. 1963(4): 45–54.

TYLER, V. E., JR., *et al.* 1966. Occurrence of amanita toxins in American collections of deadly amanitas. J. Pharm. Sci. 55: 590–593.

WEIR, J. K., AND V. E. TYLER, JR. 1963. Quantitative determination of serotonin in *Panaeolus* species. J. Pharm. Sci. 52: 419–422.

WIELAND, T. 1968. Poisonous principles of mushrooms in the genus *Amanita*. Science 159: 946–952.

## DISCUSSION

MILLER: You made a rather strong statement on the quality of biochemical and morphological characters. On what taxonomic level are we talking? Genus? Family? Species?

TYLER: I think my statement was that we are ignorant as to whether one kind of taxonomic character is more significant than another at the present time. I will stand on my last statement, that in the end chemical taxonomy will win out, so to speak, on the basis of the determination of the sequence of nucleotides in DNA. Once this has been catalogued for all living things, then we can tell what they are and to what extent they are related, regardless of what taxonomic level.

I think at the present time that it is much more useful at the species and genus level. I don't want to preclude something from happening in the future, and that is why I answered as I did at first. At the present time there is no question but that its principal application is at the species and genus level, and groups of genera.

ROGERS: It seems to me the criteria of chemotaxonomy have to be evaluated in terms of other character correlations—their taxonomic distribution and the hypothesis for the derivation of the compounds concerned—and this evaluation is just the same as we give other taxonomic characters. There isn't any reason why chemical characters should be of any less or any different importance from morphological ones. I can't see why equivalence here is not entire.

SMITH: Of the members of the Strophariaceae you mentioned, it seems to me that *Stropharia hornemanii* and *S. rugosa-annulata* were really very closely related stirpes—a very small group that is related both by morphological and chemical characters. In my present thinking I think of *Psilocybe*, *Stropharia*, and *Naematoloma* as one genus because of the particular pattern of distribution of morphological characters and the arbitrariness of decisions one must make to segregate a set of genera. I think in chemotaxonomy that we have related groups of all sizes. It all depends on how the characters are used.

FERGUS: It would be ideal if we could analyze the DNA in an organism and I think ultimately we will do so. That would be the final answer. Yet you say it is extremely precarious to analyze the chemical constituents of the secondary somatic mycelium and to compare them with those of the mycelium of the carpophore, which is the tertiary mycelium. Yet the DNA must be the same. We are going to have to know something besides just the order of the messengers of the DNA that operate from the DNA. We are going to have to know what is repression and suppression, induction and shutting off the genes, and the sequence thereof. I am struck by the incongruity of our present knowledge concerning DNA content of secondary mycelium and the tertiary mycelium, as reflected by the difference in the chemical composition of somatic, vegetative, and assimilating hyphae versus the carpophore hyphae.

TYLER: We must also consider the ecological situation in which the organism finds itself, because we know that ecology modifies form and structure. In a sense it is like the blind man and the elephant. Everything depends on what part of the creature you are examining. If you are looking at the cuticle of *Amanita muscaria* you would say that it is red, but if you take a section from the stipe it is white.

SMITH: We are talking about a sequence of metabolic events. When a mycelium starts to grow, it is primarily in a food-gathering state. When it starts to fruit it is primarily in a food-utilization state. I think it is well known that there are many metabolic changes that take place in one of these fruit bodies from youth to age and that this is the essence of what we are talking about. At a certain age the DNA dictates that a certain compound shall be made.

PETERSEN: Even when the mycelium is growing in the ground and not fruiting it has not lost the gene for the capacity to form a fruiting body. The DNA has to be the same in the nucleus of the vegetative hypha as it is in the nucleus of the basidium. If we analyze DNA of that taxon, it should be the same throughout.

HEIM: I was very interested in your very excellent exposition. I was interested also because some matters are very close to me. I have four items I would like to discuss. The first is the problem of *Panaeolus*. You mentioned a publication by Heim and Hofmann (1958) in which we reported the isolation of psilocybin from *Panaeolus* in chromatographic columns. The presence of psilocybin is frequent in *Panaeolus*. These problems are being investigated by Dr. G. Ola'h.

Point two will perfectly underline this problem. The example of the hallucinogenic mushrooms is one of the most interesting of all the problems concerned with the relations between chemical composition and taxonomic position. We can say that in *Psilocybe*, all species of Dr. Singer's group *Caerulescentes*, which is a perfect group within *Psilocybe s.l.*, contain psilocybin and frequently psilocin. *Stropharia cubensis*, which is very close, contains psilocybin, psilocin, and a substance which is toxic. Finally, some of these compounds are found in *Panaeolus*, some *Stropharia*, all *Psilocybe* gr. *Caerulescentes*, and a few *Conocybe* which are very close.

The third point is concerned with reports that bufotenine is hallucinogenic. The molecular configuration (OH on position 5) dictates to the contrary. Bufotenine is not hallucinogenic.

The fourth point is quite different. I wonder if there is any *Amanita muscaria* in the eastern United States. I have seen many specimens from Vermont and New Hampshire, but they are not exactly those of Europe. Moreover, the problem of color is not connected with muscarine. There are strong indications that it is the presence of a vanadium complex which gives color to *A. muscaria*. This is quite like *Craterellus cornucopioides*, whose color is related to a manganese salt. I suppose that in certain regions of eastern North America there also may be many varieties which possibly cannot assimilate a vanadium complex. In the red variety, it is possible to do this. I think this is a very perplexing problem, but I reserve my position due to the lack of evidence of eastern American *A. muscaria* and other forms or subspecies of this group. Of course, all the varieties of *A. muscaria* with yellow color are different.

TYLER: It appears from what we know now about *Panaeolus* that there will be a large number of chemical races involved and that these may change under cultural conditions and according to the part of the country in which the material is collected. All of our material was from the Pacific northwest, and all I can do is to report the results as we found them.

I agree with you entirely about bufotenine. Bufotenine is not a psychotomimetic drug unless it is injected into the bloodstream, and we are not talking about that kind of thing here. *Clitoycbe gallinacea*, which I believe you mentioned, is a psychotomimetic mushroom and is indeed an interesting species. I believe there is a report from Hungary concerning the fact that it contains lysergic acid derivatives. We have never been able to verify this because we have never been able to collect the species, but it may well contain something like lysergic acid or at least some other psychotomimetic compound.

With respect to our *Amanita muscaria*, I am unable to say much about the eastern United States varieties because those specimens with which we worked were all from the Puget Sound area in western Washington where this species occurs exactly as it occurs in Europe. I have collected it in Germany as well, and the specimens are identical, at least chemically. We do have these varieties of pileus color in which there is a yellow form and a white form as well as the very dark red form.

MILLER: In your work with *A. muscaria* and *A. pantherina* on the characterization of ibotenic acid, did you intimate that you found that you could not work with dried material? Must you use fresh specimens immediately secured in the field?

TYLER: We tested for ibotenic acid in herbarium samples up to eight years old and found that in the specimens quantities declined with age until after about three years it was very uncertain as to whether we could detect its presence or not. In some of the four- and five-year-old samples, ibotenic acid could not be detected by our method, so we concluded that in normally stored herbarium samples the acid does disappear over a period of a very few years. We tried extracting young buttons as well as older material and found essentially no difference on a dry-weight basis.

DUBUVOY: The chemical races you obtained in the laboratory could also lead to ecological races. Have you done studies on the same species collected in several ecological areas? Is there a qualitative or quantitative chemical difference?

TYLER: I can answer your question only in terms of those species with which I have worked. It would be very dangerous for me to try to generalize. In the case of *A. muscaria* which I had collected from Seattle to Germany and in many places between, all were very similar. We also have data on differences between fruiting bodies and mycelial cultures. Our cultures of *Psilocybe cubensis*, for example, produce psilocybin very nicely in fermentative shake cultures. In our *Panaeolus* species, however, although the mycelium grows very well, it contains none of the hydroxytryptamines in which we are interested. So a lot depends first on your ability to cultivate the mycelium of the organism in culture, then the ability of that mycelium to produce the secondary constituents under the specific conditions being studied. Where produced, the quantity of the compounds is comparable on a weight-to-weight basis with that found in the carpophore. As a matter of fact, with *P. cubensis*, some of our better yields in a good medium went up to more than 1% in one week, which is probably a higher concentration than in the carpophore.

POMERLEAU: Did you say that *A. brunnescens* would have to be separated from *A. citrina*?

TYLER: *A. citrina*, *A. porphyria*, and *A. tomentella* produce very large amounts of bufotenine and related hydroxytryptamines. These are not present in *A. brunnescens*, nor are the peptide toxins.

SINGER: How do you explain the results of the classical studies of Ford, who evidently worked with *A. brunnescens* or what he called *A. phalloides* and who reported the presence of amanita toxins in this group?

TYLER: Ford worked with preparations and extracts which were usually administered by injection into small animals. There might be all kinds of materials that would be toxic used in this way that would not act the same if ingested, and that might in this way lead us astray. I think that in order to be sure one has to look at the phenomenon uniformly. We also know there are differences in animal sensitivity.

HASCKAYLO: Is there any correlation between the accumulation or utilization of urea and the mode of existence of the fungus—whether it is a purely saprophytic organism and fruits saprophytically or whether it is a symbiotic fungus which requires a mycorrhizal association to produce a carpophore?

TYLER: The accumulation of urea is widely distributed sporadically in many of these groups. I have not been able to correlate it with mycorrhizal relationships.

NOËL ARPIN and JEAN-LOUIS FIASSON

*Faculté des Sciences, Université de Lyon, Lyon, France*

# THE PIGMENTS OF BASIDIOMYCETES: THEIR CHEMOTAXONOMIC INTEREST[1]

## Introduction

The color of living things has always been used by man to categorize them; this has been particularly so in the case of mushrooms (Pastac, 1942), whose colors present a much clearer and more varied range in certain groups than morphology. However, pigmentation has long been used by the mycologist, not only to define species but also to classify them. It is useful to recall that the first major classification of the Agaricales was based on the mass coloration of their spores, and in more than one case the color of the fructification is homogeneous enough in a group, species, or subspecies for the taxon to be classified by it. Such classifications based on a similar coloration imply a common origin, a common chemical composition involved. In that way, Fries's classification of the Agaricales was already chemotaxonomic. The color, however—the macroscopic and subjective quality—is a weak element. Fries's classification, therefore, has often been criticized, and Heim (1941) enumerated cases where it was mistaken. On one hand, the colors arranged in one group were in fact the most varied; on the other, in the Leucosporae one finds species with really colorless spores side by side with those whose spores are extremely pale in color. To understand the sources of mistakes which can arise from the crude use of the color factor, and, consequently, to warn mycologists against hasty taxonomic generalization, we believe it useful to make the following comments:

First, the same coloring can be the result of two totally different structures, which, far from demonstrating a relationship, indicate on the contrary a divergence in orientation of the chemical composition. Thus *Clitocybe venustissima* (Fries) Sacc. and *Hygrophoropsis au-*

[1] No. 17 of a series, "Recherches chimiotaxinomiques sur les Champignons. Preceding publication: Gluchoff, K. 1970. Etude chimiotaxinomique des pigments des Russules. Bull. Soc. Natur. Archeol. Ain (in press).

*rantiaca* (Wulfen ex Fr.) R. Maire present orange colors which are very similar and could *a priori* support a grouping of the two species. Whereas the coloring agent of the first is carotenoid, however, that of the second is entirely different (Arpin, 1966).

Conversely, a difference in color can arise from a simple variation in overall richness in the same pigment complex. *Phillipsia carminea* (Pat.) Le Gal, colored bright red, and *Cookeina sulcipes* (Berk.) Kuntze, pinkish-fawn in color, have some pigmentary stocks which are extremely close in nature and proportion to the various carotenoids present. These are present in much greater pure proportions, however, in the former (Arpin & Liaaen Jensen, 1967a, b).

Similarly, two taxa clearly distinct in color can possess extremely closely related pigments from a chemical or biogenetic point of view: for this to be so it is sufficient for them to differ only at the level of the enzyme which determines the conversion of one compound to another. From *Polyporus leucomelas* Pers. ex Fr., Akagi (1942) extracted a pigment called leucomelone, and, in a slightly greater quantity, a colorless derivative of this called protoleucomelone. It is reasonable to suppose that, in certain colorless species related to this mushroom, all the pigment might be found in the colorless form.

All that is necessary for this to be so is a single gene change, a minimal modification without great taxonomic importance. The study of mutations, experimental or natural, should make us cautious in this sense. Easily used for the elucidation of the biosynthetic chains of pigments, they have furnished some strains with very varied colors which, nevertheless, clearly belong to the species from which they are produced. Let us emphasize that frequently the pigment in question is present, but in the form of a precursor of a different color, or even with no color.

Finally, a color sometimes results from the superposition of two types of very different pigments: *Sowerbyella unicolor* (Gill) Nannfeldt possesses some carotenoid pigments and others entirely different (Arpin, 1968). It is similar in this way to *Leotia lubrica* Pers. and other Geoglossaceae (Arpin, 1968) where, thanks to their preferential solubility, it is easy to separate some bluish-green hydrosoluble pigments from some yellowish-orange lipophilic compounds and carotenoid pigments.

These remarks about the natural coloring of the mushroom (or of some parts of it) apply equally to operations carried out by the mycologist to arrive at his taxonomic decision. We are thinking here both of spontaneous color reactions produced by bruising or sectioning, and coloration appearing by the action of chemical reagents of one kind or another on the macro- or microscopic level, as is so usual in mycology. Their value depends obviously on their frequently unknown specific features; even when the compounds in question are identified, the meanings of negative results can be very varied. Thus, the blueing of certain Boletaceae upon breaking corresponds to the oxidation of the "boletol" by the oxygen of the air in the presence of a phenoloxydase. It may not take place if either an enzyme or the boletol is missing. This second possibility can mean that there is an alteration, perhaps small, or that there is a complete absence of the chain of necessary biogenetic reactions. The taxonomic value of these various possibilities is clearly very different.

All this shows that the apparent common identity of two colors (or their difference) is only a very superficial indication which must be supported with much more precise data. Before the physico-chemical techniques had reached the mycology laboratories, the characteristics of solubility and the cytologic localization of pigments, in direct relation to their structure, had already furnished precious indications from which systematists were able to draw great benefit (Kühner, 1934). Let us note, however, that these data were not in themselves conclusive. The carotenoids, for instance, are not the only yellow, liposoluble compounds which can color cytoplasmic droplets. Conversely, particularly in the case of oxygenated pigments, the successive elements of a single biosynthetic pathway can have different solubilities and hence different topographies. Only chemical identification brings strict criteria. But it can furnish more than that. Indeed, if a distribution table of defined (but not identified) chemical compounds brings taxonomic elements of great value, the knowledge of the structure of these molecules and of their biogenetic relationship can demonstrate some unsuspected relationships between compounds with very different properties, particularly of color, thus opening the door to some fundamental phyletic hypotheses. Indeed, if we confirm that compounds A, B, and C are derived biochemically from one another in a certain sequence, it is legitimate to postulate their appearance in the same order in the course of biological evolution. Analysis, then, furnishes not only an oriented but literally quantified view of evolution. Care must be taken in the application of this principle of recapitulation of phylogeny by biogenesis—a regeneration here fully justified by Haeckel's law, where the superposition of chemical evolution and biological evolution permits the analysis of the latter by the former. A phyletic derivation can be accompanied by a loss as well as a gain of biogenetic capacity and, following some chain reactions, can regress.

## Summary of the Principal Chemical Structures Encountered

### *Pigments with a Fundamentally Hydrocarbon Skeleton*

These compounds are clearly the hydrocarbons and their derivatives. Their lipophilic nature explains their presence in cytoplasm where they are found in lipoid droplets or in the form of granulations or crystals.

#### *Pigments of Lactarius deliciosus*

The skeleton of these pigments, at present known only in *Lactarius deliciosus* (Willstaedt, 1946 a,b), results from the joining of a heptagonal and a pentagonal ring. The structures of two molecules are now known: that of lactarazulene (fig. 1), a blue pigment, and of lactaroviolin (fig. 2), forming reddish-violet crystals.

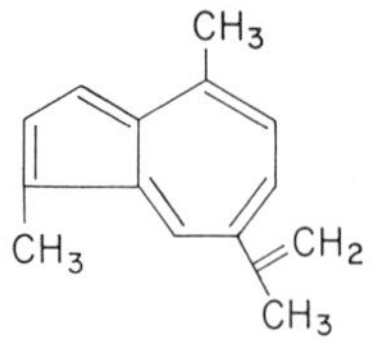

FIGURE 1. Lactarazulene.

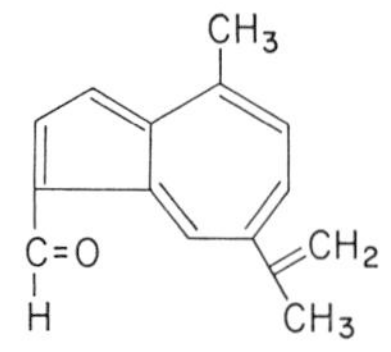

FIGURE 2. Lactaroviolin.

From a biosynthetic point of view the aldehyde function of lactaroviolin, whose structure was revealed by Plattner *et al.* (1954) and Heilbronner and Schmidt (1954), originates from the oxidation of the corresponding methyl group of lactarazulene, whose structure was previously ascertained by Sorm, Benesova, and Herout (1954).

*Pigments of Certain Corticiaceae*

Erdtman (1948) isolated from *Corticium croceum* Bres. (= *C. sulphureum* [Fr.] Fr.) a pigment of a polyethylene nature which he named corticrocin (fig. 3).

FIGURE 3. Corticrocin.

Some years later, from *C. salicinum* Fr. growing on the dead branches of *Salix*, Gripenberg (1952) extracted a new pigment of ethylenic type, which he called cortisalin (fig. 4). This differed from corticrocin by possessing two extra carbon atoms in the aliphatic chain, and replacement of one of the two carboxylic groups by a *p*-hydroxyphenyl group.

FIGURE 4. Cortisalin.

*The Carotenoids*

The yellow, orange, or red pigments which we call carotenoids are liposoluble substances possessing for the most part 40 carbon atoms. Their hydrocarbon skeleton presents a system of conjugated double bonds–(lycopene)–fig. 4[1], more-or-less important, which are responsible for various colors. This skeleton results from the "head to tail" joining of eight isoprenic units with a central symmetry of the molecule. The differences between the very numerous known carotenoids (over 150) come largely from modifications occurring at the extremities of the molecule: cyclization, dehydrogenation, and oxygenation. All these molecules are derived biosynthetically from colorless, saturated compounds such as phytoene, with the number of conjugated double bonds slightly raised.

Carotenoid pigments are present in the most varied groups of both superior and inferior fungi (see Lederer, 1938); however, it is at the level of the Deuteromycetes and Ascomycetes that they appear to be proportionally most prevalent.

## *Pigments of a Quinonoid Nature*

The quinones, widespread throughout the vegetable kingdom, are particularly prevalent among fungi. In 1960, out of 80 known anthraquinones, more than half had been isolated from mycelial or lichen cultivation (the fact that none were isolated from algae, but only from lichens, indicates that, among the latter, it is the mycosymbiont which is responsible for the synthesis of the molecules).

The quinones, derived from aromatic, colorless skeletons, owe their color to the presence in the molecule of the chromophore C=O; some groups, such as -OH,-$OCH_3$ without being chromophores, can modify the molecule's color.

### *Benzoquinones*

Extremely abundant in *Penicillium* and *Aspergillus*, their structural study was dealt with particularly by Raistrick, Robinson, and Todd (1933). Two representatives of this type of quinone have been isolated among the Basidiomycetes: (*a*)methoxy-4, either toluquinone or coprinin (fig. 5), from *Coprinus similis* Berk. & Br. and from

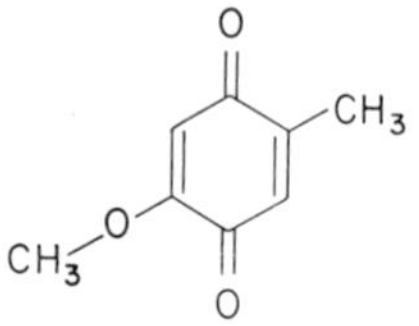

FIGURE 5. Methoxy-4 toluquinone.

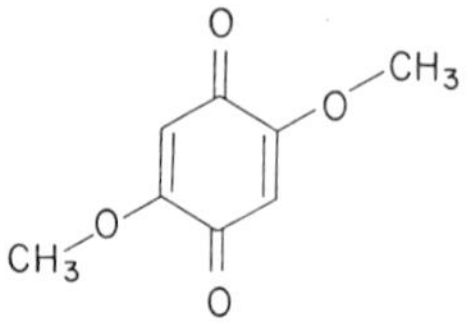

FIGURE 6. Dimethoxy-1,4 benzoquinone.

*Lentinus degener* Kalchbr., sowed in a partly natural environment for cultivation (Anchel *et al.*, 1948); (*b*)dimethoxy-1, 4 benzoquinone (fig. 6) from *Polyporus fumosus* (Pers.) Fr. cultivated in the same conditions (Anslow, Ashley, & Raistrick, 1938).

### *Pigments Derived from P-diphenyl Benzoquinone*

Our principal knowledge of these molecules is derived from the old work of Kögl *et al.* (1928) and from the more recent work of Gripenberg (1958 a,b, 1960). The credit must go to Kögl for having ascertained the structure of polyporic acid (fig. 7), which appears in the form of a violet color following the reaction of ammonia on the carpophore of *Phaeolus rutilans* (Pers.) Fr., and in certain lichens (Murray, 1952).

The structure of another acid (fig. 8) has recently been ascertained by Gripenberg (1960). He has, moreover, effected its synthesis.

Other colorants, atromentin (fig. 9), leucomelone (fig. 10), aurantiacin (fig. 11), and muscarufin (fig. 12), possess the same basic

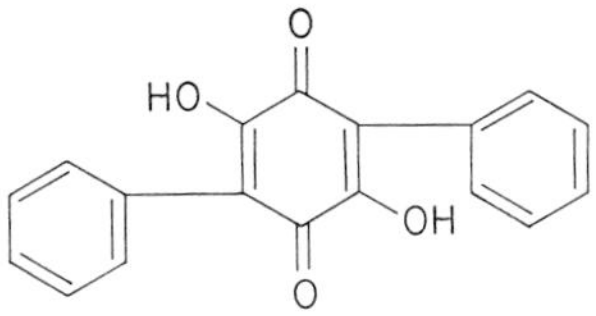

FIGURE 7. Polyporic acid.

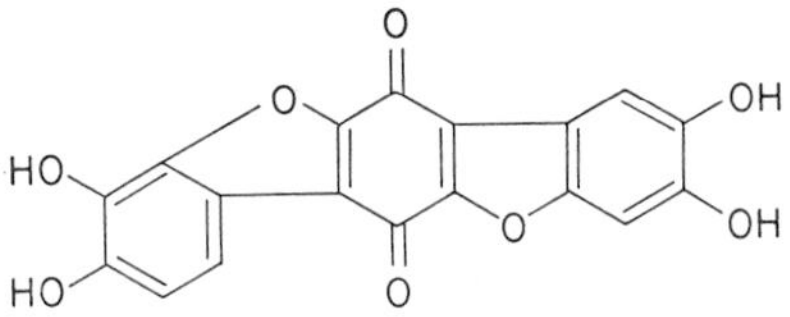

FIGURE 8. Thelephoric acid.

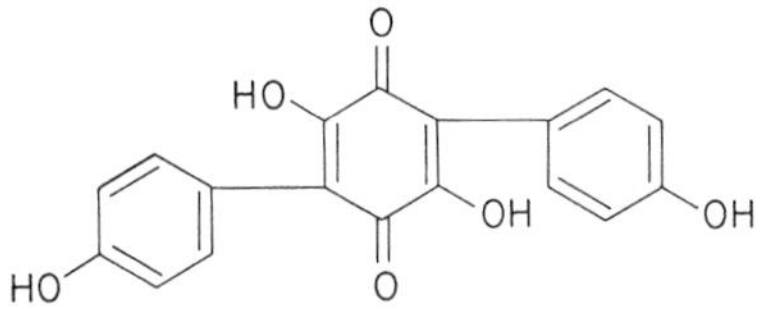

FIGURE 9. Atromentin.

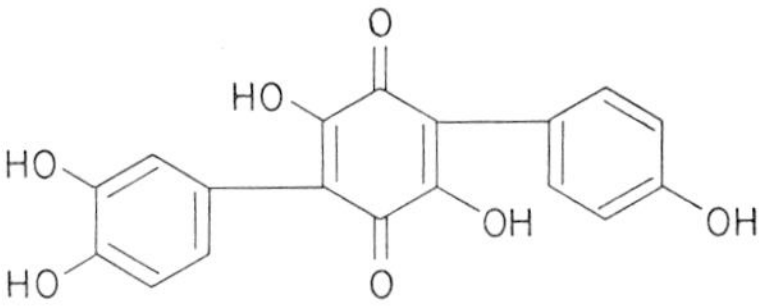

FIGURE 10. Leucomelone.

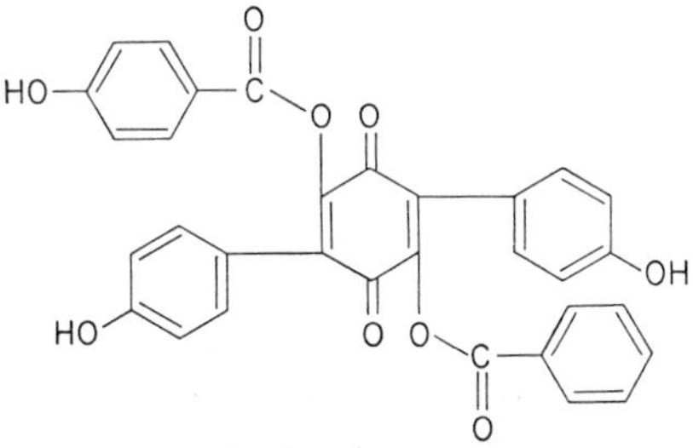

FIGURE 11. Aurantiacin.

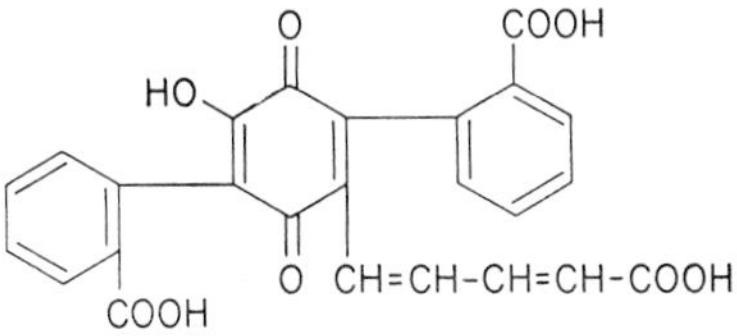

FIGURE 12. Muscarufin.

structure onto which are grafted various smaller parts: OH, COOH, $C_6H_5$.

*Naphthoquinones*

These are encountered in numerous fungi and other microorganisms, but little known among the Basidiomycetes. Bendz (1948), however, isolated methyl-6 naphthoquinone (fig. 13) from *Marasmius graminum* Lib.

O
CH3
O

FIGURE 13. Methyl-6 naphthoquinone.

*Anthraquinones*

In spite of the classic usage of these molecules in dyeing because of their exceptional stability, anthraquinone structures (figs. 14–16) remain poorly understood. In fact, although it is relatively easy to know the nature and number of substituents, their position is most

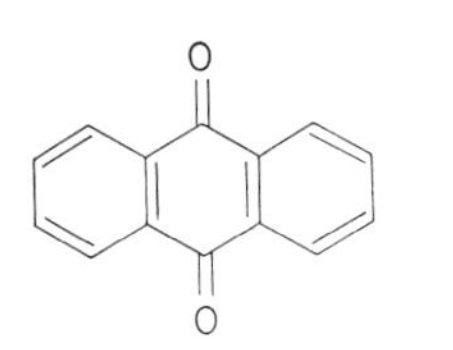


FIGURE 14.
Anthraquinone structure.

(α)8
1(α)
(β)7
2(β)
(β)6
3(β)
(α)5
4(α)

FIGURES: 15 & 16.
Isomeric positions
on anthraquinone model.

difficult to establish: for an anthraquinone with five substituents the number of possible isomers is 210. Work at the school of Raistrick has been a decisive help in elucidating the structure of these molecules found among the lower mushrooms. Two specific molecules isolated are emodin (fig. 17) and dermocybin (fig. 18).

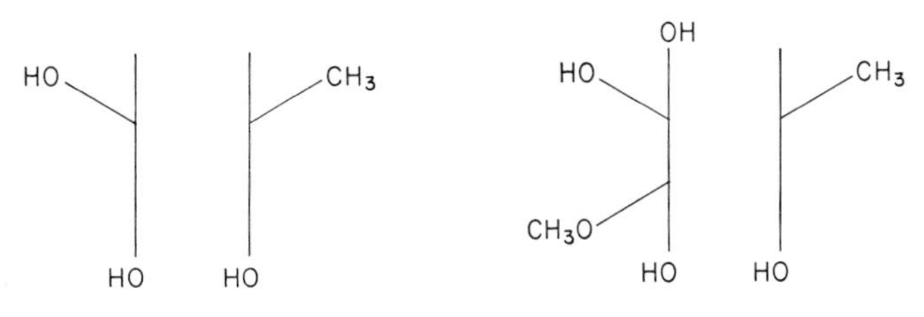


FIGURE 17. Emodin.

FIGURE 18. Dermocybin.

*The Nitrogenized Pigments of the Phenoxazine Type*

It is again to Gripenberg (1963) that we owe the isolation and structural classification of the red pigments of certain *Trametes*. Even though it was already known that an alkaline hydrolysis of those pigments would liberate an ammonia molecule, it was only in 1958 that the structure of the three compounds was established.

Cinnabarin (fig. 19) and cinnabaric acid (fig. 20) were isolated

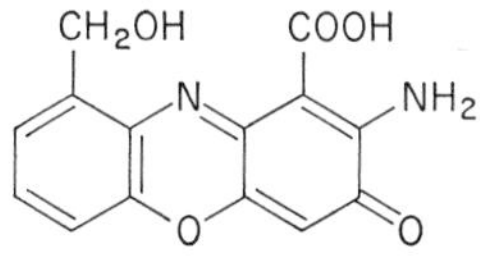

FIGURE 19. Cinnabarin.

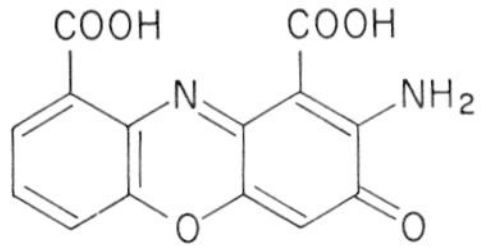

FIGURE 20. Cinnabaric Acid.

from *Trametes cinnabarina* (Jacq.) Fr. (= *Polystictus sanguineus* L.). An African sample from the former Belgian Congo yielded tramesanguin (fig. 21) from *T. cinnabarina* var. *sanguinea* (L.) Pilát. Ac-

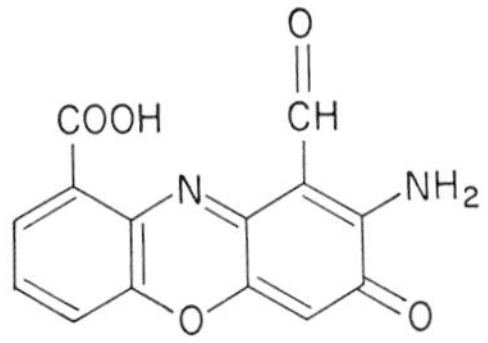

FIGURE 21. Tramesanguin.

cording to Gripenberg, three different types of mushrooms exist able to synthesize the nucleus phenoxazine: (*a*) those which synthesize only cinnabarin; (*b*) those which synthesize cinnabarin and cinnabaric acid; and (*c*) those which synthesize cinnabarin and tramesanguin.

The last two types belong to African species; *T. cinnabarina* as gathered in North America (Pennsylvania) and in Finland totally lacks cinnabaric acid.

## The Cortinarii

Although the presence and distribution of the anthraquinonoid pigments of numerous species of cortinarii were studied by Mme. Gabriel (1965) it is Kühner (1949) to whom we owe the microtopographic study of the colorants of this vast group of Ochrosporae. We shall first consider the localization of pigments; then we shall examine in more detail the case of anthraquinonoid colorants. They will provide us with an excellent example of the systematic interest it is possible to find in such research.

### *Different Types of Pigments Met among the Cortinarii*

We envisage them by decreasing generality of systematic distribution.

#### *Membrane Pigments*

Like most Ochrosporae the cortinarii possess certain yellowish-brown pigments that are very hard to extract. The pigments are positioned most often on the surface of the pileus but can also be found in the various parts of the fruit body.

#### *Cytoplasmic Pigments*

These are generally blue pigments, which unlike those above are found almost exclusively within this group. Except for *C. violaceus*, which owes its color to the presence of dissolved compounds in the vacuole, the pigments are in the cytoplasm. One characteristic is their very great lability. Because of it they are only easily distinguished in the carpophore while it is growing. It has been confirmed that carotenoids do not enter into the question.

*Extracellular Pigments*

Epicellular or intercellular, yellow, yellowish-green, and changing easily to red when acted on by alkalis, they are present in the subgenus *Dermocybe*, principally in various species of the *Cinnamomei* section of Kühner and Romagnesi (1953) as well as in the subgenus *Phlegmacium.* These pigments, anthranolic in nature, are greatly responsible for the pigmentation of *Phlegmacium* and for the flourescence of these species under the Wood light.

*Vacuolar Pigments*

Frequent in the subgenus *Dermocybe* Fr. *sensu* Orton (section *Sanguinei* and *Cinnamomei* of Kühner & Romagnesi, 1953) and in the subgenus *Cortinarius* (=*Inoloma* Fr.), they color the different parts of the carpophore yellow, pink, red, sometimes light brown in the case of *C. melanotus* and violet in the case of *C. violaceus.* They are much less widespread in the other subgenera. Nevertheless, we have discovered these pigments in certain *Hydrocybe* Fr. of the *Miniatopodes* section, such as *C. bulliardi,* or again in some species (*C. elegantior* and *odorifer*) of the subgenus *Phlegmacium.*

We must emphasize that a carpophore can present simultaneously pigments in very different localizations. Thus, according to Kühner, in the case of *C. concinnus,* the yellow color of the lamella corresponds to the anthranols (interhyphoid pigments) whereas the brick-red color of the pileus is a result of the superposition of the reddish-pink color of vacuolar pigments and the brown of the membrane pigments.

### *Correlations between Nature and Localization of Pigments*

The extracellular and vacuolar pigments are hydrophiles and easily extracted by ethanol. They are either anthraquinones, anthranols, or xanthones(?), the latter having hitherto only been met in glycosidoid form in the subgenus *Cortinarius* (= *Inoloma* Fr.). The membrane pigments, like those of the cytoplasm so common in the cortinarii, have a structure which as yet remains unknown. In any case, let us occupy ourselves only with hydrosoluble compounds in order to sum up the taxonomic conclusions which Mme. Gabriel (1965) has drawn from pigment study.

## *Hydrosoluble Pigments of the Cortinarii: the Systematic Significance in their Study*

*Subgenus Dermocybe Fr. sensu Orton* (table 1)

Fourteen hydrophile pigments have been isolated from this group. Twelve of these give the characteristic reactions of the anthraquinones, and one of an anthranol, although the exact nature of the latter

TABLE 1. SUBGENUS *Dermocybe*. DISTRIBUTION OF PIGMENTS.

| | Group *Sanguinei* | | | Group *Cinnamomei* | | | | | | |
|---|---|---|---|---|---|---|---|---|---|---|
| | *sanguineus* | *semi-sanguineus* | *phoeniceus* | *croceus* | *conformis* | *malicorius* | *cinnamolutescens* | *sphagneti* | *uliginosus* | *olivaceo-fuscus* |
| 1 | o o o | | | | | o o | | | | |
| 2 | o | | | | | | | | | |
| 2′ | | | | o o o | o o o | | o o o | o o | o o | |
| 3 | o o o o | | | | | o o o o | | | | |
| 3′ | o | o o | o o | | | | | | | |
| 4 | o o o | o o o | o o o | | | | | | | |
| 5 | o o o o | o o o | o o o | o o o o | o o o o | o | o o o o | o o | o o o o | |
| 6 | o | o o o o | o o o | o o | o o o | o o o | o o o o | o o o | o o o o | |
| 6′ | | | | | | | | | | o o |
| 7 | | | | o o | o | | o o | o o | o | o o o o |
| 8 | | | | o | o o | | o o | o | o o o o | |
| 9 | | | | o | o o | o | o o o | o o o o | o o o o | |
| 10 | | | | | | | o o | | | |
| 11 | o | | o | | | | | | | |
| 12 | | o | o | | o | o | | | | |
| 13 | | o o | | | | | | | | |

does not appear to be known. The number of pigments varies in each species, the average being six or seven. Even though their anthraquinonoid nature is recognized, the structure of each pigment is not entirely clear, with two exceptions, namely emodin and dermocybin. The first is present in *C. sanguineus* and *C. malicorius* Fr., the second only in *C. sanguineus*. Table 1 summarizes the presence of the different pigments of each species and permits us to group them into three categories as Mme. Gabriel has indicated:

*1. Pigments Characteristic of One or Two Species.* Ex. emodin for *C. sanguineus* and *C. malicorius*, dermocybin for *C. sanguineus*, pigment n⁰ 10 for *C. cinnamolutescens* Hemy.

2. *Pigments Characteristic of One Section.* The *Sanguinei* section includes pigments 3′ and 4, the *Cinnamomei* section is completely without them. However, it is within the latter that one encounters anthranoles, pigment 2′ forming bright yellow lumps, and in most cases, pigments 7, 8, and 9.

3. *Pigments Characteristic of a Subgenus.* These are clearly pigments 5 and 6, which abound in all species of *Dermocybe* with the exception of *C. olivaceofuscus,* which in terms of pigments appears exceptional.

Mme. Gabriel noted that compounds 5 and 6 had an $R_f$ intermediate between those raised from *Sanguinei* and those which were low in the same chromatographic system from *Cinnamomei.* The fact that pigment 6 was present in other subgenera of *Cortinarius* allowed us to consider it to be the type pigment of the group.

Certainly all these compounds appear closely linked to one another, probably only differing in the nature and position of one or several small parts in the anthraquinonic nucleus. It would, nevertheless, be interesting to know precisely the chemical structure of each of them. Perhaps then, we could understand why pigment 3 appears only in the two emodin species, and why only the anthranol species (pigment 2) contain pigments 7 and 8, which are absent from *C. malicorius* which is also devoid of anthranol.

*Subgenus Hydrocybe Fries*

This subgenus includes species which show hygrophanous morphology. In the *Miniatopodes* group, some species like *C. bulliardi* Fr. ex Pers. are colored by vacuolar compounds. Kühner and Romagnesi (1953) also noted that *C. cinnabarinus* Fr., placed in *Dermocybe* (*Sanguinei*), had a hygrophanous pileus. Moreover, Kühner underlined the great resemblance of the color of the stipes of *C. cinnabarinus* and *C. bulliardi.* From that basis it was interesting to compare the pigments of the two species. Results obtained by paper chromatography allow us to suggest the probability of a very close relationship between the two species. Indeed, the four anthraquinonic pigments isolated from *C. cinnabarinus* differ from those isolated from *Dermocybe* in general, but on the other hand, prove to be identical to those of *C. bulliardi.* Also, it seems right to consider *C.*

*cinnabarinus* "as a *Hydrocybe* of section *Miniatopodes* in which the red pigment, instead of being located at the base of the stem, covers the entire carpophore."

*Subgenus Phlegmacium*

The color of the four species of the *Scauri* group (*C. elegantior sensu* Moser, *C. erichalceus* Batsch ex Fr. *sensu* Moser, *C. odorifer* Britz., *C. atrovirens* Kalchbr.) as well as two others of the *Cliduchi* section (*C. percomis* Fr. and *C. nanceiensis* R. Maire) is essentially due to anthranols. Although these anthranols are very unstable, those of *C. percomis* and *C. nanceiensis* appear to be very close, if not identical. Let us point out as well that *C. elegantior* and *C. odorifer* have an anthraquinonic pigment very near or identical to the pigment 6 of *Dermocybe*.

*Subgenus Cortinarius*

It is to Henry (1937) that we owe the proof of the action of the silver nitrate on *C. cotoneus* Fr. *sensu* Quélet. The flesh of this species turns bright red in the presence of this reagent. Kühner (1960) stated that the same reaction occurred with two neighboring species, *C. melanotus* Kalchbr. and *C. venetus* Fries. This characteristic reaction supports anatomic characters (subglobose spores, very broad hyphae in the trama of the pileus) in separating very exactly *Olivascentes* of Kühner and Romagnesi from *Dermocybe*. Mme. Gabriel (1962) also noted the "beautiful yellow-scarlet fluorescence of the stipe, the lamellae and the flesh" as distinctive features. *C. cotoneus*, *C. venetus*, and *C. melanotus* have undergone a comparative analysis from which we can draw the following conclusions: (*a*) The pigment composition of the carpophores of the three species is very similar; it includes anthraquinonic pigments and for two of them the pigment no. 6 of *Dermocybe*. (*b*) The compound responsible in the three cases for the reaction of the silver nitrate is a hydrosoluble pigment with the ultraviolet spectrum characteristic of a xanthone. It is the first time that a pigment of this sort has been discovered in a mushroom, and its glycosidic form makes the discovery even more original. It must be stated, nevertheless, with Gatenbeck (1960) that from the biogenetic standpoint it is perfectly plausible to accept a relationship between the xanthones and the anthraquinones.

## The Boletes

Among the Basidiomycetes, the boletes (table 2) stand out with a property known for a very long time—the common bluing that takes place in the flesh on contact with the air. G. Bertrand (1902) showed the appearance of the blue coloration results from the oxidation of a yellow pigment which he named boletol. This oxidation is enzymatic (phenol-oxydase) and needs the presence of oxygen from the air.

Boletol, which is very lightly colored (yellow crystals), presents an absorption spectrum characteristic of an anthraquinone ($\lambda_M$ ethanol : 215, 257, 395 nm.) The presence of a carboxylic group explains the solubility of this compound in bicarbonate solutions. Adding a few drops of potato juice to a solution of boletol produces a marked change in color from yellow to green, then blue. Under the influence of the enzyme, boletol (fig. 22) becomes boletoquinone (fig. 23).

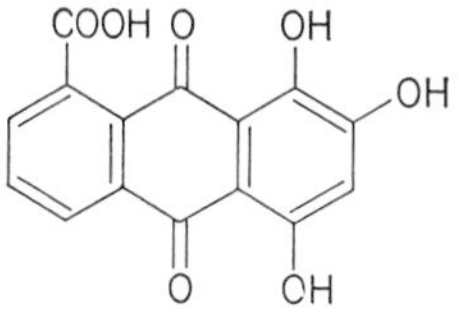

FIGURE 22. Boletol.

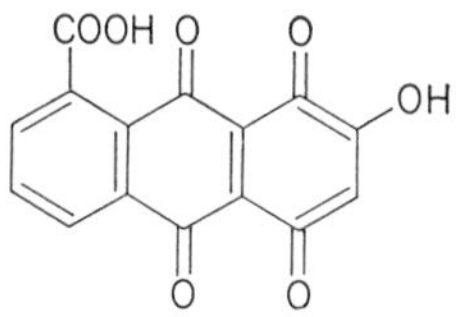

FIGURE 23. Boletoquinone.

In fact, the paradiquinone formed is red, but the presence of alkaline-earths makes it turn blue. The carboxylic moiety being possible either in position 5 or 8, there exist both a boletol and a pseudoboletol (=isoboletol). Their spectrum properties are identical, but their chromatographic properties differ slightly, permitting the separation of the two isotopes, which can be synthesized by the same carpophore, as is the case with *Xerocomus piperatus* and *X. parasiticus.*

Although the mechanism of coloration of the boletes has long been clear and the formulae of the compounds responsible known since the work of Kögl and Deijs (1935a, b), much less has been known about the distribution of the substances until recent years.[2] We are

[2] Recent work (Edwards et al. 1967. Chem. Conn. 373; 1968. J. Chem. Soc., 2968; Steglich et al. 1968. Z. Naturforschung 23:1044) has shown that the basic skeleton of the substances responsible for the blueing of bolete fruit bodies is not anthraquinonic but tetraonic-variegatic and xerocomic acids, not boletol and isoboletol respectively.

TABLE 2. BOLETES AND RELATED SPECIES. DISTRIBUTION OF BOLETOL AND PSEUDO-BOLETOL.

| Genera | Species | Blueing in air | Boletol | Pseudo-boletol |
|---|---|---|---|---|
| *BOLETUS* Dill. ex Fr. (= *TUBIPORUS* Karst.) | *edulis* | | * | |
| | *reticulatus* | | * | |
| | *pinicola* | | * | |
| | *calopus* | bl. | * | |
| | *appendiculatus* | bl. | * | |
| | *pulverulentus* | bl. | * | |
| | *luridus* | bl. | * | |
| | *erythropus* | bl. | * | traces |
| | *rhodoxanthus* (*purpureus*) | bl. | * | |
| *TYLOPILUS* Karst. | *felleus* | | | |
| *LECCINUM* S.F. Gray (= *KROMBHOLZIA* Karst.) | *nigrescens* (*crocipodius*) | | | |
| | *aurantiacum* | | | |
| | *carpini* | | | |
| | *scabrum* (*leucophaeus*) | | | |
| *PHYLLOPORUS* Quel. | *pelletieri* (*rhodoxanthus*) | | * | |
| *XEROCOMUS* Quel. | *chrysenteron* | bl. | * | |
| | *subtomentosus* | | * | |
| | *parasiticus* | | trace | * |
| | *badius* | bl. | * | |
| *BOLETINUS* Kalchbr. | *cavipes* | | * | |
| *SUILLUS* Micheli ex S.F. Gray (= *IXOCOMUS* Quel.) | *aeruginascens* (*viscidus*) | bl. | * | |
| | *tridentinus* | | | |
| | *grevillei* (*elegans*) | | | |
| | *flavidus* | | | |
| | *luteus* | | trace | |
| | *granulatus* | | | |
| | *placidus* | | * | trace |
| | *bovinus* | bl. | * | |
| | *variegatus* | bl. | * | |
| | *piperatus* | | * | * |
| *GYRODON* Opat. | *lividus* | bl. | * | |
| *GYROPORUS* Quel. | *castaneus* | | | |
| | *cyanescens* | bl. | | |
| *STROBILOMYCES* Berk. | *floccopus* (*strobilaceus*) | | | |
| *PORPHYRELLUS* Gil. | *pseudo-scaber* (*porphyrosporus*) | | | |
| *GOMPHIDIUS* Fr. | *helveticus* | | | * |
| | *viscidus* | | | * |
| | *glutinosus* | | | * |
| *PAXILLUS* Fr. | *atrotomentosus* | | | |
| | *panuoïdes* | | | |
| | *involutus* | | | |

now much better informed since the work of Mme. Gabriel (1965) who, working from a chemotaxonomic view, completed the sparse data furnished in this field by Bertrand (1901, 1902), Kögl and Deijs (1935a, b), and Gilbert (1931). This work, summarized in table 2, brings some very useful knowledge concerning the system of this group.

In most cases those results concerning only the presence or absence of boletol or pseudo-boletol confirm clearly the value of former taxonomic restrictions made by mycologists. Among the Boletoideae all nine species viewed as being of the *Boletus* type contain boletol. On the other hand, the *Leccinum* type is totally lacking in it. We believe, however, that the species of the latter type should possess some molecules of a related structure, which would explain why they blacken on reaction to phenol-oxydases, exactly as the boletol species react to these same enzymes by turning blue. One could question whether Singer (1962) was right to place the *Tylopilus* species among the Boletoideae, since *T. felleus* lacks boletol and does not blacken on exposure to the air.

The Xerocomoideae appears homogeneous from the point of view which concerns us. Nevertheless, the species *B. parasiticus* is distinguished by containing only a trace of boletol, but also by having a more significant quantity of pseudo-boletol, a pigment encountered in abundance in the Gomphidiaceae.

In the Suilloideae boletol is unequally distributed: two species (*B. elegans* and *B. tridentinus*) of section *Larigni* Singer (comprising the species of Meleze and those with a ringed stalk) and two others (growing under pine) of section *Granulati* (*B. luteus* and *B. flavidus*) contain neither boletol nor pseudo-boletol. On the one hand, Mme. Gabriel (1965) extracted yellow pigments from these species, one common to *B. flavidus* and *B. luteus*, the other common to *B. elegans* and *B. tridentinus*. These yellow pigments are of an anthraquinonic nature as is revealed by their chemical, chromatographic, and spectral properties. The spectral properties lead one to think that the anthraquinones possess several hydroxyls in position β, unlike boletol.

At the level of the Gyrodontoideae we note the presence of boletol in *Gyrodon lividus* and its absence in *Gyroporus*, in particular in *Gyroporus cyanescens*. Although former experimenters have noted boletol in *G. cyanescens*, it seems certain that this species in fact has none. The results of Mme. Gabriel thus would confirm the data provided by Gilbert (1931) and Singer (1962). Finally, in the two representatives of the Strobilomycetaceae studied, the total absence of boletol and pseudo-boletol has been confirmed.

Singer (1962) placed beside the Boletaceae two other families representing lamellar formation: the Paxillaceae and the Gomphidiaceae, as had Kühner and Romagnesi (1953) in their *Flore analytique des champignons supérieurs*. We know that the tubes of the boletes, following the gelification of their linings, lift easily from the flesh; besides, the reticulum of the tubes presents a clear two-sidedness. One finds both this gelification and the two-sidedness among the paxilli and the gomphidii. The affinity of certain lamellae to the boletes is confirmed by Mme. Gabriel's results. It appears from table 2 that the three species of gomphidii studied, *G. helveticus*, *G. viscidus*, and *G. glutinosus*, contain pseudo-boletol and therefore are seen to be related to the typical boletes. On the other hand, the paxilli differ from the gomphidii and from the true boletes by the total absence of boletol and of pseudo-boletol. Moreover, if we consider the spore form, the fusiform spores of the gomphidii are much more like those of the boletes and unlike the much more rounded spores of paxilli. In addition, we know from the work of Kögl *et al.* (1928) that in *P. atrotomentosus* (Batsch) Fr., the colorant is in fact a dihydroxyl derivative of polyporic acid, atromentin, of an entirely different nature from that of boletol. Extremely important in this species (2%), atromentin is found mainly in the form of leuco-derivative oxidizing on contact with the air. The fresh mushroom possesses white flesh and is colored only on the surface.

One species poses in itself the whole problem of the link between the typical and lamellate Boletaceae—*Xerocomus* (= *Boletus*) *subtomentosus* and *Phylloporus pelletieri* (= *P. rhodoxanthus*). These have sometimes been considered as two extreme aspects of the same species, but their boletol contents are unknown.

The distribution of boletol and pseudo-boletol must provide mycologists with excellent information for taxonomic classification of the boletes and neighboring groups. There is no doubt that future chemical study of such mushrooms will still prove very fruitful. Numerous other pigments, anthraquinonic and nonanthraquinonic, also exist in these species, and knowledge of their nature and distribution will doubtless make classification of this group more exact.

## The Lactarii and Russulas

These two genera, placed side-by-side since Persoon (1801), are defined as Agaricales with vesiculous flesh due to the presence of sphaerocysts. These elements are found in the laminae of the russulas, but not of the lactarii, and are considered evolved structures by Kühner. Many other characteristics link the lactarii and the russulas, however: many are mycorrhizal species; the mycelium always lacks clamps; both have representatives with bitter taste; their amyloid spores have a uniform decoration, reticulate or echinulate, which has led to the common group name of "asterosporeés."

Observation of various agarics, particularly "lactario-russulae" under the Wood lamp, showed Josserand and Netien (1938–1939) that whereas the russulas examined were brightly flourescent, all the lactarii were dull. These observations, confirmed later by the work of Deysson (1958), led Josserand and Netien to the following conclusions: "This maintains a division, at least small, between the two genera which modern systematics tends to unite." [tr.]

Kühner, following precise observations of pigment topography, stated that among the lactarii the pigments were only rarely vacuolar, usually being intercellular or on the membranes. He concluded, on the other hand, in favor of the almost general vacuolar nature of the pigments of the russulas. This explained the frequent discoloration of the pileus of certain russulas, a phenomenon never found among the lactarii. This gives us a glimpse of another difference resulting, it would appear, from the chemical nature of the coloring substances. Indeed, although our knowledge is very limited, Willstaedt (1946a, b) isolated pigments from *Lactarius deliciosus*, six of them lipophilic, and three hydrocarbons. Thanks to the work of Plattner *et al.* and Sorm *et al.* we know the structures of lactaroviolin and lactarazulene. A hydrocarbon, verdazulene ($C_{15}H_{16}$), has likewise been isolated from *L. deliciosus*. To our knowledge, these are the only chemical results that have been published about the pigments of the lactarii, and these obviously do not allow us to extend the lipophilic nature of the pigmentation observed in the one species studied to the whole group.

Unfortunately, we are hardly more advanced in the study of the pigmentation of the russulas (table 3), in spite of the importance of this character in distinguishing a number of them. The name "russularhodin" was proposed by Balenovic *et al.* (1955) to represent the major red pigment isolated from *Russula emetica*, but no structural hypothesis has been proposed as yet, in spite of the work currently going on in our laboratory (Gluchoff's unpublished results).[3]

TABLE 3. PIGMENTS EXTRACTED FROM *Russula* SPP.

| Group | Species | No. | coll. | $Y^1$ | $R^1$ | $R^2$ | B | $Y^2$ | $Y^3$ | $Y^4$ | $Y^5$ | $Y^6$ | I | $R_3$ | $Y_7$ | other |
|---|---|---|---|---|---|---|---|---|---|---|---|---|---|---|---|---|
| | R. emetica | 5 | | + | ++ | + | o | o | o | — | — | — | — | — | — | 2 R |
| | R. mairei | 5 | | + | ++ | + | o | o | o | — | — | — | — | — | — | 2 R |
| | R. betularum | 5 | | ++ | + | + | (+) | o | o | — | — | — | — | — | — | 2 R |
| | R. sardonia | 5 | | + | ++ | + | + | o | o | — | — | — | — | — | — | 1 R |
| | R. fragilis | 5 | | + | ++ | + | + | o | o | — | — | — | — | — | — | — |
| | R. erythropus | 2 | | + | ++ | + | + | o | o | — | — | — | — | — | — | 1 R |
| | R. xerompalina (purple) | 2 | | + | ++ | — | + | o | o | — | — | — | — | — | — | — |
| RB (red-blue group) | R. xerompalina (brown) | 1 | | ++ | + | — | + | o | o | — | — | — | — | — | — | — |
| | R. caerulea | 5 | | + | ++ | + | + | o | o | + | + | + | — | — | — | — |
| | R. versicolor | 5 | | + | ++ | — | + | o | o | + | + | + | — | — | — | — |
| | R. velenovskyi | 4 | | + | ++ | + | + | o | o | + | + | + | — | — | — | 1 R |
| | R. rosea | 5 | | + | ++ | + | (+) | o | o | + | + | + | — | — | — | — |
| | R. lapida | 3 | | + | ++ | + | (+) | o | o | + | + | + | — | — | — | 1 R |
| | R. atropurpurea | 5 | | + | + | ++ | + | o | o | — | — | — | — | — | — | 2 R |
| | R. lutea (apricot) | 3 | | o | — | + | o | ++ | + | — | — | — | — | — | — | — |
| Y (yellow group) | R. lutea (yellow) | 2 | | o | — | (+) | o | ++ | + | — | — | — | — | — | — | — |
| | R. claroflava | 2 | | o | o | o | o | ++ | + | — | — | — | — | — | — | — |
| | R. ochroleuca | 5 | | o | o | o | o | ++ | + | — | — | — | ? | ? | — | — |
| | R. vesca | 5 | | o | o | o | o | o | o | — | — | — | + | + | ? | 1I, 1Y |
| IP (indigo-pink group) | R. cyanoxantha | 5 | | o | o | o | o | o | o | — | — | — | + | + | ? | 1I, 1Y |
| | R. grisea | 5 | | o | o | o | o | o | o | — | — | — | + | ? | ? | — |
| | R. heterophylla | 5 | | o | o | o | o | o | o | — | — | — | + | ? | ? | 1I, 1R |
| | R. amoena (yellow cap) | 2 | | o | — | (+) | o | o | o | — | — | — | — | — | ++ | — |
| | R. amoena (purple stalk) | 2 | | o | (+) | + | o | o | o | — | — | — | — | — | — | — |
| | R. fellea | 5 | | o | o | o | o | o | o | o | o | o | — | — | — | — |
| O (nil) group | R. foetens | 3 | | o | o | o | o | o | o | o | o | o | — | — | — | — |
| | R. sororia | 5 | | o | o | o | o | o | o | o | o | o | — | — | — | — |
| | R. nigricans | 4 | | o | o | o | o | o | o | o | o | o | — | — | — | — |
| | R. virescens | 2 | | o | o | o | o | o | o | o | o | o | — | — | — | — |

++ = large amounts of pigment present; + = small amounts present; (+) = traces present; — = not detected; o = almost certainly not present; ? = pigment of appropriate color and appropriate $R_f$ present but identity not certain. R = red; Y = yellow; I = indigo.

The major pigment ($R_1$) has been noted in several species by Watson (1966) among others, as it is the most frequently encountered of a certain number of other pigments. The table presented by Watson summarized the results obtained by thin-layer chromatography on about thirty species of *Russula* studied. The species *R. fellea, R. foetens, R. sororia* (of the section *Ingratae* Quélet), *R. virescens* (of section *Palumbinae* Kühner & Romagnesi), and *R. nigricans* (section *Compactae* Fries) possess pigments fundamentally different from the other russulas: incrusted pigments inextractable by aqua

[3] Gluchoff, K. 1969. Étude chimiotaxinomique des pigments des Russules. Thèse 3<sup>eme</sup> cycle, Lyon; Gluchoff, K. and Ph. LeBreton. C. R. Acad. Sci. Paris (in press); Gluchoff, K. Bull. Soc. Natur. Archéol. Ain (in press).

picolin or pyridin, solvents which are very effective as far as the colorants of other russulas are concerned, among which Watson distinguished three groups: (*a*) The I.P. group (indigo and red $R_3$) which includes those species of section *Palumbinae* Kühner & Romagnesi; (*b*) The Y group, with yellow pigments $Y_2$ and $Y_3$, which are very widespread among the russulas of Kühner and Romagnesi. However, those writers recognized that *R. claroflava* was closely related to *R. ochroleuca*, though the former was placed in section C of the *Genuinae* and the latter among the *Ingratae*. (*c*) Finally, the R.B. group included some species with red, blue, and yellow pigments, certain of them being distinguished particularly by the presence of yellow pigments $Y_4$, $Y_5$, and $Y_6$.

The uncertainty which exists concerning the structure of the various pigments and their possible connections to one another and the relatively small number of species studied makes a more profound discussion of the subject difficult at present. Also, the problem raised by Heim (1941) as to the identity of the red pigment of the russulae (rubein) and its comparability to that of the gastromycete mushroom *Elasmomyces russuloïdes* does not yet seem to have been solved.

## The Cantharelloid Fungi

*Cantharellus* seems to be intermediate between the Agaricales and the agaricoid Aphyllophorales. This morphological definition was obviously too wide to embrace a natural group—some indisputable but retrogressive agarics belonged, in fact—and had to be restricted. The character held as being essential was the stichobasidial manner of the nuclear division in the hymenium. This character seems clearly to separate the chanterelles, rich in carotenoids, from the agarics, which until recently were considered to be devoid of these pigments. Two problems appear here: relationships within the group thus defined (distinction and true degree of affinity between *Cantharellus* and *Craterellus*) and the position of the group as a whole, or at least of the genus *Cantharellus*.

### *Relations among the Cantharellaceae*

Kühner and Romagnesi (1953), studying European chanterelles, placed them in a single genus *Cantharellus*. Based on the presence

(*Cantharellus*) or absence (*Craterellus*) of clamps, they distinguished at the most two subgenera, the second including among others *C. cornucopioides* and *C. cinereus*. Corner (1966), attributing an importance of the first order to the development and structure of fructifications, and considering the stichobasidium as a secondary characteristic, separated *Craterellus* (restricted to the species with most infundibuliform fruit bodies) with *C. cornucopioides*, from *Cantharellus* (where *C. cinereus* is placed near *C. tubaeformis* and *C. lutescens*), the development of the first appearing to him relatively nearer to the clavarias than to true *Cantharellus*. Donk (1964), on the other hand, according a greater importance to the manner in which the basidial nuclei were divided, considered them very close and joined them in the single family Cantharellaceae.

Both by its carotenoid pigments and by compounds of another sort, *C. cornucopioides* links closely to the group *C. lutescens–C. tubaeformis*, from which it differs only in having a weaker carotenogenesis, with correspondingly relatively strong development of dark pigments of another sort. These data argue very strongly in favor of the homogeneity of *Cantharellus* (*sensu* Kühner-Romagnesi).

We can note that Corner (1966) subdivided the genus *Cantharellus* into several subgenera based on general aspect. *Cantharellus* (with *C. cibarius*, *C. cinnabarinus*, and *C. friesii*) accumulates bicyclic carotenoids (β-carotene and its ketonic derivatives). *Phaeocantharellus* (with *C. lutescens* and *C. tubaeformis*) accumulates exclusively aliphatic compounds. In view of the order of appearance of these compounds in the course of biogenesis, one is tempted to classify as more primitive the species with aliphatic carotenoids which also contain brownish pigments of another nature. From them, one could postulate an evolutionary divergence leading either to *C. cinereus* or to *C. cibarius* and from the latter to *C. friesii*, and in both cases loss or regression of one of the two types of pigment initially present. It would appear very unlikely, however, that the production of the bicyclic carotenoids in certain chanterelles should be a repeat performance. On the other hand, it seems that neurosporene only accumulates when carotenogenesis undergoes a partial restriction. Third, one might think that it was from *C. cibarius*, whose pigmentation appears little differentiated (and whose pigmentation, as we shall see, is found in a species near to the chanterelles), that an equally divergent evolution should have proceeded and led either to *C. friesii* or *C. cinereus.* In this scheme the *C. tubaeformis–C. lutescens* group must mark a regression of the carotenoids along the line toward *C. cinereus.*

### *Relations between the Cantharellaceae and Other Groups*

Kühner (1928), in a description of *Omphalia chrysophylla* Fr., noted how much this agaric resembled the real chanterelles by its yellow spores and its hymenium indistinctly separated from the subhymenium, as well as its color and general aspect. Because the exclusion of the Cantharellaceae from the Agaricales was often considered as established, however, it was among the Aphyllophorales that writers looked for affinities to the chanterelles.

R. Heim (1949) underlined the characteristics which, according to him, linked *Cantharellus* and various simple clavarias, in particular the exotic carotenogenic species *Clavaria cardinalis* (=*Clavulinopsis miniata* [Berk.] Corner). Although more recent authors do not seem to have retained this relationship between the genera *Cantharellus* and *Clavulinopsis,* the common presence of carotenoids in these two groups is interesting to consider. It must always be remembered that the mere possession of pigments by this family is only a preliminary indication whose significance must be confirmed by the determination of their chemical structure. From this standpoint the two carotenogenic clavarias hitherto analyzed (*Clavaria helicoides* and *Pistillaria micans*) are characterized by a metabolic orientation toward the monocyclic carotenoids, a fairly original feature and as yet unknown among the chanterelles. On the other hand, although Corner noted that in the genus *Clavulinopsis* the red color seemed to be due in many cases to carotenoids, the true nature of these pigments ought to be verified by analysis. In fact, in *Cl. miniata,* the carotenoid nature of whose pigments seems established only on cytological observations, it is the medullar hyphae which are the most colored. The orange pigments are concentrated in the subhymenium, and if their cytological localization recalls carotenoids, they are carotenoids with a yellow color and of a different nature than those found in *Cl. corniculata.* Finally, although these pigments appear to turn green in reaction with iodine,[4] they do not undergo any modification with con-

[4] One of us (NA) has recently shown that the green reaction of lipid droplets in which carotenoids are dissolved is not due to the presence of the carotenoids, but apparently to the polyosidic nature of the membrane surrounding the inclusions.

centrated sulphuric acid. Contrary to their position in the cell, their position in the mushroom seems to reduce the likelihood that they are carotenoids.

Corner (1950) linked *Cantharellus* to the large simple clavarias of *Clavariadelphus. Cl. pistillaris* possessed the veined hymenium of the chanterelles; *Cl. truncatus* had a truncate sterile top which was not far removed from the origin or suppression of a pileus. According to Corner, apart from the way in which the basidial nuclei are divided (a characteristic to which he does not attach great importance) the two genera differ only in that *Cantharellus* has a real pileus, whereas *Clavariadelphus truncatus* acquires a marginally greater size and has a fructification strongly resembling that of *C. cuticulatus.* On the other hand, if *Cantharellus* were to lose its marginal size, the result would be *Cl. truncatus.* The way in which this evolution has taken place remains to be determined.

Donk (1964), on the contrary, accorded great importance to the mode of division of basidial nuclei and entirely separated *Cantharellus* and *Clavariadelphus* (the morphological similarity only resulted for him from a phenomenom of convergence). He placed the latter instead with *Gomphus,* the fruit bodies of which, like *Clavariadelphus,* but unlike the chanterelles, turn green on reaction with ferric sulphate. The views of Petersen (1967a, b) have been closer to those of Corner as to the importance to be given to the manner of the division of the basidial nuclei, and his study of the coloration of spores in cotton blue goes entirely against the opinion of Donk, clearly dividing *Clavariadelphus* from *Gomphus* and linking it to *Cantharellus* (in the midst, it is true, of a vast group containing also the *Clavaria-Clavulinopsis* series).

In fact, it appears to us that *Cantharellus,* rich in carotenoids, cannot be derived from *Clavariadelphus,* deprived of these pigments as it is. If there is a relationship between these two genera, evolution must have proceeded from a still chiastobasidial pre-*Cantharellus* toward *Clavariadelphus* (if, like Corner [1966], Boidin [1958], and Petersen [1967a], one considers the stichobasidial manner of division as a secondary character). Against this, the identity of the pigmentation and the macroscopic and microscopic elements described by Kühner (and to which we can add the odor) add up to a quantity of characters common to *Gerronema chrysophyllum* (Fr.) Singer (=*Omphalia chrysophylla* Fr.) and to the chanterelles which these share with no other group. This leads us to believe that it is around *Omphalia* that we should look for the origin of the chanterelles, the exclusion of which from the Agaricales seems to us excessive. The real evolutionary target in the Cantharellaceae has been recognized; it seems to be *Cantharellus,* which appears to us to be very near *Gerronema.* Two species have already been shown to be carotenogenic (*G. chrysophyllum*—Fiasson [1968a], and *G. venustissimum* [=*Clitocybe venustissimum* (Fr.) Sacc.]—Arpin [1966]) and in these species the yellow spores of many chanterelles are also often found. Microscopically, giving a secondary value to the stichobasidial character, *Cantharellus* appears to us as a deviation originating from the omphalias and carotenogenic clitocybes.[5]

## The Aphyllophorales

Among the polyporoid fungi, nitrogen-containing pigments of various structures have been discovered; some cinnabarin, cinnabaric acid, and tramesanguin, as we have seen, have been isolated from *Trametes* (Gripenberg, 1963).

Another sort of colorant much more widespread among the Basidiomycetes is also encountered in a large number of polyporoid fungi. We knew since Kögl (1928) the nature of the acid responsible for the violet coloration obtained by the action of the ammonia on *Phaeolus rutilans* (Pers.) Pat. (=*Polyporus rutilans*). It was without question polyporic acid, a colorant derived from *p*-diphenyl benzoquinone, a basic structure often met in the Aphyllophorales. The fruit bodies of other species of *Phaeolus* (*P. croceus* . . . ) change color in the presence of ammonia, apparently expressing a chemistry related to, if not identical to, that of *P. rutilans* (Fallahyan, 1964).

Thelephoric acid, a molecule derived from polyporic acid, was recognized by Zopf (1890) in many species of mushrooms which for the most part fall in the series Phylactériae Patouillard—*Thelephora palmata* Scop., *T. flabelliformis* Fr., *T. caryophyllea* Schaeff., *T. terrestris* Ehrb., *T. coralloides* Fr., *T. crustacea* Schum.,[6] *T. intybacea*

[5] See also: Fiasson, J. L., R. H. Petersen, M. P. Bouchez, and N. Arpin. 1970. Recherches chimiotaxinomique sur les Champignons. XIV. Contribution biochimique à la connaissance taxinomique de certains Champignons cantharelloïdes et clavarioïdes. Rev. Mycol. (in press).

[6] It is interesting to note that although this species, unlike the others which were placed among the Phylactériae by Patouillard (1900), was put in *Tomentella,* Bourdot and Galzin (1928) thought it better placed among the Phylactériés.

Pers., and *T. laciniata* Pers. Other phylactériae also possess this acid, especially the genus *Calodon* Quélet (=*Hydnum* Fr.), including *C. ferrugineum* (Fr.) Pat. (Zellner, 1915), and *C. suaveolens* (Scop) Quél. Gripenberg (1960) elucidated the chemical structure of this compound from this last species. The structure had been already partly established by Kögl, Erxeleben, and Janecke (1930). Gripenberg (1958a, b) likewise showed that the coloration of *Hydnum aurantiacum* Batsch was due, besides thelephoric acid, to aurantiacin (dibenzoate of atromentin).

The corticiums, like the "pores," synthesize colorant materials derived from polyporic acid (Gripenberg, 1965). The specific acid has been distinguished in *Peniophora filamentosa* as xylerythrin, resulting from the copulation of polyporic acid and *p*-hydroxyphenylacetic acid. Xylerythrin has also been found in *Peniophora sanguinea* Bres. (sect. *Radicatae* Bourdot & Galzin). In the merulii some molecules of the same type have been met in the genus *Phlebia*. The cultivation of *P. strigozonata* has furnished phlebiarubrone (McMorris & Anchel, 1963), whose synthesis from polyporic acid was recently accomplished by Gripenberg (1966).

Among the corticiums some pigments fundamentally different from the preceding share in the coloring of various species. On one hand, corticrocin is responsible for the coloring of *Corticium sulphureum* (Fr.) Fr. (=*C. croceum* Bres.) and cortisalin for *C. salicinum* Fr. (=*C. rutilans* [Pers.] Quél.). The strong resemblance between the two structures makes one think of the possible existence of a common precursor, and the presence of the *p*-hydroxybenzoic nucleus is not surprising in cortisalin as this molecule has been found in cultures of *Corticium sasakii*. Some carotenoids are also present in *Peniophora s. str.* (=section *Coloratae* Bourdot & Galzin). This section whose autonomy and homogeneity have been demonstrated by Eriksson (1950) and Boidin (1958, 1965) therefore is distinguished by the nature of its pigmentation, which is entirely original among the Aphyllophorales (excepting *Cantharellus*). The pigmentation of one of them, *P. aurantiaca* Bres., has just been studied by Arpin, Lebreton, & Fiasson (1966). The most striking observation is the great richness of oxygenated carotenes derived from β-carotene: echinenone and especially astaxanthin, which appears to us to indicate a certain evolution of pigmentary metabolism. Some closely neighboring, and in some cases identical, pigments had only been isolated from the American chanterelle, *Cantharellus cinnabarinus* by Haxo (1950). The existence of such pigments has recently been demonstrated in *Phyllotopsis nidulans* (Fr. ex Pers.) Donk by Fiasson (1968b).

With the mention of the presence of carotenoids in various groups of Heterobasidiomycetes—Calocerales (*Calocera viscosa, C. cornea, Dacrymyces palmatus*), Tremellales (*Tremella mesenterica*), and Uredinales (many *Gymnosporangium* and *Puccinia* species)—and also Homobasidiomycetes like the Phallales (*Anthurus, Clathrus, Lysurus,* and *Mutinus*), we finish the revue of our present knowledge of the pigments of Basidiomycetes.[7]

## Conclusion

It is still too soon for the results obtained to lead to real enrichment of our knowledge as too much remains unknown. Both the structural chemical plans (anthraquinones of the cortinarii, pigments of the russulas) are poorly understood. Finally, we must emphasize that pigment analysis has hitherto been principally concerned with European and to a lesser degree with North American species. The present general tendency in other branches of mycology to devote to the exotic species all the interest they deserve should appear in the same way in our field. The few analyses carried out on such species have been shown to be most fruitful.

In spite of the importance of our knowledge gaps, from now on we can establish a certain "balance" between carbon pigments, whether reduced (carotenoids) or of a fundamentally oxygenated nature. For example, the pigmentation of *Tremella mesenterica* is of a carotenoid nature, whereas that of *Guepinia helvelloides* is not, and species of *Peniophora s. str.* are the only corticiums which possess carotenoids, which appear to be very rare at the level of the Agaricales.

It is plausible to think that on the chemical level evolution proceeded together with a growing oxidation of the molecules, progression along biosynthetic pigment pathways occurring generally

[7] Concerning the carotenoids of the various groups cited above, the reader is referred to the publications of Mme. P. Heim (1947), G. Turian (1960), and Fiasson, Le Breton, and Arpin (1968).

through progressive oxidation. In a similar way one can imagine that at a higher level of evolution there corresponds a structure of a more oxygenated state. According to this hypothesis it is not, therefore, surprising to establish that in the Agaricales carotenoids are present only in exceptional cases whereas, on the other hand, the anthraquinones are widely distributed. Within other groups of fungi, one can likewise observe a "balance" between oxygenated pigments and carotenoids, as among the discomycetes, but here the proportion of carotenogenic species is much higher than that established among the Agaricales. Among the more evolved Discomycetes one can distinguish two quite distinct types of pigmentation, which taken outside of taxonomy seem to cancel one another: oxygenated pigments of the morels, and carotenoids—very oxygenated themselves—of the Sarcoscyphaceae.

Finally, at the level of the "link" groups, the simultaneous presence of two types of pigmentation is very common. We must entertain the idea that in the course of evolution a choice was made between the two metabolic systems initially present, with perhaps even a competition at some level for a common precursor.

## References

Akagi, M. 1942. J. Pharm. Soc. (Japan) 62: 129 (*cited in* Miller, p. 247).

Anchel, M., *et al.* 1948. Antibiotic substances from Basidiomycetes. III. *Coprinus similis* and *Lentinus degener*. Proc. Nat. Acad. Sci., U.S.A. 34: 498–502.

Anslow, W. K., J. N. Ashley, and H. Raistrick. 1938. The isomeric dimethoxy-2:5 toluquinones and certain related compounds. J. Chem. Soc. 1938: 439–442.

Arpin, N. 1966. Recherches chimiotaxinomiques sur les champignons. Sur la présence de carotenès chez *Clitocybe venustissima* (Fries) Sacc. C. R. Acad. Sci. Paris 262: 347–349.

———. 1968. Les caroténoïdes des Discomycetes; essai chimiotaxinomique. Thése, Lyon. 169 p.

———. 1968. Recherches chimiotaxinomiques sur les champignons. X. Nature et distribution des caroténoïdes chez les Discomycetes operculés (Sarcoscyphaceae exclues). Conséquences taxinomiques. Bull. Soc. Mycol. France 84: 427–474.

ARPIN, N., PH. LEBRETON, AND J. L. FIASSON. 1966. Recherches chimiotaxinomiques sur les champignons. II. Les caroténoïdes de *Peniophora aurantiaca* (Bres.). Bull. Soc. Mycol. France 82: 450–459.

ARPIN, N., AND S. LIAAEN-JENSEN. 1967a. Recherches chimiotaxinomiques sur les champignons. IV. Les caroténoïdes de *Phillipsia carminea* (Pat.) Le Gal; isolement et identification d'une xanthophylle naturelle nouvelle. Bull. Soc. Chim. Biol. 49: 527–536.

———. 1967b. Recherches chimiotaxinomiques sur les champignons. Sur le présence de l'ester de la torularhodine chez *Cookeina sulcipes* (Berk.) Kuntze (Ascomycètes). C. R. Acad. Sci. Paris 265: 1083–1085.

BALENOVIC, K., *et al.* 1955. The chemistry of higher fungi. III. Contributions to the chemistry of the genus *Russula*. Archiv Kem. Yug. 27: 15–20.

BENDZ, G. 1948. An antibiotic agent from *Marasmius graminum*. Acta Chem. Scand. 2: 192.

BERTRAND, G. 1901. Sur le bleuissement de certains champignons. C. R. Acad. Sci. Fr. 133: 1233–1256.

———. 1902. Sur l'extraction du bolétol. C.R. Acad. Sci. Fr. 134: 124–126.

BOIDIN, J. 1958. Essai biotaxinomique sur les Hydnés résupinés et les Corticiés. Etude spéciale du comportement nucléaire et des mycéliums. Rev. Mycol., Mem. Hors-Ser. no. VI, in 8°, 390 p., 103 f. (These Fac.).

———. 1965. Le genre *Peniophora sensu stricto* en France (Basidiomycètes). Bull. Soc. Linn. Lyon 34: 161–169, 213–219.

BOURDOT, H., AND A. GALZIN. 1928. Hyménomycètes de France. Paris. 761 p.

CORNER, E. J. H. 1950. A monograph of *Clavaria* and allied genera. Ann. Bot. Mem. 1: 1–740.

———. 1966. A monograph of cantharelloid fungi. Ann. Bot. Mem. 2: 1–255.

DEYSSON, G. 1958. Sur l'examen des champignons en lumière de Wood. Son intérêt pour l'étude du genre *Russula*. Bull. Soc. Mycol. Fr. 74: 207–215.

DONK, M. A. 1964. A conspectus of the families of Aphyllophorales. Persoonia 3: 199–324.

ERDTMAN, H. 1948. Corticrocin, a pigment from the mycelium of a mycorrhiza fungus. II. Acta Chem. Scand. 2: 209–219.

ERIKSSON, J. 1950. *Peniophora* Cke sect. *Coloratae* Bourd. et Galz. A taxonomical study with special references to the Swedish species. Symb. Bot. Upsaliensis 10: 1–76.

FALLAHYAN, F. 1964. Étude culturales et anatomique de quelques Polypores conidiophorés à trame molle ou coriace. Thèse Doct. Univ. Paris. 89 p.

FIASSON, J.-L. 1968a. Recherches chimiotaxinomiques sur les champignons. VI. Les carotènes de *Omphalia chrysophylla.* Fr. C.R. Acad. Sci. Paris 266: 1379–1381.

———. 1968b. Les caroténoïdes des Basidiomycètes; survol chimiotaxinomique. Thèse 3ème cycle, Lyon.

FIASSON, J.-L., PH. LEBRETON, AND N. ARPIN. 1968. Les caroténoïdes des champignons. Bull. Soc. Nat. Archeol. Ain 82: 1–67.

GATENBECK, S. 1960. Studies on the biosynthesis of anthraquinones in lowest fungi. Svensk Kem. Tidsk. 72: 188–203.

GABRIEL, N. 1962. Recherches sur les pigments des Agaricales. VI. Pigments des Cortinaires du groupe *Olivascentes.* Bull. Soc. Mycol. Fr. 78: 359–366.

———. 1965. Contribution à la chimiotaxinomie des Agaricales. Pigments des Bolets et des Cortinaires. Thèse, Lyon. 70 p.

GILBERT, E. J. 1931. Les Bolets. Librairie E. Le Francois, Paris. 254 p.

GRIPENBERG, J. 1952. Fungus pigments. II. Cortisalin, a new polyethenoid pigment. Acta Chem. Scand. 6: 580–586.

———. 1958a. Fungus pigments. VIII. The structure of cinnabarin and cinnabaric acid. Acta Chem. Scand. 12: 603–610.

———. 1958b. Fungus pigments. IX. Some further constituents of *Hydnum aurantiacum* Batsch. Acta Chem. Scand. 12: 1411–1414.

———. 1960. Fungus pigments. XII. The structure and synthesis of thelephoric acid. Tetrahedron 10: 135–143.

———. 1963. Fungus pigments. XIII. Tramesanguin, the pigment of *Trametes cinnabarina* var. *sanguinea* (L). Pilat. Acta Chem. Scand. 17: 703–708.

———. 1965. Fungus pigments. XVI. The pigments of *Peniophora sanguinea* Bres. Acta Chem. Scand. 19: 2242–2259.

———. 1966. Fungus pigments. XVII. The synthesis of phlebiarubrone. Tetrahedron Letters 7: 697–698.

HAXO, F. 1950. Carotenoids of the mushroom *Cantharellus cinnabarinus.* Bot. Gaz. 111: 228–232.

HEILBRONNER, E., AND R. W. SCHMIDT. 1954. 238. Zur Kenntnis des Sesquiterpene und Azulene. 113, Mitteilung. Azulenaldehyde und Azulenketone: die Struktur des Lactaroviolins. Helv. Chim. Acta 37: 2018–2039.

HEIM, MME. P. 1947. Études sur la localisation des pigments carotiniens chez les champignons. Rev. Mycol. 12: 104–125.

HEIM, R. 1941. Les pigments des champignons dans leur rapport avec la systematique. Bull. Soc. Chim. Biol. 29: 48–79.

———. 1949. Une Clavaire cantharelloïde australienne à pigment carotinien cristallisé. Rev. Mycol. 14: 113–120.

HENRY, R. 1937. Révision de quelques Cortinaires. Bull. Soc. Mycol. Fr. 53: 49–80.

JOSSERAND, M., AND G. NETIEN. 1938, 1939. Observations sur la fluorescence de 175 espèces de champignons charnus examinés en lumière de Wood. Bull. Soc. Linn. Lyon 7: 283–292; 8: 14–23.

KÖGL, F. 1928. Untersuchungen über Pilzfarbstoffe. VII. Die Synthese des Atrementins. Zur Kenntnis der Atromentinsaure. Ann. Chem. 465: 243–256.

KÖGL, F., AND W. B. DEIJS. 1935a. Untersuchungen über Polzfarbstoffe. XI. Uber Boletol den Farbstoffe der blau anlaufenden Boleten. Ann. Chem. 515: 10–23.

———. 1935b. Untersuchungen über Pilzfarbstoffe. XII. Die Synthese von Boletol und Isoboletol. Ann. Chem. 515: 24–33.

KÖGL, F., H. ERXELEBEN, AND L. JANECKE. 1930. Untersuchungen über Pilzfarbstoffe. IX. Die Konstitution der Thelephorsaüre. Ann. Chem. 482: 105–119.

KÖGL, F., *et al.* 1928. Untersuchungen über Pilzfarbstoffe. VI. Die Konstitution des Atromentins. Ann. Chem. 465: 211–242.

KÜHNER, R. 1928. *Omphalia chrysophylla* Fr. Bull. Soc. Mycol. Fr. 44: pl. XXVIII.

———. 1934. Observation sur la localisation cytologique des substances coloreés chez les Agarics et les Bolets. Le Bot. 26: 347–369.

———. 1949. Remarques sur quelques caractères cytologiques habituellement négligés des Cortinaires et particulièrement sur la localisation de leurs substances colorées. Bull. Soc. Natur. Oyonnax 3: 1–8.

———. 1960. Notes descriptives sur les Agarics de France. I. *Cortinarius*. Bull. Soc. Linn. Lyon 29: 40–56, 219–227, 260–266.

KÜHNER, R., AND H. ROMAGNESI. 1953. Flore analytique des Champignons supérieurs. Masson Cie, Paris. 557 p. + 677 figs.

LEDERER, E. 1938. Sur les caroténoïdes des Cryptogames. Bull. Soc. Chim. Biol. 20: 611–634.

MCMORRIS, T. C., AND M. ANCHEL. 1963. Phlebiarubrone, a basidiomycete pigment related to polyporic acid. Tetrahedron Letters 5: 335–337.

MILLER, M. W. 1961. The Pfizer handbook of microbial metabolites. McGraw-Hill, New York, 772 p.

MURRAY, J. 1952. Lichens and Fungi. I. Polyporic acid in Stictae. J. Chem. Soc. G. B. 1952: 1345–1350.

PATOUILLARD, N. 1900. Essai taxonomique sur les familles et les genres des Hyménomycètes. Lons le Saulnier. 184 p.

PASTAC, I. A. 1942. Les matières colorantes des Champignons. Rev. Mycol. Mem. Hors-Ser. No. 2. 88 p.

PERSOON, C. H. 1801. Synopsis methodica fungorum. (Reprint, New York, 1952.) 708 p.

PETERSEN, R. H. 1967a. Notes on clavarioid fungi. VI. Two new species and notes on the origin of *Clavulina.* Mycologia 59: 39–46.

———. 1967b. Evidences of interrelationships of the families of clavarioid fungi. Trans. Brit. Mycol. Soc. 50: 641–648.

PLATTNER, P. A., *et al.* 1954. The structure of Lactaroviolin. Chem. and Ind. Sept. 25: 1202–1203.

RAISTRICK, H., R. ROBINSON, AND A. R. TODD. 1933. A synthesis of helminthosporin. J. Chem. Soc., 1933: 488–489.

SINGER, R. 1962. The Agaricales in modern taxonomy. 2nd ed. Cramer, Weinheim. 915 p., 73 pls.

SORM, F., B. BENESOVA, AND V. HEROUT. 1953. Chem. Listy 47: 1856 (*cited in* Miller, p. 161).

TURIAN, G. 1960. Identification des caroténoïdes majeurs de quelques champignons Ascomycètes et Basidiomycètes. Neurosporène chez *Cantharellus infundibuliformis.* Arch. Mikrobio. 36: 139–146.

WATSON, P. 1966. Investigation of pigments from *Russula* spp. by thin layer chromatography. Trans. Brit. Mycol. Soc. 49: 11–17.

WILLSTAEDT, H. 1946a. Pilzfarbstoffe. V. Zur Natur des Sauerstoffs im Lactaroviolin. Svensk Kem. Tidskr. 58: 23–26.

———. 1946b. Pilzfarbstoffe. VI. Uber zwei neue lipoidlösiche Farbstoffe aus dem echten Reizker. Svensk Kem. Tidskr. 58: 81–85.

ZELLNER, J. 1915. Zur Chemie der höhreren Pilze. XI. Mitteilung über *Lactarius scrobiculatus* Scop., *Hydnum ferrugineum* Fr., *Hydnum imbricatum* L., und *Polyporus applanatus* Wallr. Monatshefte f Chem. 36: 611–632.
ZOPF, W. 1890. Die Pilze. Trewendt, Breslau.

## RESUMÉ

Après avoir rappelé que la pigmentation a été depuis longtemps utilisée par le Mycologue pour définir et regrouper les espèces, l'accent est mis sur la nécessité de connaître la structure chimique de la matière colorante pour déboucher sur une systématique naturelle valable.

Suit un bref aperçu des principales structures chimiques des pigments rencontrés chez les Basidiomycètes. Parmi les pigments à squelette fondamentalement hydrocarboné figurent les pigments du Lactaire délicieux, ceux de certains Corticiés et les caroténoïdes. En ce qui concerne les pigments quinoniques, quelques exemples sont donnés relatifs aux benzoquinones, napthtoquinones et anthraquinones, ainsi qu'aux dérivés de la *p*-diphényl-benzoquinone. Enfin sont exposées les structures des pigments azotés isolés des *Trametes.*

Il est ensuite tenté de tirer partie, dans un but taxonomique, de l'étude des pigments dans quelques groupes de Basidiomycètes; sans entrer dans la discussion au niveau des taxons inférieurs, les quelques problèmes suivants ont été abordés:

Validité des divers sous-genres de Cortinaires: les résultats concernent la distribution des anthraquinones et des anthranols.

Relations entre Bolétacées typiques et lamellées, envisagées sous l'angle particulier de la répartition de deux anthraquinones: bolétol et isobolétol.

Relations au niveau des Lactaires et des Russules.

En tenant compte de la pigmentation caroténoïdique, le problème des relations des Cantharellacées avec d'autres groupes de Basidiomycètes (Agarics et Clavaires) est discuté; le probleme de l'homogénéité des Cantharellacées est également posé.

Au sein des Aphyllophorales, et tout particulièrement des Corticiacées, la nature hétérogène de la pigmentation est manifeste: des groupes à pigments de nature fondamentalement hydrocarbonée co-existent avec d'autres à pigments dérivés de la *p*-diphényl-benzoquinone.

Il est encore trop tôt pour que les études chimiques entreprises sur le pigmentation des Champignons apportent tous les enrichissements que l'on peut attendre de telles recherches: beaucoup trop d'inconnues subsistent, d'une part sur le plan structural (anthraquinones des Cortinaires, pigments des Russules . . .), d'autre part sur le plan de la distribution de molécules connues (acides polyporique, théléphorique . . .).

Il n'en demeure pas moins qu'une certaine "balance" entre pigments hydrocarbonés réduits (et pouvant de ce fait être considérés comme relativement primitifs) et pigments oxygénés, semble deja se dégager, soit au niveau des grands groupes, soit au niveau de taxons plus réduits.

## Discussion

Singer: I might make a comment on the *Cortinarius* question. I think our study of the genus *Dermocybe* shows the extraordinary importance that the investigation of the pigmentation in this group and other groups has for the taxonomist. When we invited Dr. Moser to study the cortinarii of temperate South America, he was able to collect more than 300 species of *Cortinarius*, most of them new. Among all these, there was not one that could be referred to the genus *Dermocybe*. Through our knowledge of pigments the genus has now been accurately delimited, with this further phytogeographic evidence as a new argument in favor of this delimitation.

Dubovoy: If all of your pigment analyses were made on mature stages of the fruit body, then during development you could get an undetected ontogenetic series. Have you done pigment studies during the development of the fruit body?

Arpin: In *Sarcoscypha coccinea* I found the same pigment in the same concentration in the small carpophore or fruit body as in the mature fruit body. In this and others, I don't think there is much difference from the immature to the mature fruit body. In some Discomycetes, however, it is possible to find the evolution of sophistication of the pigment molecule. I studied many species of Discomycetes for carotenoids, and in those species I have seen very excellent carotenes. Some pigments differ by the location of double bonds. Aleuriaxanthine, very abundant in *Aleuria aurantia,* is one-hydroxy. In *Sarcoscypha coccinea* the opposite is found: 2′-dehydroplectaniaxanthine. Some pigments we can consider quite advanced. By pigment analysis we may get some clue to the lines of evolution.

By biochemical analysis we can comprehend the phylogenetic state of an organism. The synthesis of compounds is the direct consequence of enzymes or enzymatic systems. However, although an index of the phylogenetic state, all molecules are not interesting. The molecules for which we know the determined role, chlorophyll for instance, are not interesting for chemical taxonomy. In our laboratory, those on which we cannot put any determined role are the most interesting for the taxonomist.

OLEXIA: It would seem difficult to be able to hypothesize a phylogenetic pathway for compounds which apparently have no great selective value. Dr. Arpin is apparently working with compounds which have no selective value or apparently we don't know what the function of these compounds is.

PETERSEN: But there is a great difference between having no significance and our not knowing what that significance might be. The work of Goldstrohm and Lilly[1] on *Dacryopinax* might be mentioned in this regard. Cultures grown in the dark did not produce carotenoid pigments and were UV-sensitive. Cultures grown in the light did produce carotenoids and were highly UV-resistant. The mortality rate for those grown in the dark and without carotenoid pigments was much higher. We found identical results when we irradiated *Dacrymyces palmatus* to enhance antibiotic production.

HEIM: I am of exactly the same opinion as Dr. Arpin about the importance of certain compounds and the lack of importance of others. We can say the same thing for every living form, every living example, every group of products.

About the question of boletol, I remember that Mrs. Gabriel found boletol in nearly all the boletes, but when she repeated, the results were not exactly the same. In the genera you have mentioned, the

[1] Goldstrohm, D.D. and V. G. Lilly. 1965. The effect of light on the survival of pigmented and nonpigmented cells of *Dacryopinax spathularia.* Mycologia 57:612–623.

presence of boletol does not seem to indicate a phylogenetic position. For instance, you state that *Gyrodon* has boletol and *Gyroporus* has none. I think of *Gyroporus* as evolved and *Gyrodon* as primitive. You say that *Xerocomus* has boletol, and *Xerocomus,* as far as I know, is advanced. But on the other hand, *Xerocomus* yes, *Phylloporus* yes, and *Paxillus* no—all this creates a complex evolutionary puzzle. In short, the presence or absence of boletol does not necessarily indicate generic relationships.

ARPIN: Up to this time, only boletol and isoboletol have been studied, but other things should be studied very carefully, and then we will make better decisions. Technically, much depends on the size of the sample, then the quantity in each sample.

In *Aleuria aurantia,* it is possible to characterize some products with one or two fruit bodies. To find a new alcohol, on the other hand, you need at least two kilos of fresh material.

BURDSALL: Can pigment studies be made on dried material or do they have to be done at the time the specimen is collected?

ARPIN: This depends on the molecules involved. The carotenes are pigments which are very labile. It is possible to work with dried material, but often in nature the carotenes are in the *trans* form, but when material is dried in air in light the pigment is transformed to *cis.* It is more difficult to study a *cis* pigment than a *trans* pigment.

PETERSEN: I'm impressed with the phylogenetic chart on *Cantharellus,* for with the insertion of about three or four other species, not only did I have the same species, but in the same places. We seem to agree totally on *Cantharellus.*

ARPIN: It is evident that in America you have many more species than we have in Europe. Because the wealth is greater than in Europe, much more can be accomplished.

PAUL L. LENTZ

*Research Mycologist*
*Crops Research Division, ARS, Plant Industry Station, Beltsville, Maryland*

# ANALYSIS OF MODIFIED HYPHAE AS A TOOL IN TAXONOMIC RESEARCH IN THE HIGHER BASIDIOMYCETES

❦

The anatomical versatility of hymenomycetous hyphae is such that the term "modified hyphae" is hardly definitive. Hyphal modifications range from commonplace features, such as distinctive branching or wall thickening, to the production of bizarre elements either within the hyphal mass or at its surface. Modified hyphae may develop both in the reproductive basidiocarp and in the vegetative mycelium, and under natural or laboratory conditions. They appear not only in dikaryotic mycelium, but also in haploid monokaryotic mycelium. Such monokaryotic mycelium is now commonly termed the "primary mycelium," as contrasted with the secondary mycelium of the dikaryotic state.

Until recently, mycologists thought that generic traits of hymenomycetes could be determined readily only by study of basidiocarps, and not by examination of cultures. However, intensive microscopic reinvestigation of many basidiomycetes has revealed that generic traits may, indeed, be recognized in the microscopic anatomy of fungus cultures. Nobles (1958), for example, has shown that a considerable number of polyporaceous species may be arranged in groups separately homogeneous for cultural characteristics, but some of these groups may not be entirely natural. Gross (1964), for example, denied that *Echinodontium tinctorium* (Ell. & Ev.) Ell. & Ev. and *Fomes juniperinus* (Schrenk) Sacc. & Syd. are closely related, even though Nobles placed them together. However, other groups are natural affiliations readily referable to recognized genera such as *Ganoderma* and *Phellinus–Inonotus*.

Donk (1956, 1957, 1958, 1962) and others have called attention to the feasibility of redefining genera of basidiomycetes in such a way that each genus may be homogeneous for many significant characteristics, including microscopic anatomical details of the basidiocarps. Because cultural characteristics also reveal significant generic features, natural genera should consist of species having in common a broad range of attributes, including those of basidiocarps and mycelial cultures. Examples of these natural genera may be contrasted with artificial genera, in which various characteristics are heterogeneous. Because the validity of such a recitation must depend heavily on the significance of various kinds of hyphae and their modifications, a discussion of hyphal modifications in the Hymenomycetes should be useful.

In the order Aphyllophorales, which includes most of the hymenomycetes other than mushrooms, Corner's (1932a, 1932b, 1953) concept of hyphal systems is the basis for nearly all recent investigations of structural development. Cunningham's (1946) work on the Polyporaceae was the earliest attempt to establish a comprehensive taxonomic system on the basis of Corner's method. Basically, Corner visualized at least three principal kinds of hyphae as constituents of the basidiocarp, but only one as absolutely necessary for basidiocarp development. This essential element is called the "generative hypha" (Plate I, figs. 1–3). The generative system of hyphae is the only one capable of producing clamp connections and basidia. For this reason, any clamp-bearing hypha either of basidiocarp or mycelial culture must be regarded as generative. Hyphae of this kind usually are freely branched. In basidiomycetes that do not have clamp connections, generative hyphae may be recognized as multiseptate, freely branched elements.

Skeletal hyphae (Plate I, figs. 4, 5), one of the two other principal elements, originate from generative hyphae. In orthodox instances, development is not by transformation of a generative to a skeletal unit, but by apical outgrowth of the skeletal element from the generative. For this reason, mycologists who have a fundamentalist concept of hyphal systems would not agree that a skeletal hypha may develop as an intercalary portion of a generative hypha.

Skeletal hyphae usually are not freely branched, although certain kinds may have a few branches. Characteristically, a skeletal hypha is a thick-walled, nonseptate, relatively straight element that grows from its point of origin rather directly toward the growing margin of the basidiocarp or, often by bending, toward some other surface.

Tips of skeletal hyphae commonly protrude into the hymenium or subhymenium of the basidiocarp, or the skeletals may have a longitudinal orientation in the spines or dissepiments of fungi that have spiny or poroid hymenial surfaces.

The third of the principal elements are "binding hyphae" (Plate I, figs. 7–9). Pouzar (1966) suggested replacing the word "binding" by "ligative" because, he claimed, Corner's original term is not easily transferred to all languages. However, Nybakken (1959) translated the Latin verb *ligare* as meaning "to bind, tie." Actually, "ligative" and "binding" are synonymous, so "binding" can be translated as "ligative" when necessary.

Binding hyphae originate from the generative elements, commonly as side branches. They are profusely branched and relatively slender, with a rather limited potential for elongation. Typical binding hyphae are rarely septate. Differences in their origin and branching in basidiocarps of various genera result in many more kinds of modifications than one would expect of elements characterized almost solely by their branching habit.

Pouzar (1966) suggested that skeletal and binding hyphae are morphologically and functionally similar and that they should be grouped as "vegetative hyphae," in contrast with generative hyphae. Instead, if a distinction is to be made between two classes of basidiocarpous hyphae, the terms "generative" and "derived" or "derivative" would seem more aptly to express the nature of the relationship.

Several additional kinds of hyphae have been described, some as minor intermediates between the three principal systems. Teixeira (1956, 1962) designated certain skeletal-like hyphae as "aciculiform," "arboriform" (Plate I, fig. 6), and "vermiculiform." Generative hyphae, especially, are able to undergo almost limitless modifications, in many instances without losing potential for resumption of unmodified growth originating from modified portions of the hyphae. Donk (1964) proposed the term "sclerified generative hyphae" for long, thick-walled sections of hyphae that grow as intercalary units of otherwise thin-walled generative hyphae. In fungi with clamp connections, a clamp may appear at the basal end of the sclerified cell, and one also may develop at the terminal end where growth of the thin-walled element is resumed. The existence of sclerified generative hyphae poses a troublesome problem for mycologists, some of whom would count such sclerified hyphae as skeletals.

When Corner described generative, skeletal, and binding hyphae, he also suggested (1932b) that the number of hyphal systems in the basidiocarp should be indicated by the terms "monomitic," "dimitic," and "trimitic." In addition, Corner (1932a) discussed "mediate hyphae," which he regarded as a fourth system, and "mycelial hyphae," which are remnants of the assimilative phase. Later (Talbot, 1954a), the mediate system was recognized as transitional between either generative and skeletal or generative and binding hyphae.

Several other intermediate kinds of hyphae also have been described. Maas Geesteranus (1963b), for example, described "connecting hyphae" of *Beenakia dacostae* Reid as those that branch at right angles from the main hyphae. In *Steccherinum fistulatum* (Cunn.) Reid, he found that the connecting hyphae are either short or very long and tortuous, in the latter instance assuming the form of interweaving hyphae. In discussing *Donkia pulcherrima* (Berk. & Curt.) Pilát, Maas Geesteranus (1962) advocated the use of Teixeira's (1960) term "bridge hyphae" for short connecting hyphae. Then, he (1967) introduced the term "hyphes sarmenteuses" (tendril hyphae) (Plate II, fig. 10) as a substitute for Corner's (1950) "interweaving laterals." According to Butler (1958), Falck had earlier applied "tendril hyphae" as a name for hyphal branches which become closely associated with their parent hyphae and grow either alongside or in contact with them.

Other kinds of hyphae may appear in basidiocarps of many hymenomycetes, but these also usually are regarded as modifications of the systems described by Corner. Conducting, oleiferous, and laticiferous hyphae, for example, may be modified generative hyphae. Terms such as "fundamental tissue" and "connective tissue," which Fayod (1889) applied in his studies of mushrooms, apparently have little meaningful relationship with Corner's terminology. Fundamental hyphae are said to form the main body elements, or framework, of the fleshy basidiocarp. The hyphal cells usually are large in comparison with those of the connective hyphae. Existing descriptions give the impression that a considerable number of mushrooms consist entirely of generative hyphal elements, many of which show modifications such as inflation or formation of special hyphal vessels. Mushrooms such as these actually have a monomitic structure on the basis of Corner's terminology. According to Thind (1961), most of the clavariaceous fungi also have only generative hyphae. Thus, a monomitic series, in which the hyphae commonly inflate, is characteristic of many fleshy basidiomycetes.

Generative hyphae of basidiocarps are easily demonstrated when the hyphae are of the monomitic series. In di- and trimitic examples, the generative hyphae may be obscured by others, but normal basidia cannot develop in the absence of generative hyphae. This fact is our assurance that generative hyphae are present from the moment that the dikaryon originates until after production of the basidia.

Generative hyphae also can be expected to develop in dikaryotic mycelial cultures. Moreover, vegetative structures such as rhizomorphs and sclerotia also have generative hyphae. In fungi which form clamp connections, participation of generative hyphae in sclerotial development is shown by the presence of clamps on the young sclerotial cells.

The generative system is now seen as the hyphal thread of life. Teixeira (1962) wrote that generative hyphae, as well as differentiated hyphae, are not just a part of the fungus, but that they are an organism in themselves. If all nonessential elements are stripped away, only the generative hypha remains to build the basidiocarp and produce basidia. In the Aphyllophorales, "*Corticium*" *coeruleum* (Fr.) Fr. is an example of a fungus with this simple structure. Every hypha is generative and each cell bears a clamp connection. With nothing more than generative hyphae as structural units, both cultures and basidiocarps can become progressively more elaborate, until even the complicated basidiocarp of a polypore or mushroom may be formed.

*Amphinema byssoides* (Pers. ex Fr.) Erikss. (Plate II, fig. 11) is another fungus in which the basidiocarp is constructed entirely of generative hyphae with clamp connections. However, some of the hyphae grow beyond the hymenial surface and protrude as cylindrical thin-walled elements with clamps (Plate II, fig. 12). These are the least modified hyphal elements that can be called "cystidia." They seem to belong within the limits of Donk's (1964) "hyphocystidia."

*Tubulicrinis chaetophorus* (Höhn.) Donk also has slender, hyphae-like cystidia. These originate in the context and pass through the hymenium without becoming involved in the production of basidia.

However, their nature is much different from that of the cystidia in *Amphinema*. Instead of being slightly specialized generative hyphae, they are modified elements produced by the generative hyphae. As such, their relationship with the generative hyphae is similar to that of skeletal hyphae and other modified structures produced by generative hyphae. The cystidium of *T. chaetophorus* is a specialized cell having a limited potential for additional growth and modification and with characteristics decidedly different from those of the subtending hypha. Whether the specialized cell has its origin in the context or in the subhymenium is of much less consequence than the fact that the nature of the cell has been greatly altered from that of the parent hypha.

In the development of a single basidiocarp, the abundance, positions, and even the forms of particular hyphal elements and products are partly hereditary and partly accidental. Hyphae elongate at their apices (Smith, 1923), so the particular relationships that hyphal elements have with one another are determined by the developmental interrelationships among the various hyphal tips. Partly because of this, the forms and abundance of modified elements produced by generative hyphae may show variation even in different areas of the individual context or hymenium.

Some of the larger genera of Hymenomycetes now are well enough characterized to serve as testing standards for the theorem that generative hyphae reveal generic attributes. *Stereum* is an example of such a genus, for it has been subjected to intensive study during the past fifteen years. Among the species in Burt's (1920) monograph are *S. ostrea* (Blume & Nees ex Fr.) Fr. (as *S. fasciatum* [Schw.] Fr.), *S. ochraceo-flavum* (Schw.) Ell., *S. purpureum* (Pers. ex Fr.) Fr., and *S. roseo-carneum* (Schw.) Fr. Of these, *S. ostrea* and *S. ochraceo-flavum* still are referred to *Stereum*, but *S. purpureum* now is placed in *Chondrostereum*, and *S. roseo-carneum* in *Laeticorticium*. These latter two are easily distinguishable from *Stereum* on the basis of hyphal characteristics alone.

In *Stereum*, clamp connections are not found on hyphae of the basidiocarp, but both single and multiple clamps develop on the hyphae of cultures. Moreover, the multiple clamps most often appear on broad hyphae, single clamps on hyphae of intermediate width, and clampless septa on slender hyphae (Plate II, fig. 13). In contrast, the generative hyphae of *C. purpureum* (Pers. ex Fr.) Pouzar and *L. roseo-carneum* (Schw.) Boidin have only single clamp connections on hyphae of both cultures and basidiocarps. Other hyphal characteristics also are distinctive, as well as spore features and physiological qualities.

Additional examples of distinctive homogeneity contrasted with heterogeneity may be found in several genera which, at least partially, remain taxonomically artificial. In these examples, heterogeneity is shown both in the generative hyphae and in the structures that they produce. An instructive case is that of *Lopharia*, which includes *L. mirabilis* (Berk. & Br.) Pat. as the type, *L. cinerascens* (Schw.) Cunn. as a closely related species (Boidin, 1959) or possibly the same species (Talbot, 1954b), *L. crassa* (Lév.) Boidin, *L. papyrina* (Mont.) Boidin, and a few other species. *L. crassa* has no clamp connections on the hyphae of basidiocarps and relatively few in culture; generative hyphae are branched at broad angles to form a subiculoid context lacking unidirectional orientation; fiber hyphae are not formed in culture; basidia and spores are of moderate size. In contrast, *L. cinerascens* has clamp connections on hyphae of the basidiocarps and cultures; orientation of generative elements has a tendency to follow the unidirectional orientation of the skeletals; fiber hyphae develop in culture; basidia and spores are large. Other characteristics also militate against the likelihood that both species belong within a single genus. Many of the characteristics of *L. crassa* are shared by *L. papyrina*, so two or more groups of species may be found in *Lopharia*.

*Vararia* also may be a heterogeneous aggregation of several homogeneous lines. Some species have clamp connections in basidiocarps and cultures, but others do not. When other characteristics also are compared, several rather vaguely defined groups emerge. One such group includes *V. effuscata* (Cke. & Ell.) Rogers & Jacks., *V. peniophoroides* (Burt) Rogers & Jacks., and *V. pallescens* (Schw.) Rogers & Jacks. All produce globose basidiospores with strongly amyloid warts or ridges (Plate II, fig. 14); generative hyphae have clamp connections; cultures form chlamydospores, oedocephaloid conidiophores, and oil-filled hyphae with prominent papillae. *V. granulosa* (Fr.) Laurila is somewhat aberrant, with small, ellipsoidal basidiospores marked by discontinuous, slightly raised, amyloid striae.

A second group in *Vararia* includes *V. investiens* (Schw.) Karst., *V. pectinata* (Burt) Rogers & Jacks., and probably *V. fusispora* Cunn.

All have curved, subfusoid basidiospores (Plate II, fig. 15); generative hyphae form clamp connections; oedocephaloid conidiophores do not appear in culture. The basidiospores of both *V. investiens* and *V. pectinata* have small amyloid plaques near their apiculi, but these have not been found on spores of *V. fusispora*. *V. racemosa* (Burt) Rogers & Jacks., with cylindrical spores, also may belong in this group.

Clamp connections are absent from hyphae of a third group in *Vararia*. In this group—which includes *V. luteopora* (Bondarzew) Singer, *V. ochroleuca* (Bourd. & Galz.) Donk, and a species described as *V. tropica* Welden (1965)—basidiospores are smooth and broadly ellipsoidal to subglobose (Plate II, fig. 16). Spore amyloidity in this group requires further study, but spores of *V. tropica* resemble those of *V. investiens* in their development of amyloid plaques. Discussion of other species would add complications to this brief outline, as would any attempt to interpret the taxonomic implications of different forms of dichohyphidia and gloeocystidia.

If this discussion may seem to have overemphasized clamp connections, perhaps perspective may be added by a few additional comments. According to Donk (1964): "The value of clamps as a taxonomic feature differs from group to group, and may even appear erratic within rather small taxa of lower rank such as species." Then Donk concluded with the comment that the typically trimitic Polyporaceae seem to have clamps invariably formed at their primary septa, but that the Septobasidiales and Hymenochaetaceae apparently never form clamp connections.

In an earlier paper (Lentz & McKay, 1966), I discussed the development of clamps in a considerable range of genera in the Aphyllophorales. At about the same time, Furtado (1966) wrote a paper of 20 pages entirely on the subject of clamp connections. One of his conclusions was that "coexistence of species with different type of septa within the same genera has not been confirmed in the taxonomic studies of the Polyporaceae" by Teixeira, O. and M. E. P. K. Fidalgo, and Furtado.

Meaningful records of the presence or absence of clamp connections in the Aphyllophorales are not so easily obtainable as may be imagined. One major difficulty results from the great number of monotypic genera. Another arises from lack of agreement as to which species belong in particular genera, even when the genera are so well known as *Phlebia*, for example. Even so, relative numbers may be helpful: records of approximately 160 miscellaneous nonpolyporaceous genera of the Aphyllophorales and peripheral groups show that about 100 genera have clamp connections on at least some of the generative hyphae in every recognized species; approximately 25 genera are entirely without clamp connections; and approximately 35 genera include some species with and others without clamps. In the last of these groups, relatively few genera have an equal balance between the species with and those without clamps. Instead, most of these genera include only a few species, as few as one, which are discordant elements in relation to presence or absence of clamps.

According to Cunningham (1965), clamp connections appear in 151 of the 242 polyporaceous species that he recorded from Australia, New Zealand, and vicinity. Smith (1966) reported that clamps are not found in "the astrogastraceous line of the Gastromycetes," that they are present "in fewer than five of 137 species" of *Rhizopogon*, that they are absent from *Boletus* and other larger genera of the Boletaceae but present in some groups, and that they are regularly present in nearly all of more than 700 species of *Cortinarius*. Maas Geesteranus (1963a) found that constant presence of clamps is a family characteristic in the Auriscalpiaceae. In a study of mycelial growth of 110 isolates in 57 species of the Agaricaceae, Semerdzieva and Cejp (1966) found clamp connections on hyphae of 69%.

In *Stereum*, the relatively unusual pattern of clamp formation is a useful generic taxonomic characteristic. However, several other genera also have multiple clamps either in basidiocarps or in cultures, or both. Multiple clamps are well known in cultures of the common species of *Coniophora*, for example, but they do not develop in the basidiocarps (Lentz, 1957; Cunningham, 1963; Aoshima & Hayashi, 1967). According to Maas Geesteranus (1962), multiple clamp connections develop on the hyphae of cultures and basidiocarps in the genus *Donkia* (Plate II, fig. 17); Cooke (1957) reported the presence of multiple clamps on hyphae in basidiocarps of *Serpula lacrimans* Pers. ex S. F. Gray; and Boidin (1958) illustrated double clamps on hyphae in cultures of *Merulius corium* Fr.

In *Phanerochaete*, including "*Peniophora*" *cremea* (Bres.) Sacc. & Syd. and related species, the pattern of clamp formation is somewhat similar to that in *Stereum*, but not quite the same. Occasional basidiocarps possibly are devoid of clamp connections, and most descriptions (Christiansen, 1960; Slysh, 1960) emphasize that clamps are rare or absent. However, both single and multiple clamps usually can be found in the marginal subiculum of the basidiocarps, and multiple clamps are common on the larger hyphae of cultures (Plate II, fig. 18). Biggs (1938) reported multiple clamps on hyphae of "*Peniophora*" *gigantea* (Fr.) Mass. in culture. This species has been referred to both *Phanerochaete* (Donk, 1962) and *Phlebia* (Donk, 1957; Christiansen, 1960). In "*Peniophora*" *cremea*, false clamps and bridge hyphae (or H-connections) (Plate III, fig. 19) are relatively common. Furtado's (1966) review paper on clamp connections includes a discussion of cause and mechanism in the formation of pseudoclamps.

Several other generative hyphal attributes may be correlated with characteristics of septation, and the sum of these very often constitutes a complex of features which are distinctive for a particular genus. In addition to hyphal width, cell length, and wall thickness, some other attributes include modification of hyphal or cell shape as the result of inflation, characteristic branching patterns as the result of the particular angles at which branches originate from the parent cells, and branching patterns which reveal tendencies of the branches to develop from particular areas of the parent cells.

Many of the Aphyllophorales in the *Ceratobasidium-Pellicularia* complex show a distinctive pattern of wide-angled hyphal branching, which is coupled with a tendency for the individual hyphae to be very broad. This group of fungi is the subject of nomenclatural and taxonomic controversy, but some mycologists, at least, refer the various species to *Botryobasidium* (with the well-characterized subgenera *Dimorphonema* [Plate III, fig. 20] and *Brevibasidium*) and to other genera such as *Botryohypochnus*, *Thanatephorus*, *Waitea*, and *Uthatobasidium*. *Ceratobasidium* also is very close, and even *Tulasnella*—as represented by *T. violea* (Quél.) Bourd. & Galz.—shares some of the hyphal characteristics of this broad group. Talbot's (1965) review of these genera may be consulted for many details of morphology and classification.

Of all these taxa near and including *Botryobasidium*, only *Botryobasidium* subgenus *Botryobasidium* has clamp connections on the hyphae in all recognized species. With the exception of *Tulasnella*, none of the others has clamps, except in a few species doubtfully included. However, several other genera may be related more-or-less closely to the *Botryobasidium* group, and clamps appear in these related genera. *Suillosporium*, which was referred to the Coniophoraceae by Pouzar (1958), is another genus which has broad hyphae that branch at wide angles. According to Oberwinkler (1965), *Suillosporium* belongs among the genera showing a rather close relationship with *Botryobasidium*. The single recognized species, *S. cystidiatum* (Rogers) Pouzar, has generative hyphae with small clamp connections.

Various species in or close to *Athelia* have moderately slender hyphae which, nevertheless, also have a tendency to branch at the wide angles reminiscent of the *Botryobasidium* group. Oberwinkler (1965) even believes that *Athelidium*, with broad hyphae, constitutes a bridge between *Botryobasidium* and *Athelia*. Much more remote, but still characterized by hyphae in which the angle of branching is wide, at least in the context, is *Phanerochaete*.

Highly individualistic hyphal branching patterns also are characteristic of various other genera. *Coniophora* and *Asterostroma* share a rather unusual sort in which a hyphal cell becomes apically broadened and produces at least three or four branches from a single level of the broadened portion. In *Hyphodontia*, the majority of branches develop opposite (and at slightly lower levels than) the clamp connections; however, a branch occasionally grows directly from a clamp. Hyphae of many other hymenomycetes also characteristically branch opposite clamp connections, but some have a definite tendency to branch directly from the clamps. Still others show no distinct pattern.

Generative hyphae are directly responsible for most of the modified hyphal products that appear in basidiocarps. Skeletal and binding hyphae are responsible for very few of consequence. Instead, skeletal and binding hyphae, as modified products of generative hyphae, are disposed at one end of a series which, at the other end, consists of elements such as true cystidia and hyphidia. In a particular genus, the presence of skeletal hyphae in basidiocarps of a single species is an excellent indication that skeletal hyphae will be found in basidiocarps of all other species of the genus. The same is true of binding hyphae, and it is also true for particular kinds of cystidia and hyphidia and, to a lesser degree, for practically all other specialized products of generative hyphae.

In this light, skeletal and binding hyphae are seen not so much as special elements placed in certain groups of fungi especially for the edification of taxonomists. Instead, they are just two of many elements which generative hyphae are able to form, and which can reasonably be expected to appear in relatively similar form and position in all basidiocarps of a single genus. The same, essentially, may be said for the various specialized products that generative hyphae form in culture.

Skeletal hyphae may appear in all parts of the basidiocarp, or they may be limited to certain areas. By means of a diagram, Teixeira (1956, 1962) has shown the areas which should be examined. When the examination is made, a highly developed special understanding of hyphal modifications must be brought to bear on interpretation of the elements which are found. Some mycologists, including Corner, regard hyphal systems as relative, or descriptive, categories. Others regard them as absolute morphological entities. For Corner, skeletal hyphae apparently are hyphae that constitute the framework of the basidiocarp. They grow in the main direction of growth of the fruiting body, a fact which distinguishes them from binding hyphae and, to a lesser degree, from generative hyphae.

A considerable number of mycologists are much less permissive in their definitions. Skeletal hyphae, in their view, could never assume the function of generative hyphae. One of the most reasonable brief expositions on this subject was written by Maas Geesteranus (1967). As freely translated, his statement is as follows:

> Concerning skeletal hyphae, we tend to adhere when possible to the definition given by Corner (1932): "thick-walled, unbranched, aseptate, straight or slightly flexuous, longitudinal . . . with the lumen more-or-less obliterated in mature parts, but the apices thin-walled with dense contents."
>
> The generative hyphae pose a problem for us because they appear in a great variety of sometimes bizarre forms. The definition by Corner is not applicable in all instances, and a new classification seems desirable. For the moment, it is sufficient to say that generative hyphae are elements which may not be classified as skeletal hyphae, binding hyphae, or tendril hyphae. It follows that the hyphae of which Boidin (1960) wrote ("May one use the term 'skeletals' for those hyphae which are not solely terminal elements?") and which without the least hesitation were referred to as skeletals by Corner and Thind (1961) ("as intercalary parts of generative hyphae . . .") are not skeletals at all. They are sclerified generative hyphae, a term proposed

by Donk (1964). It then follows that probably some of the species of Aphyllophorales that have been described as having a dimitic context will be revealed as monomitic.

Various other elements of the basidiocarp may appear similar to skeletal hyphae. Thick-walled cystidia that develop in monomitic basidiocarps, or which develop as elements apparently separate from the skeletal hyphae of dimitic or trimitic basidiocarps, may have greatly elongated pedicels, as in *Lopharia crassa*. In such instances, these thick-walled elements may present difficulties in interpretation. In the case of *Hydnum stereoides* Cke., Maas Geesteranus (1964) found that the context is monomitic, but the trama of the spines "neither truly monomitic nor dimitic." The cystidia in the basal parts of the spines were said to have thick walls and to resemble skeletal hyphae. However, in very many instances, if not practically always, thick-walled cystidia which have heavily incrusted apices or which are otherwise considerably differentiated can, by very careful examination of their bases, be found to be independent of true skeletal hyphae, even when the two occur together in the subhymenial region.

In a study of *Tulostoma*, Wright (1955) described some "conspicuously thickened and colored terminal hyphae" on the endoperidium of *T. berteroanum* Lév. They were not similar to the setae of the xanthochroic basidiomycetes, so Wright called them "mycosclerids." Although this term recently was recommended as a substitute for the name "skeletal hyphae" (Smith, 1966), illustrations and Wright's description give no indication whatsoever that they have even the most remote resemblance to well-developed skeletal hyphae. Perhaps they are similar to the thick-walled pileocystidia or caulocystidia of *Podoscypha*, as illustrated by Reid (1965).

In relation to the genus *Trogia*, Corner (1966) introduced the terms "sarcodimitic" and "sarcotrimitic," which are based on Corner's functional concept of hyphal systems. Sarcodimitic construction is dependent on the presence of very long and greatly inflated thick-walled cells that serve the function of skeletals, but which normally have clamp connections. In the sarcotrimitic construction, thick-walled but septate, outgrowths from the thin-walled generative hyphae simulate the binding hyphae of the trimitic polypores. Reid (1967) objected to Corner's terminology as confusing, since the terms "sarcodimitic" and "sarcotrimitic" are not morphologically comparable with "dimitic" and "trimitic." Similarly, Smith (1966) used the term "physalomitic" for a system of inflated hyphal cells. This terminology has no relationship at all with the numerical basis of Corner's mitic systems.

A discussion of hymenomycetous hyphal elements should include at least brief comment on a considerable number of conspicuously modified hyphal forms and products that constituted the subject matter for earlier discussions of hyphal elements (Lentz, 1954; Talbot, 1954a). Most of these elements are either modifications of generative hyphae or modified products of generative hyphae.

Oil-bearing hyphae (Plate III, fig. 21) are well known in the fungi, and they are relatively common both in basidiocarps and cultures of the Aphyllophorales. Simple oil-filled hyphae may be generative hyphae, often with clamp connections. In cultures of *Vararia* and of at least a few other genera, oil-filled hyphae may produce cylindrical or mammillate protuberances that usually bear oily globules. Generative hyphae also may contain solid, yellowish material having a resinous appearance.

Substances of an oily or waxy appearance, often colored, may be found in hyphae with few if any septa (Plate III, figs. 22–24). Sometimes called "conducting hyphae" (Talbot, 1954a) or "vascular hyphae" (Lentz, 1954), these elements were referred to by Singer (1962) as "gloeo-vessels" and by Donk (1967) as "gloeoplerous hyphae." Similar hyphae, in *Lactarius*, may carry a kind of latex, so they are called "laticiferous hyphae." Oleiferous and laticiferous hyphae, including those that resemble branched tubes, were discussed by Heim (1931, 1936a, 1936b), Talbot (1954a), and Singer (1962). In most instances, they probably are transformed generative hyphae.

In his conspectus of the families of Aphyllophorales, Donk (1964) listed two principal kinds of cystidia as "hymenial" and "tramal." According to Donk, tramal cystidia are the ends of tramal hyphae, including generative, skeletal, and gloeoplerous hyphae. In place of "false setae" and "pseudocystidia" (Lentz, 1954), or "cystidia" (Burt, 1920; Bourdot & Galzin, 1928; Pilát, 1930), Donk used the term "skeletocystidia" (Plate III, fig. 25) for the modified ends of skeletal hyphae that grow into the hymenial region. Skeletocystidia are defined without reference to their lengths or basal septa. A skeletocystidium is simply the apical portion of a skeletal hypha, usually slightly modified and inflated, protruding into or even beyond the hymenium.

Skeletocystidia of the sanguinolentous basidiocarps in certain species of *Stereum* may, in addition, be classed as pseudocystidia (Plate III, fig. 26) because they are terminations of skeletal hyphae that bear a dark red liquid material which oozes out of the injured hyphae (Bergenthal, 1933). According to Singer and Gamundi (1963), pseudocystidia may originate in any infrahymenial layer, and they either have an excretory function or, at least, they are continuations of the conducting system. Prominent hymenial terminations of oleiferous and laticiferous tubes such as those illustrated by Heim (1936b) for *Mycena atro-violacea* Heim and *Collybia lacrimosa* Heim also may be classed as pseudocystidia. In addition, Singer (1962) includes coscinocystidia (Singer, 1947), hymenial terminations of coscinoids, which appear in *Linderomyces* as nonseptate hyphae with pitted walls and spongy insides.

Gloeocystidia (Plate III, fig. 27) also have been included within the category of pseudocystidia (Lemke, 1964). However, Donk (1964) and others (Lentz & McKay, 1966) insist on using the term "gloeocystidium" for the inflated, often ventricose, or irregularly shaped hyphal endings that commonly have waxy or waxy-vitreous contents. They may occur either in the hymenium or in other areas of the basidiocarp. Perhaps some gloeocystidia are formed as portions of gloeoplerous hyphae, but a great number originate directly from typical generative hyphae. In *Asterostroma* and *Vararia*, gloeocystidia are borne on ordinary generative hyphae which, in *Vararia*, have clamp connections in the species characterized by formation of clamps. Similarly, Reid (1965) found that the gloeocystidia of *Cymatoderma* and *Podoscypha* originate from thin-walled generative hyphae with clamp connections. Boidin and Ahmad (1963) illustrated gloeocystidia produced by clamped generative hyphae of *Duportella tristicula* Pat., and Maas Geesteranus (1963a) reported that clamped generative hyphae of *Auriscalpium* give origin to gloeocystidia as terminal hyphal cells.

In several genera of hymenomycetes, various kinds of cells that occur have in common the fact that they tend to assume a spherical shape. Some probably are elements of the generative system; others may be gloeocystidial. In the basidiocarps of *Russula*, many cells, presumably of the fundamental tissue, assume more-or-less spherical shapes somewhat altered by the fact that the cells are pressed together in rosette-like clusters. These cells, called "spherocysts" (Plate IV, fig. 28), are said to be responsible for the brittle texture of the Lactario-Russulae (Josserand, 1952).

Spherical cells also appear in the context or subhymenium of *Chondrostereum purpureum* and *Cystostereum murraii* (Berk. & Curt.) Pouzar. These cells, known as "vesicular bodies" or "vesicles" (Plate IV, fig. 29), are not grouped in regular patterns, and they apparently develop by enlargement of scattered hyphal tips. In *C. murraii,* they probably are transitional to yellowish, refractive gloeocystidia that appear in the hymenium (Lentz, 1955).

The hyphae in basidiocarps and cultures of *Hyphoderma* produce lateral outgrowths that develop into peculiar subspherical bodies known as "stephanocysts" (Plate IV, fig. 30). The flattened bases of the stephanocysts are surrounded by distinctive tongue-shaped flaps. The formation of stephanocysts by hyphae of *"Peniophora" pubera* (Fr.) Sacc. (Boidin, 1958) may be a good indication that this species belongs in or very near *Hyphoderma.*

In the Basidiomycetes, the xanthochroic series, or family Hymenochaetaceae (Donk, 1948, 1964), is an outstanding example of a group that cuts directly across the traditional Friesian system of classification. This series includes fungi which, in older classification systems, might have been placed into the families Thelephoraceae, Clavariaceae, Hydnaceae, and Polyporaceae. The homogeneous nature of the Hymenochaetaceae is well substantiated by the many key characteristics shared by all, or substantially all, members. Production of brown setae or various kinds of setiform elements is especially prevalent.

Recently (Smith, 1966), objection has been registered against use of the term "seta" for the kind of brown, spinelike elements so highly characteristic of the Hymenochaetaceae. The objection apparently is based partially on the fact that use of the general term "setae" for the elements in the Hymenochaetaceae would seem to require that some other term be used for an assortment of setiform elements found in a few other hymenomycetes.

With the exception of setae in the deuteromycetous genus *Colletotrichum,* those of the Hymenochaetaceae are probably the best-known examples of setae in the fungi. The term "setae" was applied to these elements by Massee (1890), Burt (1918), Rea (1922), Shope (1931), Overholts (1929, 1953), Lowe (1934, 1957, 1966), Bessey

(1950), Talbot (1954a), O. and M. E. P. K. Fidalgo (1967), and many other authors. Ainsworth and Bisby's (1961) illustration of a seta is based on an example from a xanthochroic basidiomycete.

Talbot (1954a) described, discussed, and illustrated practically every conceivable setiform element of the Hymenochaetaceae. Little can be added. The several kinds of elements may be defined somewhat as follows (Lentz, 1954):

1. *Seta*, a brown, spinelike, sterile hyphal end appearing in various parts of xanthochroic basidiocarps, never very conspicuously incrusted, usually thick-walled, darkening conspicuously when moistened with potassium hydroxide solution.

2. *Embedded seta* (including macroseta, tramal seta, setal hypha), an enlarged, elongated, brown, thick-walled tube embedded in the context or trama of a xanthochroic basidiocarp, and generally following the course of the ordinary hyphae of the region in which it occurs.

3. *Stellate seta* (asteroseta), a compound seta having several radiating arms originating from an enlarged center.

Setae of all categories apparently originate from generative hyphae, although Cunningham (1946) asserted that those in basidiocarps with a dimitic hyphal system arise from skeletal hyphae. Simple setae (Plate IV, fig. 31) appear in *Hymenochaete* and in most other members of the family. Often they may be seen to originate beneath the hymenium and to project beyond the hymenial surface. Embedded setae (Plate IV, figs. 32, 33), according to Cunningham (1965), are found in basidiocarps of a few species of *Phellinus* and *Inonotus*, including *P. lamaensis* (Murr.) Heim and *I. glomeratus* (Pk.) Murr. Stellate setae (Plate IV, figs. 34, 35) are well exemplified in the basidiocarps of *Asterodon* and *Asterostroma*. Development of the asterosetae of *Asterostroma* begins with development of a characteristic pyramidal enlargement at the end of a hypha, with subsequent production of slender tubes which eventually become spinelike arms of the asteroseta. *Asterodon* is a typical member of the Hymenochaetaceae, but *Asterostroma* exhibits several characteristics that differ significantly from those of the xanthochroic fungi.

With the exception of spores, probably no microscopic elements of the basidiocarp have been utilized more for taxonomic purposes than the miscellany of structures styled as cystidia (Plate IV, fig. 36; Plate V, fig. 37). Almost any nonsporiferous cell of noteworthy form or position in the basidiocarp, and especially at any surface of the basidiocarp, may be classed as some kind of cystidium. However, the majority of elements regarded as true cystidia appear in the hymenium and very commonly protrude beyond the other hymenial structures. In a brief discussion, little more than a list of cystidial names can be given. Cystidia have been classified according to origin, position, form, and contents. For more-or-less extensive discussions of cystidia, reference is made to papers by Romagnesi (1944), Josserand (1952), Lentz (1954), Talbot (1954a), Singer (1962), Donk (1964), Reid (1965), and Smith (1966), in addition to pertinent notes and comments by many other recent authors.

The importance of the generative hypha in the production of essentially all drastically modified structures of the basidiocarp has already been emphasized. True cystidia almost certainly are produced only by generative hyphae. Heavily incrusted, thick-walled cystidia may appear to be modifications of skeletal hyphae, but very careful dissection of the subhymenium or other appropriate regions will reveal that thick-walled cystidia originate from septate hyphae, usually thin-walled, and very often bearing clamp connections.

In my introductory paragraph, I said that bizarre elements may be produced either within the fungus hyphal mass or at its surface. Some of the more unusual elements could serve as the subject for further extensive discussion. In addition to pseudo-, gloeo-, skeleto-, hypho-, and coscinocystidia, which have already been mentioned, and besides Donk's (1967) hymenial and tramal cystidia, various other cystidiiform elements have been described.

Pileocystidia develop on the pileus surface and have a potential for assuming a great variety of forms. Some may intergrade with other kinds of cells (Plate V, figs. 38, 39) so that their characterization as pileocystidia is not stable. Romagnesi (1944) differentiated between pileocystidia and surface hairs, but the distinction is not always clear. Surface hairs often resemble terminal portions of skeletal hyphae, but in *Panus rudis* Fr. they are generative hyphae with clamps (Plate V, fig. 40). Caulocystidia (Plate V, fig. 41) develop on the stems of many stipitate fungi, and these cystidiiform elements also may resemble hairlike or inflated hyphae, some having clamp connections. Cheilocystidia are found on the gill edges of various mushrooms, and other kinds of cystidia are recognizable as structures that grow in particular areas of the basidiocarp, have distinctive shapes, or contain noteworthy substances.

In some fungi of both the Heterobasidiomycetes and Homobasidiomycetes, peculiar peglike hyphal aggregations (Plate V, figs. 42, 43) protrude beyond the hymenial surfaces much in the manner of cystidia. However, these pegs are compound structures often rather easily observable without magnification. According to Cunningham (1963), the hyphal pegs of *Mycobonia* develop from the basal region and grow through the context before protruding from the hymenium. Examination of the type specimen of *M. flava* (Sw. ex Berk.) Pat. (Plate V, fig. 43), which is the type species of *Mycobonia*, shows that the pegs actually originate in or near the hymenium and do not grow through the context.

Finally, any discussion of modified hyphae must turn to the structures that Donk (1956) called "hyphidia." Just a list of these elements would be extensive, so remarks concerning hyphidia will be restricted to mention of dendrohyphidia (Plate V, fig. 44) (in the hymenia of *Laeticorticium* and various other Basidiomycetes) and dichohyphidia (Plate II, fig. 16; Plate V, fig. 45), which are characteristic of *Vararia*. The dichohyphidia of *Vararia*, both in basidiocarps and cultures, may assume so many different forms that one kind is hardly comparable with another. Reid (1965) used the term "dichophytic binding hyphae" for the dichohyphidia of *Dichopleuropus spathulatus* Reid, and he also suggested that the dichohyphidia of *Vararia* and *Lachnocladium* may be specialized forms of binding hyphae.

This brings us to the point at which we introduced the subject: The anatomical versatility of hymenomycetous hyphae is such that the term "modified hyphae" is hardly definitive. With this in mind, we may wish to compare the modified hyphae of the Heterobasidiomycetes with those of the Homobasidiomycetes. When we find that the same kinds of highly differentiated structures may appear in both groups, we begin to comprehend some of the problems associated with consideration of modified hyphae as taxonomic tools.

## References

Ainsworth, G. C. 1961. Ainsworth and Bisby's dictionary of the fungi. 5th ed. Commonwealth Mycol. Instit., Kew. 547 p.

AOSHIMA, K., AND Y. HAYASHI. 1967. Notes on the genus *Coniophora* of Japan. Mycol. Soc. Japan Trans. 8: 80–86.

BERGENTHAL, W. 1933. Untersuchungen zur Biologie der wichtigsten deutschen Arten der Gattung *Stereum*. Centralbl. Bact., II, 89: 209–236.

BESSEY, E. A. 1950. Morphology and taxonomy of fungi. McGraw-Hill Book Co., New York. 791 p.

BIGGS, R. 1938. Cultural studies in the Thelephoraceae and related fungi. Mycologia 30: 64–78.

BOIDIN, J. 1958. Essai biotaxonomique sur les Hydnés résupinés et les Corticiés. Rev. Mycol., Mém. hors-sér. 6. 387 p.

———. 1959. Hétérobasidiomycètes saprophytes et Homobasidiomycetès résupinés. VII.—Essai sur le genre "Stereum sensu lato" (Troisième contribution). Soc. Linn. Lyon Bull. Mensuel 28: 205–222.

———. 1960. Le genre *Stereum* Pers. s.l. au Congo Belge. Jard. Bot. État Bruxelles Bull. 30: 283–355. [Cited in the quotation from Maas Geesteranus, 1967.]

BOIDIN, J., AND S. AHMAD. 1963. The position of *Duportella tristicula* Pat. (Basidiomycetes:Thelephoraceae). Biologia 9: 33–38.

BOURDOT, H., AND A. GALZIN. 1928. Hyménomycètes de France. Marcel Bry, Seine. 761 p.

BURT, E. A. 1918. The Thelephoraceae of North America. X. *Hymenochaete*. Missouri Bot. Gard. Annals. 5: 301–372.

———. 1920. The Thelephoraceae of North America. XII. *Stereum*. Missouri Bot. Gard. Annals 7: 81–248.

BUTLER, G. M. 1958. The development and behaviour of mycelial strands in *Merulius lacrymans* (Wulf.) Fr. II. Hyphal behaviour during strand formation. Annals Bot., n.s., 22: 219–236.

CHRISTIANSEN, M. P. 1960. Danish resupinate fungi. Part II. Homobasidiomycetes. Dansk Bot. Arkiv 19: 63–388.

COOKE, W. B. 1957. The genera *Serpula* and *Meruliporia*. Mycologia 49: 197–225.

CORNER, E. J. H. 1932a. The fruit-body of *Polystictus xanthopus* Fr. Annals Bot. 46: 71–111.

———. 1932b. A *Fomes* with two systems of hyphae. Trans. Brit. Mycol. Soc. 17: 51–81.

———. 1950. A monograph of *Clavaria* and allied genera. Oxford Univ. Press, London. 740 p.

———. 1953. The construction of polypores—1. Introduction: *Polyporus sulphureus, P. squamosus, P. betulinus,* and *Polystictus microcyclus.* Phytomorphology 3: 152–167.
———. 1966. A monograph of cantharelloid fungi. Oxford Univ. Press, London. 255 p.
CORNER, E. J. H., AND K. S. THIND. 1961. Dimitic species of *Ramaria* (Clavariaceae). Trans. Brit. Mycol. Soc. 44: 233–238. [Cited in the quotation from Maas Geesteranus, 1967.]
CUNNINGHAM, G. H. 1946. Notes on classification of the Polyporaceae. New Zealand Jour. Sci. Technol. 28, sec. A: 238–251.
———. 1963. The Thelephoraceae of Australia and New Zealand. New Zealand Dept. Sci. Industr. Res. Bull. 145. Wellington. 359 p.
———. 1965. Polyporaceae of New Zealand. New Zealand Dept. Sci. Industr. Res. Bull. 164. 304 p.
DONK, M. A. 1948. Notes on Malesian fungi, I. Bot. Gard. Buitenzorg Bull., ser. III, 17: 473–482.
———. 1956. Notes on resupinate Hymenomycetes—III. Fungus 26: 3–24.
———. 1957. Notes on resupinate Hymenomycetes—IV. Fungus 27: 1–29.
———. 1958. Notes on resupinate Hymenomycetes—V. Fungus 28: 16–36.
———. 1962. Notes on resupinate Hymenomycetes—VI. Persoonia 2: 217–238.
———. 1964. A conspectus of the families of Aphyllophorales. Persoonia 3: 199–324.
———. 1967. Notes on European polypores—II. Persoonia 5: 47–130.
FAYOD, V. 1889. Prodrome d'une histoire naturelle des agaricinés. Ann. Sci. Nat.. 7 sér., 9: 181–411.
FIDALGO, O., AND M. E. P. K. FIDALGO. 1967. Dictionário micológico. Rickia, suppl. 2. Instit. Bot. São Paulo Publ. 232 p.
FURTADO, J. S. 1966. Significance of the clamp-connection in the Basidiomycetes. Persoonia 4: 125–144.
GROSS, H. L. 1964. The Echinodontiaceae. Mycopath. Mycol. Appl. 24: 1–26.
HEIM, R. 1931. Le genre *Inocybe. In* Encyclopédie mycologique. Paul Lechevalier & Fils, Paris. 429 p.
———. 1936a. Sur trois agarics à latex de la flore malgache. Acad. Sci. [Paris] Compt. Rend Séances 202: 1450–1452.

———. 1936b. Observations sur la flore mycologique malgache. Rev. Mycol., n.s., 1: 223–256.

JOSSERAND, M. 1952. La description des champignons supérieurs. *In* Encyclopédie mycologique. Paul Lechevalier, Paris. 338 p.

LEMKE, P. A. 1964. The genus *Aleurodiscus* (*sensu stricto*) in North America. Canad. Jour. Bot. 42: 213–282.

LENTZ, P. L. 1954. Modified hyphae of Hymenomycetes. Bot. Rev. 20: 135–199.

———. 1955. *Stereum* and allied genera of fungi in the upper Mississippi Valley. USDA Agric. Monogr. 24. 74 p.

———. 1957. Studies in *Coniophora*. I. The basidium. Mycologia 49: 534–544.

LENTZ, P. L., AND H. H. MCKAY. 1966. Delineations of forest fungi: Morphology and relationships of *Vararia*. Mycopath. Mycol. Appl. 29: 1–25.

LOWE, J. L. 1934. The Polyporaceae of New York State (Pileate species). N.Y. State College Forestry. Syracuse Univ. Tech. Publ. 41. 142 p.

———. 1957. Polyporaceae of North America. The genus *Fomes*. N.Y. State College Forestry. Syracuse Univ. Tech. Publ. 80. 97 p.

———. 1966. Polyporaceae of North America. The genus *Poria*. N.Y. State College Forestry. Syracuse Univ. Tech. Publ. 90. 183 p.

MAAS GEESTERANUS, R. A. 1962. Hyphal structures in Hydnums. Persoonia 2: 377–405.

———. 1963a. Hyphal structures in Hydnums. II. Koninkl. Nederl. Akad. Wetenschappen–Amsterdam Proc., ser. C, 66: 426–436.

———. 1963b. Hyphal structures in Hydnums. III. Koninkl. Nederl. Akad. Wetenschappen–Amsterdam Proc., ser. C, 66: 437–446.

———. 1964. Notes on Hydnums–II. Persoonia 3: 155–192.

———. 1967. Quelques champignons hydnoides du Congo. Jard. Bot. National Belgique Bull. 37: 77–107.

MASSEE, G. 1890. A monograph of the Thelephoreae. Part II. Linnean Soc. Bot. Jour. 27: 95–205.

NOBLES, M. K. 1958. Cultural characters as a guide to the taxonomy and phylogeny of the Polyporaceae. Canad. Jour. Bot. 36: 883–926.

NYBAKKEN, O. E. 1959. Greek and Latin in scientific terminology. Iowa State Univ. Press, Ames. 321 p.

OBERWINKLER, F. 1965. Primitive Basidiomyceten. Revision einiger

Formenkreise von Basidienpilzen mit plastischer Basidie. Sydowia 19: 1–72.

OVERHOLTS, L. O. 1929. Research methods in the taxonomy of the Hymenomycetes. Internatl. Congr. Pl. Sci. Proc. 2: 1688–1712.

———. 1953. Polyporaceae of the United States, Alaska, and Canada. Univ. Michigan Press, Ann Arbor. 466 p.

PILÁT, A. 1930. Monographie der europäischen Stereaceen. Hedwigia 70: 10–132.

POUZAR, Z. 1958. Nova genera macromycetum II. Ceská Mykol. 12: 31–36.

———. 1966. Studies in the taxonomy of the polypores I. Ceská Mykol. 20: 171–177.

REA, C. 1922. British Basidiomycetae. Cambridge. 799 p.

REID, D. A. 1965. A monograph of the stipitate stereoid fungi. Nova Hedwigia 18(Beihefte). 388 p.

———. 1967. *Review*, A Monograph of Cantharelloid Fungi, E. J. H. Corner, 1966. Trans. Brit. Mycol. Soc. 50: 333–335.

ROMAGNESI, H. 1944. La cystide chez les Agaricacées. Rev. Mycol., n.s., 9 (Suppl.): 4–21.

SEMERDZIEVA, M., AND K. CEJP. 1966. Investigation of mycelial growth in some gill fungi under laboratory conditions. Folia Microbiol. 11: 146–154.

SHOPE, P. F. 1931. The Polyporaceae of Colorado. Missouri Bot. Gard. Annals 18: 287–456.

SINGER, R. 1947. Coscinoids and coscinocystidia in *Linderomyces lateritius*. Farlowia 3: 155–157.

———. 1962. The Agaricales in modern taxonomy. J. Cramer, Weinheim. 2nd ed. 915 p.

SINGER, R., AND I. J. GAMUNDI. 1963. Paraphyses. Taxon 12: 147–150.

SLYSH, A. R. 1960. The genus *Peniophora* in New York State and adjacent regions. N.Y. State College Forestry. Syracuse Univ. Tech. Publ. 83. 95 p.

SMITH, A. H. 1966. The hyphal structure of the basidiocarp, pp. 151–177, Vol. 2. *In* G. C. Ainsworth, and A. S. Sussman. The fungi. Academic Press, N.Y.

SMITH, J. H. 1923. On the apical growth of fungal hyphae. Annals Bot. 37: 341–343.

TALBOT, P. H. B. 1954a. Micromorphology of the lower Hymenomycetes. Bothalia 6: 249–299.

———. 1954b. On the genus *Lopharia* Kalchbrenner & MacOwan. Bothalia 6: 339–346.
———. 1965. Studies of '*Pellicularia*' and associated genera of Hymenomycetes. Persoonia 3: 371–406.
TEIXEIRA, A. R. 1956. Método para estudo das hifas do carpóforo de fungos poliporáceos. Instit. Bot. São Paulo Publ. 22 p.
———. 1960. Characteristics of the generative hyphae of polypores of North America, with special reference to the presence or absence of clamp-connections. Mycologia 52: 30–39.
———. 1962. The taxonomy of the Polyporaceae. Biol. Rev. 37: 51–81.
THIND, K. S. 1961. The Clavariaceae of India. Indian Council Agricult. Res., New Delhi. 197 p.
WELDEN, A. L. 1965. West Indian species of *Vararia* with notes on extralimital species. Mycologia 57: 502–520.
WRIGHT, J. E. 1955. Evaluation of specific characters in the genus *Tulostoma* Pers. Michigan Acad. Sci. Arts Lett. Papers 40: 79–87.

## DISCUSSION

DUBUVOY: I do not see why you make a distinction between sclerified generative hyphae and skeletal hyphae. You must give a special morphogenetic potentiality to the apical cell of any hypha. You are taking skeletals as different because they are clampless, yet you have said that cystidia which originate from skeletal hyphae are clamped. I would like to know if you have some other, perhaps chemical, criterion which will give you a difference in the apical cell from the other cells of the hypha so you can think that these are different elements.

LENTZ: I do not believe that I would have made any such statement. In fact, I distinguish between skeletocystidia, which are noteworthy ends of skeletal hyphae, and true cystidia which never arise from skeletal hyphae. True skeletal hyphae do not form clamps, so the modified end of a skeletal hypha could not be clamped. True cystidia and skeletocystidia may occur in the same basidiocarp, but they are separate elements.

I tried to present both sides of this, however. I follow what I assume is the point of view taken by Teixeira and Fidalgo and some others, and apparently to some degree by Maas Geesteranus, that if the sclerified generative hypha subsequently becomes thin-walled, produces clamp connections, and ultimately forms basidia, it should be distinguished from the kinds of hyphae that do not have the faculty of resuming growth as generative hyphae.

I do not think we have reached the point where we can really attach significance to some of these names and say, "now we have it for all time." Dr. Smith has stated in print that perhaps the most important feature of Corner's hyphal systems, and of the work that others have done in describing hyphae, is that such a scheme causes us to examine the microscopic structure much more carefully than we have in the past.

LARSEN: How common, in your experience, are "H"-connections, and what is their function?

LENTZ: I have seen some of them, but I don't know how common they are. I would refer you to Furtado's paper (1966), which is rather detailed.

GRAND: In my study of the cultural characteristics of some *Suillus* species, I found the same "H"-connections. I have watched them develop, and it is, in the case of several *Suillus* species at least, a true anastomosis. It is not a branching hypha as such.

HANLIN: Your definition of a seta seems to exclude the structure found in *Colletotrichum*.

LENTZ: The structures in the Hymenochaetaceae have been called setae by a great number of mycologists for a very long time. These are setae special to the Hymenochaetaceae, and I think that should be recognized. In *Colletotrichum*, I assume that there are setae of a different nature. In my own mind, I keep them completely separate. A greater difficulty, as I understand it, is that some other basidiomycete structures seem to have a close resemblance to the *Hymenochaete* setae, but they do not have some of the special properties of the setae of *Hymenochaete*. In those other cases in the Basidiomycetes, it seems to me that the attachment of some prefix or the use of some other nomenclature would be desirable. At this late date I just don't feel that a useful purpose will be served by using another name for setae so well known in the Hymenochaetaceae.

ROGERS: The setae of *Colletotrichum* would be comparable to the paraphyses of the Discomycetes. The setae of *Hymenochaete* would be comparable to paraphyses in Basidiomycetes. That is, they belong to different cytological generations. Whatever their morphological similarities, I do not think we could call them homologous structures at all. I see no objection to calling them both setae, because they are so obviously separated in the life cycles of the fungi.

SINGER: We have discussed with Dr. Donk the question of setae in the Agaricales. Maybe you would care to comment on this. I understand that you think we should maintain the term "setae" as a general term in the Basidiomycetes and perhaps in other fungi and then subdivide them into subtypes. I think in the Agaricales we have at least two quite different things which are termed setae according to the general definition that we have now. For instance, in *Marasmius* the so-called setae look entirely like setae in the Aphyllophorales, but they do not darken in KOH. They originate from generative hyphae that are clamped. And in *Marasmius* they originate from broom cells (which you showed in your illustrations very nicely) whereby one of the setulae of the broom cells is elongated very strongly, and the base becomes difficult to observe. Conversely, in *Boletochaete* setae are actually modified cystidia of the *Boletus* type, but they happen to have colored walls, either green or brown.

LENTZ: My suggestion, if I have to make a suggestion, would be to use the term "setae" for these structures in the Hymenochaetaceae and use a prefix elsewhere. In some cases I am sure that entirely different words would be appropriate for some of the setaform elements in the mushrooms. If I were doing it, I would say that the structures in the Hymenochaetaceae only are setae, and the others would have to be called something else.

REID: I am under the impression that you have said that the thick-walled cystidia in the hymenium arise from thick-walled septate hyphae and not usually from skeletals. It occurs to me that this is precisely the same thing that skeletal hyphae do. In these thick-walled hymenial cystidia with very deep-seated origin, there is a very long drawn-out basal portion. If you search back far enough you come to a septate thick-walled hypha. But likewise, if it were a skeletal hypha it might originate from precisely the same type of hypha. So we return again to deciding more-or-less how differentiated the hyphae are.

LENTZ: I spent three weeks once, as an example, working with one basidiocarp to find that out to my satisfaction. This is a fungus in which a mycologist described the thick-walled cystidia as being the ends of skeletal hyphae. There were many skeletal hyphae which did end in the vicinity of the hymenium, and these were all—as many as I could trace back into the trama—very long. Far down the thick-walled cystidia in that region—you know the subhymenial region where things are so difficult to tease apart without injuring the fruiting body—it is very difficult to make a connection. However, the thick-walled cystidia were attached to thin-walled subhymenial hyphae, and I felt, as I say, that there was a rather decided break and discontinuity. The skeletal hyphae did not seem to originate in or near the subhymenium. In most or perhaps all of these fungi there perhaps isn't any way to reach an absolute determination.

DONK: The terminology for hyphae and cystidia is rapidly becoming bewildering. One way out of the confusion is to find a reasoned basis for the terminology. Since the fruiting body is a growing thing and the hyphae develop during the process of growth I think we might as well divide the fruiting body into "surfaces." There is the growing margin of the cap which constitutes a morphogenetic "line" (a narrow strip) and the growing ends of the hymenophore which may be "points" (spines) or "edges" (tubes, gills). Moreover, there are the surfaces of the cap and the stalk; these surfaces are bi-dimensional, so to say, to contrast them to the growing "lines" and "points" of the first category. One of these surfaces is that remarkable structure, the hymenium. Most mycologists do not bring the products formed by the growing margins and the growing surfaces into connection with each other. I believe that since both categories are parts of the same fruiting body it should be kept in mind that both may be expected to form basically the same elements. Thus, typical skeletal hyphae formed by the margin become longer and longer because they keep pace with the advancing margin, but when these elements are formed in the hymenium they look quite different because they are subjected to the rules of a different morphogenetic field. Therefore, it should not be surprising when in certain taxa thick-walled hymenial cystidia are to be interpreted as modified skeletal hyphae, which assume a strongly abbreviated form because the rules of the hymenium prevent them from developing otherwise. Elements of a euhymenium tend to be strongly abbreviated and to become more-or-less inflated; corresponding elements arising in margins are longer and more slender.

As a very peculiar morphogenetic field, the hymenium not only influences the development of the strictly hymenial elements, but it may also act on the growth of the elements of an adjacent region that has its own morphogenetic tendencies. For instance, skeletals formed in the growing margin of a dissepiment of a tube or tip of a spine may originate so close to the side where the hymenium will develop that it will soon be included into the sphere of influence of the subhymenium and react according, viz., assume certain characters to be expected of thick-walled *hymenial* cystidia. What we need are additional characters that will aid in establishing the homology between the two extreme forms (skeletal and hymenial cystidia); in such cases chemical tests will often be indispensable.

The views just sketched might become a basis for a renewed appreciation of the situation in the Lachnocladiaceae. The dendrohyphidia (*Scytinostroma*) and dichohyphidia (*Vararia, Lachnocladium*) are surface structures in origin; they are formed first, then the basidia (and eventual other hymenial elements) may follow. The hyphidia are rather a strongly abbreviated and branched type of hyphae comparable to skeletals; they make out part of the catahymenium, which induces some of their characteristics.

Skeletals and the like are formed as end cells. They soon become embedded in the context which exercises its own morphogenetic forces and these may redirect the course of development of the skeletals. These forces may also induce the change of generative hyphae into sclerified ones, and so on. Such secondary influences on preformed hyphal elements may be called epigenic.

BURDSALL: Would you comment on what purpose the terms monomitic, dimitic, and trimitic really serve at this point? Do they serve any, other than forcing mycologists to look at hyphae more carefully?

LENTZ: I think they are useful terms to build something on. You realize, I am sure, that binding hyphae have many different natures, especially when we include such things as dichohyphidia which occur in the context of *Vararia*. Even leaving these out because they are rather extreme and specialized structures, the variety of binding hyphae is considerable. We also have the sclerified generative hyphae as opposed to strictly skeletal hyphae and so on. At this point I would feel uncertain about how many monomitics, dimitics, and trimitics there were, and whether that particular kind of emphasis should be given to the hyphae.

JACQUES BOIDIN
*Professeur*
*Faculté des Sciences, Université de Lyon, Lyon, France*

# NUCLEAR BEHAVIOR IN THE MYCELIUM AND THE EVOLUTION OF THE BASIDIOMYCETES

The complete study of nuclear behavior from the germination of the spore until its formation on the basidium has only been accomplished for some species distributed among the lignicolous Aphyllophorales. For a number of other Basidiomycetes our knowledge remains very fragmentary. Let me start by reporting the results of our studies of almost 400 corticiums *sensu lato* and pore fungi.

## Nuclear Behavior

Five types of behaviors can be distinguished:

1. *Normal* behavior: the uninucleate spore germinates into a mycelium composed of uninucleate cells; the diplont is regularly binucleate.

2. *Subnormal* behavior: after the germination of the binucleate spore, a brief plurinucleate state precedes the appearance of mycelium which rapidly becomes uninucleate; the diplont is binucleate.

3. *Heterocytic* behavior: some binucleate or exceptionally uninucleate spores germinate into primary mycelia which remain, always —at least within the growing cells—pluri- to multinucleate, but the diplont is still regularly binucleate.

4. *Astatocenocytic* behavior. This series is composed of the species with inconstant clamp connections herein called "variable" because, although constant on the entirely aerial hyphae, they become inconstant and then totally lacking on the asphyxiated hyphae where the septa themselves are very widely spaced. The germination, even if the spore is uninucleate, is strongly coenocytic, and this state is maintained during all of the haplophase. As to the difference from the heterocytic species, the secondary mycelium is only binucleate and clamped under conditions of sufficient aeration; or else it re-

turns in a reversible manner to the coenocytic condition of youth, losing partially, then totally, the clamps. The fructification is binucleate. The factor responsible for the progressive loss of the binucleate clamped state is the accumulation of carbon dioxide in the environment (Gabriel, 1967).

5. *Holocenocytic* behavior: The spore with one or two nuclei germinates into a mycelium with plurinucleate cells. This coenocytic state is maintained during all the life of the mycelium and extends into the fructification where, in certain extreme cases, the basidiole alone is binucleate.

The progressive importance of the multinucleate phase is illustrated in table 1. As you can see, the plurinucleate state, nonexistent

TABLE 1. NUCLEAR BEHAVIOR CLASSED ACCORDING TO THE GROWING IMPORTANCE OF THE PLURINUCLEATE PHASE

| | Germination | Haplont | Diplont | Subiculum | Subhymenium | Basidia |
|---|---|---|---|---|---|---|
| Normal | ------ | ------ | ====== | ====== | ====== | ====== |
| Subnormal | ××× --- | ------ | ====== | ====== | ====== | ====== |
| Heterocytic | ×××××× | ×××××× | ====== | ====== | ====== | ====== |
| Astatocenocytic | ×××××× | ×××××× | =×=×=× | ====== | ====== | ====== |
| Holocenocytic | ×××××× | ×××××× | ×××××× | =×=×=× | =×=×=× | ====== |

--- : Uninucleate    === : Binucleate    ××× : Multinucleate

in the mycelium of the "normal" species, exerts itself at the germination of the spores of the "subnormal" species and invades more and more the life cycles of the heterocytic and particularly the holocenocytic species.

These schemes recapitulate almost wholly the events known among the Aphyllophorales. Only certain shortened cycles remain difficult to classify: *Hypochnus terrestris* (according to Kniep, 1913) whose binucleate spores germinate into mycelia with dicaryons cannot be called subnormal or heterocytic because their behavior is differentiated to the level where haploidy is suppressed. One finds the same thing in the Agaricales in the genus *Octojuga*. The parthenogenetic species—rare in this group—are easier to classify. Notice, however, that the species without clamps and with plurinucleate monosporic mycelium from which we have not been able to obtain the fructification in the laboratory would be called holocenocytic, behavior that conceals also a haploid parthenogenesis as the absence of haplophase. They are—by the impossibility of distinguishing two kinds of successive mycelia (primary and secondary, mono- and poly-spored)—doubtfully considered as homothallic;[1] it is not within the true definition of this term.

Although within the pore fungi and corticiums (*s.l.*) the appearance of the cenocyty is always early in the cycle, frequent exceptions to this pattern are found among the Agaricales. The species of *Mycena* with "normal" mycelial behavior possess multinucleate fundamental hyphae within the stipe. Likewise, the plurinucleate cystidia growing within the hymenium of the *tenacella* group of *Collybia* arise from elements elsewhere binucleate and without clamps. One finds these same facts in the heterocytic Agaricales where the secondary mycelium is therefore dicaryotic: the fundamental hyphae of *Conocybe*, the cystidia of *Agrocybe praecox*, etc., are rich in nuclei.

## Correlations between Nuclear Behavior, Clamps, and Number of Nuclei in Spores

For the corticiums *s.l.* (Corticiaceae, Punctulariaceae, Hymenochaetaceae [polyporoid forms not included], Coniophoraceae, residual Hydnaceae, resupinate Lachnocladiaceae, resupinate Auriscalpiaceae, Hericiaceae, Stereaceae and Podoscyphaceae) table 2 indicates the number and percentages of differentiated cases by comparing clamps, number of spore nuclei, and nuclear behaviors.

The normal and subnormal behaviors are very close and only differ by the number of nuclei within the spore at the start of the cycle. These represent 62.8% of the cases. Next comes holocenocytism (23.9%). A more detailed observation showed that the more statis-

[1] Cf. table 4 where it is written "(hom.)," the parentheses indicating probable data.

tically important cases, in decreasing order, are: (*a*) normal behavior (58%), with constant clamps: 50%; (*b*) holocenocytic behavior (23.9%), with verticillate clamps: 12.8%; (*c*) heterocytic behavior (7%) associated very often with binucleate spores; 6.4% (more than 90% of the cases); (*d*) astatocenocytic behavior (6%), associated

TABLE 2. NUCLEAR BEHAVIOR OF THE CORTICIUMS *s. l.* IN ASSOCIATION WITH CLAMPS AND THE NUMBER OF NUCLEI OF THE SPORES.

| | Nuclear Behavior | Normal | Subnormal | Heterocytic | | Astatocenocytic | | Holocenocytic | | Total |
|---|---|---|---|---|---|---|---|---|---|---|
| | *Spores with* | *1 nucleus* | *2 nuclei* | *1 nucleus* | *2 nuclei* | *1 nucleus* | *2 nuclei* | *1 nucleus* | *2 nuclei* | |
| CLAMPS | Constant | 148<br>50 % | 7<br>2.3% | 2<br>0.6% | 19<br>6.4% | | | | | 176<br>59.4% |
| | Inconstant | 5<br>1.6% | 1<br>0.3% | | | | | 3<br>1 % | 14<br>4.7% | 23<br>7.7% |
| | Variable | | | | | 14<br>4.7% | 4<br>1.3% | | | 18<br>6 % |
| | Absent | 19<br>6.4% | 6<br>2 % | | | | | 6<br>2 % | 10<br>3.3% | 41<br>13.8% |
| | Verticillate | | | | | | | 8<br>2.7% | 30<br>10.1% | 38<br>12.8% |
| | Totals | 172 | 14 | 2 | 19 | 14 | 4 | 17 | 54 | 296 |
| | | 186 | 62.8% | 21 | 7 % | 18 | 6 % | 71 | 23.8% | |

The number of representatives and the percentages are indicated by each known combination.

more often with uninucleate spores (in 77.7% of the cases), which should be emphasized because the spores with one nucleus generally give rise (84% of the cases) to the mycelia with normal behavior (Yen's rule, cf. Boidin, 1958, pp. 297–298).

The empty spaces are not without interest. Undoubtedly with reason one can consider that some of these will remain because the characters which are matched in them are incompatible: holocenocytic and constant clamps; verticillate clamps with normal, heterocytic, and astatocenocytic nuclear behaviors (cf. hatched areas of table 2). Other spaces probably will become filled; heterocytic behavior with inconstant clamps and binucleate spores is known beyond the corticiums, in *Coprinus impatiens, C. lipophilus, C. hiascens, C. subdisseminatus.*

Likewise, the heterocytic behavior without clamps appears to be a fact in *Coprinus congregatus*, remarkable since among the corticiums the "regularization" which is accomplished in the heterocytic species by the passage of the primary mycelium to the secondary mycelium appears to necessitate the action of the clamp connections.

Table 3, more schematic, groups the corticiums *s.l.* and polypores.

TABLE 3. NUCLEAR BEHAVIOR OF THE CORTICIUMS *S. l.* AND THE PORE-FUNGI.

| Nuclear Behavior | | Normal and Subnormal | Heterocytic | Astatocenocytic | Holocenocytic | Totals | Percentages |
|---|---|---|---|---|---|---|---|
| CLAMPS | Constant | 198<br>50.1% | 27<br>6.8% | | | 225 | 56.9% |
| | Inconstant | 6<br>1.5% | | | 18<br>4.5% | 24 | 6 % |
| | Variable | | | 28<br>7% | | 28 | 7 % |
| | Absent | 32<br>8.1% | | | 48<br>12.1% | 80 | 20.2% |
| | Verticillate | | | | 38<br>9.6% | 38 | 9.6% |
| Totals | | 236 | 27 | 28 | 104 | 395 | |
| Percentages | | 59.7% | 6.8% | 7% | 26.3% | | |

One notices that the introduction of 100 polyporoid fungi does not modify the percentages appreciably. Notice the absence of species with verticillate clamps among the true polyporoid fungi (the decrease from 12.8 to 9.6%) and especially the abundance among the poroid Hymenochaetaceae of holocenocytic species without clamps (the percentage changes from 5.3 to 12.1).

Table 4, more complex, introduces an additional factor called schematically "sexuality," that is, the factor which accounts for the homo- or heterothallic behavior, and which is more precisely bi- or tetrapolarity. Many of the species of which only the polarity, or only the number of spore nuclei is unknown, cannot be represented in the table; the data appearing on this table concerns, however, 248 species.

One can state the dominant correlations as follows: (*a*) normal behavior, constant clamps, tetrapolarity, spores uninucleate; (*b*) heterocytic behavior, constant clamps, bi- or tetrapolarity, spores binucleate; (*c*) astatocenocytic behavior, clamps variable, bipolarity, spores uninucleate; (*d*) holocenocytic behavior, homothallic, verticillate clamps, spores binucleate.

TABLE 4. NUCLEAR BEHAVIOR, CLAMPING, AND SEXUALITY IN THE CORTICIUMS *s. l.*, AND THE PORE-FUNGI.

| Behavior | Clamps on the Mycelium | Sexuality | Nuclei in Spores | Number of Cases | Examples |
|---|---|---|---|---|---|
| Normal | Constant | IV | 1 | 102 | *Hyphodontia, Steccherinum, Podoscypha, Coriolus,* etc. |
| | | II | 1 | 24 | *Hyphoderma, Coriolellus* |
| | | Hom. | 1 | 6 | " " |
| | Inconstant | IV | 1 | 3 | *Chondrostereum* |
| | Absent | Het. | 1 | 2 | *Dendrothele griseo-cana,* |
| | | Hom. | 1 | 14 | *Hymenoch. rubiginosa, Ceratob. cornigerum, Uthat. fusisporum* |
| Subnormal | Constant | IV | 2 | 4 | *Pen. incarnata, P. aurantiaca* |
| | | II | 2 | 1 | *Galzinia pedicellata* |
| | Absent | Het. | 2 | 6 | *Scytinostroma* spp. |
| Heterocytic | Constant | IV | 2 | 9 | *Vuilleminia, Funalia* |
| | | II | 1 | 1 | *Hapalopilus aurantiacus,* |
| | | II | 2 | 10 | *Punctularia, Gloeophyllum* |
| | | Hom. | 2 | 1 | *Gloeocystidiellum leucoxanthum* |
| | Absent | Het. | | 2 | *Vararia ochroleuca* |
| Astatocenocytic | Variable | IV | 1 | 1 | *Tyromyces albellus,* |
| | | IV | 2 | 1 | *Bjerkandera imberbis,* |
| | | II | 1 | 12 | *Phlebia, Mycoacia, Hapalopilus nidulans, H. croceus* |
| | | II | 2 | 3 | *Cerrena unicolor, Loph. spadicea,* |
| | | Hom. | 1 | 2 | *Mycoacia stenodon, Phl. pendula* |
| Holocenocytic | Inconstant | Hom. | 2 | 6 | *Xylobolus* spp., *Lopharia crassa* |
| | Absent | (Hom.) | 1 | 3 | *Hym. tabacina, Dimorph. pruinata,* |
| | | (Hom.) | 2 | 6 | *Heterobasidion annosus, Tulasn. violacea, Peniophora erikssonii* |
| | Verticillate | Hom. | 1 | 7 | *Phanerochaete* spp., *Donkia pulcherrima, Meruliopsis taxicola* |
| | | Hom. | 2 | 22 | *Stereum, Coniophora, Phanerochaete* spp. |

Parentheses indicate a controversial result.

In the Agaricales, the data only bear upon certain favored genera, some important aggregates not germinating or growing badly in culture (*Cortinarius s.l., Inocybe,* Lactario-russulas); they remain frequently too incomplete for the species easily cultured. One can, however, say that the majority of the behaviors observed among the Aphyllophorales can be found: normal behavior is frequent in the "leucosporae" where it is also tied more frequently to the presence of clamps and to tetrapolarity; heterocytic behavior widely distributed among the "chromosporae" (*Pholiota*, *Hypholoma*, etc.); astatocenocytic behavior rather exceptional (*Flammula gummosa p. ex.*). Within *Coprinus*, a greater variety is observed, from heterocytic species with clamps and tetrapolarity (*C. lagopus*) or inconstant clamps and bipolarity (*C. impatiens*, *C. comatus*), to heterocytic species without clamps (*C. congregatus*), and finally holocenocytic species with inconstant clamps (*C. ephemerus*, *C. subpurpureus*) or without clamps and homothallic (*C. pellucidus*, *C. heptemerus*). This genus would contain also one species with verticillate clamps.

## Nuclear and Sexual Behavior and Systematic Groups

For the systematist, it is refreshing to encounter families or genera of which all the species have identical nuclear behavior: the genera *Stereum*, *Phanerochaete*, *Coniophora*—holocenocytic; *Phlebia*—astatocenocytic; *Vuilleminia*, *Gloeophyllum*, and numerous genera of the colored-spored Agaricales—heterocytic; Auriscalpiaceae, Hericiaceae, and the genera *Steccherinum*, *Cymatoderma*, *Hyphoderma*, etc.—normal. This seems to confirm the validity of these alliances. But, on the contrary, for the phylogeneticists, there are the families or the genera of which all, in being indisputably natural, show within themselves a certain diversity of behavior which has more interest.

No one disputes the autonomy of the Hymenochaetaceae, nor the relationship of *Ceratobasidium* with the genera created from *Pellicularia sensu* Rogers. We may establish in the first group (table 5) a transition of species without clamps but with dicaryons to the species with plurinucleate cells within several genera. Within the second group (table 6) this same phenomenon seems to be preceded by the progressive loss of clamps. One can summarize the evolution: clamped-binucleate; binucleate with variable clamps; binucleate without clamps; plurinucleate; multinucleate (divisions not or imperfectly conjugated).

The genus *Peniophora*, well characterized by its pink spores, the presence of oxygenated bicyclic carotenes (Arpin, Lebreton, & Fiasson, 1966; Fiasson, 1968), the hemichiastobasidial arrangement of the meiotic spindles, etc., can be cut up into sections on the basis of morphological and microscopic characters. While the very great majority of the species have constant clamps with normal nuclear behavior and tetrapolarity, the "*cinerea* group" contains two representatives with clamps partially lacking; the "*polygonia* group" one subnormal species and particularly the "*incarnata* group," which contains on one hand subnormal species like *P. aurantiaca*, and on the other holocenocytic species without clamps, morphologically very

TABLE 5. HYMENOCHAETACEAE.

| Nuclear Behavior | | Diplont with 2-3 Nuclei | Diplont Multinucleate |
|---|---|---|---|
| Clamps | Absent | *Hym. mougeotii*<br>*Hym. rubiginosa*<br>*Phellinus gilvus*<br>*Ph. torulosus*<br>*Ph. nigrolimitatus* | *Hym. tabacina*<br>*Hym. corrugata*<br>*Ph. pini*<br>*Ph. igniarius*<br>*Ph. conchatus*<br>*Inonotus rheades*<br>*I. cuticularis*, etc. |

TABLE 6. *Botryobasidium* AND RELATED GENERA.

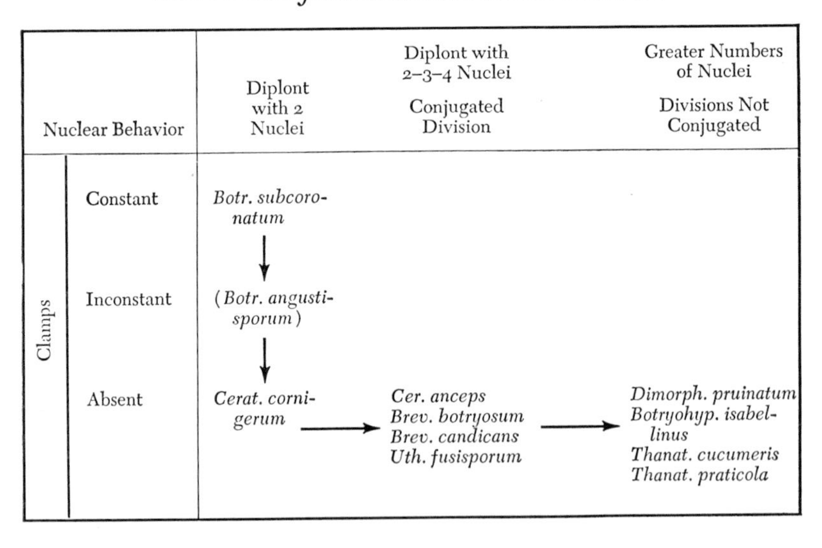

| Nuclear Behavior | | Diplont with 2 Nuclei | Diplont with 2–3–4 Nuclei<br>Conjugated Division | Greater Numbers of Nuclei<br>Divisions Not Conjugated |
|---|---|---|---|---|
| Clamps | Constant | *Botr. subcoronatum*<br>↓ | | |
| | Inconstant | (*Botr. angustisporum*)<br>↓ | | |
| | Absent | *Cerat. cornigerum* → | *Cer. anceps*<br>*Brev. botryosum*<br>*Brev. candicans*<br>*Uth. fusisporum* → | *Dimorph. pruinatum*<br>*Botryohyp. isabellinus*<br>*Thanat. cucumeris*<br>*Thanat. praticola* |

near the latter with which it has been long since confused—*P. erikssonii.* It is impossible not to see everything in *Peniophora,* beginning with a group of normal behavior, the species differentiated either by progressive loss of clamps with preservation of the dicaryons (*P. limitata* and *P. piceae*) or by the introduction of subnormal behavior (binucleate spores) which suddenly changes afterwards (total loss of clamps, plurinucleate condition during all the life of the mycelium) to the holocenocytic condition (table 7). We should speak

TABLE 7. *Peniophora* (*Coloratae*).

| | Nuclear Behavior / Spores with | Normal / 1 nucleus | Subnormal / 2 nuclei | Holocenocytic / 2 nuclei |
|---|---|---|---|---|
| CLAMPS | Constant | *polygonia*<br>*laeta*<br>*cinerea . . .* | *lilacea*<br>*incarnata*<br>*aurantiaca* | |
| | Inconstant | *limitata*<br>*piceae* | | |
| | Absent | | | *erikssonii* |

here of neoteny, of the crude generalization that the "larval" state becomes capable of sexual reproduction. *Irpex lacteus* and *tulipiferae,* long since confused, illustrate a case rather similar; the first is normal, with constant clamps and uninucleate spores; the second, morphologically similar, has a majority of binucleate spores and a mycelium without clamps and plurinucleate.

Table 8 represents the diverse behaviors within the genus *Laeticorticium* and within the subgenus *Aleurodiscus.* In the first one sees a transition: normal–subnormal–heterocytic, the clamps remaining constant. In the second, one witnesses a heterocytic–holocenocytic transition in which the clamps become scarce (*A. wakefieldiae* with inconstant clamps, *A. oakesii* with rare clamps located within the carpophore) and then disappear (*A. aurantius, A. amorphus*).

The genus *Gloeocystidiellum* (table 9) is the one which displays the greatest diversity of behavior. One finds together some normal clamped species showing tetrapolarity and holocenocytic species

TABLE 8. *Laeticorticium* AND *Aleurodiscus* SUBG. *Aleurodiscus*.

| Nuclear Behavior | | Normal | Subnormal | Heterocytic | Holocenocytic |
|---|---|---|---|---|---|
| CLAMPS | Constant | *L. ussuricum* (IV) → | *L. ionides* ↘ | *A. atlanticus* (IV)<br>*A. penicillatus* (h.)<br>*L. polygonioides* (IV)<br>*L. rosco-carneum* (IV)<br>*L. roseum* (II) | ↘ |
| | Inconstant | | | | *A. wakefieldiae* (H)<br>↓<br>*A. oakesii* (H)<br>↓ |
| | Absent | | | | *A. aurantius* (H)<br>*A. amorphus* (H) |

Abbreviations: h = heterothallic; IV = tetrapolar; II = bipolar; H = homothallic.

TABLE 9. *Gloeocystidiellum*.

| Nuclear Behavior | | Normal | Subnormal | Heterocytic | Holocenocytic |
|---|---|---|---|---|---|
| | Spores with | 1 Nucleus | 2 Nuclei | 2 Nuclei | 2 Nuclei |
| CLAMPS | Constant | *porosum* (IV)<br>*convolvens* (II?) → ↘ | | *luridum*<br>n° 6063 (II)<br>*leucoxanthum* (H) ↘ | |
| | Absent | | *lactescens* (h)<br>*citrinum* (h)<br>(*humile*) - - - → | | *friesii* (H)<br>↓ |
| | Verticillate | | | | *flammeum* (H)<br>*heimii* (H) |

with verticillate clamps. These latter could have been derived from species without clamps by overevolution. It appears to me in effect that the verticillate clamps have not been derived but are an extravagant reinvention because they have relation to neither the dicaryon nor nuclear fusion. They do not exist in the hymenium and have been detected only at the margin or in culture; that is to say, they are found at the farthest place from karyogamy and meiosis.

Another very peculiar behavior which has not been encountered in the above-mentioned genera is astatocenocity. Contrary to Yen's rule, this behavior is typically associated with uninucleate spores and characterizes some well-defined genera: *Phlebia, Mycoacia, Hapalopilus*, etc., or isolated species, and to our knowledge rarely coexists with another behavior than heterocity. The most evident case is that of *Hapalopilus* (table 10) where one finds a hetero-

TABLE 10. *Hapalopilus.*

| | Nuclear Behavior | Heterocytic | Astatocenocytic |
|---|---|---|---|
| | Spores with | 1 nucleus | 1 nucleus |
| CLAMPS | Constant | *aurantiacus* (II) → | |
| | Variable | | → *nidulans* (II)<br>*croceus* (II) |

cytic species (*H. aurantiacus*) with uninucleate spores, an exceptional character for heterocyty. It also appears true for certain tropical *Mycoacia* and *Phlebia sensu* Donk, but the knowledge of their life cycles is incomplete. Therefore, we easily accept the idea that the heterocytic species with uninucleate spores are some "preastatocenocytics" according to the scheme: heterocytic with uninucleate spore–astatocenocytic with uninucleate spore.

A more detailed knowledge of the species of group 54 of Nobles presumes astatocenocyty, and some related species can contribute to the argument for or against this hypothesis.

Table 11 diagrams all the relationships indicated above.

In reference to table 4, one can verify that tetrapolarity, bipolarity, and homothallism are not evenly distributed. There is a decrease in the percentage of the tetrapolar species from the top to the bottom of the table, from normal behavior to holocenocytic behavior. Bipolarity progresses inversely: normal and subnormal behavior = 16.2% of bipolar species; heterocytic behavior = 52.3% of bipolar species; astatocenocytic behavior = 78.9% of bipolar species.

TABLE 11. SCHEMATIC REPRESENTATION OF THE DIVERSE RELATIONSHIPS SHOWN IN TABLES 5–10.

| | Nuclear Behavior | Normal | Sub-normal | Heterocytic | | Astatocenocytic | | Holocenocytic | |
|---|---|---|---|---|---|---|---|---|---|
| | Spores with | 1 nucleus | 2 nuclei | 1 nucleus | 2 nuclei | 1 nucleus | 2 nuclei | 1 nucleus | 2 nuclei |
| CLAMPS | Constant | | | | | | | | |
| | Inconstant | | | | | | | | |
| | Variable | | | | | | | | |
| | Absent | | | | | | | | |
| | Verticillate | | | | | | | | |

The dotted line represents the close relationship of the *Peniophora aurantiaca–erikssonnii* and the *Irpex lacteus–tulipiferae* complexes.

Holocenocytic behavior is always associated with homothallism, but the term "homothallism" very certainly includes different phenomena. Recall the reservation made above about the "homothallic" species without clamps. Notice, moreover, that all of the holocenocytic species with rare or verticillate clamps are homothallic. But homothallism, notably primary homothallism, is found in connection with the other types of nuclear behavior. It only seems to be an accident, only a failing of the incompatibility phenomenon. It is found especially among certain "races" of heterothallic, bipolar species: for example, certain *Hyphoderma tenue*, *H. setigerum*, or among certain species of *Phlebia*, *Mycoacia*, *Coriolellus*, of which the other representatives are bipolar. Bipolarity seems to be, apparently, the access to homothallism.

## Evolution, Phylogeny

The existence of different behaviors within the natural assemblage permits a connection of the main cases observed (table 11), but nothing indicates the direction of evolution. Are the more primitive hymenomycetes the binucleate species or the multinucleate? Are the more evolved *Aleurodiscus* heterocytic, clamped, and tetrapolar or, on the contrary, homothallic without clamps?

### *Morphological and Structural Criteria*

#### *Among the Aphyllophorales*

It is exceedingly difficult within the framework of the Aphyllophorales to find the decisive arguments. The fact that corticioid forms are encountered within diverse families: *Ramaricium* in the Gomphaceae, *Laxitextum* in the Hericiaceae, *Hymenocheate* in the Hymenochaetaceae, *Clavulicium* in the Clavulinaceae, *Tomentella* in the Thelephoraceae, can be interpreted in favor of an evolution by morphological reduction, and the heterogeneity of the Corticiaceae would partially agree; but we should not generalize from these examples. One can easily see relapsed evolution, as in the cases of over-evolution shown us by paleontology (the unrolling of the cretaceous Ammonites, etc.).

#### *Among the Agaricales*

Others here will discuss the problem of the Agaricales for which, without paleontological proof, it appears much too difficult to create more than an opinion. It is common knowledge that many of the modern authors (Heim, Kühner, etc.), following Fayod, see evolution proceeding from simple toward veiled forms, thinking that the chanterelles or mycenas have preceded in time the lactarii and amanitas, or that the marasmii or collybias have preceded the coprini with exhibited differentiation and biological adaptations to some very special environments (i.e., animal excrement). The other authors following Brefeld (notably R. Singer) have defended an opposite argument, in which veils and structural complication would prove an origin among the Gastromycetes, evolution having proceeded toward morphological simplification.

## *"Multinucleate State" Criterion*

### *Among the Basidiomycetes*

As R. Kühner had very clearly explained, morphological complexity is correlated with an augmentation of the number of nuclei (binucleated spores of the amanitas and lepiotas) and with total or partial loss of clamps (lactario-russulas, amanitas, etc.). Among the species with colored spores, the differentiated spore is very generally binucleate and the nuclear behavior is very often heterocytic (*Stropharia*, *Hypholoma*) or sometimes holocenocytic when clamps are lacking (some coprini). In brief, "abnormal" behaviors develop together with morphological complexity. This interesting relationship doesn't permit, however, a decision in the choice between morphologically progressive evolution (Fayod) or regressive (Brefeld), because we don't know evolution in the time of the nuclear behavior. If one compares the Agaricales and Aphyllophorales, one discovers this time that the lower Aphyllophorales (Corticiaceae, Stereaceae, Coniophoraceae, etc.), that is, the less morphologically complex Homobasidiomycetes, are richer in "abnormal" behaviors (and practically the only ones to possess verticillate clamps). This proves without doubt that evolution here has not followed a one-way road and would favor a regressive evolution of at least a part of the corticiums.

### *Elsewhere within the Plant Kingdom*

If one looks outside the Basidiomycetes, the frequency of the multinucleate state among the lower fungi can be seen. Some will wish to argue this as proof that the cenocytic condition is primitive. This is not what is shown in the evolution of prothalli among Pteridophytes, or the embryo sac of the angiosperms.

### *Ontogenetic Proof*

One of the laws stated by the geologists (Pavlov, etc.) has been summarized for the plants by Gaussen in these terms: "A character susceptible to evolution appears more evolved among the juvenile than among the adults, the juveniles indicate, therefore, the direction of the future evolution. . . ." Table 1 can be a magnificent example of this law; it shows very clearly how the plurinucleate state when it is poorly developed is located in the germination of the spores, that is among the juveniles, and thereafter, can gain the haplontic and then the diplontic state.

### *Personal Position*

The homology between the ascogenous crozier and the clamp and the impossibility of seeing homothallism or possession of verticillate clamps as primitive characters compels me to believe that the primitive Basidiomycetes were binucleate, clamped, and heterothallic (and in spite of this unusual characteristic, tetrapolar rather than bipolar like everything else). This group of characters is found among more than 25% of the Aphyllophorales and is seen in all groups of saprophytic Basidiomycetes, with a high frequency in the Auriculariales and Tremellales, in the white-spored Agarics, etc.

The loss of clamps and the frequently concommitant appearance of plurinucleate elements (in the diplont) would, in the Basidiomycetes, indicate evolution. One cannot consider as primitive the multinucleate cystidia of the *tenacella* group of *Collybia* or the plurinucleate fundamental hyphae of *Mycena*, whereas their connective and generative hyphae remain binucleate.

The multinucleate state among the Basidiomycetes would indicate evolution, or rather overevolution, that is, the recurrence of the very early ancestral state (lower fungi) by final evolution. Likewise, the simple physiognomy of the corticiums can be a primitive character for some fungi while for others it is the effect of a recurrence, of a pseudo-cyclic evolution.[2]

## Conclusions

These are probably not unique lines. Evolution and overevolution are associated in order to confuse the picture that we are actually handed by the world of Basidiomycetes. In the morphological domain, evolution must go from the simple to the complex, but in certain cases must return from the complex to the simple (morphological overevolution). Within the life cycle, I think that the primitive Basidiomycetes generalized at the same time the crozier and the

[2] We have previously indicated the same hypothesis for the hemichiastic disposition of the meiotic spindles and then for the stichic disposition. Verticillate clamps perhaps also are a regressive overevolution made by the mycelia (and within the phyla) momentarily deprived of clamps; their lack of function and their location far from the basidia are explained.

dikaryon were invented by the Ascomycetes. Their revolution has been to abandon the ascus in order to produce the basidium with its four external, uninucleate spores with different polarity (tetrapolarity). This behavior is "normal," therefore, at the outset.

The evolutive force goes forward, according to the example, preferentially either in structural complexity (stalk, gill, veil, etc.) or chemistry (colored spores, etc.), in the Agaricales, or toward disorderly nuclearity, the loss of clamps, homothallism in the case of corticiums *s.l.* In the latter case two processes are possible, both illustrated at their inception by the genus *Peniophora.* (*a*) Progressive loss of clamps, then subsequent establishment of the plurinucleate state with homothallism (at least apparent); (*b*) utilization of the four residual nuclei abandoned in the basidium at its invention (8 nuclei are formed, in effect, from the three series of divisions of the fused nucleus in the Basidiomycetes, just as in the Ascomycetes) from which binucleate spores are established; whence the progressive establishment of cenocyty in the haplont (heterocyty), followed by loss of the clamps in the diplont (holocenocyty).

For those who absolutely must see cenocyty as a primitive character—which I personally doubt a great deal—this evolution to an alleged ancestoral state again becomes an overevolution.

## References

Arpin, N., Ph. Lebreton, and J. L. Fiasson. 1966. Recherches chimiotaxinomiques sur les Champignons. II. Les caroténoïdes de *Peniophora aurantiaca* (Bres.) Hoehn. et Litsch. (Basidiomycetes). Bull. Soc. Mycol. France 82: 450–459.

Boidin, J. 1958. Essai biotaxonomique sur les hydnés résupinés et les corticiés. Rev. Mycol. Mem. 6: 1–387.

Fiasson, J. 1968. Les Caroténoïdes des Basidiomycètes. Survol Chimiotaxinomique. Thesis, Université de Lyon. 84 p.

Gabriel, M. 1967. Recherches sur la Physiologie du mycélium des Basidiomycètes en aérobiose et anaérobiose. Second thesis, Université de Lyon. 110 p.

Kniep, H. 1913. Beitrage zur Kenntnis der Hymenomyceten. I. Die Entwicklungsgeschichte von *Hypochnus terrestris,* nov. spec. Z. Bot. 5: 593–609.

## Resumé

Le comportement nucléaire des mycéliums n'est suffisamment connu que chez les Porés et les Corticiés (sensu lato); de très grands ensembles (la plupart des Agaricales . . .) sont très mal connus à ce propos. Si l'on se rapporte aux Aphyllophorales, on constate que, à coté d'un comportment dit "normal" (spore, germination et haplonte à éléments uninucléés, mycélium secondaire binucléé) intéressant 56% d'entre elles, on peut distinguer les comportements subnormaux (3:5%), hétérocytiques (7%), astatocénocytiques (7%), et holocénocytiques (26.5%) pour lesquels un état plurinucléé se prolonge de plus en plus loin au cours du cycle.

Si l'on recherche les liens pouvant exister entre ces comportements et la présence des boucles (constantes 56.9%, inconstantes 6%, variables 7%, absentes 20.2%, verticillées 9.6%), on constate que le plus souvent, le comportement nucléaire normal va de pair avec la constance des boucles et la tétrapolarité. A l'opposé, les espèces holocénocytiques sont le plus souvent homothalles, sans boucles ou à boucles rares, parfois opposées ou verticillées autour d'une même cloison. Ces comportements extrêmes peuvent caractêriser certains genres ou certaines familles. Cependant à l'intérieur d'ensembles de valeur indiscutable, une variation peut s'observer qui pourrait aller, pour des espèces bouclées par exemple, dans l'ordre suivant: Normal–Subnormal–Hétérocytique–Astatocénocytique–Holocénocytique.

Le problême épineux est de savoir dans quel sens se fait l'évolution. Part-on d'espèces au comportement nucléaire normal, tétrapolaires et à boucles constantes, ces 3 caractères étant alors des critères primitifs? Ou doit–on au contraire considérer,–les champignons inférieurs étant en majorité cénocytiques,–que l'holocénicytie est la caractère primitif et par conséquent l'homothallie, les spores binculéées et bien souvent les boucles verticillées?

Considérant entre autres l'homologie entre crochet ascogène et anse d'anastomose, nous serions tenté de croire que l'évolution des Basidiomycètes es partie d'un type bouclé et par suite "normal," que l'on retrouve fréquemment chez les Hétérobasidies (Auriculariales, Tremellales) et dans tous les ensembles d'Homobasidiomycètes. Le genre *Coprinus,* indiscutablement évolué est le seul parmi les Agaricales à posséder une espèce à boucles verticillées, à coté d'espèces holocénocytiques, hétérocytiques . . . .

Cette manière de voir placerait les Agaricales chromosporées aux spores généralement binuclées (et de ce fait au comportement le plus souvent subnormal ou hétérocytique) au dessus des Leucosporeés, et parmi ces derniéres, les Amanites et les Lépiotes au dessus des Marasmes ou des Collybies. Parmi les Porés, le mycélium des *Coriolus* trimitiques serait plus primitif que celui des *Gloeophyllum* et des *Hapalopilus*. Parmi les Corticiés s.l., l'évolution mycélienne maximum est réalisée par les *Stereum* s.str., les *Phanerochaete*, les *Coniophora*; elle dépasse ce qui a pu être à ce jour remarqué dans tous les autres groupes. Cela signifie–t'il que ce sont les plus évolués des Basidiomycètes. Nous nous contenterons de conclure qu'ils ont les mycéliums les plus évolués.

## Discussion

Raper: Do you make any distinction in the homothallic forms that you have between primary and secondary homothallism, i.e., between those forms in which the assimilative will give rise to the complete sexual cycle and those in which binucleate spores have actually a primitive incompatibility—bipolar or even tetrapolar involved?

Boidin: In the body of table 4 there were 66 cases of apparent homothallism; you might consider some of them primary homothallism (i.e., those exhibiting normal, subnormal, heterocytic, and astatocenocytic behavior).

In 22 cases the basidiospore received one reduced nucleus and the dicaryotization appears when the daughters of this single nucleus begin to combine by two; the clamps begin to form at this time, if the species is clamp-forming.

Where it is no longer possible to see the dicaryotic situation (holocenocytic behavior, especially if the fungus is either non-clamp-forming or has verticillate clamps), it is difficult to say if the homothallism is primary or nonprimary. Sometimes one can obtain fruiting bodies and yet one has not done the genetic studies to show whether real sexuality is taking place.

Dick: In *Schizophyllum* we commonly see certain hemi-compatible situations. It is very interesting for those of us who work with *Schizophyllum* to know something about the origin of this split control over incompatibility and the dicaryon in this situation. Do you see in the species of the "lower" higher Basidiomycetes any cases of this hemi-compatible growth? Do you know of any correlation in the number of alleles that are present in the population of the particular species that are tetrapolar or bipolar? Is there any certain correlation between the extension of the number of alleles in the species and the advanced or primitive stage of these species?

BOIDIN: The hemi-compatible situation has not been noticed, and it is difficult to know the number of alleles in all the species.

SINGER: A poor indication was given of the Agaricales. Will you tell us whether you have repeated the observations of Kühner and Lange, whether your data are based on your observations, or whether they were built from the data given by other authors?

BOIDIN: The Agaricales data was mostly from Yen, Kühner, and Lange, and other literature, but many of them investigated at Lyon are still unpublished.

GILBERTSON: In regard to the distribution of these various types of mating systems throughout these groups, the figures that we usually have were given by Dr. Raper. You may correct me if I have them down wrong. Approximately 65% of the Homobasidiomycetes which have been investigated are tetrapolar, 25% are bipolar, and 10% homothallic. Would that be fairly accurate?

RAPER: That was the figure that I gave, and it was rough.

GILBERTSON: I wonder if these figures might not be revised as some of the more advanced resupinates are investigated. Perhaps both the tetrapolar and bipolar species will go down while the homothallic will go up.

BOIDIN: From our study of about 400 species, I would say that the Corticiaceae are more advanced. In the case of the Aphyllophorales, where the nuclear behavior has been completely studied, there is tetrapolarity (50.5%), bipolarity (21.5%), and homothallism (28%).

GILBERTSON: This 65–25–10 ratio is derived primarily from the polypores and the agarics which are very easy to isolate.

RAPER: I would like to call attention to Dr. Boidin's chart in which the very large samples are tetrapolar. In the normals there are 102 tetrapolar compared with 24 bipolar. Of course, that bears out what we said.

NOBLES: Do you find any evidence of relationship between the poroid and resupinate forms, and things that have been put into the Thelephoraceae? Do you find any evidence of close relationship between a poroid form, a resupinate form, and a smooth form?

BOIDIN: Nuclear behavior correlates quite well with other data which served as a basis for defining the taxa. But above all, it is the other characteristics which tend to justify these relationships.

DUBUVOY: In the astatocenocytic species you have said that the culture medium plays a major role in the nuclear condition. Are there other cases in the other series in which conditions of culture medium play a role in changing nuclear behavior? We have found in *Psilocybe* that oxidation-reduction potential, beside carbon dioxide, also plays a major role in controlling the astatocenocytic condition. In the heterocytic what importance has the culture medium?

BOIDIN: The astatocenocytic fungi are a very peculiar group, especially because of anaerobic conditions. The astatocenocytic species are wood decaying; they penetrate deeply into the wood where the concentration of carbon dioxide and oxygen is critical.

PETERSEN: You have studied many species of resupinates. You have data on other people's work on the Agaricales, conjecturing that there is some trend from uninucleate spore and constant clamps all the way to the other degenerate end of the chart. In all of the Friesian groups could the same projection be made, conjecturally—for the clavarioid fungi, the cantharelloid fungi, the thelephoroid fungi, and others? If this trend is general, then we cannot project backward to any given form for the primitive Basidiomycetes. In other words, we cannot go back to a single morphological form as *the* primitive Basidiomycete, using nuclear behavior or the nuclear condition as the means to do so.

BOIDIN: There are underevolved species and overevolved species. All the phylla which passed through the corticioid to the lamellate form have the same "normal" nuclear behavior (e.g., Auriscalpiaceae). We cannot recognize here if morphologic evolution is related to the evolution of nuclear behavior.

JOHN R. RAPER and ABRAHAM S. FLEXER

*Department of Biology, Harvard University*
*Cambridge, Massachusetts*

# MATING SYSTEMS AND EVOLUTION OF THE BASIDIOMYCETES

Pronouncements about the evolution of any group of organisms for which there is no extensive fossil record are extraordinarily hazardous. This is, in general, no less true of the group of higher fungi known as the Basidiomycetes than for other organisms. This particular group of organisms, however, is rich in features that are unique among the fungi or, for that matter, among all organisms. These include the basidium in its various morphological forms, the dolipore septum, the dikaryon, the clamp connection, and the several mating systems that occur among the Basidiomycetes. These features, if they could be correctly read, should indicate the course of evolution with some degree of accuracy. The extant groups, however, represent only those assemblages of forms best adapted to present environments and in all probability do not reflect the details of the evolutionary history of the Basidiomycetes any more accurately than the extant vertebrates reflect their evolutionary history as we know it from the fossil record.

In the absence of rather specific information about numerous transient forms that were ideally suited to penetrate a wide range of different niches and to give rise to a variety of divergent lines that ultimately overwhelmed their progenitors, we can say little. The prospects for a single definitive evolutionary history of the fungi are indeed dim. It has been pointed out (Raper, 1968) that perhaps the best we can hope for is a consensus of plausibility concerning the evolutionary significance of the several components that together compose the constellations of characteristics by which we define the major groups of the fungi. One difficulty, of course, is that no two students of the Basidiomycetes will agree entirely on the relative significance of the various features involved. This, however, should discourage no one: that one's own ideas cannot be unequivocally proved to be correct is a small price to pay for the certainty that they will not be proved to be entirely wrong.

With these general considerations in mind, we would like to discuss here some ideas concerning the evolution of the Basidiomycetes that follow from a study of their sexual phenomena. There are three basic parameters of sexuality, (*a*) life cycles, (*b*) patterns of sexuality or mating systems, and (*c*) sexual mechanisms (Raper, 1966b). When the Basidiomycetes are examined in the light of these three parameters, only one is found to be subject to significant variation or divergence. Consider first the matter of life cycles. To equate the life cycle of a long-cycle rust, with its numerous spore forms, to that of a smut or to that of any of the higher Basidiomycetes may seem odd, but in principle this is the case: the life cycle of each consists of a haploid phase, a dikaryotic phase, and a single diploid nuclear generation prior to meiosis. In many of the smuts, the sporidia fuse while still attached to the basidium; this, however, is not an obligatory process and hardly constitutes a distinct exception to the basic type of life cycle characteristic of the Basidiomycetes.

Next, consider sexual mechanisms. Sexual fusions in Basidiomycetes are typically somatogamous. This may involve either the fusion of meiospores prior to any propagation in the haploid phase or the fusion of undifferentiated vegetative cells. The direct fusion of meiotic products commonly occurs among fungi only in two very disparate groups—the sporidia of many species of smuts and the ascospores of numerous species of the Hemiascomycetes. The fusion of vegetative cells in heterokaryosis, however, is very widespread, if not universal, throughout the Euascomycetes and Basidiomycetes, but it is only in the Basidiomycetes that such fusions are an integral step in the sexual progression. The only exceptions to somatogamous fusions in the Basidiomycetes occur in the rusts, in which very simple differentiated sexual elements are sometimes present.

It is only in the matter of patterns of sexuality or mating systems that significant evolutionary divergence seems to have occurred in the history of the Basidiomycetes. Aside from homothallism, which occurs in all groups, there are four basic mating systems that occur among the Basidiomycetes, and some significance can be read into the distribution of these different systems. A simple incompatibility system comprising two alternate alleles at a single locus characterizes the heterothallic species of rusts. In some forms, this simple incompatibility system is superposed on functional hermaphroditism. In such species, the flexuous hypha serves as a receptive or female organ that is fertilized by a spermatizing agent, usually a uninucleate pycniospore. This pattern of sexuality, essentially similar to that in the Euascomycetes as typified by *Neurospora*, permits reciprocal cross-fertilization between mycelia of the two different mating types. In many other rusts that lack these differentiated sexual elements, the sexual fusions are somatogamous.

Most of the smuts also have a simple incompatibility system consisting of two alternate alleles at a single locus—designated *A*—but they, like all of the higher Basidiomycetes, lack any sexual differentiation whatever. In a very few species, such as *Ustilago longissima* and *U. maydis*, a more complex system has evolved that superposes a second locus—the *B* locus—with multiple alleles upon the basic one-locus-two-allele system characteristic of most smuts (Rowell, 1955). The function of the multiple-allelic *B* locus, however, is somewhat obscure (Halisky, 1965; Day, 1966). Fusions occur between two sporidia or two vegetative cells only if they carry different *A* factors, whether the alleles at the *B* locus are the same or different. The subsequent development of the dikaryotic phase, however, depends upon the presence of two different *B* factors. Because the dikaryon is an obligate parasite on the host plant, however, it is a moot point whether the *B* factor is to be considered a second locus determining mating competence, i.e., an incompatibility factor, or simply as a factor determining pathogenicity. In any event, this system with its two loci—one locus with two alleles, the other with multiple alleles—appears to be quite distinct from the two-factor system known in the Homobasidiomycetes.

In the remaining groups of Heterobasidiomycetes, the predominant mating system is governed by a single factor with multiple alleles, a pattern that has long been termed "bipolar sexuality" but which might better be termed "bipolar incompatibility." The natural population of a bipolar species contains a large number of mating types, each determined by a specific allele of a single factor. The number of distinct alleles of the incompatibility factor in such species is of the order of 30 to 50, and any strain is compatible with any other strain that carries a different allele. Multiple-allelic incompatibility here constitutes a very real innovation, in that it tremendously increases the outbreeding potential as compared to a two-allele system.

One species has been reported to deviate significantly from this characteristic pattern: for *Tremella mesenterica*, Bandoni (1965) has reported a mating system almost identical to that of *Ustilago longissima* and *U. maydis*. Such a mating pattern is quite unexpected in this group. It should be noted, however, that throughout the higher Basidiomycetes, the bipolar nonparasitic Heterobasidiomycetes included, fusion per se is not dependent upon sexual compatibility, and the restriction on initial fusion may have a biologically trivial basis unrelated to the sexual progression. Numerous mutations are known in the higher Basidiomycetes that have such effects.

The one-factor-multiple-allelic system is also widely distributed throughout the Homobasidiomycetes, and, in these forms, the operation of this system is indistinguishable from that in those Heterobasidiomycetes in which it is known to occur. In addition to this mating system, however, a more highly evolved system involving multiple alleles of two factors is the predominant mating system in the Homobasidiomycetes. This system, "tetrapolar sexuality" or more properly "tetrapolar incompatibility," occurs in a majority of species of the higher Basidiomycetes and constitutes a most efficient outbreeding system. Tetrapolar incompatibility involves two multiple-allelic factors, *A* and *B*, which together regulate dikaryosis. Different *B* factors are required for nuclear migration, while different *A* factors are required for nuclear pairing and the formation of clamp connections. Thus, two homokaryons are fully compatible when they carry different *A* and different *B* factors (fig. 1; Raper & Raper, 1968). A typical tetrapolar species, *Schizophyllum commune*, has approximately 450 *A* factors and approximately 90 *B* factors (Raper, Baxter, & Ellingboe, 1960; Koltin, Raper, & Simchen, 1967). Each factor, however, has been resolved into two loci, $\alpha$ and $\beta$, each with multiple alleles. Each specific combination of $\alpha$ and $\beta$ alleles specifies a distinct factor-phenotype: e.g., 9 $A\alpha$'s and approximately 50 $A\beta$'s thus account for 450 *A* factors. This system can generate >40,000 distinct mating types, and these are interfertile to the extent of 98+% (Raper, 1966a).

Tetrapolar incompatibility is certainly the most complicated pattern of sexuality known among the fungi; actually, it is a very strong contender for the most complex of all mating systems. The complexity of its genetic basis and of its physiological manifestations suggests it to be a highly evolved system that can only have originated in many successive stages occurring over long periods of time and probably under a very wide range of selective pressures. In so complex a system, one might well expect considerable variation on the basic tetrapolar scheme among subgroups of the Homobasidiomycetes and possibly even that such variation would reflect evolutionary progression. This, however, has not proved to be the case;

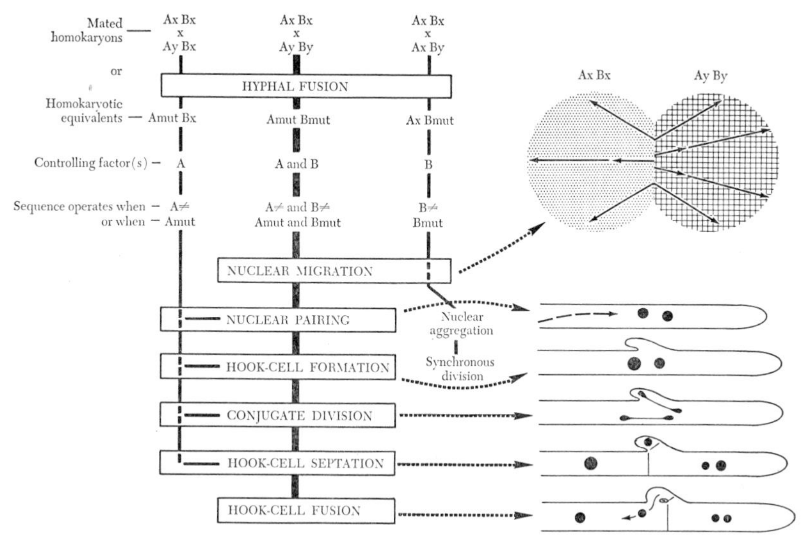


FIGURE 1. Control of sexual morphogenesis in *Schizophyllum commune*. (From Raper, 1968.)

wherever the tetrapolar system has been carefully examined, only very minor differences have been found, either in the genetic basis of the system or in its physiological manifestations (Raper, 1966a). This fact strongly suggests not only that the Homobasidiomycetes constitute a monophyletic line, but also that the ancestral form possessed a mating system practically identical to that of well over half of the present members of the subclass.

This interpretation calls attention to what appears to be a conspicuous contradiction: an examination of the patterns of sexuality among species composing the various families and even some of the larger genera throughout the Hymenomycetes reveals a more-or-less random mix of 10% homothallic, 25% bipolar, and 65% tetrapolar species (Esser, 1967). The Gastromycetes are not as well known, and of the dozen or so species that have been analyzed all are tetrapolar except for a single bipolar species (Esser, 1967). If the distribution of homothallism, bipolarity, and tetrapolarity were at the ordinal or higher taxon-levels, polyphyletic origins of the subclass might be indicated. Distribution by species or even subspecies (see below) certainly indicates a more recent differentiation of mating systems among very closely related forms. This prompts a consideration of possible mechanisms by which bipolar and homothallic systems have been derived—and derived repeatedly—from tetrapolar forms. Recent genetic work with representative and typical tetrapolar species, preeminently *Schizophyllum commune* and *Coprinus* spp., gives some hints of how such derivation might have come about. Mutations in any of the loci of the incompatibility factors in these forms essentially removes the incompatibility barrier as it is determined by the factor in which the mutation is located. For example, in *S. commune*, a mutation in the *B* factor renders the mutant strain compatible with all *B* factors, itself included. As a consequence, such a mutant strain will mate with any strain that carries a different *A* factor, and the result is essentially a conversion from the tetrapolar to the bipolar type of incompatibility (Day, 1963; Parag, 1962). By the same token, a strain that has a mutation in the *A* factor as well as a mutation in the *B* factor is a perfectly good mimic of a fertile dikaryon, down to such details as binucleate cells and the presence of clamp connections. Such doubly mutant strains fruit reasonably well, and each spore gives rise to a self-fertile, "dikaryotic" mycelium (Raper, Boyd, & Raper, 1965). Two mutations will thus remove the incompatibility barrier completely and convert a tetrapolar form into an essentially homothallic one. Such "homothallic" strains, however, are unstable in culture, and the competence to fruit is rapidly attenuated within a few generations. Such mutations may be expected (Raper, 1966a) to have slight selective advantage over their wild alleles, and this could well be sufficient to permit their establishment in natural populations, under which conditions there would be strong selective pressure toward the accumulation of stabilizing modifying factors.

Were this the case, one might well expect different types of bipolar systems as the result of different mutative routes from tetrapolarity, i.e., somewhat different end results might well be expected from the loss of discrimination in the *A* factor as compared to the loss in the *B* factor. Minor differences have been reported in, for example, the bipolar species of *Coprinus* (P. R. Day, personal communication), but the differences are minor and of such nature that they cannot be correlated with certainty to the different and recognizable functions of the *A* and *B* factors of tetrapolar forms.

If, as now appears to be the case, the *A* and *B* factors of all tetrapolar species are each constituted of two separate loci—to date no exceptions have been found among the half-dozen or so adequately investigated species—this same reasoning can be carried one step further to the homothallic species. Mutations in both loci of the *B* factor, $B\alpha$ and $B\beta$, of *S. commune* have been characterized (Koltin, in press) and although the basic discrimination due to the *B* factor is lost in both cases, there are minor differences in the effects of mutations in the two loci. Homothallic species might thus be expected to differ in very minor respects. To our knowledge, this has been put to the test in only a single case with rather surprising and, we feel, significant results. A group of very simple, resupinate, arachnoid forms comprising homothallic, bipolar, and tetrapolar races is included under the binomial *Sistotrema brinkmanni.* Lemke (1966) examined this complex in respect to mating systems and interfertility. Matings between wild-type strains of the three different races, homothallic, bipolar, and tetrapolar, gave no discernible interactions, and it is very likely that in nature the three races are genetically isolated. In similar matings of auxotrophic mutant (i.e., biochemically deficient) strains, however, some interactions were observed. No fertile crosses were achieved between tetrapolar strains and either bipolar or homothallic strains, but one successful fertile cross was achieved between a bipolar strain and a homothallic strain. Genetic isolation between bipolar and homothallic races of the *S. brinkmanni* complex is not absolute. One may further speculate that this observation suggests the interconvertibility of homothallic and bipolar forms—it does not specify, however, the direction of the change.

More significant to present considerations, however, were the interactions among three homothallic collections that were investigated. Matings between auxotrophic strains derived from two collections were fertile in all combinations, i.e., in all cases, fertile dikaryons were established. These dikaryons fruited readily and produced spores among which all characters of both parents segregated normally. In contrast to this, strains of either of these two homothallic collections interacted with strains of the third homothallic collection to yield, not fertile dikaryons, but instead infertile heterokaryons that had many of the characteristics of the common-*A* or hemicompatible heterokaryon of a tetrapolar form, such as *Schizophyllum*. It is not yet known whether the same sort of situation might obtain in other "species" in which homothallic and heterothallic forms have been described (Lange, 1952; Boidin, 1958).

Mating systems in and of themselves thus serve primarily to reinforce the present definitions of major groupings of Basidiomycetes which were originally established on the basis of far more conspicuous features, principally morphological characters. Mating systems, when taken in isolation, thus add little information. The story is somewhat more revealing when mating systems are considered in connection with certain morphological and genetic characteristics with which they are associated. Considered from this bias, it is possible to trace what we consider a reasonable sequence through which the gradual accession of features could have resulted in the constellations of characters that delimit the major groups of Basidiomycetes. We present this sequence as a highly speculative and provocative view of fungal evolution, which, however, is quite consistent with what is known of sexuality in contemporary fungi.

The proposed sequence (fig. 2) must start from a very, very ancient beginning, far enough in antiquity to encompass the divergence of the main lines of the Ascomycetes. Most of the forms that are considered to constitute the most primitive Ascomycetes are homothallic. There are among them, however, a few heterothallic species, and in these, self-sterility, cross-fertility is determined by two alleles at a single locus. This type of mating system, or others that appear to be derived from it, are characteristic of all of the more highly evolved fungi.

Two major lines were early differentiated, one giving rise to the Hemiascomycetes, a group in which nuclear fusion immediately follows the fusion of sexual elements, the nuclear phase thus immediately passing from haploid to diploid. In the other line, a few to many mitotic divisions were interposed between plasmogamy, i.e., the fusion of sexually compatible cells, and karogamy, i.e., the fusion of nuclei. The intervening mycelial stage contains pairs of compatible nuclei, the dikaryon, a characteristic of all higher fungi. This primitive dikaryotic line must then soon have diverged into two separate lines, in one of which the dikaryon is parasitic on the homokaryotic mycelium of the maternal parent. This line led to the Euascomycetes,

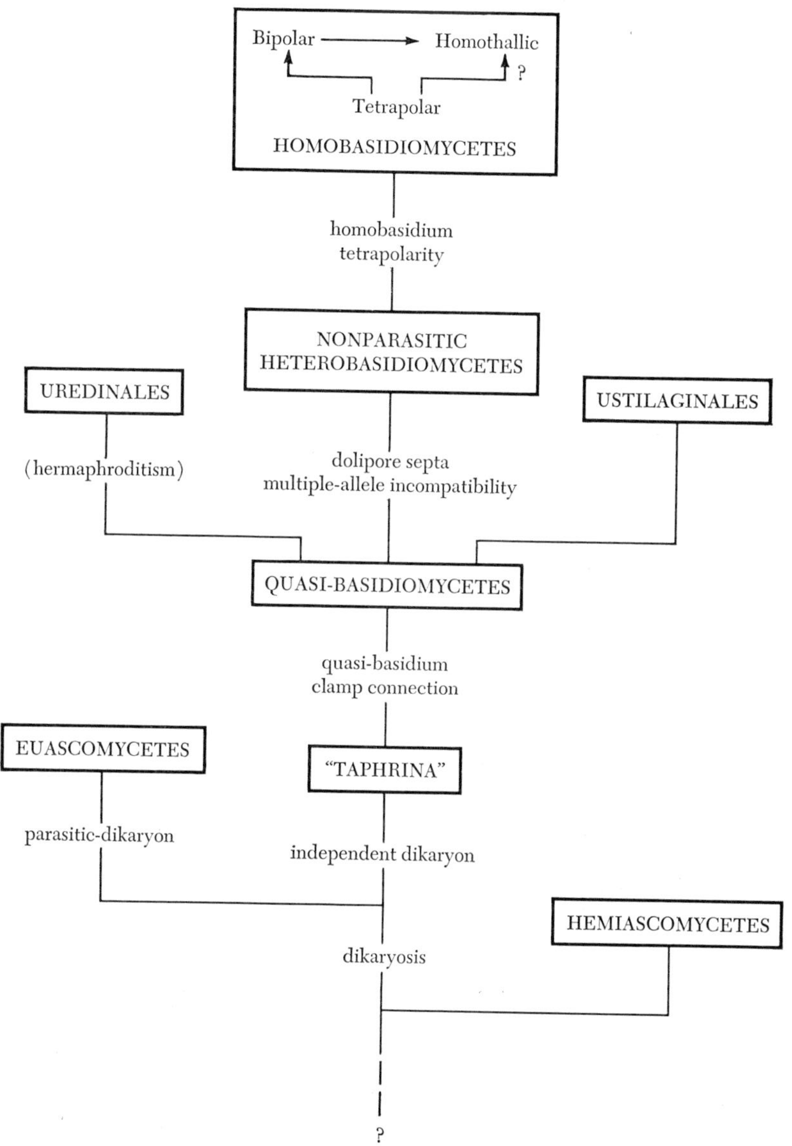


FIGURE 2. Proposed derivation of major groups of the higher fungi.

with its characteristic pattern of sexuality, one-locus-two-allele incompatibility superposed on hermaphroditism. The other line, characterized by a vegetatively independent dikaryon, would seem to be the proper forerunner of the Basidiomycetes.

The free-dikaryotic line would thus have, at this stage, the following characteristics: an extended vegetative dikaryotic mycelium constituted of cells separated by simple septa and having an incompatibility system constituted of two alleles at a single locus. Two additional features, a low level of morphological specialization and a parasitic habit, make the description of this hypothetical organism closely resemble that of the heterothallic species of the present-day genus, *Taphrina* (Wieben, 1927), a form that Savile (1968) has suggested as possessing the most probable combination of characteristics befitting the ancestor of the Basidiomycetes.

As Savile argues, evolutionary theory dictates that it is in relatively unspecialized, primitive, plastic forms such as this "*Taphrina*-like" organism that minor changes in phenotype can lead to the maximal adaptive radiation into new niches and the maximal degree of evolutionary divergence. Two features differentiate such a form from all Basidiomycetes: the ascus instead of the basidium *and* the absence of clamp connections. There can be no question that the ascus and the basidium are strictly homologous. The difference between endogenous and exogenous production of spores by the two types of meiosporangia, however, has been interpreted to constitute a formidable barrier—at least, in the mind of the mycologist—in the derivation of Basidiomycetes from Ascomycetes. This interpretation is in all probability correct, and there is no reason to think that this conversion has occurred repeatedly. On the other hand, there is no reason to preclude the evolution of a long series of very minor changes in the process of spore production, some of which, in effect, achieved the transition from the ascus to a number of intermediate types of quasi-basidia from which the range of present-day basidia could have been derived.

Clamp connections probably also first appeared in these primitive quasi-Basidiomycetes. We are quite unable to specify the adaptive significance of clamp connections and of the association of clamp connections and the exogenous production of meiospores, but these characteristics are both universal throughout the Basidiomycetes in groups that otherwise have little in common. But whatever the detailed basis of their adaptive value, the ancestral stock with these characters apparently gave rise to three major divergent lines.

1. One line could have given rise to the present-day rusts. This would have involved no change in the parasitic habit, in the basic pattern of sexuality, and only a slight specialization in respect to the sexual mechanism. The receptive hypha is strictly homologous only to the trichogyne of the female apparatus in the Ascomycetes. The other functions of the ascogonium are assumed in the rusts by elements in the aecidium, a structure that has no homologue in the Ascomycetes, while the spermatizing elements resemble the asexual spores of many forms more than they resemble the spermatia of the Ascomycetes.

2. A second divergent line from our hypothetical quasi-Basidiomycete could have given rise with very little change to the smuts. Certain characteristics of the smuts are comparable to features of *Taphrina*: (*a*) the fusion of sporidia and ascospores prior to their germination by budding to form vegetative colonies consisting of haploid unicells and (*b*) the extensive development of the dikaryon as a parasitic, vegetative stage.

3. The third line would have involved two very significant developments. In the first place, the incompatibility factor became multiply allelic. Even with a very few alleles, the tremendously increased outbreeding potential would be subject to strong positive selection. For example, with ten alleles at one locus, the outbreeding efficiency is 90% as compared with 50% for the original two-allele condition. A second significant development was the innovation of the dolipore septum. Again it is impossible to specify the adaptive value of the dolipore septum, but its occurrence in those forms in which dikaryotization is achieved by nuclear migration—and this is the case in all Basidiomycetes exclusive of the rusts and smuts—and in no other forms suggests a critical role in the process of nuclear migration. Such a role for the dolipore septum appears rather strange in view of its structural complexity, but it should be noted that there is no necessary relationship between structural complexity and resistance to enzymatic degradation. In any event, very rapid nuclear migration and the dolipore septum constitute associated features in the higher Basidiomycetes. This third line, with some divergence in respect to basidial character, could have given rise to the nonparasitic Heterobasidiomycetes. There is relatively little information about the de-

tails of sexuality among the members of this group, but insofar as it is known, bipolar incompatibility appears to be characteristic, with the possible exception of *Tremella mesenterica*.

Two additional developments would then be required for the origin of the Homobasidiomycetes: (*a*) the development of the homobasidium and (*b*) the duplication of the multiple-allelic *A* factor.

Of these two developments, the derivation of the homobasidium from heterobasidia is being considered in more detail elsewhere herein. The duplication of the incompatibility factor and the subsequent functional differentiation of the two series of factors, *A* and *B*, constitutes by any criteria a very improbable development. The system does exist, however, and we are stuck with the problem of how it *could* have developed, even if we have very little hope of determining how it *did* develop. Although no one has yet been able to resolve the genetic structure of the *A* factor of bipolar forms, we suspect it of having a two-locus structure corresponding to that of the *A* and *B* factors of tetrapolar forms. If this is so, the origin of tetrapolarity need have involved only duplication of this already dual character. This could well have occurred in numerous ways, translocation and recombination being the simplest, hence the most probable mechanism. This development would promote a slightly increased outbreeding efficiency, as the altered phenotype would be expected to be universally compatible with the original type. Of greater selective advantage, the conversion from bipolarity to tetrapolarity reduces potential inbreeding from a maximum of 50% to a maximum of 25%.

At a later time, with the duplicated factor well established in the population, a gradual assumption of different functions by the two factors would be expected (Darlington, 1939; Grant, 1963). These developments apparently gave rise to a plastic group, capable of wide dispersion, from which originated the various divergent lines that compose the present-day Homobasidiomycetes.

There has often been question about the biological utility of tetrapolar incompatibility: what important biological function does it serve that could account for its evolution and for its maintenance? In respect to outbreeding efficiency, its value has been termed qualitatively obvious but quantitatively puzzling (Raper & Esser, 1964). A possible solution to this puzzle has recently been advanced by Simchen (1967), who made a comparison of the relative effectiveness of two-locus versus one-locus incompatibility factors. Simchen showed that for a given total number of alleles, the partitioning of these alleles between two loci provides greater outbreeding efficiency than maintaining the same number of alleles in a single locus. Perhaps more important from the evolutionary point of view is his estimate of the minimal population required to maintain a given number of factor-phenotypes. For S. *commune*, the approximately 450 phenotypes determined by the *A* factor could be maintained in 174 homokaryotic strains with a two-locus factor as compared to the 4,822 homokaryons that would be needed if the factor had only a single locus. The same reasoning with some modification would apply to the two-factor or tetrapolar system as compared with the one-factor or bipolar system. The advantages inherent in the four-locus, tetrapolar system could very well be decisive in survival under "catastrophic" circumstances, where new populations and eventually new lines emerged from small populations. Such small populations would be subject to genetic drift, and entirely new evolutionary lines might emerge from tiny populations according to the founder principle (Mayr, 1942). In the long history of fungal evolution, extremely severe environmental conditions covering wide areas of the earth could well have reduced fungal populations of certain types to near extinction. The origin of the major groups of Basidiomycetes from our hypothetical quasi-Basidiomycete could well have occurred under such circumstances.

This, then, is our view of a feasible phylogeny for the Basidiomycetes. Admittedly, primary emphasis has been placed on sexuality to the near-exclusion of many other biological parameters of perhaps equal importance. Nevertheless, the proposed phylogeny is at least consistent with the facts known to us and at best may have heuristic value for further exploration of the fascinating area of fungal evolution.

## References

Bandoni, R. J. 1965. Secondary control of conjugation in *Tremella mesenterica.* Canad. J. Bot. 43: 627–630.

Boidin, J. 1958. Essai biotaxonomique sur le hydnés résupinés et les corticiés. Rev. de Mycol. (mem. hors–sér.) No. 6 (Paris). 388 pp.

Darlington, C. D. 1939. The evolution of genetic systems. Cambridge. 265 p.

DAY, P. R. 1963. Mutations affecting the *A* mating type locus in *Coprinus lagopus*. Genet. Res. Camb. 4: 55–65.

———. 1966. Recent developments in the genetics of the host-parasite system. Ann. Rev. Phytopathology 4: 245–268.

ESSER, K. 1967. Die Verbreitung der Incompatibilität bei Thallophyten, pp. 321–343. *In* W. Ruhland, ed., Handb. Pflanzenphysiol. 18.

GRANT, V. 1963. The origin of adaptations. Columbia Univ. Press, New York. 606 p.

HALISKY, P. M. 1965. Physiologic specialization and genetics of the smut fungi. III. Botan. Rev. 31: 114–150.

KOLTIN, Y. in press. The genetic structure of the incompatibility factors of *Schizophyllum commune*: comparative studies of primary mutations in the *B* factor. Mol. Gen. Genet.

KOLTIN, Y., J. R. RAPER, AND G. SIMCHEN. 1967. Genetics of the incompatibility factors of *Schizophyllum commune*: the *B* factor. Proc. Nat. Acad. Sci. (U. S.) 57: 55–62.

LANGE, M. 1952. Species concept in the genus *Coprinus*. Dansk. Botan. Ark. 14: 1–164.

LEMKE, P. A. 1966. The genetics of dikaryosis in a homothallic basidiomycete, *Sistotrema brinkmanni*. Thesis, Harvard Univ., Cambridge, Mass.

MAYR, E. 1942. Systematics and the origin of species. Columbia Univ. Press, New York. 334 p.

PARAG, Y. 1962. Mutations in the *B* incompatibility factor in *Schizophyllum commune*. Proc. Nat. Acad. Sci. (U. S.) 48: 743–750.

RAPER, J. R. 1966a. Genetics of sexuality in higher fungi. Ronald Press, New York. 283 p.

———. 1966b. Life cycles, basic patterns of sexuality, and sexual mechanisms, vol. 2, pp. 473–511. *In* G. C. Ainsworth and A. S. Sussman, eds., The Fungi. Academic Press, New York.

———. 1968. On the evolution of fungi, vol. 3, pp. 677–693. *In* G. C. Ainsworth and A. S. Sussman, eds., The Fungi. Academic Press, New York.

RAPER, J. R., M. G. BAXTER, AND A. H. ELLINGBOE. 1960. The genetic structure of the incompatibility factors of *Schizophyllum commune*: the *A* factor. Proc. Nat. Acad. Sci. (U. S.) 46: 833–842.

RAPER, J. R., D. H. BOYD, AND C. A. RAPER. 1965. Primary and secondary mutations at the incompatibility loci in *Schizophyllum.* Proc. Nat. Acad. Sci. (U. S.) 53: 1324–1332.

RAPER, J. R., AND K. ESSER. 1964. The Fungi, vol. 6, pp. 139–245. *In* J. Brachet and A. E. Mirsky, eds., The Cell. Academic Press, New York.

RAPER, J. R., AND C. A. RAPER. 1968. Genetic regulation of sexual morphogenesis in *Schizophyllum commune.* J. Elisha Mitchell Sci. Soc. 84: 267–273.

ROWELL, J. B. 1955. Functional role of incompatibility factors and an in vitro test for sexual compatible haploid lines of *Ustilago zeae.* Phytopathology 45: 370–375.

SAVILE, D. B. O. 1968. Possible inter-relationships between fungal groups, vol. 3, pp. 649–675. *In* G. C. Ainsworth and A. S. Sussman, eds., The Fungi. Academic Press, New York.

SIMCHEN, G. 1967. Genetic control of recombination and the incompatibility system in *Schizophyllum commune.* Genet. Res. Camb. 9: 195–210.

WIEBEN, M. 1927. Die Infektion, die Myzelüberwinterung und die Kopulation bei Exoasceen. Forsch. a.d. Geb. d. Pflanzenkr. u.d. Imm. i. Pflanzenr. 3: 139–176.

## DISCUSSION

BOIDIN: It is remarkable that the figures given by you are so close to mine, but I can speak only for the Homobasidiomycetes. I am quite struck by the correspondence of our main hypotheses of sexual evolution.

RAPER: This is quite comforting. I have not documented my sources fully here, and I admit that I have gotten them from wherever possible, and I owe Dr. Boidin quite a debt.

BOTH: I was particularly interested in your mention of the primitive ascomycete in the distant past. Has any work been done on the effect of extremes in radiation on these fungi?

RAPER: Most fungi are horribly x-ray resistant. For *Schizophyllum*, for instance, a 90% mortal dose is around 250,000 roentgens. I think particle radiation resistance is pretty much the same. In short, they are extremely resistant to ionization or ionizing radiation. But in reference to genetic drift and the founder principle, I was thinking of such things as ice ages and other longer range climatic and ecological changes. Under such conditions, very tiny populations could have been isolated and maintained for long periods of time.

SINGER: Do you derive the Hemiascomycetes and the main group of the Ascomycetes from what you called "primitive ascomycetes of taxonomic importance"?

RAPER: The Hemiascomycetes might well have diverged earlier from some common stem, but the Euascomycetes—with a number of shared characteristics with all "higher forms"—seem clearly to have been derived from these "primitive Ascomycetes." I can think of something like *Taphrina*, for instance, as possibly being very primitive in certain characters but not in others. I also think it is safer to assume that these primitive Ascomycetes were actually independently derived.

HAYNES: I am interested in hearing your comments on this parasitic dikaryon in the Ascomycetes.

RAPER: This idea is lifted rather largely from an article from D. B. O. Savile[1]. I see no reason to disagree with his evaluation of the situation, that the ancestral form was something of the general type of the heterothallic taphrinas—parasitic, dikaryotic, relatively unspecialized. This would be the sort of gene pool from which you might expect an adaptive radiation to occur that might have led to the Basidiomycetes.

HAYNES: Are you saying, then, that in these Ascomycetes, the dikaryotic mycelium is parasitic on the female ascogonium?

RAPER: No. It is not so in *Taphrina*, for instance, nor in some others. It is so in the Euascomycetes, but I don't think ascogenous hyphae in those fungi have ever been isolated in culture. For instance, Bistis removed the fertilized ascogonia of *Ascobolus* at the time that they began to push out the beginnings of ascigerous hyphae, but was unable to induce them to develop beyond this stage. When he put them back in their sheaths of haploid mycelium from the maternal parent, though, they developed normally, so there is a very significant dependence there.

BAKER: Do you see the crozier as arising distinctly twice?

[1] Savile, D. B. O. 1968. Possible interrelationships between fungal groups. *In* The Fungi, vol. 3, pp. 635–649. Academic Press, New York.

RAPER: I really do not have any very definite ideas about it. I suggest that the crozier is a reasonably simple development compared to many of the other things we are discussing, and I do not see any absolute necessity for strict homology. On the other hand, I do not hold this view with any deep conviction.

LUTTRELL: Could you comment on these mating-type systems as evolutionary mechanisms. Do they function as regulating mechanisms?

RAPER: I think we know too little about too few systems in sufficient detail to make any generalizations. To discuss the role of mating systems or any sort of sexual phenomena in the process of speciation and the like, we need to know very much more than we presently know about these forms.

SMITH: In how many species of the Basidiomycetes characterized by lack of clamp connections has the dolipore apparatus been studied and what is the longevity of this apparatus in the hyphae? I am speaking about only the dolipore septa in vegetative mycelium. This would include both the homokaryotic and dikaryotic hyphae.

RAPER: Royall Moore would be more competent than I to answer that.

MOORE: I would say that wherever the somatic hypha forms a cross-wall it seems to be a universal characteristic, but there is a distressing failure of it to show up in the Uredinales and in the Ustilaginales. It seems to be universal in the Homobasidiomycetes. I would say it was more characteristic of the Basidiomycetes than basidia, in that it appears in the imperfect mycelium where you have no fruiting body. This apparatus (Plate VI) would distinguish Ascomycetes from Basidiomycetes and is present wherever you find a septum.

JESSOP: Does the dolipore apparatus we are talking about include the parenthesomes or just the aperture?

RAPER: The whole structure. I do not know of any cases where part of the apparatus is present and not all.

MOORE: There is one interesting enigma here that is hard to resolve. In some species the parenthesomes appear to be continuous, and in other taxa—*Schizophyllum* is one—the parenthesomes are perforate. What significance can be attributed to this is for future resolution.

SMITH: When you look at the hyphae of the fruiting structure of certain boletes, you find that the cross-wall is a "doughnut" with a hole 3–5 microns in diameter. There is no evidence of a dolipore under light microscopy. How long in the life of the hypha does the dolipore function? Herbarium material of *Boletus calopus* lacks the dolipore apparatus. In fact, what is the function of the dolipore?

MOORE: I get the impression that the primary hyphae have plastic septa which are subject to dissolution in nuclear migration, but that the subsequent septa in secondary hypha are permanent structures because they have the same appearance as many cell walls. The dolipore is also present in the tertiary hyphae in the fruiting body. At what point it ceases to function in its role I don't know. I would say that it is part of the architecture of the secondary and tertiary mycelium.

Surprisingly, it is apparently extremely difficult to resolve the dolipore with light microscopy. There is a degree of inflation and other factors. It would be very interesting for someone to do some electron microscopy on herbarium material.

RAPER: Actually I think the dolipore apparatus is transient under certain circumstances and just might not survive herbarium treatment.

SMITH: In *Boletus calopus* we have a case in which the cross-walls are obviously different from the longitudinal hyphal walls. They stain differently so that the cross-walls stand out extremely like a doughnut with a hole. It bothers me that we are putting so much emphasis on the dolipore apparatus in the Basidiomycetes.

FERGUS: How many exceptions to the dolipore are there? Moore says it is distressingly absent in some groups. How many species have we actually looked at?

MOORE: Going through the literature, I would say probably somewhere around 25 to 50 species of Basidiomycetes have been reported. In the primary mycelium the septum seems to be composed of an amorphous electron-dense material which Bracker has shown can dissociate. The membrane that is constraining this material or forming it can be ruptured or dissociated. Day has shown that in nuclear migration the septum also seems to dissociate. It seems that when the dikaryotic mycelium is formed, septa are found which are electron-transparent like the wall. One gets the impression, therefore, that this secondary septum is made up of a fill of chitinaceous material, which prohibits nuclear migration. The attempts that Snider made to get nuclear migration in a dikaryotic *Schizophyllum* got negative results. I don't know whether anything further has been done or not.

RAPER: I would like to point out that in the common-A heterokaryon in which facile nuclear migration occurs and in the mutant-B homokaryon, we get what looks like dissolution of the septa. It is precisely the same sort of thing that Giesy and Day saw in primary mycelia through which nuclei were migrating, so it is beginning to add up to a fairly consistent pattern. Whether it is universal is again a matter of question. I am not sure that all major groups of the Homobasidiomycetes have actually been examined.

Koltin and I have just reported on a single gene in *Schizophyllum* which we call "dik." It is a dominant gene and it is necessary for the maintenance of the dikaryon. The homoallelic recessive immediately diploidizes. So despite the tremendously important role the dikaryon plays in this whole group, in this particular form its maintenance is due to a single gene. I suggest that it is a single gene that has tremendous selective value in the peculiar biology of these forms. In this common-A or mutant-B material we frequently get interseptal septa and these are usually more-or-less simple (i.e., like the septa of Ascomycetes) as seen under the electron microscope. I do not pretend to know what their role is, though.

SMITH: I think it would be very interesting to study these secondary septa under the electron microscope. Corner has been using these as generic characters, for instance.

ROGERS: There are some reports of fusion of pycnospores to pycnospores, of infection by pycnospores, and reports of examination of mycelia which seem to indicate that any haploid cell in a rust is potentially receptive. I wonder what you would think of extending your generalization.

RAPER: I would make whatever accommodations are needed. Many of the rusts are certainly somatogamous, but some actually have a differentiated sexual system. The sorts of things you have mentioned are not at all surprising to me, and I will make whatever modifications are necessary to accept them.

MILDRED K. NOBLES

*Principal Mycologist*
*Plant Research Institute, Ottawa, Canada*

# CULTURAL CHARACTERS AS A GUIDE TO THE TAXONOMY OF THE POLYPORACEAE[1]

Although wood-inhabiting species of Hymenomycetes develop mainly within the host, where their complex physiological activities break down the components of the wood and cause typical brown or white rots, their classification has been based solely on the characters of their fruit bodies, structures which are highly specialized for the protection of basidia and the dispersal of basidiospores. Early systems of classification, in which genera and higher taxonomic units were segregated on the basis of gross morphological characters of fruit bodies and especially the hymenial configuration, are being replaced by systems in which the types of hyphae and their arrangement in fruit bodies are accorded major significance. Corner (1932), in emphasizing the importance in taxonomy of the structure of the tissues of fruit bodies and the nature of the hyphae composing them, recognized three main categories of hyphae, generative hyphae and their modifications to form skeletal and binding hyphae, and three hyphal systems, the monomitic, dimitic, and trimitic. To follow Corner's methods it is necessary to observe the form and arrangement of hyphae composing fruit bodies, a task which becomes very laborious in fruit bodies with complex structure such as those of many species with poroid hymenial surfaces. Fidalgo (1967) recently stressed the difficulty of preparing satisfactory mounts for examination when he resorted to the use of ultrasonic vibrations "to release unbroken hyphae from the plectenchyma and provide better information about the microstructures of the sporophore." Meanwhile, in cultures of all the species, including those whose fruit bodies are so difficult to dissect and whose hyphal systems are difficult to determine, generative hyphae and hyphae modified to form the characteristic microstructures of the species may be observed with ease.

[1] Contribution No. 688 of the Plant Research Institute.

Pinto-Lopes (1952) recognized this when he reported the occurrence of similar types of hyphae in fruit bodies and cultures of 90 species of Polyporaceae and proposed a classification of the family based on these microstructures. In 1958 I too suggested that certain cultural characters, including the types of generative hyphae and their modifications, could provide a guide to relationships in the Polyporaceae. Ten years later I am still convinced that cultural characters have taxonomic significance and I wish to present the basis for this conviction.

## Methods

I have worked with cultures of wood-inhabiting species of Hymenomycetes for many years, at first for the practical purpose of identifying large numbers of cultures isolated from decays of forest trees in Canada during extensive surveys and more recently in a search for the possible contribution to taxonomy of cultural characters. Cultures of over 600 species of Hymenomycetes have been accumulated, many accompanied by voucher specimens in the Canada Department of Agriculture, Ottawa, Mycological Herbarium, descriptions of over 300 species have been prepared, of which 150 have been published, and the species have been arranged in diagnostic keys to facilitate comparisons between isolates to be identified and named cultures. The standard procedure is to grow cultures for identification or description on agar containing 1.25% Difco Bacto malt extract and 2% Difco Bacto-Agar in Petri dishes from a side inoculum, at room temperature or at 25° C., in the dark, for 6 weeks. At weekly intervals, records are made of the radius of the colony, the rate of growth being expressed as the number of weeks required by the culture to cover the surface of the agar in the Petri dish, the form and character of the advancing zone, the color, topography, and texture of the mat, the presence or absence of fruiting areas, the color changes in the agar induced by the growth of the fungus, and the odor. At suitable intervals, mycelium from the advancing zone, from the older parts of the colony, both aerial and submerged, and from fruiting areas is mounted in aqueous KOH and stained with an aqueous solution of phloxine for the observation and description of hyphal characters.

The presence or absence of extracellular oxidase is determined by the Bavendamm method, in which the cultures are grown for 1 week on malt agar containing 0.5% gallic or tannic acid, when the appearance of brown diffusion zones indicates the formation of extracellular oxidase; the absence of such diffusion zones indicates lack of extracellular oxidase. A rapid method giving comparable results consists of dropping an alcoholic solution of guaiacum on the mat of a growing culture, where the rapid appearance of a blue color indicates the occurrence of extracellular oxidase; no change or a tardy appearance of a pale blue color indicates its absence. The color changes are probably caused by one or more laccases released by the cells into the medium.

## Diagnostic Characters

Whenever possible, my descriptions of the cultural characters of a species are based on several isolates since these may show some variation in their macroscopic characters, topography, texture, color, and ability to fruit, although they are remarkably uniform in their growth rate at constant temperatures and in their microscopic characters. Since the microstructures produced in culture are constant for each species, major divisions in my diagnostic keys are based on hyphal characters, including the type of septation in generative hyphae and the special structures produced by the differentiation of those hyphae. Four types of generative hyphae occur (Nobles, 1965). In the largest number of species all thin-walled hyphae are nodose-septate. In 17 of the species treated here but in a number of species in other families, thin-walled hyphae in the advancing zone and frequently in the submerged mycelium are simple-septate and these hyphae, usually broad, give off branches which are narrower and nodose-septate, the older part of the mat being composed of such nodose-septate hyphae. In four species treated here but in larger numbers of species of the Coniophoraceae and Stereaceae, the thin-walled hyphae are mainly simple-septate but have occasional single or multiple clamp connections, these occurring most frequently on the hyphae of the advancing zone. In the remaining species, the thin-walled hyphae are simple-septate. Since nearly 80% of the species in this study (exclusive of the 48 species with poroid, xanthochroic fruit bodies more properly assigned to the Hymenochaetaceae) have nodose-septate hyphae throughout or in all regions except the advancing zone and since a number of species in which the septation is usually simple occasionally form clamp connections, it may be postulated that in this group, which represents the family Polyporaceae as it is commonly accepted, the nodose-septate condition was originally universal and that certain species have lost the property of forming clamp connections.

In about 50 of the species treated here the thin-walled generative hyphae remained undifferentiated throughout the period of examination, which extends over 6 weeks in my studies but over only 2 weeks in the studies of some other authors whose data I have used in preparing this paper. There is always the possibility that the apparent absence of differentiated hyphae results from the failure to observe them. However, in many species there is definitely no differentiation of hyphae in culture and it is of interest to observe that the hyphal systems in fruit bodies of many of these species are described as monomitic.

The differentiation of generative hyphae to form fiber hyphae occurs in the cultures of many species. In my terminology, fiber hyphae are hyphae with thick, refractive walls, hyaline or brown, and lumina narrow or apparently lacking, arising as the elongated terminal cell of a hypha and thus aseptate, rarely branched in some species, frequently branched in others. In cultures of many species the fiber hyphae are numerous, long and flexuous, and interwoven to form a thick mat, often coherent and tough. These fiber hyphae are homologous with the skeletal and binding hyphae of fruit bodies and, in general, occur in those species whose fruit body hyphal systems are described as dimitic or trimitic. In a few species fiber hyphae are produced tardily and may not be observed during the 2-week period of observation used by some investigators or even during the 6-week period that I use, but they are readily observed in most species.

In 28 of the species included in the present study, hyphae are differentiated to form nodose-septate hyphae with irregularly thickened walls or with scattered thick-walled, refractive areas on the walls. In some species the walls appear to swell in aqueous KOH while the lumen is reduced to a narrow line with occasional expansions, often running from side to side of the hypha. The clamp connections may become enlarged, thick-walled, and in appearance suggest a ball-and-socket bone structure. In other species, the first indication of this type of differentiation is the formation of hyphae with numerous short branches or protuberances whose walls are thick or solid and refractive. These thick-walled nodose-septate hyphae occur unaccompanied by fiber hyphae in cultures of *Leptotrimitus semipileatus* (Peck) Pouz., *Poria placenta* (Fr.) Cke, and others and in association with numerous fiber hyphae in *Daedalea* spp., *Fomes cajanderi* Karst., etc. They were recorded in fruit bodies of *Coriolellus* (*Daedalea*) spp. by Sarkar (1959) and in those of *Osteina obducta* (Berk.) Donk by Donk (1966b), and were included in the diagnostic characters of the genus *Daedalea* by Aoshima (1967).

In a number of species, generative hyphae differentiate to form cuticular cells, which originate as terminal or intercalary swellings, at first hyaline with contents staining in phloxine, finally with walls slightly thickened and remaining hyaline or becoming brown. These are crowded to form pseudo-parenchymatous areas which appear skinlike or crustose on the surface of mycelial mats or form translucent sheets in the submerged mycelium. Cuticular cells accompanied by fiber hyphae are characteristic of cultures of *Ganoderma* spp. and may be homologous with the inflated hyphal ends that form the upper crust of fruit bodies of species in that genus. In poroid species of the Hymenochaetaceae with xanthochroic fruit bodies, cuticular cells interwoven with fiber hyphae occur in *Fomes igniarius* (L. ex Fr.) Kickx and its varieties and in some isolates of *Poria obliqua* (Pers. ex Fr.) Karst. Cuticular cells without accompanying fiber hyphae form coherent surface layers or sheets in the agar in cultures of *Heterobasidion annosum* (Fr.) Bref. and *Rigidoporus* spp.

In a number of other species dark-colored skinlike surface layers are composed of hyphae with numerous short, hooked, or recurved branches, as in *Polyporus vulpinus* Fr., *Poria tsugina* (Murr.) Sacc. & Trott., and *Polyporus radiatus* Sow. ex Fr. of the Hymenochaetaceae or of hyphae with irregular, thick-walled branches, nodules, or protuberances as in cultures of species of *Polyporus s. st.*

Hyphae with contorted tips, heavily incrusted, are conspicuous in mounts of aerial mycelium from cultures of *Climacocystis borealis* (Fr.) Kotl. & Pouz., *Leptotrimitus semipileatus* (Peck) Pouz., and *Incrustoporia* spp. They appear to be specialized structures, differing from the lightly incrusted hyphae or hyphal segments that occur in cultures of many species. Similar incrusted hyphal tips in the ends of dissepiments of fruit bodies of *Poria* (*Incrustoporia*) *stellae* Pil. and *P. tschulymica* Pil. were described by Eriksson (1958).

Other microstructures formed by the differentiation of hyphae in cultures include cystidia and gloeocystidia, setae and setal hyphae, swellings of various types, and conidia, chlamydospores, and oidia. Such structures are included in descriptions of cultures of individual species and are useful in diagnosis but they appear to have little or no taxonomic significance above the species level and so are not discussed here. Similarly, growth rates and the appearance of cultures may provide clues for the identification of individual cultures but they play no part in segregating groups of related species and so are omitted from the discussion.

Earlier I remarked that the vegetative phase in the life cycle of wood-rotting fungi has not been considered in taxonomy although the results of its enzymatic activity can be readily observed. Decays in wood caused by these fungi are divided into two main types, the brown rots, in which the cellulose content is reduced while the lignin component remains practically constant, and the white rots in which both cellulose and lignin are attacked. It seems reasonable to assume that the species that cause both brown and white rots possess an enzyme system capable of reducing cellulose and that those species associated with white rots have acquired a second enzyme system involved in the degradation of lignin. The question is whether the enzyme system involved in lignin breakdown was acquired independently by each of the species associated with white rots or whether, in the course of evolution, this enzyme system was developed on one or only a few occasions. If lignin-degrading enzyme systems were developed independently by the species causing white rots, then such species do not necessarily demonstrate kinship by their possession of these enzymes and may be related to different parent species that cause brown rots. I believe that the evidence supports the alternate hypothesis, that the enzyme systems involved in lignin breakdown developed on one or only a few occasions and that the species in which these enzymes occur are more closely related among themselves than with the species that cause no appreciable lignin degradation.

When the Bavendamm or alcoholic gum guaiacum test for extracellular oxidase is applied to cultures of wood-rotting Hymenomycetes, with few exceptions cultures of species that cause brown rots show no color change, whereas those that cause white rots show the color changes that indicate the presence of extracellular oxidase. There is no claim that this enzyme, probably a laccase, is responsible for lignin degradation but its occurrence is apparently linked with that of the enzyme system that is involved.

Information on interfertility phenomena is available for 85 of the 275 species included in the present study, three species being homothallic and 82 species being heterothallic; 27 showing the bipolar type of interfertility governed by alleles at one locus, 52 the tetrapolar type of interfertility governed by alleles at two loci, with 3 heterothallic species for which the type of interfertility was not determined. Of species whose cultures do not produce extracellular oxidase, 3 are homothallic, 2 are heterothallic but the type is not known, 16 are bipolar, and 4 are tetrapolar. Of the species whose cultures produce extracellular oxidase and whose generative hyphae are regularly nodose-septate, 48 are tetrapolar, 2 are bipolar, whereas 9 species with generative hyphae simple-septate in the advancing zone and nodose-septate elsewhere are bipolar. This suggests strongly that, in general, bipolarity is the usual type of interfertility in species that lack extracellular oxidase and are associated with brown rots and that tetrapolarity is the usual type of interfertility in species associated with white rots, whose cultures produce extracellular oxidase and are regularly nodose-septate.

The classification of the Polyporaceae based on cultural characters that I am now presenting is essentially the same as the one that I published in 1958 and presented in a demonstration paper at the Ninth International Botanical Congress in 1959. In that scheme, the groups of species brought together in the final divisions were listed alphabetically according to genus and species, the generic names being those used by Overholts, Lowe, and other conservative taxonomists. I am now attempting to correlate the information on relationships within the Polyporaceae provided by cultural studies with recent systems of classification. The suggested classification is shown diagrammatically in figure 1 in which only species that have been designated as types of genera are listed and in table 2 where the species that compose each of the final groups are listed, with the type species preceded by an asterisk. Following a system of notation used in my key for the identification of cultures of wood-inhabiting

PRIMITIVE GROUP SPECIAL GROUP I SPECIAL GROUP II

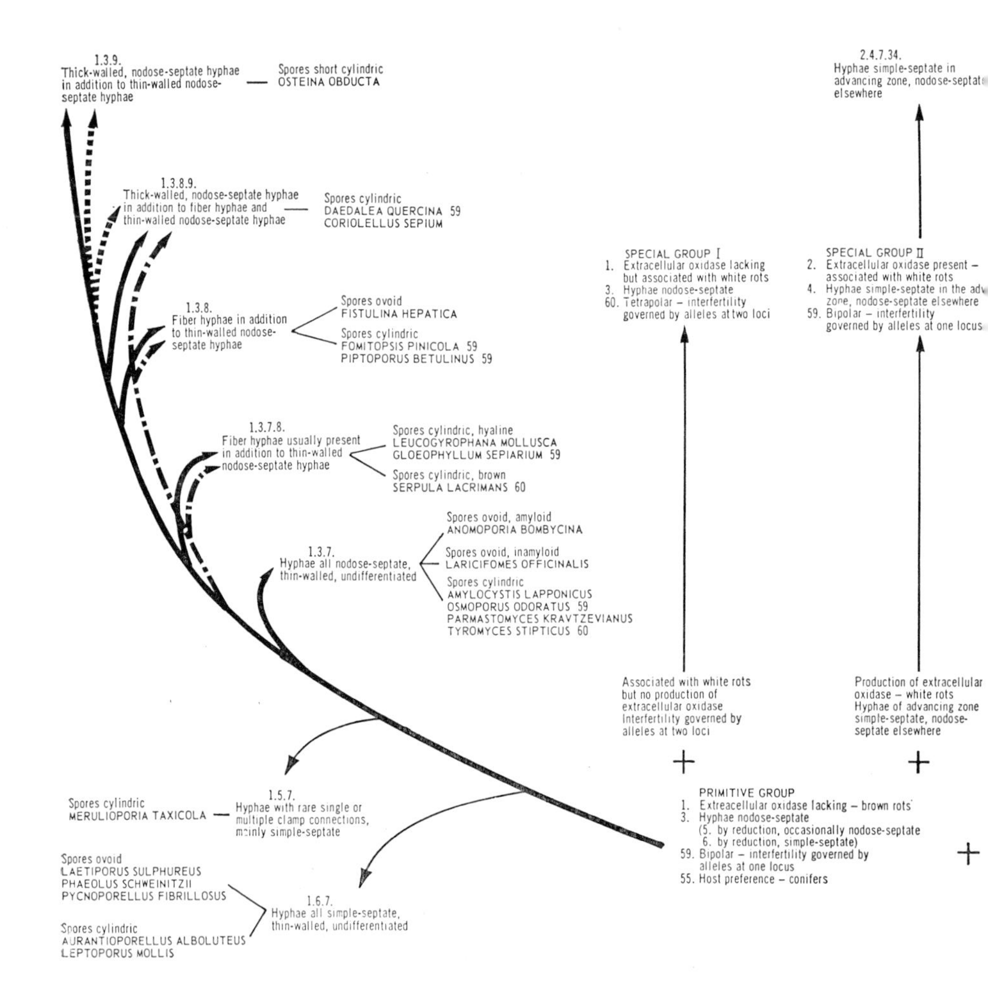

1.3.9.
Thick-walled, nodose-septate hyphae in addition to thin-walled nodose-septate hyphae
Spores short cylindric
OSTEINA OBDUCTA
1.3.8.9.
Thick-walled, nodose-septate hyphae in addition to fiber hyphae and thin-walled nodose-septate hyphae
Spores cylindric
DAEDALEA QUERCINA 59
CORIOLELLUS SEPIUM
1.3.8.
Fiber hyphae in addition to thin-walled nodose-septate hyphae
Spores ovoid
FISTULINA HEPATICA
Spores cylindric
FOMITOPSIS PINICOLA 59
PIPTOPORUS BETULINUS 59
1.3.7.8.
Fiber hyphae usually present in addition to thin-walled nodose-septate hyphae
Spores cylindric, hyaline
LEUCOGYROPHANA MOLLUSCA
GLOEOPHYLLUM SEPIARIUM 59
Spores cylindric, brown
SERPULA LACRIMANS 60
1.3.7.
Hyphae all nodose-septate, thin-walled, undifferentiated
Spores ovoid, amyloid
ANOMOPORIA BOMBYCINA
Spores ovoid, inamyloid
LARICIFOMES OFFICINALIS
Spores cylindric
AMYLOCYSTIS LAPPONICUS
OSMOPORUS ODORATUS 59
PARMASTOMYCES KRAVTZEVIANUS
TYROMYCES STIPTICUS 60
1.5.7.
Hyphae with rare single or multiple clamp connections, mainly simple-septate
Spores cylindric
MERULIOPORIA TAXICOLA
1.6.7.
Hyphae all simple-septate, thin-walled, undifferentiated
Spores ovoid
LAETIPORUS SULPHUREUS
PHAEOLUS SCHWEINITZII
PYCNOPORELLUS FIBRILLOSUS
Spores cylindric
AURANTIOPORELLUS ALBOLUTEUS
LEPTOPORUS MOLLIS
2.4.7.34.
Hyphae simple-septate in advancing zone, nodose-septat
elsewhere
SPECIAL GROUP I
1. Extracellular oxidase lacking but associated with white rots
3. Hyphae nodose-septate
60. Tetrapolar – interfertility governed by alleles at two loci
SPECIAL GROUP II
2. Extracellular oxidase present – associated with white rots
4. Hyphae simple-septate in the adv
zone, nodose-septate elsewhere
59. Bipolar – interfertility governed by alleles at one locus
Associated with white rots but no production of extracellular oxidase
Interfertility governed by alleles at two loci
Production of extracellular oxidase – white rots
Hyphae of advancing zone simple-septate, nodose-septate elsewhere
+
+
PRIMITIVE GROUP
1. Extreacellular oxidase lacking – brown rots
3. Hyphae nodose-septate
(5. by reduction, occasionally nodose-septate
6. by reduction, simple-septate)
59. Bipolar – interfertility governed by alleles at one locus
55. Host preference – conifers
+

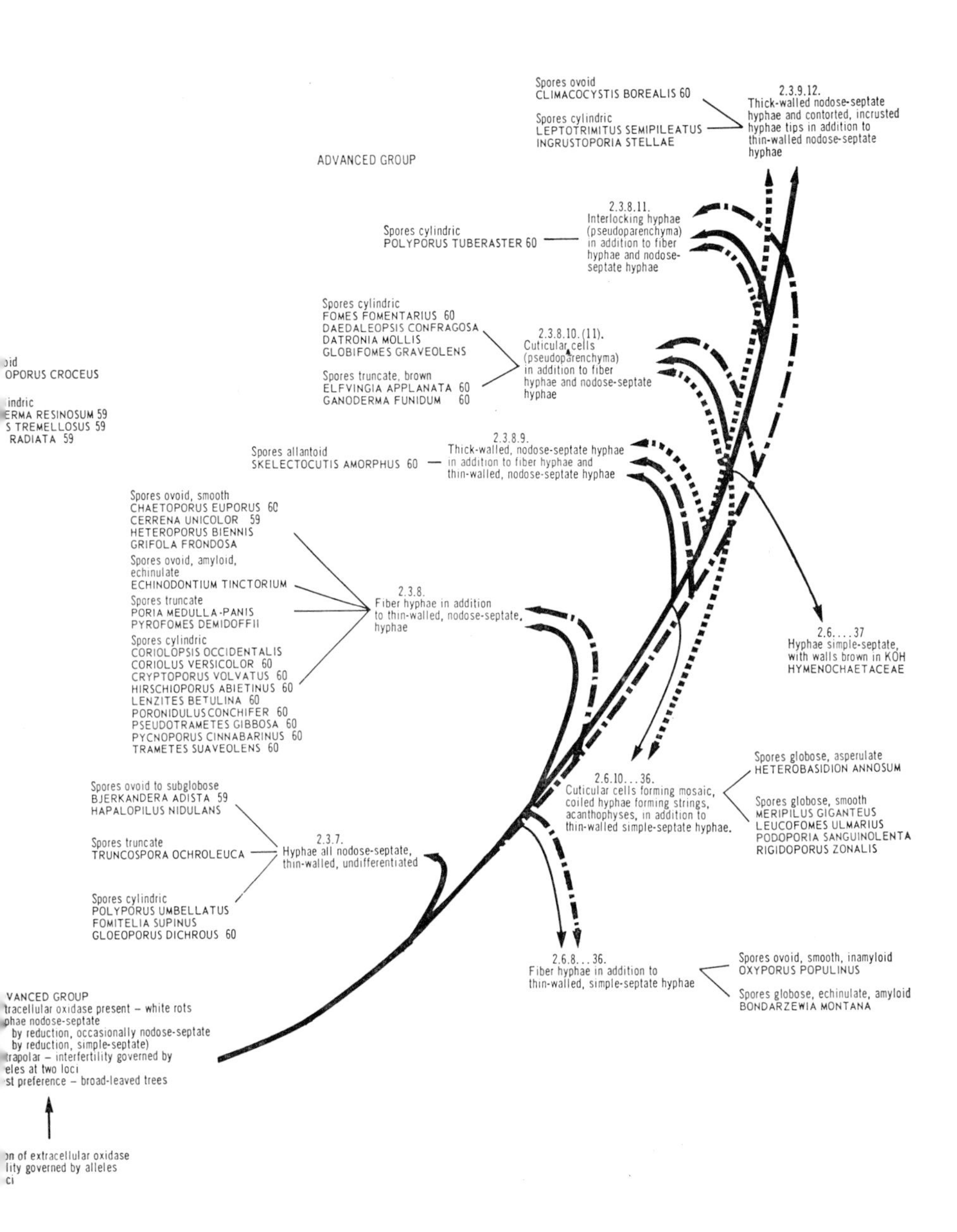


FIGURE 1. Cultural characters as a guide to the taxonomy of the Polyporaceae.

TABLE 1. DIAGNOSTIC CHARACTERS AND THEIR CODE SYMBOLS.

| | |
|---|---|
| Code Symbol 1. | Results negative in tests for extracellular oxidase |
| Code Symbol 2. | Results positive in tests for extracellular oxidase |
| Code Symbol 3. | Thin-walled hyphae consistently nodose-septate |
| Code Symbol 4. | Thin-walled hyphae simple-septate in the advancing zone, nodose-septate elsewhere |
| Code Symbol 5. | Thin-walled hyphae mainly simple-septate, with occasional single or multiple clamp connections |
| Code Symbol 6. | Thin-walled hyphae consistently simple-septate |
| Code Symbol 7. | Hyphae remaining thin-walled and undifferentiated |
| Code Symbol 8. | Hyphae differentiated to form fiber hyphae |
| Code Symbol 9. | Hyphae differentiated to form nodose-septate hyphae with irregularly thickened walls or with scattered thick refractive areas on walls |
| Code Symbol 10. | Hyphae differentiated to form cuticular cells, closely packed to form a pseudo-parenchyma |
| Code Symbol 11. | Hyphae differentiated through formation of numerous short branches, hooked or recurved, or thick-walled nodules, interlocked to form a plectenchyma. |
| Code Symbol 12. | Hyphae differentiated to form numerous contorted incrusted hyphal tips |
| Code Symbol 36. | Hyphae hyaline and mats white or pale in color |
| Code Symbol 37. | Hyphae yellow or brown when mounted in KOH solution and mats yellow or brown, at least in part |
| Code Symbol 57. | Species homothallic, completing the life cycle from a single basidiospore |
| Code Symbol 58. | Heterothallic, but type of interfertility not known |
| Code Symbol 59. | Heterothallic, showing the bipolar type of interfertility |
| Code Symbol 60. | Heterothallic, showing the tetrapolar type of interfertility |

TABLE 2. A CLASSIFICATION OF THE "POLYPORACEAE" BASED ON CULTURAL CHARACTERS.

Key Code 1.3.7.

Spores ovoid, amyloid

*Anomoporia albolutescens*
**Anomoporia bombycina*

Spores, ovoid, inamyloid

**Laricifomes officinalis*
**Leucogyrophana molluscus*
*"Polyporus" anthracophilus*
*"Polyporus" australiensis*
*"Polyporus" balsameus*
*"Polyporus" guttulatus*
*"Polyporus" portentosus*
*"Poria" aneirina*
*"Poria" crassa*

Spores cylindric, hyaline

**Amylocystis lapponicus*
**Osmoporus odoratus*–59

Spores cylindric, hyaline—continued

**Parmastomyces kravtzevianus*
*"Poria" carbonica*
*"Poria" gossypium*
*"Poria" radiculosa*
*"Poria" sequoiae*

Spores cylindric, brown

**Meruliporia incrassata*
*"Merulius" aureus*
*"Merulius" lignicola*
*Serpula pinastri*

Spores allantoid, small, hyaline

*"Merulius" niveus*
*"Polyporus" caesius*
*"Polyporus" fragilis*
*"Poria" pannocincta*

Key Code 1.3.8.36.

Spores ovoid

**Fistulina hepatica*
*"Polyporus" spraguei*
*"Poria" vaillantii*
*"Polyporus" meliae*–59
*"Polyporus" palustris*–59
*"Polyporus" rubidus*
*"Trametes" dickinsii*

Spores cylindric

**Fomitopsis pinicola*–59
*Fomitopsis rosea*–59

Spores allantoid, small

**Piptoporus betulinus*–59

Key Code 1.3.8.37.

Spores ellipsoid, brown

*"Poria" nigra*
*Serpula himantioides*–60
**Serpula lacrimans*–60

Spores cylindric, hyaline

*Gloeophyllum abietinum*–59
**Gloeophyllum sepiarium*–59
**Phaeocoriolellus trabeus*–59
*"Lenzites" striata*
*"Trametes" carbonaria*

TABLE 2. A CLASSIFICATION OF THE "POLYPORACEAE" (*Cont'd*).

Key Code 1.3.8.9.

Spores cylindric

*Coriolellus heteromorphus*–59
*Coriolellus malicola*–57
**Coriolellus sepium*–58
*Coriolellus serialis*–59
*Coriolellus sinuosus*–58
*Coriolellus variiformis*–59
**Daedalea quercina*–59

Spores cylindric–continued

*"Fomes" cajanderi*–59
*"Polyporus" durus*
*"Trametes" lilacino-gilva*

Spores allantoid, small

*"Poria" alpina*
*"Poria" xantha*

Key Code 1.3.9.

Spores oblong or short cylindric

**Osteina obducta*
*"Poria" oleracea*
*"Poria" placenta*–59
*"Poria" sericeomollis*

Spores allantoid, small

*Tyromyces albo-brunneus*
**Tyromyces stipticus*–60
*Tyromyces tephroleucus*–59
*Tyromyces undosus*

Key Code 1.5. (multiple) 7.

Spores short cylindric or allantoid

*"Merulius" ambiguus*
*"Merulius" corium*–57
*"Poria" griseoalba*
*"Poria" taxicola*–57

Key Code 1.2.6.7.

**Aurantioporellus alboluteus*
**Laetiporus sulphureus*
**Leptoporus mollis*
*"Merulius" confluens*
**Phaeolus schweinitzii*
**Pycnoporellus fibrillosus*
*"Poria" aurea*
*"Poria" cocos*
*"Poria" inflata*
*"Poria" metamorphosa*
*"Poria" reticulata*
*"Poria" rhodella*
*"Poria" spissa*

Key Code 2.3.7.

Spores globose, echinulate

*"Poria" candidissima*

Spores ovoid, smooth–continued

*"Polyporus" fumidiceps*

TABLE 2. A CLASSIFICATION OF THE "POLYPORACEAE" (*Cont'd*).

Spores truncate

**Truncospora ochroleuca*
*Truncospora ohiensis*

Spores ovoid, smooth

**Bjerkandera adusta*–59
*Bjerkandera fumosa*–60
**Hapalopilus nidulans*
*"Polyporus" consors*
*"Polyporus" delectans*–60
*"Polyporus" fissilis*
*"Polyporus" peckianus*
*"Poria" overholtsii*
*"Poria" versipora*

Spores cylindric

**Fomitella supina*–60
*"Polyporus" hirtus*
*"Polyporus" umbellatus*
*"Poria" mappa*

Spores allantoid

**Gloeoporus dichrous*–60
*"Poria" crustulina*

Key Code 2.3.8.

Spores subglobose to ellipsoid, chlamydospores very numerous

**Grifola frondosa*
**Heteroporus biennis*
*"Polyporus" obtusus*–60
*"Polyporus" robiniophilus*
*"Polyporus" spumeus* var. *mongolica*
*"Trametes" cingulata*
*"Trametes" ljubarskyi*–60

Spores ellipsoid, amyloid, echinulate

**Echinodontium tinctorium*

Spores truncate, hyaline

*"Fomes" ellisianus*–60
*"Fomes" fraxinophilus*–60
*"Polyporus" compactus*–60
**Poria medulla-panis*
*Poria subacida*
*Poria tenuis* var. *pulchella*

Spores cylindric

**Chaetoporus euporus*–60
*Chaetoporus rixosa*
*Chaetoporus variecolor*
**Coriolopsis occidentalis*
*Coriolus hirsutus*–60
*Coriolus pinsitus*
*Coriolus pubescens*–60
**Coriolus versicolor*–60
*Coriolus zonatus*–60
**Cryptoporus volvatus*–60
*"Daedalea" ambigua*
**Dichomitus squalens*–60
*Favolus alveolaris*–60
**Globifomes graveolens*
**Hirschioporus abietinus*–60
*Hirschioporus fusco-violaceus* –60
*Hirschioporus laricinus*–60
*Hirschioporus pergamenus*–60
**Lenzites betulina*–60
*Lenzites flaccida*
*Lenzites palisoti*
*Lenzites reichardtii*–60
*"Polystictus" leoninus*
*"Polystictus" xanthopus*–60

TABLE 2. A CLASSIFICATION OF THE "POLYPORACEAE" (*Cont'd*).

Spores oblong—ellipsoid

**Cerrena unicolor*—59
"*Poria*" *aestivale*
"*Poria*" *albidofusca*—60
"*Poria*" *fissiliformis*

Spores truncate, brown

**Phaeotrametes decipiens*—58
"*Polyporus*" *megaloporus*—60
**Pyrofomes demidoffii*
**Poronidulus conchifer*—60
**Pseudotrametes gibbosa*—60
**Pycnoporus cinnabarinus*—60
*Pycnoporus coccineus*—60
*Pycnoporus sanguineus*—60
*Trametes flavescens*—60
*Trametes hispida*—60
*Trametes meyenii*
*Trametes proteus*
**Trametes suaveolens*—60
*Trametes trogii*—60

Key Code 2.3.8.9.

Spores allantoid

**Skeletocutis amorphus*—60

Key Code 2.3.8.10.

Spores subglobose, hyaline, smooth

"*Fomes*" *fraxineus*—60

Spores truncate, brown, appearing echinulate

*Amauroderma rude*

Spores truncate, etc.—continued

**Elfvingia applanata*—60
*Ganoderma colossum*
*Ganoderma lobatum*
**Ganoderma lucidum*—60
*Ganoderma oregonense*
*Ganoderma tsugae*
*Ganoderma resinaceum*

Key Code 2.3.8.10.11.

Spores cylindric

**Daedaleopsis confragosa*
*Daedaleopsis confragosa* var. *Lenzites tricolor*
**Datronia mollis*—60
**Fomes fomentarius*—60
*Fomes sclerodermeus*
*Fomes scutellatus*
"*Lenzites*" *eximia*
"*Trametes*" *campestris*—60

TABLE 2. A CLASSIFICATION OF THE "POLYPORACEAE" (*Cont'd*).

Key Code 2.3.8.11.

Spores cylindric

*Polyporus arcularius*–60
*Polyporus brumalis*–60
*Polyporus elegans*
*Polyporus melanopus*
*Polyporus platensis*

Spores cylindric–continued

*Polyporus radicatus*
*Polyporus squamosus*–60
*Polyporus tuberaster*–60
*Polyporus varius*

Key Code 2.3.9.12.

Spores ovoid

* *Climacocystis borealis*–60

Spores allantoid

* *Incrustoporia stellae*
*Incrustoporia subincarnata*

Spores allantoid–continued

*Incrustoporia tschulymica*
* *Leptotrimitus semipileatus*
"*Polyporus*" *albellus*–60
"*Poria*" *odora*

Key Code 2.4.7.

Spores ovoid

* *Aurantioporus croceus*
*Merulius porinoides*–59
*Phlebia atkinsoniana*
"*Polyporus*" *galactinus*–59
"*Polyporus*" *pulcherrimus*
"*Poria*" *mutans*
"*Poria*" *rivulosa*–59
"*Poria*" *salmonicolor*
"*Radulum casearium*"–59

Spores allantoid

*Ischnoderma benzoinum*
* *Ischnoderma resinosum*–59
* *Merulius tremellosus*–59
*Phlebia albida*
* *Phlebia radiata*–59
"*Poria*" *cinerascens*–59
"*Poria*" *subvermispora*–59
"*Poria*" *zonata*

Key Code 2.6.8.

Spores ovoid, smooth, inamyloid

* *Oxyporus populinus*

Spores cylindric

* *Irpex lacteus*

Spores globose, echinulate, amyloid

*Bondarzewia berkeleyi*
* *Bondarzewia montana*

TABLE 2. A CLASSIFICATION OF THE "POLYPORACEAE" (*Cont'd*).

Key Code 2.6.10.

Spores globose, asperulate

**Heterobasidion annosum*

Spores globose, smooth

**Meripilus giganteus*
*"Polyporus" catevatus*
*"Polyporus" rigidus*
*"Polyporus" rugulosus*
*"Poria" adiposa*

Spores globose, smooth—continued

*"Poria" hypolateritia*
*"Poria" vincta*
*Rigidoporus latemarginatus*
*Rigidoporus lignosus*
*Rigidoporus nigrescens*
*Rigidoporus sanguinolentus*
*Rigidoporus ulmarius*
*Rigidoporus vitreus*
*Rigidoporus zonalis*

Key Code 2.3.37.

Hymenochaetaceae 48 species

Hymenomycetes (Nobles, 1965), subdivisions are headed by key codes, a series of numbers (referred to as Code Symbols), each of which denotes a cultural character. The Code Symbols for those cultural characters which I have discussed above and to which I attribute taxonomic significance above the species level are listed in table 1.

A Species Code, the series of Code Symbols that denote the cultural characters shown by cultures of a species, was prepared for each of the 275 species in the study from descriptions made in my laboratory or published by a number of authors. With the exception of 12 species belonging to *Merulius, Phlebia, Echinodontium*, and *Radulum*, all the species have fruit bodies with poroid hymenial configuration, the mark of the family Polyporaceae as it is commonly accepted. By sorting the Species Codes an arrangement of species that may indicate relationships was obtained.

As shown in figure 1, the first division is based on the results of tests for extracellular oxidase, the species whose cultures produced no extracellular oxidase being shown on the left side of the chart under Code Symbol 1, those whose cultures produced extracellular oxidase being on the right side under Code Symbol 2. The cultures of 70 species gave consistently negative results, 183 gave consistently positive results, and 22 species gave negative results in some tests, positive in others. On the basis of the types of their associated rots, 13 of the 22 species were assigned to the group whose cultures lack extracellular oxidase, 9 to the group whose cultures produce oxidase. In the chart, the heavy solid line indicates that the generative hyphae are nodose-septate so that all the species above these lines in groups designated by Key Codes 1.3. and 2.3. have generative hyphae that are regularly nodose-septate whereas species in the group designated by Key Code 2.4. have generative hyphae that are simple-septate in the advancing zone, nodose-septate elsewhere. The groups below the heavy solid line, with Key Codes 1.5., 1.6., 2.6., consist of species whose generative hyphae are simple-septate or have rare single or multiple clamp connections, which I interpret as a form of reduction. The species in whose cultures the generative hyphae remain undifferentiated or at least show none of the modifications considered to have taxonomic value above the species level fall under Key Codes 1.3.7. and 2.3.7. Above these, the dash-dot lines indicate the differentiation of nodose-septate hyphae to form fiber hyphae, which appear tardily or may be lacking in species under Key Code 1.3.7.8., but are present in all species under 1.3.8. and 2.3.8. The dotted lines indicate the differentiation of generative hyphae to form thick-walled nodose-septate hyphae, present in association with fiber hyphae in groups under Key Codes 1.3.8.9. and 2.3.8.9. and alone in groups designated by Key Codes 1.3.9. and 2.3.9. In the left-hand group, this is the highest degree of differentiation observed. On the right side, cuticular cells, indicated by a dotted line and the Code Symbol 10, and interlocking hyphae, indicated by dash-dot-dot and Code Symbol 11, represent the highest degree of differentiation observed.

The decision to introduce basidiospore morphology as the basis for the final divisions in a classification based on cultural characters resulted from two observations. In well-defined genera composed of species whose fruit bodies and cultures show clear evidence of relationship, such as *Polyporus s. st.*, *Daedalea*, and *Hirschioporus*, basidiospore form and even size are uniform. Hence, genera whose component species show a wide range in basidiospore characters may be suspect. Also, when preparing my paper on classification of the Polyporaceae in 1958, I compared the cultures (in tubes) for the species in each group and even by this crude method was able to recognize similarity among species in some groups while other groups were heterogeneous in respect to characters other than oxidase production and hyphal differentiation. When cultures of species in these heterogeneous groups were sorted into groups of species with similar basidiospore characters, groups of species more uniform in appearance were obtained. Hence, the decision to include a noncultural character was made.

With few exceptions, all the species that fall under Key Code 1.3. on the left side of the chart above the solid line are alike in lacking extracellular oxidase, in being associated with brown rots, in having the bipolar type of interfertility governed by alleles at one locus, and in showing a host preference for conifers. The 17 species with rare clamps or simple septa also lack oxidase, are associated with brown rots, and in at least two species are homothallic. I interpret lack of oxidase, association with brown rots, bipolarity, and a host preference for conifers as primitive characters and therefore describe the group as a primitive one. As noted earlier, differentiation of hyphae in this group results in formation of only fiber hyphae and nodose-septate hyphae with thick walls.

In contrast, all the species that fall under Key Code 2.3. on the right side of the chart above the solid line are alike in producing extracellular oxidase, in causing the breakdown of lignin and the formation of white rots in wood, in having the tetrapolar type of interfertility governed by alleles at two loci, and in showing a host preference for broad-leaved trees. In addition, their hyphae show a high degree of differentiation, with formation of pseudo-parenchymatous areas in cultures, formed of microstructures which may be homologous with microstructures in fruit bodies that characterize the stipitate species of *Polyporus s. st.* and the hard upper surfaces of *Ganoderma* spp. Even in species reduced in the sense that their hyphae lack clamp connections, fiber hyphae occur in *Oxyporus populinus* and cuticular cells and other highly specialized structures in *Rigidoporus* spp. It is of interest also that ovoid to cylindric spores occur in both major divisions but that truncate spores occur only in species on the right side of the chart, where species with truncate spores occur under Key Code 2.3.7. in *Truncospora* spp., under Key Code 2.3.8. in *Poria* spp. and *Pyrofomes demidoffii*, and under Key Code 2.3.8.10. in *Elfvingia applanata* and *Ganoderma* spp., which may suggest a sequence in development within the main group. The number of advanced characters (extracellular oxidase production, ability to degrade lignin and cause white rots, tetrapolarity, host preference for broad-leaved trees, and highly differentiated hyphal structures) common to all species on the right of the chart under Key Codes 2.3. and 2.6.36. (excluding Hymenochaetaceae) suggests that these are related species and represent a monophyletic line of development.

Special Group II, denoted by Key Code 2.4.7.34., is composed of 17 species which may be considered advanced in their production of extracellular oxidase and association with white rots, primitive in their bipolar type of interfertility, and reduced in that hyphae of the advancing zone and submerged mycelium lack clamp connections. In discussing this group in 1958, I wrote

> The cultures show other remarkable similarities: in all, the rate of growth is rapid, the surface of the agar in Petri dishes being covered in one to two weeks; in all but *Polyporus resinosus* and *Radulum casearium*, many of the broad, simple-septate hyphae become thick-walled, the walls appearing to be gelatinously modified; characteristic globose chlamydospores are present and usually numerous in all species except *Merulius rufus* and *Phlebia merismoides*; in these two species and in *Poria rubens*, oidia are produced in abundance, and they occur sparingly in *P. albipellucida* and *P. cinerascens*; and finally, the eight species in the group whose type of interfertility is known are bipolar, in contrast to other species that give a positive test for extracellular oxidase and are tetrapolar. The number and nature of the correlated characters common to all the species suggest that the group is of monophyletic origin. The independent development of the same complex of diverse characters in unrelated species, with the formation of such a group through convergent evolution, seems unlikely.

Boidin (1958) recognized the same group of correlated characters in cultures of certain species of *Merulius*, *Phlebia*, and *Corticium* Sect. *Ceracea*, and in addition similarities in the distribution of nuclei in germinating spores, monocaryotic and dicaryotic mycelia. He too suggested close relationship or convergent evolution as possible explanations of the similarities. Along with the *Corticium* spp. that Boidin included, the *Phlebia* spp. and *Merulius* spp. that we both included, and the species with poroid hymenial configuration in my list, at least one member of the Agaricaceae, *Panellus serotinus* (Pers.) Kühner, has similar cultural characters. To me the group represents a single line of evolution, in which the various types of hymenial configuration and growth habit have developed independently.

In a survey such as this it is possible to observe general lines of development that may reflect phylogeny but quite impossible to look at each of the 32 groups into which the species have been sorted and assess the significance of these groups in taxonomy. In general, species in genera clearly defined by the characters of their fruit bodies show equally striking similarities in their cultures. For example, in table 2 under Key Code 2.3.8.11. nine species of *Polyporus s. st.*, including its type, are listed. The cultures are alike in hyphal characters and in macroscopic appearance, the dark brown, wrinkled, pseudo-parenchymatous areas contrasting sharply with the white cottony or woolly mats. The fruit bodies are alike in being stipitate with at least the basal portion of the stipe covered with a dark-colored cuticle similar to that produced in culture. The basidiospores are large, cylindric, or ellipsoid-cylindric. *Polyporus umbellatus*, which Pouzar (1966) designated as type of *Polyporus* subg. *Dendropolyporus*, and *P. hirtus* lack the fiber hyphae and interlocking hyphae which occur in cultures of *Polyporus spp.* and should, on the evidence from cultural studies, be excluded from *Polyporus s. st.*

Under Key Code 2.3.8.10. (spores truncate, brown, appearing echinulate), one species of *Amauroderma*, one of *Elfvingia*, and six of *Ganoderma* all appear similar in cultural characters, which indicates that they are closely related if not congeneric. Under this Key Code and under Key Code 2.3.8.10.11., which is an admission that the plectenchyma formed in the cultures was so compact that I could not determine whether it was composed of cuticular cells or interlocking hyphae, are eight species with cylindric spores, including the types of *Daedaleopsis*, *Datronia*, and *Fomes*, and one species, *Fomes fraxineus*, with subglobose spores. The similarities in cultural characters suggest relationship and indicate the need for correlated studies of cultures and fruit bodies of the species in the group. The cultural characters of one species, *Trametes campestris*, were described recently by David (1967) and by Domanski and Orlicz (1966). In both papers, the Species Code following my system of notation was given. Code Symbol 11, indicating the presence of interlocking hyphae forming a plectenchyma appeared in both and Domanski and Orlicz described an extensive "layer of plectenchymatous structure" on the surface of cultures. These authors compared this Species Code with that of *T. squalens*, the type of *Dichomitus* Reid, which lacks Code Symbol 11, and apparently on this evidence transferred *T. campestris* to *Dichomitus*. I do not know fruit bodies of *T. campestris* and it may well be that similarities in fruit bodies and especially in the form of skeletal hyphae may warrant the transfer, but the transfer cannot be justified on the basis of cultural similarities, since *Dichomitus squalens* cultures lack the interlocking hyphae that occur in cultures of *T. campestris*.

In my publication on cultural characters as a guide to the taxonomy of the Polyporaceae (Nobles, 1958), I brought together a group of species alike in a number of characters including the presence of microstructures described as nodose-septate hyphae with irregularly thickened walls and fiber hyphae. The species included *Daedalea quercina*, the type of that genus, and *Coriolellus sepium*, the type of *Coriolellus*. After a comparative study of cultures and fruit bodies of six of the species, including the type of *Coriolellus*, Sarkar (1959) presented an emended description of the genus *Coriolellus* giving its cultural characters as well as fruit body characters and listing all six species under the genus *Coriolellus*. In table 2, under Key Code 1.3.8.9., I follow her disposition of the species. Meanwhile, Donk (1966a) transferred the species to *Antrodia*, typified by *A. serpens*, (which I know has similar cultural characters although it is not included in the present study) since *Antrodia* Karst. predated *Coriolellus* Murr. In a paper that came to my attention after table 2 was prepared, Aoshima (1967) has transferred all the species except *Coriolellus sinuosus* to *Daedalea* Pers. ex Fr., a still earlier genus. This pleases me very much, for I have long known that the cultural similarities between the species of *Coriolellus* and the type species of *Daedalea*, *D. quercina*, suggested close relationship. The recognition of corresponding similarities in fruit bodies and the acceptance of *Daedalea* as the name for the group, with Aoshima's emended description of the genus including Sarkar's cultural characters, indicate the value of correlated cultural and fruit body studies. In his list Aoshima included *D. dickinsii* Yasuda, which I list under Key Code 1.3.8. on the basis of a description published by Bakshi and Bagchee (1950). It is possible that thick-walled, nodose-septate hyphae were missed in their examination of cultures and that they were observed by Aoshima in fruit bodies. Here again a correlated study of fruit bodies and cultures of the species is required. *Trametes carbonaria*, transferred to *Daedalea* by Aoshima, also occurs under Key Code 1.3.8. When she was studying the species included by her in *Coriolellus*, Sarkar carefully examined cultures of *T. carbonaria* and deliberately omitted the species from *Coriolellus* because the cultures lacked the thick-walled, nodose-septate hyphae that were present in all species that she assigned to that genus. Here again a correlated study of cultures and fruit bodies is desirable. Cultures of *T. flavescens*, also transferred to *Daedalea* by Aoshima, were described recently by David (1967), and on the basis of her description, which showed the species to have extracellular oxidase and tetrapolar type of interfertility, the species is listed under Key Code 2.3.8. along with the type and a number of other species of *Trametes*. Thus, a consideration of cultural characters may be necessary in deciding the proper disposition of certain species.

Under Key Code 1.3.8.9. (spores cylindric), in addition to the species I have mentioned are *Fomes cajanderi*, *Polyporus durus*, and *Trametes lilacino-gilva*. I am not familiar with fruit bodies of the last two species and can make no comment on their relationships, but I am concerned about the inclusion of *F. cajanderi* in the group. It has obvious similarity with *Fomitopsis rosea*, under Key Code 1.3.8.36. immediately above, but it also has thick-walled, nodose-septate hyphae in culture. Kotlaba and Pouzar (1957) suggested that *Fomes* (*Fomitopsis*) *cajanderi* seemed to occupy a somewhat transitional position between *Fomitopsis* and *Coriolellus*, so it was possible that evidence from fruit bodies also indicated relationship of this species with other species rather than with *Fomitopsis rosea*.

Obviously out of place under Key Code 1.3.8.9. is *Poria xantha*, with the tetrapolar type of interfertility and microstructures that differ from the typical thick-walled, nodose-septate hyphae denoted by Code Symbol 9, but I see no proper niche for it unless it be Special Group I.

The cultural characters of species listed under Key Code 1.3.7. (spores cylindric, brown) and under Key Code 1.3.8. (spores ellipsoid, brown) are similar, especially in the general appearance of cultures and tendency toward formation of strands and plumose mycelium. The two species for which the type of interfertility has been determined are tetrapolar, unlike the majority of species whose cultures lack extracellular oxidase. Three of the species have been assigned to the genus *Serpula*, and *Poria incrassata* has been made the type of the monotypic genus *Meruliporia.* The species may not be closely related to poroid species included in the present study but rather to the Coniophoraceae or some other group.

I would like to refer briefly to the species listed under Key Code 2.3.8. (spores cylindric). Most of the 35 species in the group were originally assigned to *Trametes* and later transferred to genera segregated from that genus. The similarities in cultural characters, including tetrapolarity, suggest close relationship but the even closer similarity among cultures assigned to *Coriolus,* to *Hirschioporus,* to *Pycnoporus,* and *Trametes s. st.* confirm the validity of these genera.

It is not possible to refer to all the groups, but these examples may indicate some practical results of cultural studies in Polyporaceae.

## References

AOSHIMA, K. 1967. Synopsis of the genus *Daedalea* Pers. ex Fr. Trans. Mycol. Soc. Japan 8: 1–4.

BAKSHI, B. K., AND B. BAGCHEE. 1950. Principal diseases and decays of oaks in India. Indian Phytopathology 3: 124–139.

BOIDIN, J. 1958. Essai biotaxonomique sur les Hydnés résupinés et les Corticiés résupinés. Rev. mycol. Mém. hors.-série No. 6.

CORNER, E. J. H. 1932. The fruit body of *Polystictus xanthopus* Fr. Ann. Bot. Lond. 46: 71–111.

DAVID, A. 1967. Caractères mycéliens de quelques *Trametes* (Polyporacées). Nat. Canad. 94: 557–572.

DOMANSKI, S., AND A. ORLICZ. 1966. *Dichomitus campestris* (Quél.) comb. nov. in Poland. Acta Soc. Bot. Polon. 35: 627–636.

DONK, M. A. 1966a. Notes on European polypores—I. Persoonia 4: 337–343.

———. 1966b. *Osteina,* a new genus of Polyporaceae. Schweiz. Zeitschr. f. Pilzkunde 62: 83–87.

ERIKSSON, J. 1958. Studies in the Heterobasidiomycetes and Homobasidiomycetes-Aphyllophorales of Muddus National Park in North Sweden. Symb. Bot. Upsal. 16: 1–172.

FIDALGO, O. 1967. Ultrasounds applied to the study of hyphal systems in Polyporaceae. Mycologia 59: 545–548.

KOTLABA, F., AND Z. POUZAR. 1957. Notes on classification of European pore fungi. Česká Mykol. 11: 152–170.

NOBLES, M. K. 1958. Cultural characters as a guide to the taxonomy and phylogeny of the Polyporaceae. Canad. J. Bot. 36: 883–926.

———. 1965. Identification of cultures of wood-inhabiting Hymenomycetes. Canad. J. Bot. 43: 1097–1139.

PINTO-LOPES, J. 1952. Polyporaceae, Contribuicao para a sua biotaxonomia. Mem. Soc. broteriana 8: 1–195.

POUZAR, Z. 1966. Studies in the taxonomy of the Polypores II. Folia Geobotan. Phytotaxon. Bohemoslovaca 1: 356–375.

SARKAR, A. 1959. Studies in wood-inhabiting Hymenomycetes IV. The genus *Coriolellus* Murr. Canad. J. Bot. 37: 1251–1270.

## DISCUSSION

DICK: Do you consider the extracellular oxidase reaction as having arisen once in the Basidiomycetes?

NOBLES: I do not think I want to be quite that definite, although I have said so on a few occasions. It is hard for me to believe that so many species in this group we call Polyporaceae could have so many characters in common without being related. It suggests to me that in this group, at least, there may be one line of development.

DICK: Does anybody know what the distribution of extracellular oxidases would be in the wood-rotting Ascomycetes?

NOBLES: I wish someone would find this out. It is so easy to drop a little guaiacum on cultures and find out what they do.

DICK: Production of these oxidases is easily gained or lost by mutations in particular strains. Wouldn't particular species differ in production as a mutational effect?

NOBLES: I am basing this report only on what I have observed in these cultures of species which are fairly closely related. But in this big group these characters seem to hang together: tetrapolarity, oxidase production, white rots, and preference for hardwoods.

TYLER: It occurs to me that these might be adaptive enzymes which could be induced as a result of a particular substrate on which the fungus is grown, and which might not be there on another substrate.

DICK: This has been shown in some of the polypores. If inducing materials such as xylamine are added to the medium, the enzymes are induced, whereas normally they are not present. The fungus has the capacity to form them provided the requirements are proper. The reactions, then, are reversible.

OLEXIA: Concerning the ideas about brown and white rots—have you been confronted with a white rot fungus which cannot use cellulose, indicating that it may possibly have lost these enzymes?

NOBLES: I am ignorant about the actual practices. I have simply observed that, in general, degradation of cellulose is common to the brown rots. In our experience these fungi always utilize cellulose. Certain species always produce brown rots. Certain other species or strains always cause white rots.

OLEXIA: I have worked to some extent with *Pycnoporus cinnabarinus*, a very attractive organism. I investigated the carbohydrate utilization capacity of the organism and got evidence that it could not degrade cellulose or any other polysaccharide which had a similar linkage, including di- and tri-saccharides. This might indicate that this species might have lost the cellulytic enzymes and merely maintained the lignitic, for I did get the positive reactions of a white rot.

NOBLES: I expect this sort of thing could be found frequently. I know little about these enzyme systems but I do know that some species on the far left side of figure I give a negative reaction for extracellular oxidases. They are actually associated with a white rot so that this reaction isn't really universal by any means, but is only the general pattern.

AMBURGEY: I am working with *Lenzites trabea*, which is a bipolar fungus and which causes a brown rot. Some of the monocaryotic cultures that I have obtained have not had the capacity to decay wood whereas many of the other monocaryotic cultures have. Some of these are still able to break down modified cellulose or cellulose products such as carboxymethyl-cellulose, indicating that there is a complex of enzymes—C-1's, C-X's, and beta-glucosidase—so a mutant obtained in one of these series of enzymes may cause the loss of the capacity to decay wood. I have crossed some of these monocaryons without this capacity and sometimes the dicaryon regains the capacity to cause to decay, but in other cases the dicaryon produced cannot.

PETERSEN: Do you have any data on how many genes may be involved or how many enzyme steps might be involved? How many places can you break it?

AMBURGEY: I have at least one monocaryon that appears to have a mutant C-1 enzyme. This means it can still reduce something like carboxymethyl-cellulose to glucose. In other cases it appears that some isolates are deficient in C-X enzymes. Whether it is one C-X or several C-X's is not quite clear right now.

DONK: I would ask if the use of the many generic names now more-or-less current in Europe means to some degree an endorsement of these genera; is it your intention to accept these genera? I think there is not much difference between *Leptotrimitus* and *Incrustoporia*; the former has effuso-reflexed fruiting bodies but in the latter these are completely resupinate.

NOBLES: I had hoped that this would be noticed because it seems to me it is an indication of relationships. I think they are so closely related as to be one genus. But in some of the others like *Trametes* and its related group and segregates I suspect that they can stand on their own in many cases. I would think certainly that *Incrustoporia* and *Leptotrimitus* might all go together.

BOIDIN: You said that 22 species gave negative results in some oxidase tests, but positive in others. Have you tested for other phenol compounds, such as gallic and tannic acid, which give brown oxidation products which could be confused with excretion? Some phenols (i.e., guaicol) give a very characteristic red color.

NOBLES: I have not followed your system, and I have attempted but not fully employed the method that Dr. Käärik suggested. This is something we should do. We have not tested for the phenol compounds.

BOIDIN: Have you tried more diluted concentrations of phenol which will allow the observation of growth and adaptive formation of enzymes?

NOBLES: I have used 0.5% only.

BOIDIN: I have not looked for possible correlation between presence or absence of oxidase and the cycle's characters. You made a very interesting observation about the Polyporaceae. For 23 species which are without oxidase, 3 are homothallic, 16 bipolar, and only 4 tetrapolar. On the contrary, the tetrapolar species are most numerous in the group with oxidase. This morning, I would have said that the absence of oxidase is an advanced character. However, I believe that you indicated that species without oxidase are primitive, is that not so?

NOBLES: After hearing your paper I wanted to revise mine. Perhaps I have to point the arrow just the other way. It seems to me that tetrapolarity, white-rot production, preference for broad-leaf trees as hosts (and this is true) all go together and would seem to be advanced characters. It is just intuitive, I presume, but it seems to me that they are advanced.

AMBURGEY: To elaborate on what Dr. Boidin has said, returning to *Lenzites trabea*: many of the monocaryons of the species will form fruiting structures (basidiocarps) in culture, which will discharge viable basidiospores and all of the basidiospores have the same mating type as the original monocaryotic culture. These spores will germinate and also will form basidiocarps which produce basidiospores, so there appears to be a homothallic situation without going through the dicaryon. In Dr. Raper's scheme, this would be somewhere between the bipolar and the homothallic, at the top of his chart, and would seem definitely to be a primitive type of fungus because it produces a brown rot, nodose-septate hyphae, and bipolar interfertility.

PETERSEN: I must ask whether you have checked the basidiospores to find out whether they are uninucleate and haploid.

AMBURGEY: I have some evidence that possibly either uninucleate or binucleate basidiospores are produced. I have some evidence that suggests that the monocaryons which fruit in culture contain two nuclei, each of which has the same factor for sexual incompatability. Clamp connections are not formed, but possibly the incompatability genes are at other loci so they may actually be partial heterocaryons.

SMITH: There is a point which concerns the nomenclature for septa of hyphae in the Basidiomycetes. The polypore specialists use the term "nodose-septum" if the cross-wall is accompanied by the specialized clamp connection. They use the term "simple-septate" for a cross-wall unaccompanied by a clamp connection or specialized branch. Since people like Royall Moore have been working, the dolipore septum, which is a truly specialized type of septum, has been described. There is also a septum which does not have a dolipore, and which is not a primary septum. This is a simple curtain-like cross-wall without nuclear or cell division as far as I can tell.

The term "simple septum" would certainly apply very admirably to it. The term "dolipore septum," of course, is self-explanatory. To me the term "nodose-septum" would ordinarily mean that the cross-wall had little granules or nodules. The "simple septum," defined in terms of polypore specialists, seems to be a misnomer. We should not try to define septa in terms of a special branch, but in terms of the characteristics of the septum itself.

DONK: I want to subscribe to Dr. Smith's criticism of the terms "nodose-septate," etc., because in the polypores themselves there are two transverse septa. One is dolipore and the other, in skeletal hyphae for instance, is a pseudo-septum and has no dolipore at all. They are both simple septa.

ORSON K. MILLER, JR.
*Research Pathologist*
*Forest Disease Laboratory, United States Forest Service, Laurel, Maryland*

# THE RELATIONSHIP OF CULTURAL CHARACTERS TO THE TAXONOMY OF THE AGARICS

This work must of necessity be regarded as a progress report. I have been culturing Basidiomycetes since 1961 but only intensively in the last four years. My objective has been to describe and photograph the sporophores from which pure cultures are obtained. In many cases killed material has been obtained for cytological study. My initial aim was to monograph a genus and include a description of the cultures obtained from each species. Interesting variation in these cultures led to my decision to obtain cultures from a number of genera in the Tricholomataceae and study them in some detail.

Unknown to me, Marta Semerdzieva from Czechoslovakia was taking a similar approach to agaric culture study. Her work, however, has either dealt with a specific complex of several species (Semerdzieva, 1965) or the selection of numerous species in many families of agarics in a very extensive manner (Semerdzieva, 1966).

In my own work I have routinely carried out growth studies by the general method followed by Davidson, Campbell, and Vaughn (1942) and Nobles (1948). However, I soon found that many fungi were very slow growers or didn't grow at all when standard methods were employed. This led to studies involving temperature variation over different periods of time and in a few cases more elaborate investigations involving factors such as light (Miller, 1967). The resultant different and often unusual morphological growth changes led to a broader study of the effects of physical factors on the growth and development of the vegetative plant. It became evident as the study progressed that certain questions needed to be resolved:

1. Does the vegetative plant have one or more characters which would strengthen the delineation of a given genus as now conceived?
2. Do isolates of a given fungus from the same or different geographical areas have the same characteristics?

3. What types of morphological structures are present in the Tricholomataceae?

4. What variation is obtained when the physical factors are altered during or at the onset of the growth of a fungus?

5. Can additional morphological characters be induced to form under special conditions?

The study to date has included 125 isolates of 70 species in the Tricholomataceae. In order to explore properly an assemblage of this size let us start in orderly fashion covering in sequence morphological structures, chemical reactions, fruiting responses, and the roll of physical factors in stimulating these responses.

## Asexual Spores

### *Chlamydospores*

Chlamydospores were found in many genera studied, including *Collybia radicata*, *Pleurotus elongatipes*, *Rhodotus palmatus*, *Pleurotus spathulatus*, *Panellus stipticus*, and *Lentinellus cochleatus*. These spores were quite variable from species to species, with walls of various thickness, variability in shape, apically oriented or intercalary but usually similar within a species.

### *Refractive Bodies*

Amorphous yellow refractive bodies were seen in many different genera. Those found in *Phyllotopsis nidulans* and *Pleurotus elongatipes* were best illustrated when mounted in phloxine and 3% KOH. The role of these bodies is unknown but they are intercalary and also apically oriented where basal clamps have been seen. They are not as numerous as the chlamydospores in most genera.

### *Conidia*

Conidia were born on simple conidiophores but were very infrequently found. *Hohenbuehelia petaloides* has thin-walled, nearly hyaline to light yellowish conidia of this type.

Conidia of a similar type were born on simple conidiophores on synnemata in a pleurotoid species which Kaufert (1935) called *Pleurotus corticatus*.[1] As Semerdzieva (1966) pointed out in her work,

[1] *Pleurotus cystidiosus* (Miller, O. K., 1969. A new species of *Pleurotus* with a coremioid imperfect stage. Mycologia 61:887–893).

it is unlikely that Kaufert correctly identified his species. However, there is little doubt from his illustrations and my studies that it was a member of the Tricholomataceae and constituted a unique type of asexual reproduction in the family.

### *Oidia*

Oidia were also not common but they were found in various genera, e.g., *Flammulina velutipes*, *Armillaria albolinaripes*, and *Pleurotus elongatipes*. They were formed on the secondary mycelium in the older portions of the culture after a period of 40–60 days.

### *Cystidia*

Cystidial end cells were either hyaline or had a yellowish granular content as seen in $H_2O$ and 3% KOH. Many shapes and sizes were found, but they were largely thin-walled with the exception of those of some species of *Panus* which have distinctly thickened walls.

## Cell Types in the Tricholomataceae

A number of differentiated tissues were observed in the mycelial mats and the question arose as to the way in which these types could be properly described. Starbäck (1895) had first proposed a system for describing fungal tissue and Korf (1958) expanded and illustrated the cell types for use in ascomycete taxonomy. The system was adopted by me for use in my study of agaric culture tissue (Plate VII).

### *Short-celled Tissue (Plate VII)*

*Textura Globosa*
Round or ovoid cells, thin-walled with intercellular spaces, often pigmented, exemplified by *Armillariella tabescens*.

*Textura Angularis*
Polyhedral cells, thin- or thick-walled with no intercellular spaces, often pigmented, exemplified by *Xeromphalina brunneola*.

### *Long-celled Types*

*Textura Intricata*
Interwoven hyphae with space between the cells, usually thin-walled and hyaline, exemplified by *Pleurotus candidissimus.*

*Textura Epidermoidea*
A tough membraneous tissue, adhering tightly together with few interhyphal spaces, usually pigmented, exemplified by *Lentinus tigrinus.*

*Textura Oblita*
Interwoven or parallel, thick-walled, usually pigmented cells, exemplified by *Armillariella mellea.*

### *Short-celled Tissue (Not Encountered in Cultures)*

*Textura Prismatica*
Small, angular, slightly thick-walled cells present in the sporophores of species of *Lentinellus* and the subhymenium of *Xeromphalina* but not encountered in culture tissue as yet.

## Consistency of a Given Structure in Isolates from Geographically Separated Fruiting Bodies of the Same Species

In general, isolates of a particular species had similar structures even when their parent fruiting bodies had been geographically separated. Cultures of *Pleurotus elongatipes* (Plate IX, figs. 17–20) had chlamydospores, amorphous yellow bodies, oidia, and clamp connections. In addition, they had a characteristic fanlike radial growth and white colonies. Isolates from various parts of the United States all showed the development of these characters.

## Variation in the Development of Morphological Characters over Equal Time Periods

Isolates of *Pleurotus dryinus* all developed similar brown-walled chlamydospores and cystidial end cells (Plate XI, fig. 52). Isolates from different geographical areas, however, developed chlamydospores and cystidia at different rates over the same period of time. The production of these structures reached a peak every day when the abundance of the spores created a distinct visual zone (see Plate XI, figs. 49–51). The Idaho isolate (Plate XI, fig. 49) had no spores at first and developed only a few zones near the point of attachment. The New York isolate (Plate XI, fig. 50) developed spores over a longer period of time whereas the Arizona isolate (Plate XI, fig. 51) had zones of spores and cystidia almost to the margin of the colony. In Plate XI, fig. 51 sectoring has occurred near the point of inoculation with a reduced production of chlamydospores within each sector. We can see from this example that much variation in the time of full development of certain cells may exist but the eventual morphology of the cells is identical. In general, this is true of the isolates that we have studied and not usually detected by short-term terminal growth studies.

## Chemical Reactions of Vegetative Hyphae

### *Extracellular Oxidase Reaction*

Gum guaiac has been used routinely to test for the presence of extracellular oxidases: the method adopted has been to place three drops in the middle of the total growth of the mycelial mat after 14 days. The majority of the species studied in the Tricholomataceae show a positive reaction. A few genera, such as the genus *Lentinus*, have no reaction.

In addition to the presence or absence of a reaction, the most promising parameter has been the time required for the reaction to occur and the intensity of the reaction when fully developed. For example, cultures of *Lentinellus ursinus* OKM 6136 developed a dark blue reaction in 3 minutes whereas *Panus rudis* OKM 4769 achieved only a pale blue reaction after 12 minutes and never became darker.

### *Amyloid Reaction*

Melzer's solution has been used along with 3% KOH and water for the microscopic study of the cultures. Amyloidity has been found in the different genera and is reported here for the first time.

*Clitocybe illudens* grew well at 12°, 18°, and 23° C. At the lower temperature, however, a well-developed "textura angularis" and "textura intricata" was differentiated after 60 days or more. This tissue exhibited scattered, clearly amyloid cells which had deep blue walls and in some cases amyloid, incrusted material deposited on the cell walls. At the higher temperatures the amount of amyloid material was greatly reduced.

*Panellus serotinus* had very deeply pigmented hyphae in culture as seen in $H_2O$ and 3% KOH. In Melzer's solution a greenish-blue reaction was observed in individual hyphae in the well-developed mycelial mat. Specific areas within the hyphal cells were greenish-blue, as indicated by the shaded areas in Plate X, fig. 46. It is possible that this may not be a true amyloid reaction, but a definite blue reaction was nevertheless present in the protoplast.

*Tricholoma imbricatum* (Wy 4227) had bands of light amyloid tissue in well-developed culture. Close examination of the cells revealed weakly amyloid cell walls. The reaction was best observed under low power where nonamyloid tissue could be observed in the same field.

*Clitocybe maxima* (OKM 4760) had light yellowish cell contents as seen in $H_2O$ and 3% KOH mounts. In Melzer's solution, however, a deep iridescent yellow reaction of the cell contents was observed.

The occurrence of amyloidity was not related to any taxonomic group in the Tricholomataceae and so far it has been found only in a few species. Genera such as *Lentinellus* which have amyloid tissue in the sporophore have exhibited no amyloid reactions in the cultures studied over the past several years. Amyloidity was found to develop slowly, however, and was often most evident in cultures which were not grown under the standard conditions usually employed to study the vegetative plant.

### *Other Microchemical Reactions*

Phloxine was also routinely employed and many species were found to have deep red to reddish-purple cells. The deep reddish-purple hyphae of *Pleurotus candidissimus* (OKM 3559) serve as a typical example of the phloxine stained hyphal cell. The staining seemed quite variable, however, very widespread, and not related to any taxonomic group. In addition, 3% KOH, ETOH, and $FeSO_4$ were routinely employed without result. The scanning of cultures with UV lamps was also carried out but no fluorescence was detected.

The general conclusion reached regarding the use of microchemical reactions was that they were species-specific. The most promising avenue of investigation was in the quantitative or colorimetric testing to determine time and intensity of the gum guaiac reaction for extracellular oxidase. For example, all the isolates of *Xeromphalina* rapidly developed deep blue reactions. Isolates of *X. campanella* required from 1.5 to 4.0 minutes to fully develop but isolates of *X. cauticinalis* were slower and a minimum development time of 6.0 minutes was the rule. This reaction combined with the unique "textura angularis" of the former and the "textura epidermoidea" of the latter was sufficient to differentiate between the species. A major obstacle, so far, has been to devise a suitable medium, preferably liquid, in which to grow these fungi in order to test them by colorimetry in a uniform way.

## Distribution of Cultural Characters in the Genera of the Tricholomataceae

From the time of Fries to the present, large genera such as *Clitocybe*, *Collybia*, *Tricholoma*, and *Pleurotus* have been subdivided to form a number of smaller genera. This has left a core of species which now compose these genera in the strict sense. The microscopic and microchemical characters which have been used to describe the new genera have often left the species which now compose the original genera with comparatively simple characters. They would now constitute the primitive genera in the family. This general pattern is also evident in the vegetative plant. The only exception found to the simple vegetative plant in *Clitocybe* (Plate X, figs. 33–36) has been *Clitocybe illudens* (Plate X, fig. 36), and the generic position of this species has long been a matter of controversy. The species of *Collybias* (Plate X, figs. 37–39) are very simple plants with occasional swollen end cells but little else. It is noteworthy that *Collybia velutipes* (Plate X, fig. 40.) is now placed in a separate genus, *Flammulina.* It differed sharply in culture from other collybias in having oidia, chlamydospores, a "textura epidermoidea," and thick-walled cells which form in age. The genus *Pleurotus* (Plate VIII, figs. 9–12) is also quite simple and has some swollen cells and occasional chlamydospores in culture. *Pleurotus palmatus* (Plate IX, fig. 32) has now been made the type of a new genus, *Rhodotus.* The unusual characters of the perfect stage were again enforced by very distinctive chlamydospores and inflated end cells in culture. *Pleurotus dryinus* (Plate VIII, fig. 13), placed in *Armillaria* by some authors, has a veil but is also an unusual member of the genus in culture with distinctive chlamydospores and cystidial end cells. *P. nidulans* (Plate VIII, fig. 14) and *P. petaloides* (Plate VIII, fig. 16) are now placed in *Phyllotopsis* and *Hohenbuehelia*, respectively. The former had amorphous yellow refractive bodies in culture, but was not distinctive in other respects. The latter had conidia on short conidiophores and is only equaled in this respect by the curious fungus which Kaufert (1935) incorrectly identified as *Pleurotus corticatus* (Plate VIII, fig. 15).[2] The production of these conidia on compact black-headed synnemata, as previously indicated, is unique in the Tricholomataceae.

The species of *Tricholoma* (Plate X, figs. 41–44) were also quite simple with only occasional chlamydospores and amorphous yellow bodies in culture. The species of *Tricholomopsis* (Plate IX, figs. 29–31) once placed in *Tricholoma* showed much variation with the development of a pigmented "textura intricata" in *T. radicata* (Plate IX, fig. 29), a pigmented "textura epidermoidea" in *Collybia (Oudemansiella) mucida* (Plate IX, fig. 31) but only inflated cells and clamp connections in *T. platyphylla* (Plate IX, fig. 30).

Species of *Panus* (Plate IX, figs. 21–24) had an array of well-differentiated characters but they did not seem similar to each other in most respects. This is also true of species of *Panellus* (Plate IX, figs. 25–28), which ranged from a well-developed "textura angularis" (Plate IX, fig. 25) or "textura intricata" (Plate IX, fig. 26) to a rather simple vegetative plant (Plate IX, fig. 28; Plate X, fig. 46) with amorphous yellow bodies.

Not enough species of *Armillaria* (Plate X, fig. 45) were cultured to make any comparison, but oidia and chlamydospores were observed. The species of *Armillariella* (Plate X, figs. 47–48), the fruiting bodies of which are so close in many respects, also had very similar cultural characters such as a "textura globosa" and "textura angularis," similar pigmentation, as well as rhizomorphs.

The species of two genera had distinctive, common characters. The first of these is *Lentinellus*, which is distinguished in the perfect stage by its minutely echinulate, amyloid spores; serrate gill

[2] *Pleurotus cystidiosus* (Miller, 1969. l. c.).

edges; lignicolous habit; and sessile or eccentric stipe. In culture, *Lentinellus* (Plate VIII, figs. 5–8) was typified by a combination of chlamydospores and/or cystidial end cells with granular contents. The second is *Xeromphalina*, which has amyloid spores, revives when moistened, has distinctive caulocystidia, and is centrally stipitate. It too had characteristic cultural characters (Plate VIII, figs. 1–4). All species had a reddish-brown "textura angularis" or "textura epidermoidea" which developed rapidly as the culture grew, forming just behind the white, simple, primary mycelium. These characteristics, combined with the very rapid deep blue reactions with gum guaiac, the strongly positive gallic and tannic acid reactions of both genera, formed a combination of distinctive generic characters.

## Induction of Other Morphological Structures

A number of Basidiomycetes were found to grow very poorly or not at all at the standard 25° C. used for routine culture study. These species were studied in more detail using a range of lower temperatures (1.7°, 7°, 13°, 18°, and 23° C.). In such cases an optional lower temperature for the species under study was selected for the routine culture study. It was soon noted, however, that many species developed different types of tissue, different pigmentation, and fruited in various ways at different temperatures.

Perhaps the most distinctive fruiting structures were those exhibited by species of *Lentinellus* (Plate XII, figs. 53–58). *L. cochleatus* (OKM 3544) grew very poorly at 25° C. When it was grown at lower temperatures normal hyphal growth usually occurred at 7°, but at 13° an occasional clavarioid sporophore was observed. A regular circle of well-developed clavarioid sporophores developed at 18° C. (Plate XII, figs. 53–55) with a cluster usually found at the point of inoculation as well. A regular cluster developed only at or near the point of attachment at 23° C. (Plate XII, fig. 56) but again it was clavarioid. The hymenium at 13° C. was composed almost entirely of chlamydospores. At 18° C. about 50% of the hymenium was made up of fertile, normal basidia and the rest of the typical thick-walled chlamydospores. The hymenium on clavarioid sporophores which developed at 23° C. was composed primarily of fertile basidia with a few scattered chlamydospores. A normal lamellate agaric was, therefore, able to complete its life cycle by means of a clavarioid fruiting structure. I wondered, at this point, if other species of *Lentinellus* could possibly be induced to form similar structures.

Subsequently, other species of *Lentinellus* were subjected to low-temperature study. Three more isolates have also produced clavarioid fruiting structures with fertile basidia and normal amyloid basidiospores produced over the apices of the branches. These were tentatively identified as *L. pilatii* AHS 73672, *L. ursinus* FS 90076 (Plate XII, fig. 57), and *L. pilatii* AHS 73305 (Plate XII, fig. 58).

The light requirement and its effect on the fruiting process has been explored (Miller, 1967) using *Panus fragilis*. It is interesting to note that a low level of light induced the gymnocarpic growth of initials. They were quite clavarioid at first and prompted one to wonder if an interruption of the fruiting process at this point in the past could have led to the production of clavarioid fruiting structures which in some cases would become the only type of sporocarp to develop. To explore this question further, exploration of the physical factors required to induce fruiting is planned.

## Phylogenetic Information Which May Be Inferred from the Stimulation of Atypical Fruiting Bodies

We cannot expect to arrive at phylogenetic conclusions without any fossil record.

The completion of the life cycle of certain fungi by a clavarioid fruiting structure, however, could indicate a type of ontogenetic evidence of the origins of the Tricholomataceae. It is not a natural family although many of the included genera appear to be related to each other.

The general form and location of the hymenium on the fruiting structures observed in *Lentinellus* are clavarioid. The fact remains that four different isolates and three species of *Lentinellus* have produced these clavarioid sporocarps *in vitro*.

Species bearing amyloid, minutely echinulate spores which are white in deposit are not found among the cantharelloid fungi, but several similar species are known in the clavarioid fungi.

Only a general study of the fruiting of species which are closely related to those which have produced atypical sporocarps can add additional insight into this interesting phenomenon.

## General Conclusions of the Studies Now Completed

Most genera do not have species with consistently similar morphological characters.

*Clitocybe*, *Collybia*, and *Tricholoma* are typified in culture by simple vegetative plants.

Several genera, including *Lentinellus*, *Lentinus*, and *Xeromphalina*, have distinctive morphological characters.

Amyloid reactions have been demonstrated for the first time in several species, but Melzer's solution is useful only in characterizing species.

Other microchemical reactions, with the exception of gum guaiac, show no logical pattern or are completely negative.

Gum guaiac has promise if the quantity and time of reaction can be accurately determined. Individual species have characteristic reaction times and all of the species of some genera have similar reaction times.

Gallic and tannic acid agars follow about the same pattern as that demonstrated by gum guaiac.

## Future Objectives

To study a wider range of species in culture so a more complete account of each genus may be obtained.

To correlate the anatomy and morphology of the perfect stage with that of the vegetative plant.

To obtain a unified species concept based on a study of the *entire* life cycle.

To study the results obtained and relate these to the genera which already have been delimited using only the perfect stage.

## References

Davidson, R. W., W. A. Campbell, and D. B. Vaughn. 1942. Fungi causing decay of living oaks in the eastern United States and their

cultural identification. Tech. Bull. 785, USDA, Washington, D. C., 65 p.
KORF, R. P. 1958. Japanese Discomycete Notes I–VIII. Sci. Rep. Yokohama Nat. Univ. II 7: 7–35.
MILLER, O. K. 1967. The role of light in the fruiting of *Panus fragilis*. Canad. J. Bot. 45: 1939–1943.
MILLER, O. K. 1969. A new species of *Pleurotus* with a coremioid imperfect stage. Mycologia 61: 887–893.
MILLER, O. K., AND R. WATLING. 1968. The status of *Boletus calopus* Fr. in North America. Notes from the Royal Botanic Garden, Edinburgh, Scotland 28: 317–325.
NOBLES, M. K. 1948. Studies in forest pathology. VI. Identification of cultures of wood-rotting fungi. Canad. J. Research, C, 26: 281–431.
SEMERDZIEVA, M. 1965. Cultivation and morphological studies of certain fungi of the Agaricaceae family in vitro. Česká Mykologie 19: 230–239.
———. 1966. Morphological observations of some *Pleurotus* mycelium. Sydowia 19: 250–258.
STÄRBACK, K. 1895. Discomyceten-Studien. Bihang Kongl. Svenska Vet.-Akad. Handl. XXI 3: 1–42.

## DISCUSSION

DUBUVOY: I assume all this work has been done with the same culture medium.

MILLER: This was all done using the same culture medium. We used the malt agar medium of Davidson and the medium that Dr. Nobles has used.

SMITH: Have you set up these cultures in a number of replicates?

MILLER: Every culture is replicated four times and repeated twice. We try to do all experiments a third time, but because of the number of cultures which we are growing, we haven't accomplished this goal as yet. Some experiments, on the other hand, have been repeated many times.

SINGER: Do you have herbarium material of all your cultures? I am thinking of *Armillariella mellea*, which is now almost a subscience in itself. It has been split up into several species which differ ecologically, at least with some small anatomical differences. Perhaps one might expect similar differences in culture characters. If the herbarium material is consistent with the culture which came from it, it is actually comparable with it in value.

MILLER: We faithfully preserve herbarium specimens so we can go back and examine the perfect stage. We also have killed material of many of the perfect stages so that they can be cytologically examined.

DUBLIN: I assume that most of these cultures are dicaryotic mycelia from tissue culture. Did you do any work with haploid mycelium?

MILLER: I should have prefaced my remarks with the fact that most of these cultures are obtained by multispore isolates or from tissue isolates.

I simply have not had the time to work with haploid cultures. It is a very worthy approach, and we have haploid cultures.

NOBLES: In my experience with the polypores in culture, some of the species will form oidia and some will not. In many cases that I have observed, the oidia are produced on haploid hyphae that have grown out from a dicaryon culture. The culture is a dicaryon mycelium, and oidial production is on a small area where there has been a reversion to the haploid condition. It is these haploid hyphae that break up to form oidia. I do not know what causes this, but it is true in some isolates. In other isolates of the same species it doesn't happen. So oidial production seems possible in all of the polypores under the haploid condition.

MILLER: Have you found this on normal dicaryotic mycelium in older portions of the mycelium?

NOBLES: Not necessarily. In *Coriollelus hirsutus*, for instance, this sectoring happens quite frequently. In the fungus we used to call *Tramates americanum* (*Osmoporus odoratus*) the whole mat reverts to the haploid. You can't keep it dicaryotic. You must know the genetic material with which you are dealing.

FERGUS: I think we are assuming that because you make a culture isolation of the mycelium from a sporophore the isolate is dicaryotic, but until more careful cytological studies are done I would question this. One of my students has completed a study on *Volvariella* and the fruit body mycelium in all cases was multinucleate up to the subhymenium, at which point it became dicaryotic. I think it is possible that the initial mycelium is dicaryotic, perhaps resulting from the germination of a mass of basidiospores and subsequent hyphal fusion, but from this point on there is no real proof that the dicaryon continues. Obviously it must dissociate or the haploid oidia mentioned by Dr. Nobles wouldn't form. Different species within a genus might have tremendous individual variability as to this nuclear dissociation and consequently perhaps what we observe in specific cases is not a general characteristic of all of the species of a single genus. All of the isolates you have produce chlamydospores, so they appear homogeneous. Considerable variation in nuclei might still exist among them, however.

I noted that your initial transfer inoculum was a plug of mycelium. Have you consistently taken mycelium from the advancing edge of the colony? If you take this from a particular portion of the mycelium and use only one plug, perhaps you are further tending to have these nuclei dissociate.

MILLER: First, the inoculum plug is never taken from morphologically highly developed cultures, but from young mycelium which has just grown. All cultures have grown approximately the same number of days and we take the plugs from the growing periphery. Second, at least 80–90% of the cells are producing clamp connections. This can usually be checked visually. Third, I know from my own experience that in the perfect stages of some of these fungi the morphologically highly developed cells, whether they are part of a rhizomorph of *Armillariella mellea* or in the sporophore of *Chroogomphus rutilus* are coenocytic. In *Xeromphalina*, I wouldn't be surprised at all if the hyphae in the early formative stages are coenocytic. In short, we do get a multinucleate condition. But how would you ever control this? We must only be aware of it.

SMITH: I think it has probably been known since the 1930's that most of the cells in an agaric fruit body having reached maturity have more than one pair of nuclei. It could be several pairs of the dicaryon. Fergus sees the possibility that you pointed out of dicaryon dissociation as equally possible. The dicaryons have merely multiplied in order to maintain a certain type of nuclear ratio. Study of this would involve counting the nuclei in each cell to ascertain whether there is an odd number or an even.

FERGUS: We did this counting and got even and odd numbers of nuclei within the cells, but I can't state the frequency of these numbers.

RAPER: I think frequency is actually more important than merely presence of even and odd numbers of nuclei. If the two are approximately equally frequent, this can mean all sorts of things, one of them being that you never had dicaryons, and another being evidence of an occasional nuclear mortality. It still doesn't mean anything really. Dicaryons do break up, but they are pretty stable things in a lot of organisms.

SINDEN: In our work with *Agaricus bisporus* we found that if we isolated from the advancing edge of the cultures we obtained very high numbers of oidium-producing cultures which were totally atypical. The variability within our cultures is extreme, depending on the part of the mycelium from which we make our vegetative isolation for transfer.

MILLER: Can you actually state what the cytological picture is in *Agaricus bisporus*?

SINDEN: We only know it in a small way at any stage. There are two nuclei in the spore, so we start with a dicaryon. The mycelium is never haploid.

MILLER: Can you establish by conventional staining procedures that this is a dicaryotic mycelium?

SINDEN: No. How can you do that with something that has paired nuclei in the basidia and in the outer part of the tramal tissue? We have taken smears of the cuticle and pellicle cells, and those cells are dicaryotic.

SMITH: If there is a breakdown in the dicaryon only when cell volume reaches a certain point, the implication is that the dicaryon is functioning throughout the fruiting body.

SINDEN: I think that is what happens. Although most investigators deal with a few isolates, we deal with thousands. We find great variations in chlamydospore production, conidiospore production, and all sorts of morphological variants of the mycelium, yet they may all come from the same original mycelial culture.

MILLER: Isn't there a potential danger that *A. bisporus*, simply from the industrial handling and the like, now has a number of abnormal mutants which have arisen? No one really has a wild type of this fungus to examine adequately and to which to ascribe a given strain name or number.

SINDEN: From what we know, most of the variants we have, have occurred in nature and have been picked up in nature and are merely perpetuated by us.

SMITH: Mr. Isaacs, who is working with me on a doctoral thesis on *Agaricus*, has some serious doubts about just what *A. bisporus* in nature really is. I am suspicious that in the cultures you have taken from one another, you may have several taxa according to his work. I would like to see the name discarded, actually.

SINDEN: I would think that there is really no single *A. bisporus*. I think that the brown strain is just as different as many species of agarics. I have never accepted the name myself. I have called it *A. campestris*, which covers both taxa except for nuclei, with *bisporus* for the bispored strain.

DUBUVOY: We have been working with cultures of *Psilocybe* for several years, and we have produced several very significant differences by changing chemicals in the medium. Even on the same medium, different strains of *Psilocybe* vary considerably even if they were collected in the same ecological area. For instance, we have found conidiospores in one strain of *Psilocybe caerulescens* but oidia in another strain. We have been able to induce chlamydospores in one strain in a medium in which all others produce oidia.

MILLER: It shows that both the ability to produce chlamydospores and oidia are in the genome of the fungus. We are doing this with *Panus fragilis* and we find the same thing. My initial attempt was to establish a uniform medium and vary the physical factors to induce morphogenic changes. Certainly a second attempt will include variations in the medium itself.

BOIDIN: Do oidium and chlamydospore production and other culture characters have any systematic or phylogenetic significance? Do you see any characters which are not consistent in the culture and fruit bodies?

MILLER: We have arbitrarily set up a series of characters in the genera of the Tricholomataceae and used them to delineate the taxa according to the perfect stage. We have talked about amyloid spores, thick-walled trámal hyphae in the lamellae, and many other characters. They do not seem to have any particular sequence. I wonder if the genera are being delineated in the right way. In other words, using both fruit body and culture characters, we see *Lentinellus* very clearly as a group of species which, in my opinion, hold together. They possess common characters all the way through the genus, in culture as well as in the perfect stage. *Xeromphalina*, with the possible exception of one species that we do not have in culture, exhibits typical cultural morphology, but no one cultural character which is present in all the species. We will not know until all of the species of *Pleurotus*, *Panus*, and *Panellus*, etc., are in culture, that the delineation of the genera themselves is correct. Not that I would base this delineation totally on the vegetative stage, but there is already much variation in opinion as to the number of genera and how they should be delineated.

GRAND: Do you ascribe any importance to the presence, absence, or relative frequencies of clamp connections in culture?

MILLER: In the tricholomatoid fungi with which I have been working, we have been dealing with multispore isolates. We do, in fact, find that a great majority have abundant clamps. It seems to me that you would have to approach the problem using haploids to really determine the relative significance of clamps.

GRAND: Have you noticed the production of clamp connections in culture where there is no indication of clamps on the hyphae of fruit body tissues?

MILLER: I have found this situation in my work on the Gomphidiaceae but I haven't run into this as yet in the Tricholomataceae. I haven't cross-examined each tissue to derive this information. Of course, it is true in the boletes and the Gomphidiaceae, in which we have the vegetative plant producing clamps but no evidence of them in the sporophore.

HARRISON: I want to call attention to the amyloid reactions that you mentioned. The same phenomenon occurs in the tramal tissue of the pilei of Hydnaceae. It is interesting to think of this being present in the pilei as well as in cultures with such exactness.

MILLER: This past winter we were able to induce thick-walled amyloid cells to develop in sporophores which formed in culture of several species of *Hericium*. It is our initial impression that this does not occur until about the third generation of fruit bodies are produced. Our initial results indicate that two or three fruit bodies must be harvested, and then when the crust seems to have matured to a certain point, the fructifications thereafter will have amyloid, thick-walled hyphae, at least near the base of the fruit body.

THIERS: In working with the amyloid reaction in the peridium of some of the gastroboletes and in the genus *Boletus* I often noticed that the cross-wall gave a positive amyloid reaction although the lateral walls apparently did not. Have you observed this at all in any of your culture work?

MILLER: No, not in culture. I too have observed it in *Boletus calopus*, on which Dr. Watling and I have a paper in press,[1] but not in the cultures themselves.

Often boletes come into culture relatively easily, as in the genus *Suillus*, for example. But in the genus *Boletus* there is great resistance. We just don't have the right nutritive medium to present to these species. For example, I can get *Boletus edulis* to grow for a brief time and then it dies. What we need is a break-through in our physiological work.

SMITH: Concerning the problem of amyloid material in and between the cells, and the amyloid cell wall, we have found a curious thing in *Rhizopogon*. When some of the peridial tissue is mounted in iodine, extracellular dark blue granules result along some of the hyphae and they do not duplicate what appears to be an amyloid reaction. I called this reaction amyloid for a while. When I mounted the tissue in KOH, everything was hyaline. Then one day I mounted the tissue in water and the dark blue granules appeared along these hyphae. It was evident, of course, that the material was not amyloid in the usual sense. One must be careful when one deals with amyloidity in hyphal content and in extracellular material. Do you mount any of these tissues in water as well as in Melzer's to get any contrast?

MILLER: I mount the tissues in KOH, alcohol, Melzer's solution, and water. I was unsure of the amyloid reaction of *Panellus serotinus* and not of the others because the hyphae in that species have an iridescent content. The color is yellow with a greenish tint, and I wondered, seeing the different reaction with Melzer's solution, if this might not be an artifact. The other reactions I reported were bona fide amyloid as far as I can tell.

[1] Miller, O. K. and R. Watling. 1968. The status of *Boletus calopus* Fr. in North America. Notes Roy. Bot. Gard., Edinburgh 28:317–325.

OLEXIA: Do those species which fruit in culture form basidia over the agar surface as well as on the fruiting body?

MILLER: I have not found basidia on the agar surface or in the mats in the Tricholomataceae. I have studied the mats of all of the species discussed. Production of fertile basidia was observed not only at the base but near the apices of the clavarioid-type structures that are produced in *Lentinellus* but not on the mat. This is true of other sporocarps produced in culture.

PETERSEN: When we grow *Sparassis crispa* we find not a fruiting body, but a completely fertile layer of basidia all the way across the agar surface.

*Clavicorona* and its relatives are branched, clavarioid fungi with amyloid, minutely echinulate spores, and are culturable and fruit in culture. Fruiting bodies in culture are small but branched. I only wish that by shining a light on a clavaria I could make a pileus appear. In the fruiting of *Panellus fragilis* does the primordium produce a pileus in the absence of light?

MILLER: If you keep the light at a low level, perhaps as low as 150 ft-c, initials are readily produced, but there will be no differentiation of a mature pileus or lamellae.

PETERSEN: Have you done any action spectrum on this? We found that in *Coprinus* we got an action spectrum for pileus production. We got another, different action spectrum for phototropism of fructifications.

MILLER: In *Panus fragilis* we noticed immediately that there was a pink pigment associated with fruiting after exposure of the mycelium to the proper level of light. If this pink pigment was pinpointed, it was at those places where the fruit body initials arose. The pigment was carried up with the initial. At the present time Dr. Collins at the University of Connecticut is attempting to identify this pigment to see if the production of the pigment is the important initiator and not the action spectrum itself, although we know that there has to be a certain intensity of light to initiate the process.

EDWARD HACSKAYLO

*Plant Physiologist*
*Forest Physiology Laboratory, Division of Timber Management Research, United States Department of Agriculture, Plant Industry Station, Beltsville, Maryland*

# THE ROLE OF MYCORRHIZAL ASSOCIATIONS IN THE EVOLUTION OF THE HIGHER BASIDIOMYCETES

We, who consider ourselves specialists on mycorrhizae, generally limit our endeavors to areas of disciplines wherein we are most familiar. We are either physiologists, morphologists, pathologists, mycologists, microbiologists, ecologists, or sometimes professional hybrids of these disciplines. Mycorrhizal research encompasses many different aspects of mycology and phanerogamic botany, so like most other plant scientists we select a somewhat narrow course and endeavor to solve some small aspect within a major research area.

Although the research in our laboratory is primarily concerned with physiology of mycorrhizae, I consider the taxonomy of mycorrhizal fungi to be of extreme importance. We must possess accurately identified organisms so that our experiments will be meaningful to others and be reproducible. Often we find ourselves in the dilemma of trying to ascertain and use accurate names for the fungi only to learn that the determination isn't easy. Sometimes a species that we have cited in a publication changes status; in other cases taxonomists cannot agree on a name, and so forth. All of you are aware of these problems. As I read papers on mycorrhizae written during the past century I cannot help wonder about the accuracy of some reports that identify particular fungi as mycorrhizal partners with specific species of phanerogams. Often it is necessary to disregard or experimentally examine those that appear to be obviously erroneous. Cautiously we attempt to synthesize broad concepts by correlating data from other papers.

I enter into this discussion on evolution of the mycorrhizal Basidiomycetes with a great deal of reservation. So little tangible evidence is available from fossil records. Attempts by mycologists or phanerogamic taxonomists to correlate mycorrhizal relationship and classification such as those by Singer (1962) are few. Certainly this would be a much better paper if it were written after this symposium is concluded. However, it is a challenge to approach the subject, and I welcome the opportunity to place before you some of our thoughts and questions.

In nearly all of the reports on mycorrhizal Basidiomycetes these fungi are cited as partners in ectotrophic association with tree species. A few orchid endophytes are exceptions. The dual organism association resulting from the ectotrophic mycorrhizal fungus and cormophyte host is sometimes referred to as an "ectotroph" (Singer & Morello, 1960), and the convenience of the term merits wider use. It is encouraging to the mycorrhiza researchers that several modern taxonomists working with the higher Basidiomycetes are attempting to interpret mycorrhizal associations as factors in classification. The number of species of Basidiomycetes included in ectotrophs emphasizes the additional consideration they will henceforth require.

Trappe (1962) compiled a listing of fungus associates of ectotrophic mycorrhizae. Within the Agaricales, 63 genera were listed. The Aphyllophorales included 14, and in other orders he listed an additional 14 genera. Thus, a total of 91 genera of Basidiomycetes have been reported as mycorrhiza-formers. That figure, of course, would be more or less, depending upon whose system of nomenclature one followed, but it gives a good approximation of what has been reported. I have not attempted to count the number of species reported, but it would amount to several hundred. Unfortunately, less than 10% of these have been positively shown to be mycorrhizal fungi in controlled experiments. The evidence is based primarily upon observation of sporophores under certain tree species. The amount of error is unknown, but it probably is considerable since fruiting may be more limited than association. Singer (1962) estimated that the mycelium of about one-half of the species within the Agaricales in temperate zones may be considered as potentially mycorrhiza-forming. Certain species within genera such as *Suillus*, *Russula*, *Lactarius*, and *Gomphidius* apparently are obligate mycorrhizal fungi. I agree with Garrett (1956) who stated that probably all mycorrhizal Basidiomycetes are ecologically obligate symbionts and that the root association is nearly always essential for completion of the life cycle of mycorrhizal fungi. The mycelium must be intimately connected with the living roots of the host as a requisite to sporophore formation under natural conditions. It seems desirable at this point to try to establish the parameters of the subject as I see them.

Ectotrophic mycorrhizal associations are limited to hosts within the following: In gymnosperms they include all the Abietineae of the Pinaceae. In angiosperms they are, with few exceptions, limited to the dicotyledonous, amentiferous families Salicaceae, Juglandaceae, Corylaceae, and Fagaceae. One major exception is the Myrtaceae containing *Eucalyptus,* many of which are known to have ectotrophic mycorrhizal associations. The other exception is the Tiliaceae, every species of which is ectotrophic. Although the literature cites many other species in ectotrophic associations, there is little conclusive evidence to establish them with certainty. The fungi associated in ectotrophs are primarily higher Basidiomycetes, and the order Agaricales is by far the largest group using numbers of genera and species as the index. To the mycorrhiza researcher Amanitaceae, Boletaceae, Cortinariaceae, Russulaceae, and Tricholomataceae are the best-known families. In the Aphyllophorales we are most familiar with the Thelephoraceae. From other orders the genera *Rhizopogon, Lycoperdon, Pisolithus,* and *Scleroderma* are most frequently mentioned. Several other groups obviously are involved in mycorrhizal associations and no doubt will receive verification by the mycorrhiza specialist and taxonomic mycologist. As one considers the broad aspects of ectotrophic associations one wonders, how did it all begin? How did it continue to evolve? What are the common denominators in the association? If we can examine a few specific examples, perhaps it will stimulate additional consideration of the problem.

The phylogenetic relationships of the host species possessing ectotrophic mycorrhizae in the gymnosperms are fairly well determined since the species are in one subfamily, Abietineae of the Pinaceae. *Abies, Tsuga, Picea, Larix,* and *Pinus* are, without exception, hosts to ectotrophic mycorrhizal fungi. However, the other subfamily, Taxodioideae, containing *Taxodium, Thuja, Chamaecyparis,* and *Juniperus,* are all associates of phycomycetous endotrophic mycorrhizal fungi.

The Pinaceae are generally considered to be among the most highly developed of the gymnosperms (Benson, 1957), but more primitive than the angiosperms. It appears, therefore, that the first

Basidiomycetes to become mycorrhizal fungi must have made the transition from a purely saprophytic to at least partially parasitic existence within this group of hosts (Garrett, 1950). Some tricholomas are modern examples of mycorrhizal Basidiomycetes that obviously possess cellulytic as well as pectolytic enzymes; thus, they are capable of utilizing cellulose as a carbon source (Norkrans, 1950). Most mycorrhizal fungi have lost this capability and can utilize only simple carbohydrates. Ritter (1964), however, reported that all the ectotrophic mycorrhizal fungi he studied could produce cellulytic enzymes. Dr. J. G. Palmer and I have not as yet substantiated this in experiments with several species of mycorrhizal fungi.

In natural environments simple carbohydrates are supplied to the fungi by roots of the host (Björkman, 1942). The intimate anatomical association apparently insures that the carbon supply is available to the mycosymbiont and not to competitive soil organisms. At present the differentiation between ectotrophic and endotrophic fungi by the Abietineae and Taxodioideae, respectively, is unsolved. Obviously, the determining factor or factors are properties of the host. Thus, from the standpoint of mycorrhizal associations the taxonomists have logically classified the Pinaceae. How does this, however, relate to the angiosperm mycorrhizal associations, if at all?

As noted before, most of the angiosperms that have Basidiomycetes as mycorrhizal associates are limited to the dicotyledonous amentiferous families Salicaceae, Juglandaceae, Corylaceae, and Fagaceae. The positions of these families in the various proposed schemes representing the phylogeny of the angiosperms are interesting. The families were considered most primitive by Engler and Prantl, who placed the catkin only a step higher than the cone of the conifers. In this scheme, then, there is a logical transition from the gymnosperms to the angiosperms, carrying along the ectotrophic mycorrhizal association. There are a couple of exceptions to this when one considers *Tilia* and *Eucalyptus,* since they are more advanced than the catkin-bearing tree species.

No unusual or obviously direct floral relationship links either *Tilia* or *Eucalyptus* to the Amentiferae or Abietineae. Whether chemical analyses within these groups indicate any phylogenetic relationships through size of molecules (Swain, 1963), class of compounds (Swain, 1963), or DNA hybridization (Dutta *et al.,* 1967) requires further investigation.

Some classification systems for angiosperms, such as those proposed by Bessey and Hutchinson, consider the Ranales most primitive. The Ranales are endotrophic in their mycorrhizal associations. In these systems the families having ectotrophic mycorrhizae are considered well advanced in the evolutionary scheme and are not grouped closely together. The catkin is regarded as an advanced inflorescence with reduced structures. Thus, it would appear that the biochemistry of the roots that essentially determines which mycorrhizal fungi can penetrate the tissues is not related to floral evolution. It is not possible for any of us at this point to do more than make the above observations. We can, however, cite some specific ecological and physiological situations and perhaps relate them to factors influencing the development and distribution of certain fungi and host species.

As I previously stated, the inherent biochemical properties of the roots of the host species determine the class from which the fungus complement will come (Melin, 1953). If species of *Pinus, Quercus,* and *Acer* are growing adjacent to each other, as is often the case in our hardwood regions, the oak and pine roots will be invaded by ectotrophic mycorrhizal Basidiomycetes but the maple roots will not. Recently we demonstrated in axenic culture experiments that *Thelephora terrestris* will readily form an ectotroph with *Pinus virginiana* but not with *Acer rubrum.* The mycelium grew equally well in the substrate and along the roots of both species. The only instance of penetration of the *Acer* roots was in the case of plants growing at very low nutrient levels where the roots had become moribund. In that instance the structure superficially appeared as an endotrophic mycorrhiza, but the fungus obviously was leading a purely saprophytic or possibly a mildly parasitic existence in the weakened cell. Undoubtedly, such observations have led some authors to suggest that several Basidiomycetes are capable of forming vesicular-arbuscular mycorrhizae. These conclusions probably are not correct, since the enzymatic capabilities of the mycorrhizal Basidiomycetes are generally limited to utilization of few relatively simple carbohydrates and not to hydrolysis of cellulose. Apparently, it is not sufficient for the mycelium of the mycorrhizal Basidiomycete only to be in superficial contact with a living root to enable it to complete its life cycle. We have also observed that in open pots in the greenhouse clamp connection hyphae obviously belonging to *T. terrestris* will grow to some extent along roots of *A. rubrum.* A mantle did not develop on the roots and there was no evidence of penetration of healthy roots by the hyphae. There was no tendency whatever for a hymenial surface to develop. With pine, however, the fungus will fruit quite readily under similar physiological conditions for the host and fungus. I do not believe the physiological specificity demonstrated by this example is limited to *T. terrestris*. The appearance of sporophores of species of *Suillus, Russula, Amanita,* or other mycorrhizal fungi only in the ecological association with certain tree species must be caused by the physiological interaction of the two symbionts.

The actual dependence of *Pinus* and other tree species on the mycorrhizal association is conclusive. One of the most recent demonstrations of this was in Puerto Rico. Attempts to establish pine on the island by seeding failed for many years. In 1953, soil containing mycorrhizal fungi was introduced and used as inoculum for pine seedlings. The results have been very dramatic. The Honduras strain of Caribbean pine now grows so successfully that it is not uncommon for the tree to grow 10 feet in one year. The doctoral dissertation by Vozzo (1968) contains a very good summary on the history of ectotrophic mycorrhizal experiments in Puerto Rico.

The geographical distribution of mycorrhizal fungi and hosts must be determined by the presence of both organisms under appropriate environmental conditions to permit establishment and function of the mycorrhizal system. These environmental conditions can indeed greatly influence the extent to which the mycorrhizal partners can survive and reproduce. An interesting paper on this subject by Singer and Morello (1960) in South America provides some data of this type. One should note, however, that recent developments would not substantiate their views that endotrophs are not a natural biological unit (see Mosse, 1963).

One of the least complex natural examples of the influence of the environment on mycorrhizal associations was very thoroughly described recently by Schramm (1966). His many years of study on establishment of vegetation on the anthracite coal banks in Pennsylvania eventually led to the conclusion that successfully establishable tree species were primarily those that could form ectotrophic mycorrhizae. Many planted species of oak and pine and spontaneously seeded birches, aspens, and willows grew on the coal banks. Schramm kept records on the appearance of sporophores of Basidiomycetes throughout many years. Invariably the sporophores belonged to just a few species that included *Inocybe* sp., *Boletus* sp., *Astraeus hygrometricus*, *T. terrestris*, *T. caryophylleae*, *Pisolithus tinctorius*, *Amanita rubescens*, and *Scleroderma aurantium.* Although natural stands of pine and the other hosts were in soil adjacent to coal banks, sporophores of other fungi did not appear on the banks. Examinations of the mycorrhizae led Schramm to the conclusion that the mycelial colonization was limited to those few species of Basidiomycetes that were in the environs. Others did not result in successful mycorrhizal associations. I spent several days with Dr. Schramm on those coal banks and was impressed with the extensive colonization of the substrate by the hyphae associated with the mycorrhizae. The yellow hyphae of *Pisolithus* were excellent for tracing through the black substrate. We collected samples of roots and the coal substrate to determine how populated the rhizosphere was by other microorganisms. With the cooperation of Dr. J. D. Menzies of the USDA at Beltsville, we found that a species of *Aspergillus* appeared consistently with all root and substrate samples almost to the exclusion of other microorganisms. The bacterial count was very low. One wonders whether the *Aspergillus* contributes to the release of ions from the substrate and thereafter to their chelation. We are continuing to examine other intricacies of the mycorrhizal association in this rather unique environment. It appears that this is one substrate that would limit colonization by certain species of Basidiomycetes regardless of the presence of appropriate hosts.

In the evolutionary development of the mycorrhizal Basidiomycetes and host plants, very likely the fungus and phanerogam flora capable of flourishing in volcanic ash would indeed have been different from that in a soil derived from recession of the seas or from weathering of parent rock materials such as in the Piedmont of the eastern United States. Those of us who live in the fall-line area are very aware of the change in vegetation over a few yards from the Coastal Plain to the Piedmont. Obviously some of the fruiting mycorrhizal fungus flora changes along with certain host species. We do not know how many other microorganisms that are under the influence of the root zones of trees are influenced by edaphic environmental changes. These physiological interactions particularly in forest soils certainly must affect the ectotrophic mycorrhizal complement.

S. A. Wilde (1968) recently expressed a broad concept of symbiotrophy in forest stands. He considers the phenomenon to include the complex biotic association of the rhizosphere within the symbiotic process. Apparently, he would not place as much emphasis on the actual penetration of roots by mycorrhizal fungi that others might. Thus, the actual existence of many mycorrhizal fungi might depend upon the phanerogam for sources of carbon and upon soil microorganisms, and perhaps roots, for solubilization and chelation of essential mineral elements. No doubt, the interaction is considerable in the soil but I am not aware of mycorrhizal Basidiomycetes developing sporophores in nature without actual penetration of the roots of the hosts. However, this brings up several other questions. Why do the mycorrhizal fungi not fruit readily in either pure culture or in axenic culture with mycorrhizal hosts? Is there a dependence upon one or more other microorganisms for completion of the life cycle of the mycorrhizal Basidiomycetes in at least some instances? What are the physiological and environmental conditions that are responsible for formation of a basidiocarp? Again, I should like to cite greenhouse and laboratory experiments that shed a little light on the matter.

Rarely has a mycorrhizal Basidiomycete been found to produce a basidiocarp and spores in axenic culture with a host. Even more rare is sporulation in pure culture. Generally we must conclude that the mycorrhizal Basidiomycetes do not consistently complete their life cycles under laboratory conditions when the substrate is free of organisms other than the symbionts. Indeed, no assurance exists that they regularly complete them in nature. This may be due to improper amounts or balances of nutrients, temperature, light, moisture, or combinations of these. Perhaps other microorganisms are more intimately involved than we now suspect. In experiments on synthesis of endotrophic mycorrhizae, Mosse (1962) found that a third member, a bacterium, was essential for establishing mycorrhizae composed of *Endogone* and a host. Again, I should like to cite some of our experiences with *T. terrestris* as a mycorrhizal fungus.

*T. terrestris* becomes established readily in the greenhouse on roots of *Pinus virginiana* and other coniferous species as a mycorrhizal associate. It very often forms sporophores on the stems of seedlings, on the substrate surface, and on the sides of flats or pots. There are always the mycelial connections between the mycorrhizae and the sporophore. In controlled experiments (Hacskaylo, 1965) we demonstrated that photosynthesis of the seedling is essential to sporophore formation of *Thelephora.* By decapitating the seedling or by covering the needles with a black bag we immediately arrested sporophore development. Later we observed that when a light-tight cover was placed over the substrate surface permitting only the needles to be exposed to light sporophores did not develop. When the pine seedlings were exposed to photoperiods of 8 hours sporophores did not form; however, on 16-hour photoperiods, they did. Also, sporophore formation was greatly influenced by variations in the levels of nitrogen and phosphorus in the medium. Details of these and other experiments will be reported elsewhere, but the development of sporophores by this mycorrhizal fungus is obviously regulated by more factors than temperature, moisture, and nutrients in the soil. Even though we performed duplicate experiments in open pot and axenic culture we have as yet to see sporulation of *T. terrestris* under axenic conditions with pine. Mycorrhizal formation is prolific and the mycelium permeates the substrate but a hymenial surface does not develop. Marx (1966) and Reid (1967) also have reported the absence of sporophores in axenic culture of *T. terrestris* and species of *Pinus.* We are continuing to study the complexity of this relationship because *T. terrestris* is the only mycorrhizal Basidiomycete for which the whole life cycle can be partially controlled within the greenhouse. There must be other fungi equally as controllable, but as yet we do not know their identity. A possibility might be *Laccaria laccata,* which frequently appears in the greenhouse on seedlings of pine and spruce.

We came upon a rather interesting aspect of fruiting by *T. terrestris* that might interest those working on its taxonomy. It shows a distinct dimorphism caused by certain as yet unknown factors.

In June, 1967, while growing seedlings of *P. virginiana* in a strictly mineral nutrient solution applied to open pots of perlite in the greenhouse, we noted that *T. terrestris* would form appressed brown patches on the surface of the perlite. We had seen these patches develop previously into characteristic frondose sporophores. Sporulation was normal. Dr. Orson K. Miller obtained spore drops from some of that material and later found that the spores germinated. In January, 1968, we established additional pots of the pine seedlings for studies on the effects of photoperiod on the mycorrhizal association and on sporulation of the fungus. As mentioned before, there was no indication of sporophore formation on the 8-hour photoperiod. In February, however, brown patches appeared on the substrate surface on the 16-hour photoperiod. They increased in size, but the characteristic frondose sporophore did not develop. When we examined the brown patches, we found that they actually were resupinate hymenial surfaces that produced the characteristic basidia and basidiospores. Later, in May, although still on a 16-hour photoperiod, the fungus began to form the characteristic collars on the stems of the seedlings and to produce fertile frondose sporophores. Thus, it was obvious that sporulation of the organism was partially regulated by its association with the host plant and environmental conditions, but the manifestation of the type of sporophore depends upon a factor or factors that we have not as yet determined. The possibility exists, of course, that a mixture of two or more species of *Thelephora* was present in the pots, but I consider this possibility minimal. Definite proof might be achieved by use of pure culture techniques, but up to the present time, as mentioned previously, fruiting has not occurred in pure culture. I presume that (*a*) it would be very possible for such an organism to receive two names, depending upon when it was collected, and (*b*) at some periods both types of fructification could be obtained in a single area. I am inclined to think that in the mycorrhizal relationship the major factor for completion of the life cycle of a symbiont is the intimate association between mycelium of the Basidiomycete and the roots of the host. Specific influences of the other organisms in the rhizosphere must be examined much more closely before their role is clearly defined.

The symbiotrophic status of mycorrhizal Basidiomycetes does not necessarily simplify the understanding of phylogenetic relationships within the fungi. I have cited examples of the influence of some specific environmental and physiological factors on the distribution and sporulation of mycorrhizal Basidiomycetes. Yet under one specialized ecological situation on the anthracite coal banks there were representatives of genera that are widely separated phylogenetically: *Amanita, Boletus, Thelephora, Pisolithus*, and *Astraeus*. Perhaps there are specialists who could point out close relationships between these organisms on a morphological basis. That is out of my realm and I shall not attempt to do it. However, on a strictly physiological basis it is apparent that there are certain common denominators among these fungi. Assuming that spores from dozens of species of mycorrhizal fungi are distributed over the coal banks from adjacent forests, only a few species are able initially to colonize the roots of pine seedlings. The requirements for temperature, moisture, and nutrients are adequate at least for the spores to germinate and for the hyphae to become associated with the tree roots. In the cases of some hosts, if not all, the system is eventually balanced so that the fungus is capable of forming a sporocarp and sexually reproducing. The ultimate in parasitism is attained. If, however, there is interference with any of the essential environmental or physiological factors, there would be total or partial inhibition of the symbiotrophic relationship to the possible demise of both organisms. It would, indeed, be convenient if we could relate the common physiological characteristics in such a manner that phylogenetic relationships of these few fungi could be correlated. At least we can say that morphologically they are all in the same class and subclass. At this point I wonder how much immediate assistance the mycorrhizal association will be to the taxonomic mycologist who is attempting to study the phylogenetic relationships of these fungi.

In reviewing the evidence that is now available we can draw a few conclusions regarding the mycorrhizal associations among the Basidiomycetes. The earliest mode of existence for the fungi must have been saprophytic in prehistoric times. The transition to reciprocal parasitism on roots took place in the development of gymnosperms, and during the evolution of certain angiosperms continued the adaptation in a balanced parasitism, particularly with species of trees. The concentration of the mycorrhizal potential within genera would indicate that physiological similarities substantiate their classification through morphological similarities. In other widely separate, morphologically different genera, the evidence seems to indicate that independent development of similar physiological and biochemical capabilities permitted transition from the saprophytic to an ecologically parasitic existence. It appears to me that for a considerable time the physiology of the higher Basidiomycetes will be of assistance more in verifying the affinities of discrete morphologically similar groups rather than in unifying the classification of all mycorrhizal Basidiomycetes.

## References

BENSON, L. 1957. Plant classification. D. C. Heath and Co., Boston. 688 p.

BJÖRKMAN, E. 1942. Über die bedingungen der Mykorrhizabildung bei Kiefer und Fichte. Symb. Bot. Upsaliensis 6: 1–190.

DUTTA, S. K., *et al.* 1967. Relatedness among species of fungi and higher plants measured by DNA hybridization and base ratios. Genetics 57(3): 719–727.

GARRETT, S. D. 1950. Ecology of the root inhabiting fungi. Biol. Rev. 25: 220–254.

———. 1956. Biology of root-infecting fungi. Cambridge Univ. Press. 293 p.

HACSKAYLO, E. 1965. *Thelephora terrestris* and mycorrhizae of Virginia pine. For. Sci. 11: 401–404.

MARX, D. H. 1966. The role of ectotrophic mycorrhizal fungi in the resistance of pine roots to infection by *Phytophthora cinnamomi* Rands. Ph.D. Diss. N. C. State Univ. 179 p.

MELIN, E. 1953. Physiology of mycorrhizal relations in plants. Ann. Rev. Plant. Physiol. 4: 325–346.

MOSSE, B. 1962. The establishment of vesicular-arbuscular mycorrhiza under aseptic conditions. J. Gen. Microbiol. 27: 509–520.

———. 1963. Vesicular-arbuscular mycorrhiza: an extreme form of fungal adaptation, pp. 146–170. *In* P. S. Nutman and Barbara Mosse, eds., Symposia of the Society for General Microbiology, No. 13, Symbiotic Associations. Cambridge Univ. Press.

NORKRANS, B. 1950. Studies in growth and cellulolytic enzymes of *Tricholoma*. Symb. Bot. Uppsaliensis 11: 1–126.

REID, C. P. P. 1967. Nutrient transfer by mycorrhizae. Ph.D. Diss. Duke Univ. 167 p.

RITTER, G. 1964. Vergleichende Untersuchungen über die Bildung von Ektoenzymen durch Mykorrhizapilze. Zeitschr. f. Allg. Mikrobiol. 4: 295–312.

SCHRAMM, J. R. 1966. Plant colonization studies on black wastes from anthracite mining in Pennsylvania. Trans. Am. Phil. Soc. Vol. 56, Part 1, 194 p.

SINGER, R. 1962. The Agaricales in modern taxonomy. J. Cramer, Weinheim. 915 p.

SINGER, R., AND J. H. MORELLO. 1960. Ectotrophic forest tree mycorrhizae and forest communities. Ecology 41: 549–550.

SWAIN, T., ed. 1963. Chemical plant taxonomy. Academic Press, London. 543 p.

TRAPPE, J. M. 1962. Fungus associates of ectotrophic mycorrhizae. Bot. Rev. 28: 538–606.

VOZZO, J. A. 1968. Inoculation of pine with mycorrhizal fungi in Puerto Rico. Ph.D. Diss. George Wash. Univ. 85 p.
WILDE, S. A. 1968. Mycorrhizae and tree nutrition. Bioscience 18: 482–484.

## DISCUSSION

SINGER: Do you have any evidence to support the theory of Moser that the role of some of the soil-inhabiting fungi in mycorrhizae is the liberation of certain substances? Perhaps this makes it possible for the mycorrhizal fungus to produce those indolic substances that provoke the morphological formation of mycorrhizal roots.

HACSKAYLO: This is still an open question. Perhaps indolic substances are being secreted by all ectotrophic mycorrhizal fungi resulting in characteristic morphological manifestations. It is not, however, something that is constant in mycorrhizal fungi as far as I know. The initial classic studies were on the effects of growth, regulatory compounds—indole acetic acid and others—on mycorrhizal roots. Morphological changes occurred in pine and other species. Many of us have avoided the subject, deferring to earlier investigators, but there is still no really clear answer. All this is going to require more study.

HANKS: Is it not possible that the fungi that are found in association with trees are there because of exudates from the trees and not from any mycorrhizal association? The fungus might be associated because of the presence of a certain carbohydrate which may be given off by the tree into the soil.

HACSKAYLO: I think all of us are aware of interdependence on such chemicals in the rhizosphere. The rhizosphere for each tree species is probably different. Therefore, promotion of growth and reproduction of many fungi occurs in very specialized conditions within the rhizosphere. Mycorrhizal fungi obviously somehow have to come in contact with the root. This may initially occur through spore germination followed by germ-tube penetration of the root. The process may be stimulated by root secretions of the host plant or by compounds present in the host. But one should consider the very distinct morphology of the fungus around the root and its regulation within a certain few tissues of the host plant. For example, the mantle formation amazes me. On the root surface a very characteristic structure forms and for each species of fungus it is fairly consistent. This is something the fungus doesn't do without the root association. If one places a stick in the soil and the fungus grows on it, a mantle does not form. Unquestionably, the fungus is being influenced by the secretions of the host plant, but the mode of penetration is unique to mycorrhizal fungi. I am convinced that the situation you describe, and which has been characterized as "peritrophic," also exists but interaction between the two organisms is not well defined.

JESSOP: Is it known what the fungus contributes to the roots of the trees?

HACSKAYLO: There is quite a bit of evidence on the translocation of the mineral elements from the soil. There have been very fine basic studies particularly in Sweden and England on that.

HESLER: What is the host range of *Pisolithus*? It is very abundant under a pin oak tree in my yard. Is oak the only symbiont that you know?

HACSKAYLO: We find it primarily with pine, and Dr. Singer reports it with *Nothofagus*, but I would not discount its capability of being a mycorrhizal fungus with oak. It is apparently a very early colonizer in some rather difficult situations. I know that as a result of inoculations with soil that was transported to Puerto Rico and contained a mixture of organisms, one of the first fungi to fruit in the nursery was *Thelephora terrestris*, and in the field it was followed by *Pisolithus*.

REID: Could you tell me about how long it is between the time that a mycorrhizal association is established by *Pisolithus* before it will produce fructifications?

HACSKAYLO: I cannot tell you that. Schramm followed this year by year, and I am sure he reported it. I know it did not fruit in our nurseries, although *Thelephora* appeared very readily. Incidentally, if you transplant a seedling with associated *Thelephora*, the fruiting body stops growing up the stem because the mycorrhizal mycelium is dissociated from the roots. In the old days they used to say the way to stop this smothering disease was to transplant the pines.

REID: In Britain *Pisolithus* is extremely rare, but last year we found it in two localities. Both here and in southern Europe, where I have also collected it, the fungus was growing in very disturbed soil, where roads had been put through, or where the top-soil had been cleared. These sites were often in very open areas away from trees.

HACSKAYLO: We also have found it fruiting along road cuts or on barren soil and in strip mine areas. In Puerto Rico I don't know what permitted it to complete its life cycle in the new habitat, but it did.

NOBLES: You spoke about a fungus probably moving from a saprophytic habit to an ectotrophic association. Do fungi ever move from an ectotrophic association to a parasitic or pathological one?

HACSKAYLO: At this point I do not believe that these fungi are pathogens. If they are pathogens they are nonmycorrhizal by definition. Furthermore, all of us who have worked since Melin, who developed pure culture techniques around 1920, have seen thousands of instances where fungi have been associated only with the roots of the higher plant in flasks under axenic conditions. If they were pathogenic they had all the opportunity in the world to show it, but they did not. If we do something which tends to weaken the system, however, then the fungus seems to fall away. For instance, if we cut off the photosynthate from the roots, the fungus appears to give up. That is how dependent it is.

BOTH: You mentioned when the host ceased its function, "the fungus falls off." What happens to it then?

HACSKAYLO: In Björkman's experiments the fungus was very well established and formed an almost complete sheet of mycelia around the periphery of the root ball. When he girdled the plant, strangled it with a wire, or covered it and reduced photosynthesis, the hyphae died and disintegrated. The fungi apparently don't continue their existence very long in mycelial form. In well-established trees, if death is not from causes which deplete the root system and kill it, the chances are the association will continue for a while until the stored carbohydrates in the root system are utilized. This can be perpetuated, as we have seen in the American chestnut. The old stumps and roots continue to send up new branches until all the stored food is depleted. This could not happen with the stumps if the root system died, but only if the root system is living. The time period would vary according to the tree species.

FERGUS: There is a tremendous amount of root grafting, particularly the *Quercus*, so even if a tree had been dead for some years, the root system could still live by being supplemented with a good nutrient supply from the surrounding trees.

GILBERTSON: Is it conceivable that a fungus might be mycorrhizal with one plant and parasitic on another species or genus? I am thinking of *Armillaria mellea*, which has been reported as mycorrhizal with orchids.

HACSKAYLO: It does not switch from a symbiont to a parasitic organism on the same plant. With orchids we are dealing not with ectotrophic mycorrhizae, but an endotrophic situation. The mycelium penetrates to the interior of the cell. Secondly, the fungi which are associated with orchid seeds and their germination phenomena can be pathogens. There are pathogenic species of *Rhizoctonia* on crop plants which will stimulate orchid seed germination in axenic cultures. The seed won't germinate without the presence of the fungus unless the medium has been supplied with certain carbon sources. Apparently, there are enzymes that the fungus contributes for the transformation of oils, etc., into utilizable carbon sources. *Armillariella* has been reported as an endophyte for orchids, but this is a little different from the regular mycorrhizal system. I like Harley's descriptions in which he refers to "ectotrophic mycorrhizae," "endotrophic mycorrhizae," and "orchid endophytes." I think there are specialized things in rhizomes.

HEIM: I would like to mention the deciduous tropical forest, where there is a tendency for ubiquitous European groups of fungi to become mycorrhizal and lignicolous and often parasitic. Among these are *Russula*, *Lactarius*, *Boletus*, *Boletinus*, and *Strobilomyces*. This seems to be a trend.

SINGER: I think we have several phenomena to discuss. I do not think that the genus *Russula*, if we include the tropical species, is entirely ectotrophically mycorrhizal. That is, I agree perfectly with Dr. Heim's saying that there are tropical species in the Russulaceae that are nonmycorrhizal according to the ecological situation. However, there is also a phenomenon described in the paper of Singer and Morello (Ecology, 1960) that Dr. Hacskaylo quoted before. We talked about "cicatrizer-mycorrhizae" in the tropics and we have followed this up further. Our system was to find the fungus in many places in a given area, make circles with a diameter more-or-less correspondent to the extension of the roots and single out the tree that could be with all of these carpophores but eliminate those trees that were not repeated in other circles. This is one of several methods we used at the same time. When we had the tree or trees identified, we dug it out and anatomically observed if the ectotrophic mycelium was there. Then, if possible, we tried to synthesize the ectotroph-association. On such occasions, I have seen tropical cicatrizing mycelia but I have not been able to make the synthesis so far. From the morphological evidence, however, I am satisfied that this relationship exists. This is work which is much more difficult than in any temperate zones because many more species are concerned, and we do not know which species are mycorrhizal. In one case we found a *Russula* species very closely related to those Dr. Heim reported from Madagascar. It forms ectotrophic mycelia only under circumstances in which the original combination of the forest has been destroyed either by fire, road construction, or agricultural activity. In the repopulation of these areas, some pioneer trees arise. So we have in *Russula* not only cases of nonmycorrhizal species, but also species that have the characteristics of cicatrizer mycelia of the tropics where the forest has been destroyed.

S. DICK: What effects are there from extracts of the host on axenic cultures of *Thelephora*, for instance?

HACSKAYLO: I cannot speak for *Thelephora*, but effects on other fungi have been reported. I think I should refer to the work of Melin on root exudates and the stimulation of spore germination, and the relation of spore germination to mycelial growth. I did some work with him on this and some of the very slow-growing fungi were remarkably stimulated by the presence of the living root. In rare instances fruiting occurred, but not as a reproducible phenomenon.

LARSEN: Would you care to elaborate on just what conditions you used to germinate the spores of *T. terrestris*?

HACSKAYLO: Orson Miller got a piece of hymenium and the spores were shed in Hagem agar tubes. He wrapped them up in foil and put them in the refrigerator. Several months later he took them out and the spores had germinated. That is all we know at the present time.

HAINES: Is there any evidence that an organism like *T. terrestris* can form its mycelial association with one plant and yet form its fruiting body on another plant?

HACSKAYLO: Yes, it does this commonly. We have it crawl up sticks, pot labels, and other objects.

VOZZO: The findings on *Populus deltoides* are seemingly anomalous. You and Schramm have found it ectotrophic in the wastelands and coal deposits of Pennsylvania, but we found it to be endotrophic in the southern United States. Are there two different mycorrhizal fungi involved? Would you speculate on a comparative physiology which might lend itself to this adaphic condition?

HACSKAYLO: An ectotrophic *P. deltoides* has been reported, but I haven't seen it personally. We did not collect *P. deltoides* in Pennsylvania, but only *P. grandidentata* and *P. tremuloides*, both of which are ectotrophic. Right now, we cannot formulate a hard and fast rule for *Populus*. Some species may have largely endotrophic associations, as *P. deltoides* apparently does. Vozzo found this and we have reports of it from Illinois. Our observations seem to indicate an endotrophic association. Perhaps the absence of ectotrophic fungi in prairie areas permits the endotrophic association. *Populus* in general is ectotrophic, and at the moment we must just let it rest there.

POMERLEAU: What is known about the morels? Are they associated, let's say, with *Populus balsamifera*?

HACSKAYLO: The best thing I know about morels is that they are good to eat. Many people have suggested that morels are mycorrhizal, but they have been reported as fruiting in many different situations. No one has really done anything to determine what their specificity is. I know some people like to collect them in old apple orchards, but apple trees have endotrophic mycorrhizae.

SINGER: I remember that morels have been brought to fructification without the interference of any other plant so it rather indicates that they are not mycorrhizal.

HEIM: In France, there is a great diversity of mycorrhizal relations between trees and morels. A great majority of morels have some mycorrhizal relation, one group with *Populus*, some with *Fraxinus*, and others with *Pinus* and *Abies*, and so on. There are some morels that are not mycorrhizal; for example, *Morchella spongida* var. *dunense*, which grows in sand only, with no relationships. *Morchella intermedia* is not associated with the same tree in Kashmir as in France; both are conifers, but not the same.

MOORE: Would you comment briefly on *Monotropa*?

HACSKAYLO: It is possible to dig out the root system of *Monotropa* and see that it is a very corralloid, clustered system covered with a sheath of hyphae. It has been known that Basidiomycetes were involved in the hyphal covering and that a mycorrhizal system exists. Clamp connections on the hyphae have been seen, and some of the boletes have been identified with it. You never find *Monotropa* except in association with trees, however, and it is only found under trees which have ectotrophic mycorrhizae. Björkman in Sweden thought that perhaps carbohydrates were transferred from the tree roots by way of a mycelium bridge into the *Monotropa*. He used isotopes and found, indeed, that the same fungus that forms mycorrhizae with the *Monotropa* forms mycorrhizae with oak or pine or whatever it is under. There is a mycelial bridge. If the radioactive phosphorus is injected into pine, it is translocated to the roots and into *Monotropa* by way of the mycelial bridge. Tagged carbohydrates as well as phosphorus are passed across. At the present time, this appears to be the answer.

REID: Only a fortnight before I came over here, I observed *Monotropa hypopitys*, which is found with beech (*Fagus sylvaticus*), fruiting in the most colossal series of "fairy rings" about 20 or 30 feet in diameter in vast quantities. These great rings were completely analogous to the fruiting of some of the mycorrhizal fungi which do the same thing. The tree trunks were at the center of the circles.

HACSKAYLO: Mycorrhizal mycelia are usually found on the root periphery on the finest roots, and that is where you would expect sporophore production. It is the area where one would expect bridges to *Monotropa*. The example you mention must be a fairly old association with the *Monotropa* because the plants aren't in isolated groups scattered around, but in perfect circles, enlarging with the root system.

GRAND: What is your feeling on the mechanism proposed by Marks and Foster (1966)[1] for invasion of the cortical region of the roots of *Pinus radiata*? They postulate that this invasion may actually be a forceful breaking apart of the cortical cells and support this with electron micrographs.

[1] Foster, R. C. and G. C. Marks. 1966. The fine structure of the mycorrhizas of *Pinus radiata* D. Don. Australian J. Biol. Sci. 19:1027–38.

HACSKAYLO: They probably have more evidence on that than I do, but we are inclined to think, until we know more about it, about the enzymatic dissolution of the middle lamella. But they seem to have come up with good electron micrographs to help clarify what is really happening.

KIMBROUGH: Has anyone looked at cycads for mycorrhizae?

HACSKAYLO: I think there have been some reports of the endotrophic type of association, but none of ectotrophic, as far as I know.

ROGERS: I believe cycads have blue-green algae associated with them, though.

POMERLEAU: Is it known that *Amanita caesaria* grows associated with oak? We have it in some places where there is no oak but *Abies*.

HACSKAYLO: It has been reported as associated with oak, but there are several species which do cross over between the gymnosperms and angiosperms as mycorrhizal fungi. They are not all as specific as some people would like them to be.

KIMBROUGH: I have noticed in Florida that Murrill tended to link certain species of higher fungi (i.e., boletes) with certain pines and certain oaks. In referring to the diagnoses of a number of rusts I have noticed that there is also an alternation between those pine and oak groups. I wonder if anyone has tried to correlate these to see if there is a common selective force operating in these tree groups for rusts, boletes, or agarics.

HACSKAYLO: Just from what you say, I think you are the one who has made the most observations on this point so far. This sounds very nice, but I know of no one who has really tried to correlate it.

PETERSEN: Is it possible that the morphological entity we pick up and classify as the fruit body of a fungus might not have been produced before the mycorrhizal association was established? If this is true, and has been forever true, then this fungus did not fruit before the tree species existed. Is this a possible way of setting up a time limit before which the fruit body of the species as we know it could not have existed? Could this be construed as an indirect "fossil" record?

HACSKAYLO: At the present time I doubt that you will find the sporophore of *T. terrestris* without a mycorrhizal association. I am not saying that its precursor necessarily was mycorrhizal, but somewhere along the line it became dependent upon the tree and you could say it might have been the initial point of its development as a species. I think without question that this evolution did take place. If it did, then much of it must have taken place in association with trees. Where it started and what the organism looked like at that time, I don't know, but mycorrhizal evolution certainly has been necessary. This may well have been started with several different groups of fungi.

VOZZO: How much weight can we place on reports of mycelial symbionts which are not substantiated or supplemented with axenic or pure culture syntheses?

HACSKAYLO: I am a bit of a purist and conservative in this area, so I usually accept these reports with a little reservation. In some instances I think many of them are distinctly probable, but in other cases I think it is possible that fungi are being reported as mycorrhiza-formers which may really be finding their livelihood in the organic matter in the soil and not necessarily connected with roots. The specificity of their association depends on certain ecological situations dictated by the chemical composition of the litter. Just because we find a certain species of fungus under pine or oak or larch, it is not necessarily a mycorrhizal fungus. I would like to see every fungus we designate as mycorrhizal verified under axenic culture conditions.

# 3

# SYSTEMATIC STUDIES OF FUNGUS GROUPS

DONALD P. ROGERS
*Department of Botany*
*University of Illinois, Urbana, Illinois*

# PATTERNS OF EVOLUTION TO THE HOMOBASIDIUM

Of the multitude of proposed generalizations concerning the science of taxonomy, two appear to be of outstanding importance and nearly universal application. The first is that in principle all characters are of equal value. It has been both the support of those who are concerned with limited groups of facts—cytotaxonomists, biochemical taxonomists, phytogeographical taxonomists, genetical taxonomists—and our first line of defense against them. It should also indicate to the taxonomist the size of his job. The second is that in practice some characters are of greater, even enormously greater, importance than others. This has always been evident to investigators concerned with major taxonomic groups—the Arthropoda rather than the genus *Drosophila*, the Basidiomycetes rather than the genus (let us say) *Phoma*. It is strikingly exemplified and confirmed by the punctilious and delightful excursion of Kendrick and Weresub (1966) into computer taxonomy as a possible source of wisdom concerning the lower Basidiomycetes.

Almost since the day it was first described (cf. Ramsbottom, 1941) the basidium has been this character of overriding importance in the taxonomy of the Basidiomycetes; and as, one after another, different types of basidia have been discovered and reported, these have become the bases of subclasses, orders, families, and lesser groups, according as they differed more or less in structure or interpretation. When basidial type is regularly correlated with other characters—when, for example, a septate basidum is borne on a gelatinous basidiocarp and produces spores that are capable of germination by repetition—this is not worth special comment. The critical cases are those where basidial type alone is the basis of the proposed classification. Take the genus *Aporpium* (Teixeira and Rogers, 1955). Although one familiar with *Aporpium caryae* can sometimes recognize the species with the unaided eye, in all respects but one it is an entirely undistinguished whitish resupinate pore-fungus, a *Poria.* But its basidia are longitudinally septate, conforming to the type characterizing the family Tremellaceae. Therefore all—or, rather, nearly all—students of the Basidiomycetes separate it from *Poria* by the highest fence available and consider it a member of the subclass Heterobasidiomycetidae, where it is unto the users of keys a stumbling-block, and unto computer-taxonomists foolishness.

Leaving aside intuition, which is a fruitful source of ideas to be investigated but a shaky support for published conclusions, there are two possible reasons for giving great weight to any character. First, it may regularly be correlated with one or several others. If they are independent, like notochord and gill-slits in the chordates, the weighting is greater than if they are dependent, like slender basidia and longitudinal meiotic spindles in the Basidiomycetes. Another way to say this is that the probable importance of a character is less the wider its distribution. Feathers occur in only one group of animals, all of whose members possess them; epidermal hairs occur in nearly all groups of plants that have an epidermis, and their presence is notoriously a matter of minor consequence.

Or a character may be given great weight for a second reason, the phylogenetic hypothesis that best explains its nature or its existence. All scientific generalizations—and any taxonomic treatment is one—are hypotheses, probable or less probable, confirmed or unconfirmed (cf. Rogers, 1958). Early mycologists, Fries, Persoon, and others of their time, worked on the hypothesis that the configuration of the hymenium was a guide to relationship. What relationship meant to Fries is not easy to understand, even though there is an exposition of the matter in more than one of his books. But in any case, fructifications having a poroid hymenium were to him related, those with teeth were related, club-shaped and coralloid fungi (that is, fructifications) were related, and so on. Later mycologists learned that many of these fruiting bodies bore their spores on basidia, and still later, these or others discovered that the basidia conformed to a limited number of strikingly different types. This is the evidence that we are confronted with. Furthermore, whatever our uncertainties about our reverend fathers in mycology, most of us know what we mean by relationship. The hypothesis of evolution is that living things now existent are descended from living things formerly existent that were different from them. And relationship to us means descent from a common ancestral group. (Most of us, that is. I do not know what hypothesis of relationship is held by those who are unable to accept the evolutionary one.)

Now, Elias Fries recognized the genera *Polyporus*, *Hydnum*, *Clavaria*, and *Thelephora*, whose hymenia were respectively poroid, toothed, claviform, and not regularly sculptured. We know that in each of these some species have club-shaped, undivided basidia, and in each at least one species has more-or-less isodiametric, longitudinally septate basidia. We can suppose, as have many who accepted as probable the evolutionary hypothesis, that those Basidiomycetes having a similar hymenial configuration had a common ancestor—in which case similar divergence in basidial type had occurred within each of the four, and the two basidial types were the result of convergence. Or else each group having a common basidial type had a common ancestor, and the similarities in hymenial configuration were the result of convergence. How do we choose? We know that a basidiocarp is fundamentally simple in structure, a complex of branching threads interwoven in various ways and subsequently more-or-less differentiated; we know that this complex is readily modifiable by circumstances. We know that although the proportions of the basidia are varied within the several types, their fundamental morphology is not. And we know also that other characters, such as germination by repetition, are usually associated with the one group of basidial types and never with the other. The hypothesis that fungi having a common basidial type are more nearly related—more recently descended from a common ancestor—than those, say, with toothed hymenia, appears the more probable hypothesis, since it requires fewer subsidiary hypotheses than the alternative. Furthermore, each time we find a known basidial type in a new group—that is, one whose basidiocarp morphology differs from those already known to possess such basidia—the hypothesis of kinship of fungi with a common basidial type is strengthened.

The basidial type is then of greater, even enormously greater, importance than other characters in determining evolutionary relationships, in developing a natural classification, of the Basidiomycetes. Whether or not we avail ourselves of the services of machines in classifying or identifying fungi, the basidium is to be given what amounts to an infinite weighting. After that come all the other characters, which in principle—that means, until we have studied and evaluated them—are of equal value. Constellations of other characters have often proved of great importance in indicating relationships, or to put it another way, have given great support to hypotheses of common descent. The characters of sterile hymenial organs and basidiocarp development by which Pilát (1926) defined his Aleurodiscineae are one example that comes to mind. The characters, progressively altered from group to group, that are shown by Malençon's (1931) Asterosporae are another; and the Phylacteriaceae of Patouillard (1900), the coniophoroid Hymenomycetes, and the *Urnigera* sections of polyporoid, hydnoid, and corticioid fungi are further examples. I am unable to recall any single character except for the basidium that has overriding importance in determining major natural groups among Basidiomycetes. Probably all Hymenomycetes with setae are related; but setae (referring to such organs as occur among the basidia of *Hymenochaete*) are always accompanied by other characters of pigmentation, texture, and reaction to alkaline solutions, and, as a matter of fact, many members of the natural group of hymenochaetoid fungi lack setae. Clamp connections are notoriously unreliable as characters for anything larger than a species; similar and presumably closely related species may show them at all septa, or frequently, or rarely, or never, as in *Corticium* sect. *Pellicularia* of Bourdot and Galzin, or only, always, or never, as in *Pellicularia* of Cooke. Sterile hymenial organs, at least as we are at present accustomed to distinguish them, are as useless or misleading: such highly characteristic cystidia as are found in *Peniophora* sect. *Tubuliferae* crop up again in *Sebacina* sect. *Heterochaetella* among the Heterobasidiomycetidae, and gloeocystidia similar to those of *Corticium* and *Stereum* occur in at least three heterobasidiomycetous genera. And as for spore color—*Coniophora*, whose spores should be brown, includes one species in which they are a rich blue. The basidium is paramount.

An acceptable hypothesis of the evolution of the Basidiomycetes requires, in addition to an idea of what it is that first of all evolved (that is, the basidium), an idea of the probable origin of the class. The importance of this state from which evolution proceeded is well illustrated by the treatments of the Ascomycetes that have appeared. For most authors they originated from the higher Phycomycetes; the first Ascomycetes were Endomycetales, and the Pezizales lie near the summit of one of the chief phyletic lines. For Bessey the Ascomycetes are modified red algae, the Pezizales are primitive, and the Endomycetales are not primitively simple but simple because they are the end term of a long series of progressive simplifications. Students of the Basidiomycetes are much better off; for most of them the Ascomycetes are the Basidiomycetes' only possible ancestors.

The facts that are best accounted for by the hypothesis of kinship between Asco- and Basidiomycetes are so well known as to need only outlining. The cytological similarity, or near identity, between ascus and basidium is the first: both are organs originally dikaryotic, in which karyogamy, meiosis, and sometimes an additional mitotic division occur, and which then proceed to the formation of spores, a process in which most, but rarely all, of the cytoplasm is included within the spore-wall. The second is the similarity or identity between the extraordinary hyphal structures called in Ascomycetes croziers and in Basidiomycetes clamp connections (Rogers, 1936). The third is the cytological life cycles, in which the higher Ascomycetes and the great majority of Basidiomycetes agree with each other and differ from all other living things: a multicellular haploid thallus, a multicellular dikaryotic thallus, and a sporogenous cell briefly diploid. The fact that, as Martens (1946) has pointed out, there is still a gap in our knowledge of the ascomycete life cycle does not for the present purpose matter in the least. In a typical ascomycete, nuclei from two parental strains are associated in the ascogonium and are paired in the crozier and, with or without croziers, in the ascus; and what happens between, in the ascogenous hyphae, is something we should like to, but need not, know at this time. There are other important similarities—metabolism and metabolic processes, the development of sporocarps, septal pores; even without them, the three groups of facts here marked have, so far as I am aware, no explanation except that of phyletic relationship. Of course, evidence of kinship is not enough; it is necessary to know, of the Asco- as of the Basidiomycetes, the direction of evolution. The origin of the Ascomycetes from Phycomycetes seems well supported, as by Atkinson and Fitzpatrick (1930), and excludes the possibility of the derivation of Asco- from Basidiomycetes. Within the Ascomycetes the system of ascogenous hyphae is a developing but physiologically dependent phase, enclosed by the haploid portion of the ascocarp. Within the Basidiomycetes the homologous system, the secondary mycelium, is dominant and independent, and in several lines becomes finally the only mycelium, with the haploid condition limited to the late basidium and the spores. The shift in relative importance of the haploid and dikaryotic phases is thus a single process in the two classes and reaches its culmination in the Basidiomycetes. We will then take it that the Basidiomycetes are not only related to but derived from the Ascomycetes. This gives us a point of departure.

The subclasses Heterobasidiomycetidae and Homobasidiomycetidae, first distinguished by Patouillard in 1887, always deserved, and within the lifetimes of most of us have received, acceptance as the first basis for discussion and treatment of the larger group. So far as the basidia are concerned, the Homobasidiomycetes are what their name implies, relatively homogeneous. The homobasidium is an undivided cell, differing in proportions but not in fundamental pattern, bearing an apical crown of outward-curving, spike-like sterigmata on which the spores are borne and from which they are discharged. The chief exceptions to this description are apobasidia, basidia whose final function of spore discharge has been lost, and whose functionless sterigmata are narrowed, thickened, lengthened, shortened, shifted in position, or suppressed, without significant alteration of the basal cell. The basidia of the Heterobasidiomycetes are not so easily characterized. As the name implies, they are different—different from what could be regarded as the typical basidium, the homobasidium, and different from each other. In fact, it was nothing less than genius in Patouillard to have assigned these fungi to one group; his contemporary, Brefeld, failed to do so, and by his prestige delayed for many years the acceptance of Patouillard's work. First of all, the basidia of many Heterobasidiomycetes are pluricellular—phragmobasidia—divided in various ways by septa. But some are undivided. Second, the basidial primordium, the dikaryotic, and then diploid, cell which corresponds to the homobasidium before the latter develops sterigmata, in most Heterobasidiomycetes gives rise to one or more outgrowths, whereby the morphology of the basidium is further developed before sterigmata are formed. These elaborations of the basidium were called epibasidia by Neuhoff (1924), in the first attempt to interpret what Patouillard had only seen; after their formation the primordium from which they arose is called the hypobasidium. I have seen no reason to adopt an alternative terminology. In some heterobasidiomycetes the primordium continues the morphological development of the basidium while still dikaryotic or, more commonly, diploid; in these fungi there is one epibasidium. In others meiosis precedes further development; these have two or four, or rarely three or perhaps as many as eight, epibasidia.

The fungi forming epibasidia before meiosis need concern us only briefly. In nearly all of them the spores, sterigmata, or sterigma-bearing filaments are borne laterally; and in spite of Linder's ingenious attempt to derive the Homobasidiomycetes from the rusts (1940), it seems gratuitous, and it certainly entails a multiplication of hypotheses, to seek the ancestry of acrosporous basidia in pleurosporous forms. Basidia with a single epibasidium characterize the Auriculariaceae and the apparently derived Septobasidiaceae, Phleogenaceae, Uredinales, and Ustilaginales. There are several notable modifications of the generalized auriculariaceous type here: the Phleogenaceae are, so to speak, heterobasidial puff-balls, and the Ustilaginales also are apobasidial—which is to say that the spores are not discharged from the basidium. In all but one of the families of this alliance the epibasidium is transversely septate and usually four-celled; in the Tilletiaceae it is a holobasidium. In some Auriculariaceae, where the basidia are deeply embedded in gelatin, the epibasidial cells bear their spores on lateral filaments that reach the surface of the basidiocarp before producing sterigmata; in others the development of the basidium is continuous, so that no distinction can be seen in the mature basidium between the hypobasidium and epibasidium. And in four of the five single-epibasidial groups there occurs a highly specialized parasitism. But there is nothing pointing in the direction of the homobasidium.

In the fungi where meiosis occurs in the probasidium and where epibasidia are formed on a cell containing plural haploid nuclei, the most obviously and characteristically heterobasidial are the Tremellaceae. Here in a short, often spherical cell three septa arise, dividing the hypobasidium into usually four quarters; regularly in some species, however, and occasionally in others, there is only one vertical septum, and the basidium is two-celled. In the great majority of species each cell then produces its own epibasidium, a stout tubular structure, usually somewhat expanded near the tip, where it bears just such a sterigma as occurs in Homobasidiomycetes. The length of the epibasidium varies almost without limit according to the depth at which the hypobasidium is formed within the basidiocarp; deep-seated hypobasidia in thick, gelatinous fructifications such as those of *Tremella* may have epibasidia over 100 μ in length; in the thin films produced by some species of *Sebacina* they may be only a few microns long, and in young fructifications or those rare species which are scarcely even waxy the epibasidia apparently disappear and the sterigmata are, or at least seem, seated on the hypobasidium. Obvious derivatives of the Tremellaceae are the Hyaloriaceae, with vertically septate apobasidia, and the Sirobasidiaceae, whose catenulate hypobasidia have been shown by Bandoni (1957) to bear not sessile basidiospores but deciduous epibasidia.

The other phragmobasidial family is the Tulasnellaceae. Here basidial ontogeny is similar to that of the Tremellaceae, but the hypobasidium is not septate. Rather, the epibasidia arise as greatly inflated bodies, four or more in number, when division of the diploid nucleus is scarcely completed. These epibasidia, which may be globose, fusiform, or claviform, have been likened to spores; and in fact their development is remarkably like that of spores, for each receives a nucleus and a share of the cytoplasm from the hypobasidium, each is then cut off from the parent cell, which is emptied and collapsed, and each produces a sterigma, or in more gelatinous fructifications a tubular filament and sterigma, on which is formed and from which is discharged a basidiospore. Where there are only four epibasidia the single nucleus, in those few species that have been studied cytologically, undergoes a postmeiotic division in the epibasidium; where a postmeiotic division has occurred in the hypobasidium and there are more than four epibasidia, no epibasidial mitosis has been observed. Once the epibasidia are separated from the hypobasidium they are physiologically independent; in gelatinous species of the Tulasnellaceae it is easy to find epibasidia wholly detached and still forming basidiospores normally.

The remaining families are holobasidiate, which is to say that their basidia are normally undivided, although under special circumstances, such as an abnormal abundance of water, septa may be formed occasionally and irregularly. The Dacrymycetaceae have basidia that resemble a tuning-fork—a club-shaped hypobasidium with two epibasidia, nearly as stout as the hypobasidium, arising at the periphery of the summit. All genera are somewhat gelatinous; the spores have their own peculiar variation of Patouillard's germination by repetition: septation, followed by production of numerous small conidia. Presumably, these characters led Patouillard to assign it to his Hétérobasidiés, where it assuredly belongs; the presence of well-distinguished epibasidia confirms it.

The remaining holobasidiate family, the Ceratobasidiaceae, was more recently recognized. Different species of *Ceratobasidium* have hypobasidia that are globose, more-or-less pyriform, nearly barrel-shaped (when the parent hypha is relatively stout), or very short claviform; the epibasidia are tubular, often somewhat distended, and two to four in number. Claviform hypobasidia with two epibasidia so closely resemble those of *Cerinomyces*, of the Dacrymycetaceae, which similarly is resupinate, that the relation of the families can scarcely be doubted; spore germination, however, is different. The Ceratobasidiaceae have germination by repetition, the production on a sterigma of a smaller secondary spore resembling the basidiospore, as do most Heterobasidiomycetes. Certain species of the genus *Ceratobasidium* had been described, before the genus or its heterobasidial character had been recognized, in the homobasidial genus *Corticium*, and these were assigned by Bourdot and Galzin to the section *Botryodea*, along with species of the genus *Pellicularia*. Although not all of them possess the characters of the section (this is notably true of *Ceratobasidium*, or *Corticium*, *cornigerum*), Bourdot's disposition of the species is entirely understandable, for there is an unbroken series of forms connecting species that are heterobasidiomycetous in all respects with loose, tufted, merely sterigma-bearing species that are unquestionable pellicularias and unquestionably homobasidiomycetous. The breaking up of this series into a number of genera has brought no advantage in the practical matter of finding names for specimens and has served only to obscure the series. It exists, nevertheless, and it is here that Homo- and Heterobasidiomycetes are united. The question over which there has been a certain amount of wrangling is the relatively inconsequential one of where to draw the line.

I am not aware of any other such bridge. It is adopted into their quite different phylogenies by both Olive (1957) and Linder. No one with any amount of either caution or judgment would assert that the very fungi before him in the twentieth century are the veritable ancestral species of nine-tenths of the class Basidiomycetes, which had representatives in the Pennsylvanian (Upper Carboniferous). On the other hand, Neuhoff has well remarked that the student of the Basidiomycetes has little need to construct hypothetical links, since he is in fact apt to be embarrassed by the abundance of existing intermediate forms. Here, then, so far as we now know, the two subclasses are joined.

Which subclass is ancestral, the Hetero- or Homobasidiomycetidae? Or, a question that should be antecedent, is there reason to suppose that the Basidiomycetes are in their origin monophyletic? I think that there is. The earlier stages in basidial ontogeny, through meiosis, are an inheritance from the Ascomycetes and could have been inherited any number of times. But the uniquely basidiomycetous structure and behavior, the curved sterigma, the asymmetrical basidiospore with its lateral apiculus, the formation of the water-drop (or, perhaps, bubble), and the ejection of the spore are so uniform among all Basidiomycetes except those obviously degenerate, and represent so complex a group of phenomena, that without direct evidence there is no justification for any supposition that they evolved more than once. Of the alternative hypotheses, that of monophyletic origin is the more probable. It is not necessary, however, to suppose that the sterigma and its spore arose *de novo*. Both spores and hyphae of Ascomycetes are known to produce conidia on what have been called sterigmata—simpler sterigmata, in that they are symmetrical and are not propulsive, and conidia simpler than basidiospores, in that they are not regularly asymmetrical nor do they possess lateral apiculi. They are analogous, and it is a part of the hypothesis now under discussion (Rogers, 1934) that they are homologous. With most of the basidium clearly homologous with the immature ascus and the sterigma and basidiospore possibly, or perhaps even probably, derived from one type of ascomycetous conidium and its support, one other structure is still left unaccounted for, the epibasidium.

If the homobasidium is the primitive type, it can be described as an ascus whose normal mode of spore production and dispersal has been suppressed (by the circumstances of its ontogeny) or lost (by genetic modification), which proceeds to the formation of spores on sterigmata, as, by hypothesis, its ascospores were able to do. With the elaboration of the means of spore ejection it has become a complete, functional sporophore. Then, according to this hypothesis, in its descendants the heterobasidia all the complexities of septation and of single or multiple epibasidia were interpolated between primordium and basidiospore, producing the several heterobasidial types as we know them.

Alternatively, the primitive basidial type is to be found among the heterobasidia. Here again, normal spore production by the ascus is suppressed or lost, perhaps both, just as spore discharge is suppressed and lost in hypogaeous Asco- and Basidiomycetes. In place, however, of forming normal but undischarged ascospores, the ascus continues its ontogeny from an earlier stage by forming evaginations of the ascus wall into which nuclei and cytoplasm migrate, or by septation, or by both. So far as I am aware no formation of evaginations from an ascus has been reported; that such a development is neither inconceivable nor improbable is shown by the well-known formation of merosporangia in the higher Mucorales. Transverse septation and subsequent fragmentation of asci have been reported and illustrated. Depending on the cytological state of the developing—or, one might say, germinating—primordium, one outgrowth, or four, or more than four, could be formed. These are forerunners of the epibasidia; each, or each segment, encloses the homologue of an ascospore protoplast, surrounded by an extension of the primordial wall as are the cells or segments of a merosporangium, and each germinates by a sterigma (ascomycete style) and conidium. Under this hypothesis the epibasidium is the organ of further development of an otherwise abortive ascus; the basidiospore is homologous not with the ascospore but with its germ conidium. The epibasidium is thus explained—that is, it finds its place within a more general hypothesis. The homobasidium is explained and its phylogeny actually exemplified by the *Ceratobasidium*—*Pellicularia* series already discussed. And if all this is probable, or even more probable than the alternative, the Heterobasidiomycetidae are to be interpreted as more primitive and the Homobasidiomycetidae as derived.

The relations among the Heterobasidiomycetes are still to be explored, and particularly the question of the primitive type among them. Whether or not the Auriculariaceae are closely related to the Tremellaceae through the fascinating and enigmatic genus *Patouillardina*, as Martin (1945) has suggested, unless we are prepared to suppose that the propulsive sterigmata and peculiar spores of the basidium arose more than once in the same form we must hypothesize a primitive type, not primitive types. I see this type best exemplified in *Tulasnella*, where the homologues of ascospores, the epibasidia, are most sporelike, most independent, best adapted to serve as organs of dissemination. (This is not to overlook *Sirobasidium*, whose catenulate hypobasidia, however, seem comparatively specialized.) It has been suggested by a number of authors, including the present one, that the *Ceratobasidium* type arose from that characterizing *Tulasnella* by the omission of the cross-walls by which *Tulasnella* epibasidia are sealed off from the hypobasidium. And this gives, as a phyletic series and the pattern of evolution to the holobasidium, (*a*) ancestral ascomycete, (*b*) Tulasnellaceae, (*c*) Ceratobasidiaceae, (*d*) *Pellicularia*.

Olive (1957) has proposed to insert in a similar series his genus

*Metabourdotia*, characterized by basidia obscurely or imperfectly cruciate-septate, like those described by Crawford (1954) in *Tremellodendropsis*. Since just such basidia are observed in established species of *Bourdotia*, along with others whose septa are well defined, it is doubtful whether either the type or the genus deserves recognition. The possibility of evolution of the ceratobasidiaceous type from the tremellaceous was never excluded. But the kinship of *Tulasnella* and *Ceratobasidium* is, at a minimum, equally real; from considerations already given *Tulasnella* seems the most likely primitive heterobasidiomycete; evolution of the Ceratobasidiaceae through both Tulasnellaceae and Tremellaceae is an unwarranted complication; the simpler hypothesis is retained as the more likely.

Nearly thirty-five years have gone by since much of what has been said here was first set down. During all that time search has been carried on for additional basidial types or modifications, or for alternative hypotheses. But our eminent predecessors, Léveillé, Berkeley, the Tulasnes, Patouillard, did their work too well. The most significant modifications of existing types of basidia since discovered seem to be what Bandoni learned about *Sirobasidium* and the other odd tremellaceous derivative that was described as *Xenolachne*; no distinct types have appeared. As to hypotheses of the derivation of the simpler Basidiomycetes, most of us have been concerned with other things, nomenclature, or reasons for breaking up old genera and describing new ones. Few have had the temerity to attempt new evolutionary hypotheses, a point on which I share the feelings of the majority. An earlier paper (Rogers, 1934) ended with a quotation from Agnes Arber: "Scientific hypotheses have in their nature no pretension to permanence, and . . . should be judged by their capacity for bringing to light further generalizations, to which, in turn, they yield place." We are still pretty much where we were, but the hopeful tenor of that sentence is still warranted.

## References

Bandoni, R. J. 1957. The spores and basidia of *Sirobasidium*. Mycologia 49: 250–255.

Bourdot, H., and A. Galzin. 1928. Hyménomycètes de France. Paris. iv & 762 p.

Crawford, D. Alleyne. 1954. Studies on New Zealand Clavariaceae. I. Trans. R. Soc. N.Z. 82: 617–631.

FITZPATRICK, H. M. 1930. The lower fungi. Phycomycetes. New York. xi & 331 p.

KENDRICK, W. B., AND LUELLA K. WERESUB. 1966. Attempting neo-Adansonian computer taxonomy at the ordinal level in the Basidiomycetes. Syst. Zool. 15: 307–329.

LINDER, D. H. 1940. Evolution of the Basidiomycetes and its relation to the terminology of the basidium. Mycologia 32: 419–447.

MALENÇON, G. 1931. La série des Astérosporées. Trav. Crypt. déd. à L. Mangin 337–396.

MARTENS, P. 1946. Cycle de développement et sexualité des Ascomycètes. La Cellule 50: 123–310.

MARTIN, G. W. 1945. The classification of the Tremellales. Mycologia 37: 527–542.

NEUHOFF, W. 1924. Zytologie und systematische Stellung der Auriculariaceen und Tremellaceen. Bot. Arch. 8: 250–297.

OLIVE, L. S. 1957. Two new genera of the Ceratobasidiaceae and their phylogenetic significance. Am. Jour. Bot. 44: 429–435.

PATOUILLARD, N. 1887. Les Hyménomycètes d'Europe. Paris. xiii & 184 p.

———. 1900. Essai taxonomique sur les familles et les genres des Hyménomycètes. Lons-le-Saunier. 184 p.

PILÁT, A. 1926. Monographie der mitteleuropäischen Aleurodiscineen. Ann. Myc. 24: 203–230. pl. 25.

RAMSBOTTOM, J. 1941. The expanding knowledge of mycology since Linneaus. Proc. Linn. Soc. London 151: 280–367.

ROGERS, D. P. 1934. The basidium. Univ. Iowa Studies Nat. Hist. 16: 160–183.

———. 1936. Basidial proliferation through clamp-formation in a new *Sebacina*. Mycologia 28: 347–362.

———. 1958. The philosophy of taxonomy. Mycologia 50: 326–332.

TEIXEIRA, A. R., AND D. P. ROGERS. 1955. *Aporpium*, a polyporoid genus of the Tremellaceae. Mycologia 47: 408–415.

## DISCUSSION

DUBUVOY: The ascus develops from a crozier and the basidium develops from an end of a hypha. You say that the crozier is like a clamp connection. Do you really see an analogy between the ascus and the basidium?

ROGERS: I would not say that there was an analogy between the ascus and the basidium but the closest possible homology, and the matter of the origin of the beak of the crozier seems to me of very little importance and, in fact, is variable. I might refer you to a paper by Rogers on *Sebacina prolifera*, in which the clamp connection is as terminal in its origin as the crozier in *Pyronema*.

SINGER: Some of my questions have to do with the problem of whether we can put out a certain hierarchy of characters. Is one character in basidiomycetes always a character of the first order, and another one of second order, and the third one subservient?

Then I thought I heard a statement that all the basidiomycetes bearing setae are related. Do you think that genera like *Boletochaete* or certain species of *Marasmius* which have setae according to me are related? And if they do have setae are they related to setae-bearing Aphyllophorales?

ROGERS: I am not familiar with either of those genera. I was thinking, of course, of Aphyllophorales and the structures that are called setae in *Hymenochaete*, not the ones that are sometimes called setae in, let us say, *Coniophora*, or *Coniophorella*. That is why I tried to typify it.

GILBERTSON: I am curious about your interpretation of the basidial type in the *Laeticorticium* group. In some of the species of *Laeticorticium*, such as *L. roseum*, one that occurs on aspen, the basidia begin development in the fall of the year. The fruiting body develops thick-walled basidial forms which remain dormant until the following spring when they continue development and produce normal basidia. Do you think that this has any possible connection with a Heterobasidiomycete ancestor?

ROGERS: I do not see any connection with the Heterobasidiomycetes, but it seems to me the same general type of development in the basidiocarp which characterizes Dr. Pilát's Aleurodiscineae. As a matter of fact I put that species there and I think this development is one of a number of excellent characters for that group of fungi.

HEIM: How would you explain the phylogeny of the morphology of the basidium in the Agaricales? There are instances which look exactly like the form of *Tulasnella* with the basidial-conidia or epibasidia, exactly as in secondary conidia which are not very rare in some Agaricales, such as *Inocybe* and so on. Sometimes the holobasidium bears no sterigmata (fig. 1). This is present in some agarics. It is exactly the same shape as *Tulasnella*. Have you considered this?

ROGERS: There is always a septum at the base of the epibasidium in *Tulasnella*. Therefore, I would say that there is no resemblance, except tangential. I don't see that it has anything to do with the *Tulasnella* basidium morphologically.

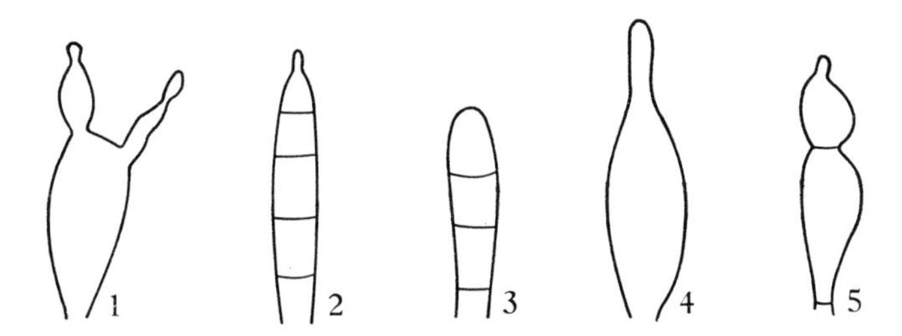


FIGURES 1–5. Rough illustrative drawings by the discussion participants.

SMITH: The picture that Dr. Heim has put on the board is not of a basidium that is characteristic of the species which he has seen, but rather something that occurs along with other more-or-less "normal" basidia. I have seen these aberrations—that is what I would call them in *Mycena*—as Dr. Heim has, in a number of agarics, but I think they are somehow connected with an abnormality in the cytology of the basidia, although I haven't proved this. It is not a characteristic of a particular species, is it, Dr. Heim?

HEIM: No, it is not characteristic of any given species, but in one species in one genus, another species in another genus and so on. Not all genera, but only some.

SINGER: I wonder if you are aware that when *Gerronema* (*Clitocybe*) *venustissimus* (which is a European species of agaric) grows in very low temperatures, it produces basidia that are septate like an *Auricularia* basidium. The basidium that looks like this (fig. 2) has usually one apical sterigma and is transversely septate. My question is almost rhetorical: would this have anything to do with *Auricularia*? My reaction is negative. It would seem like a secondary development of the homobasidium in which septation takes place.

ROGERS: I should certainly agree with you on that.

PETERSEN: I might also comment that of course Van Overeen's genus *Clavulinopsis* was based on distraught, misplaced, often subnumerary, abnormal sterigmata, so this phenomenon is widespread in several species of clavarioid fungi.

SMITH: If you really want to study "normally" abnormal basidia go to the genus *Rhizopogon.* In the first place, the hymenium is composed of cellular elements. A hymenium should be a palisade of single cells, but in *Rhizopogon* this is not often the case. But all variations occur. Some basidia are like that (fig. 3) and bear a spore at the tip. Others are like cystidia (fig. 4) but do bear spores. Some look almost like what Dr. Heim drew, with a bulge at the top (fig. 5). These often become septate and have several cells variously inflated. One of the interesting problems left in *Rhizopogon* is the study of morphological variation of the basidia.

So far as I know, a good deal of the variation in the shape of the basidium is due to particular pressure at the moment of development. The progress of development is dependent on pressure. The long narrow sterigma may project up and may have a spicule—to use the terms of Dr. Donk—on top or may not. If there is little pressure, the apical cell will simply form a basidium, and the basal cell would be about the same size.

Below the basidia there is a cellular subhymenium, so that the hymenium varies tremendously from single cells to what I would call more loosely a trichodermal hymenium. This is the only group I know where there is such a tremendous spectrum of basidial types in one section under a microscope. What the meaning of this is I don't know yet. I think that it probably has some significance, because it is not an occasional formation of an odd type of meiospore production, but is a regular occurrence.

ROGERS: I doubt that Dr. Zeller would agree with this, but I suggest that most of these things happened to the basidia in the Gastromycetes because in the Gastromycetes the basidium lost one of its principal functions and simply went wild, all in its physiognomy. There is no restraint of survival value for the normal type.

SMITH: I would more-or-less agree with that. There is a loss of emphasis.

PETERSEN: Is there a possibility that some of these kinds of aberrations could be ecological or microecological? We are all familiar, of course, with leaving collections in wax paper too many hours or days and finding basidia with epibasidia or sterigmata long and hypha-like, and sometimes hyponumerary or supernumerary. This might happen in the field due to a humidity factor. I wonder if this is in question at all?

SINGER: First, I find this in a single fruiting body that has been collected in the field. We find it under the microscope on the same slide. Second, I have looked at *Mycena* that I have brought in fresh with the dew still on it, and have found some of these greatly elongated, malformed sterigmata with spores, with the spore still attached to the sterigma but germinating and producing a hypha.

DONK: The basidium, like other organs of the Hymenomycetes, may occasionally vary in shape depending on the species. The most remarkable example I remember is perhaps *Myxomycidium*, a small clavariaceous fungus with fruiting bodies at the underside of logs in the tropics. The hymenium is amphigenous. When still young the basidia are perfectly normal holobasidia. Then the portion bearing the hymenium becomes more and more liquified and bigger until at maturity it becomes a drop which will flow around one's finger when touched. This rapid increase in the volume of the basidiiferous portion seems to account for several rigorous adaptations assumed by the not yet fully mature basidia in order to reach the more and more receding surface for the production of dischargeable spores. The hymenium becomes strongly thickened; the stalks of the basidia may elongate considerably; the basidia may even "repeat" themselves by producing an apical club-shaped outgrowth; the sterigmata may become longer and longer (and may even bifurcate) until they reach the surface where they produce their spicula and then shoot away the spores. The division of the diploid nucleus may occur in the tubular outgrowth (the "repeated" basidium). This reconstruction of what happens is based on evidence published by Martin, Linder, and Kobayasi. *Myxomycidium* has even been placed in a family of its own because it was supposed to have a special basidial type. In my opinion the basidia are of a common type, but they may assume unusual shapes because of their ability to adapt themselves.

M. P. CHRISTIANSEN

*Copenhagen, Denmark*

# VARIATION IN THE RESUPINATE HOMOBASIDIOMYCETES

From a taxonomical point of view the resupinate fungi are certainly not a well-defined group. They have, however, certain characters of gross morphology in common. The fruit body is not greatly differentiated and is appressed to the substratum and more-or-less attached to it. Rarely the margin is free or even recurved. Bracket fruit bodies are only to be found on vertical substrates and then only occasionally.

If we go for a walk in a forest on an autumn day in order to collect fungi, we can easily find the larger mushrooms, but it is very possible that we will not see a single resupinate. This is because the resupinates nearly always live a hidden existence on the lower sides of dead branches lying on the floor of the forest. If we want to collect the resupinates it is necessary therefore to search for them and we must know where we have a good chance of finding them. Many of these fungi occur on fallen or cut branches or trunks lying on the forest floor. If we find a large heap of very decayed branches we can often —if we are lucky—find specimens of many resupinates, enough for a microscopic study to last a long time. Other species occur on or inside very decayed stumps where they form a carpet on the walls of the cavities [e.g., *Cristella farinacea* (Pers. ex Fr.) Donk, *Cr. candidissima* (Schw.) Donk, *Phlebia hydnoides* (Cooke & Massee) M. P. Christ.]. Resupinates are also to be found under fallen leaves, on pine needles, under mosses [e.g., *Tylospora asterophora* (Bon.) Donk, *T. fibrillosa* (Burt) Donk, *Amphinema byssoides* (Pers. ex Fr.) J. Erikss., *Corticium byssinum* (Karst.) Massee, *C. bicolor* Peck], as a carpet on the walls of mouse holes [e.g., *Byssocorticium atrovirens* (Fr.) Bon. & Sing.], on dead stems of ferns and other plants, and on the naked earth or on stone. *Asterostroma laxum* Bres. occurs on dead wood but also at the base of living bushes such as heather (*Calluna*); other species are present on the bases of bushes also. A few species occur on dead but still attached branches of living trees [e.g., *Peniophora laeta* (Fr.) Donk on *Carpinus betulus*, *P.*

*violaceo-livida* (Sommerf.) Mass. on *Salix*-species, and *Vuilleminia comedens* (Nees ex Fr.) Maire on all kinds of deciduous trees]. *Aleurodiscus acerinus* is to be found on the bark of *Acer campestris*, both on the trunk and the branches.

We must be very careful and have a keen eye when we collect resupinates, as otherwise we shall miss many interesting species. Most of the resupinates are very clear and visible and can be seen easily from a long distance, but other species are very thin and small and form only an almost invisible film over the substrate. If we are not careful therefore, we shall miss many interesting species [e.g., *Peniophora longispora* (Pat.) Bourd. & Galz.].

The context of the resupinates is leathery, coriaceous, ceraceous, membranous, pellicular, felty, arachnoid, or gelatinous.

Generally the entire surface is fertile. Smooth or slightly verrucose hymenophores are most frequent, but tuberculate, echinulate, plicate, and porose hymenophores occur, especially on forms developed on the lower side of the substrate (e.g., *Corticium evolvens*—membranaceous; *Peniophora incarnata*, etc.—ceraceous; *Tomentella* species—felty; *Steccherinum* species—echinulate; *Merulius* species—plicate; and *Poria* species—porose).

I should like to say a few words about how to dry species of resupinates. To dry the larger mushrooms is often difficult, but it is easy to dry resupinates and quick drying (by heating) is best because many thick resupinates are attacked by larvae. In a dried condition resupinates can be kept for years and old material is often as easy to study as fresh.

## History

The foremost expert in resupinates in the early period was C. H. Persoon (1755–1837) from Holland who, like Elias M. Fries from Sweden (1794–1878), described and named many species within the resupinates. But as the use of the microscope had not been introduced in mycology it was especially species with peculiar macroscopic appearance or with specific colored fruit bodies which were collected and described, mostly species from the families Hydnaceae, Polyporaceae, Thelephoraceae, but also from the family Corticiaceae. Of other students of the resupinates, one ought to mention Giacomo Bresadola (1847–1929), a very careful collector whose descriptions were so exact that his work became a model for many later students. Other students were F. X. R. von Höhnel (1853–1920) and V. Litschauer, both from Austria, P. A. Karsten (1843–1917) from Finland and above all H. Bourdot (1861–1937) from France together with A. Galzin published the large work, *Hyménomycètes de France.*

## Identification

With few exceptions, the resupinates are all Basidiomycetes and a great majority of the species are saprophytes. Some species prefer recently deceased substrate [e.g., *Xenasma tulasnelloideum* (Höhnel & Litsch. Donk)] whereas others are confined to strongly decomposed substrates (e.g., *Tomentella* species).

Only a small minority of the species can be readily identified in the field (e.g., the few host-specific forms) and species of similar habit may prove very different under the microscope. Several species are very difficult to detect due to their habitat and inconspicuous appearance, but these are often the most conspicuous and distinct under the microscope. To arrive at a reliable identification a microscopic examination is thus generally indispensable. It is often a great help to use 3% KOH solution and 2% aqueous phloxine solution for microscopic examination, and if the stain is successful, the fungus element is a brilliant red on a clear background.

Few resupinate species belong to the Heterobasidiomycetes, but a large majority belong to the Homobasidiomycetes, mostly to the family Corticiaceae *s. lat.* and originally referred to the genera *Corticium*, *Gloeocystidium*, and *Peniophora*, all very heterogeneous assemblages. Other families are the Clavariaceae, Coniophoraceae, Hydnaceae, Hymenochaetaceae, Polyporaceae, and Thelephoraceae. Only the two last are families with large numbers of resupinate species.

In the fundamental work of Bourdot and Galzin (1928), *Hyménomycètes de France*, the authors divided *Corticium* into 15 sections and the two remaining genera into 7 sections each, but the sections have proven very distantly related in many cases. A total revision of the taxonomy was thus strictly necessary. This work was initiated by Donk, who described the new genera *Botryobasidium* and *Gloeocystidiellum*. Rogers created the genus *Ceratobasidium*, in many ways intermediate between the Heterobasidiomycetes and Homobasidiomycetes. Jackson (1948), Eriksson (1950), and others showed the necessity for further revision. Donk (1956, 1957) erected *Tubulicrinis, Laeticorticium, Scytinostroma, Tylosperma* (later *Tylospora*), *Xenasma* and emended *Gloeocystidiellum, Sistotrema, Athelia, Cristella, Hyphoderma, Peniophora,* and *Phlebia*. Eriksson (1958) added the genera *Sistotremastrum, Repetobasidium, Paullicorticium, Hypochnicium, Fibricium, Hyphodontia,* and *Membranicium* (*ad int.*).

TABLE 1. THE SUBFAMILIES AND GENERA OF THE CORTICIACEAE AS ACCEPTED BY ERIKSSON.

| | | |
|---|---|---|
| Botryobasidioideae: | Aleurodiscoideae: | Corticioideae: |
| *Cerinomyces* | *Gloeocystidiellum* | *Corticium* sect. *Laeta* |
| *Ceratobasidium* | *Laxitextum* | *Corticium* |
| *Uthatobasidium* | *Aleurodiscus* | Odontioideae: |
| *Botryohypochnus* | *Vuilleminia* | *Dacryobolus* |
| *Thanatephorus* | *Cytidia* | *Odontia* |
| *Botryobasidium* | Athelioideae: | Merulioideae: |
| *Oidium* | *Athelia* | *Merulius* |
| Sistotremoideae: | *Byssocorticium* | Stereoideae: |
| *Sistotrema* | *Tylospora* | *Stereum* |
| *Sistotremastrum* | Phlebioideae: | *Lopharia* |
| Cristelloideae: | *Phlebia* | *Podoscypha* |
| *Cristella* | *Mycoacia* | *Plicatura* |
| *Xenasma* | *Membranicium* | *Cotylidia* |
| *Paullicorticium* | Peniophoroideae: | |
| Repetobasidioideae: | *Peniophora* | |
| *Repetobasidium* | *Sterellum* | |
| Galzinioideae: | Hyphodermoideae: | |
| *Galzinia* | *Hyphoderma* | |
| *Scytinostroma* | *Hypochnicium* | |
| *Vararia* | *Hyphodontia* | |
| *Laeticorticium* | *Amphinema* | |
| Tubulicrinoideae: | *Fibricium* | |
| *Tubulicrinis* | | |

Table 1 summarizes the subfamilial classification scheme used by Eriksson. The last four subfamilies were admittedly artificial, but all the rest represent natural groupings. All these genera were based in general on microscopic details, and to arrive at a reliable identification a microscopic examination was thus indispensable. The most striking microscopic structure is the basidium and almost all forms of the basidium occur in the resupinates. Drawings of microscopic details both from the Hetero- and Homobasidiomycetes characterize the different resupinate genera in question.

Microscopic details of the resupinate species of the Heterobasidiomycetes are shown in figures 1–16. *Helicogloea farinacea* (Höhn.) Rogers (figs. 1, 2) has its fruit body effused, adnate, mealy-tomentose to submembranaceous, whitish to cream and is saprophytic on dead wood. *Platygloea peniophorae* Bourd. & Galz. (fig. 3) is parasitic on Corticiaceae. Both species belong to the family Auriculariaceae and have basidia with transverse septa. This basidium is perhaps related to the ascus of the Ascomycetes.

*Sebacina grisea* (Pers.) Bres. (figs. 4–6) from the family Tremellaceae has fruit bodies which are waxy-gelatinous to mucous-gelatinous and bluish-grey and has longitudinally septate basidia. The drawing shows a young basidium, and an older one with four long epibasidia. The genus *Sebacina* is large.

The fruit bodies of *Tulasnella griseo-rubella* Litsch. (figs. 7–12) from the family Tulasnellaceae are effused, closely adnate, when fresh almost coral-red, in drying waxy, smooth, cream with a reddish tint and occur on dead wood. The drawing shows a young basidium, a basidium with sporelike epibasidia with septation at the base, older epibasidia, and an epibasidium with basidiospore. The genus *Tulasnella* does not seem to be a proper member either of the Heterobasidiomycetes or Homobasidiomycetes. The genus and species have been placed here because the basidiospores produce secondary spores.

*Ceratobasidium pseudocornigerum* M. P. Christ. (figs. 13–16) has fruit bodies which are effused, closely adnate, very thin, arachnoid to somewhat waxy, whitish grey, and found on dead wood. The basidia of *Ceratobasidium* are morphologically somewhat like those of *Tremella* but are not septate. They are also reminiscent of the basidia of *Botryobasidium*, but these basidia, however, often have six short sterigmata. I have placed *Ceratobasidium* and *Tulasnella* in the

FIGURES 1–20.

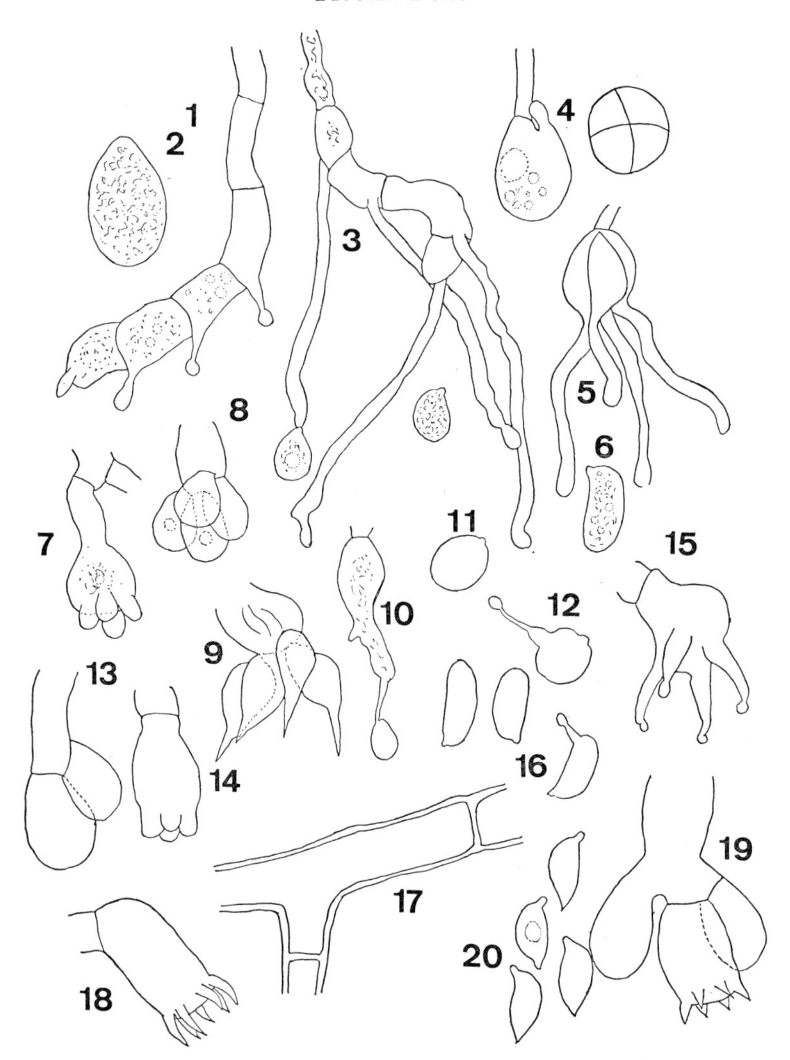


Figs. 1–16: Hyphae, basidia, and spores of Heterobasidiomycetes. Figs. 1–2, *Helicogloea farinacea*. Fig. 3, *Platygloea peniophorae*. Figs. 4–6, *Sebacina grisea*. Figs. 7–12, *Tulasnella griseo-rubella*. Figs. 13–16, *Ceratobasidium pseudocornigerum*.

Figs. 17–20: Hypha, basidia, and spores of Homobasidiomycetes, *Botryobasidium botryosum*.

Heterobasidiomycetes because the basidiospores in both genera produce secondary spores, a feature which is very characteristic and easily seen, and which does not occur in the Homobasidiomycetes. In my opinion, however, the proper limit between the Heterobasidiomycetes and Homobasidiomycetes is of smaller interest than knowledge of the many species within the two groups.

Within the large order Aphyllophorales of the subclass Homobasidiomycetes, both macroscopic and microscopic details may be used in order to come to an acceptable taxonomic conclusion concerning the placement of single species. In the genus *Botryobasidium* [e.g., *B. botryosum* (Bres.) J. Erikss., figs. 17–20] the fruit body is effuse, hypochnoid, or loosely pellicular. The basidium is short and wide, almost cylindric, with six sterigmata. The hyphae are wide and without clamp connections. Some few species of this genus have somewhat urniform basidia and hyphae with clamp connections, but cystidia are rare. As already mentioned, *Botryobasidium* has some characters in common with *Ceratobasidium.* It is known that the conidial states of some *Botryobasidium* species are species of the genus *Oidium.*

John Eriksson suggested a grouping of Scandinavian species of *Botryobasidium* into three subgenera, *Botryobasidium*, *Brevibasidium*, and *Dimorphonema*, based partly on the shape of the basidia, partly on the nature of the basal hyphae. "In the two first subgenera, the basal hyphae do not differ noteworthily from the subhymenial hyphae, but in the last subgenus there is a clear difference between the basal hyphae which are much wider and more or less pigmented, and the subhymenial hyphae which are thin walled and hyaline," Eriksson wrote. In this latter subgenus the basidium is also longer and more or less constricted.

The genus *Sistotrema* is well characterized by its urniform basidia, which are at first globose, then urniform with (4)-6-8-sterigmata. The fruit body is effuse to pellicular or fragile-membranous; the surface is even, grandinioid, or porose. The fruit body of S. *diademiferum* (Bourd. & Galz.) Donk (figs. 21–25) is even, pruinose, whitish to greyish yellow.

S. *Sernanderi* (Litsch.) Donk (figs. 26–30) differs from the latter by its narrow basidia with 3–4 sterigmata and by having cystidia. The genus *Sistotremastrum* John Eriksson is very closely related to *Sistotrema*, differing only by having nearly cylindrical basidia.

FIGURES 21–41.

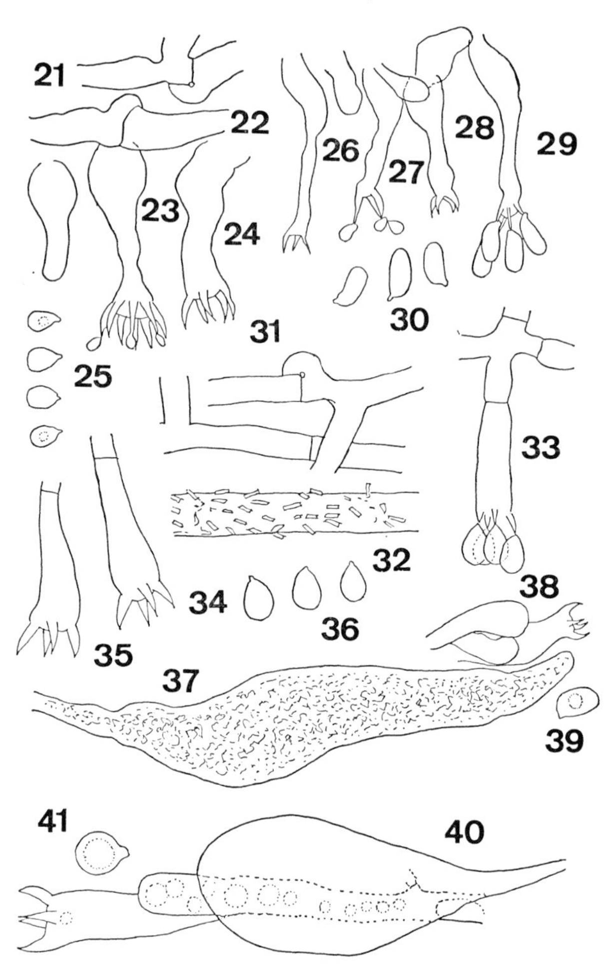


Hyphae, basidia, and spores of Homobasidiomycetes. Figs. 21–25, *Sistotrema diademiferum*. Figs. 26–30, *S. sernanderi*. Figs. 31–36, *Athelia neuhoffii*. Figs. 37–39, *Gloeocystidiellum porosum*. Figs. 40–41, *G. citrinum*.

The genus *Athelia* is large and homogeneous. *A. neuhoffii* (Bres.) Donk (figs. 31–36) has its fruit body widely effused; the subiculum loosely cobwebby, of cylindrical, thin-walled hyphae with few to many clamp connections and the hymenium pellicular, often merulioid when fresh, detachable from the substrate at least when dried and cracked. The basidia are clavate, and cystidia are lacking or rarely present.

In *Gloeocystidiellum* Donk, exemplified by *G. porosum* (Berk. & Curtis) Donk (figs. 37–39) and *G. citrinum* (Pers.) Donk (figs. 40–41), the fruit body is widely effused, the upper layer sometimes layered, waxy or nearly subgelatinous to pellicular and soft. Microscopically, the genus is characterized by its large, generally completely immersed gloeocystidia, rarely in addition to rather thick-walled cystidia. The spore wall is amyloid, smooth, rarely warted to minutely spiny.

The fruit body of *Tubulicrinis* Donk is closely adnate, thin, and pruinose, atomate to minutely granulose, and rarely waxy to membranaceous. The hyphae are indistinct or thick-walled with clamp connections, but the cystidia are often numerous, thick-walled, and usually far-protruding, more-or-less amyloid, and dissolve in 10% KOH. In *T. thermometrus* (Cunn.) M. P. Christ. (fig. 42–44) the cystidia are cylindrical and capitate, the tip thin-walled, but in *T. hirtellus* (Bourd. & Galz.) Donk (figs. 45–47) and *T. subulatus* (Bourd. & Galz.) J. Erikss. (figs. 48–50) they are more-or-less conical, becoming more-or-less abruptly thin-walled toward the pointed apex. The spores of the genus are globular to curved-cylindric and nonamyloid.

The genus *Peniophora* Cooke 1879 emend Donk 1957, has been restricted repeatedly. The fruit body of the genus is firm, closely attached to the substrate, or loosely attached at the margin, bright to dull colored, red, grey, lilaceous, blue, or brown. The basidia are clavate, the cystidia thick-walled, often with brown basal parts, incrusted toward the apices. Gloeocystidia are numerous in *P. nuda* (Fr.) Bres. (figs. 51–53) but are absent in most of the other species in the genus. Dr. John Eriksson's work on section *Coloratae* Bourd. & Galz. is excellent.

The genus *Xenasma* Donk is characterized by the broadly clavate to cylindrical, rather short pleurobasidia which are lateral extensions of creeping hyphae and which bear 2–7 sterigmata. Cystidia may be present or lacking. The spores are nearly globular or broad-ellipsoid, rather small, smooth to roughened. The fruit body is indeterminate, usually very thin and often difficult to detect, soft-waxy to gelatinous. The genus is illustrated by *Xenasma pruinosa* (Pat.) Donk (figs. 54–57).

FIGURES 42–64.

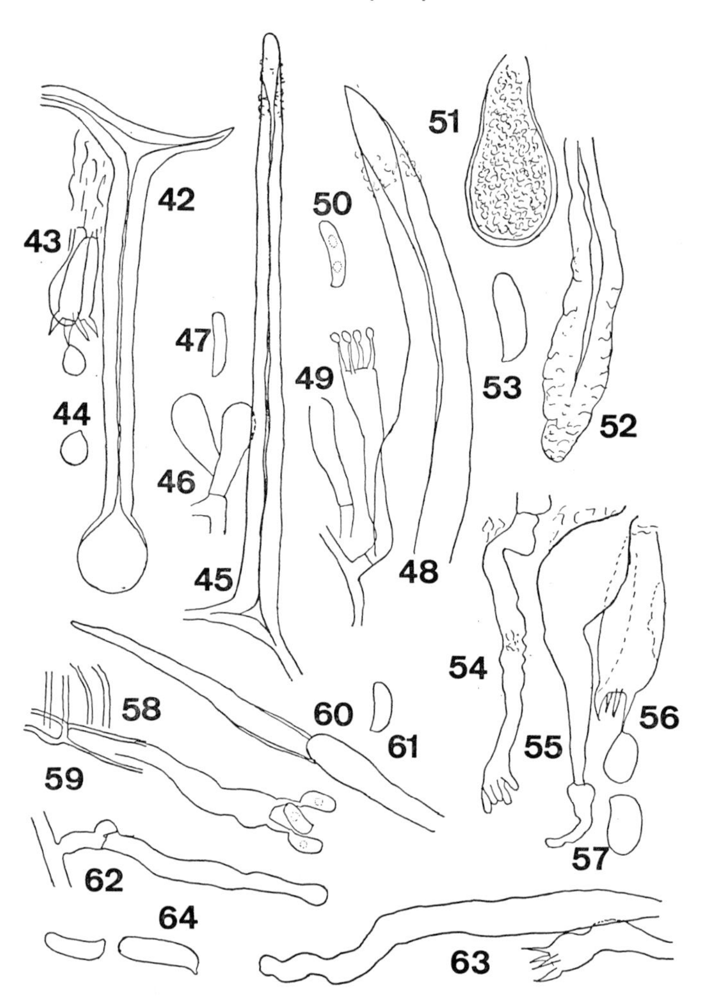


Basidia, cystidia, and spores of Homobasidiomycetes. Figs. 42–44, *Tubulicrinis thermometrus.* Figs. 45–47, *T. hirtellus.* Figs. 48–50, *T. subulatus.* Figs. 51–53, *Peniophora nuda.* Figs. 54–57, *Xenasma pruinosa.* Figs. 58–61, *Phlebia subserialis.* Figs. 63–64, *Hyphoderma definitum.*

The fruit body of *Phlebia* Fr. emend Donk is rather thin to thick, closely adnate, waxy-gelatinous to mucous, with its surface smooth, folded, or granular by crystal conglomeration. The basidia are

FIGURES 65–82.

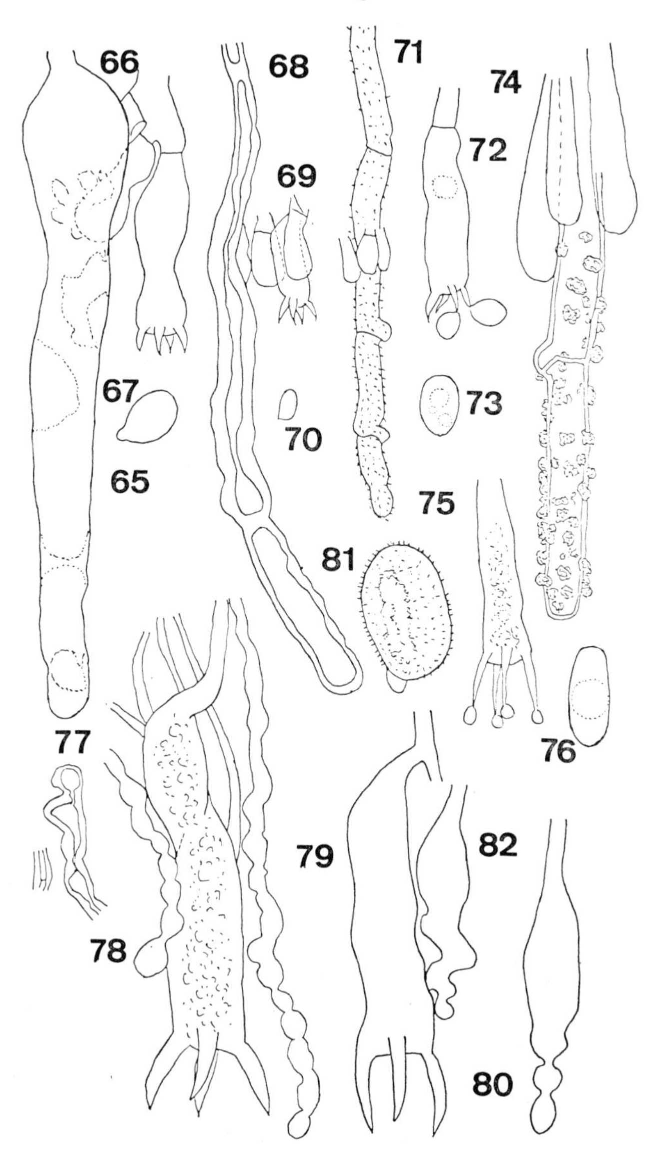


Hyphae, basidia, cystidia, and spores of Homobasidiomycetes. Figs. 65–67, *Hyphoderma magniapiculum*. Figs. 68–70, *Hyphodontia pallidula*. Figs. 71–73, *Hyphoderma polonense*. Figs. 74–76, *H. setigerum*. Figs. 77–78, *Aleurodiscus amorphus*. Figs. 79–82, *A. aurantius*.

FIGURES 83–90.

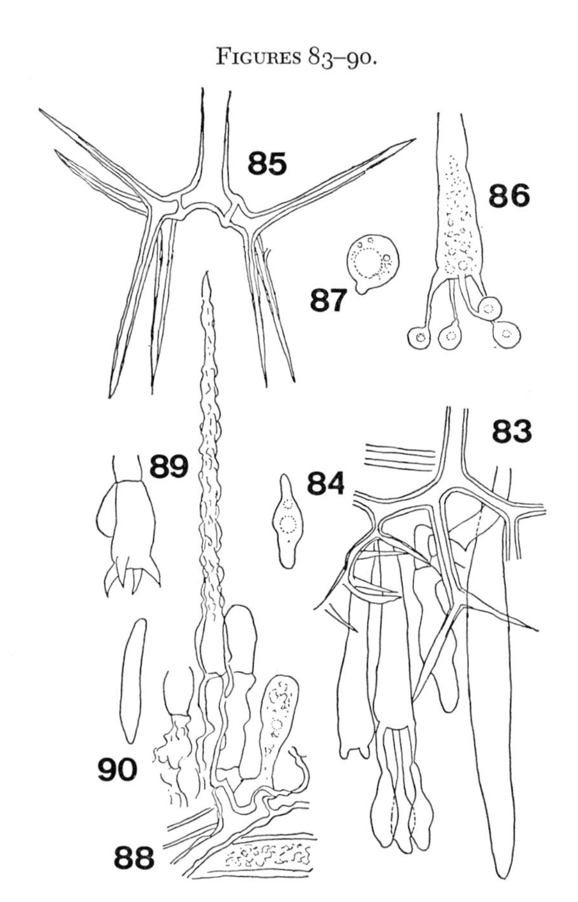


Hyphae, basidia, cystidia, and spores of Homobasidiomycetes. Figs. 83–84, *Vararia investiens.* Figs. 85–87, *Asterostroma laxa.* Figs. 88–90, *Subulicystidium longispora.*

slender-clavate. If present, the cystidia may be of different forms; slender and thin-walled (leptocystidia) [*P. subserialis* (Bourd. & Galz.) Donk, figs. 58–61] or thick-walled and incrusted (lamprocystidia). The spores are rather small, often curved, and smooth.

*Hyphoderma* Wallr. emend Donk is illustrated by four species, *H. definitum* (H. C. Jacks.) Donk (figs. 62–64), *H. magniapiculum* M. P. Christ. (figs. 65–67), *H. polonense* (Bres.) Donk (figs. 71–73), and *H. setigerum* (Fr.) Donk (figs. 74–76). The fruit body is strictly resupinate. The hymenium is rather compact, somewhat fleshy when fresh, usually smooth, toothed, or raduloid in a few species, rarely very thin and minutely floccose. The basidia are cylindrical to somewhat urniform, and the cystidia may be of different forms: somewhat vermiform and thin-walled, conical and thin-walled, cylindrical and thin-walled, septate and hairy, cylindrical and with somewhat solid walls, or septate and more-or-less incrusted with crystals. The spores are often cylindrical to oblong and flattened to somewhat depressed on one side, and rarely broad.

*Aleurodiscus amorphus* (Pers. ex Fr.) Schroet. (figs. 77–78) has a cupshaped or cupuliform fruit body, whereas *A. aurantius* (Pers. ex Fr.) Schroet. (figs. 79–82) has an effused waxy to fleshy fruit body, but both species have acanthophyses, very long basidia, and large spores which are faintly echinulate or minutely aculeate and amyloid. Other members of the genus *Aleurodiscus* have smooth spores.

The genera *Vararia* Karsten and *Scytinostroma* Donk are characterized by the dendritically ramified hyphae, dendrophyses, which are rather thick-walled, hyaline or light brown, turning reddish brown in Melzer's solution. *V. investiens* (Schw.) Karst. (figs. 83–84) has fusoid spores, which are nonamyloid. *S. portentosum* has globose spores which are amyloid. *Asterostroma laxa* Bres. (figs. 85–87) also possesses dendrophyses.

Finally, *Subulicystidium longispora* (Pat.) Parm. (figs. 88–90), with peculiar cystidia and very long spores, is not a member of the genus *Peniophora* emend Donk.

The form and size of the spores of resupinates are very different, so that a great number of species within the group have spores which are highly characteristic. The spore is of great help in the identification of numerous species. Figures 91–98 show spores of Heterobasidiomycetes. All are hyaline and smooth, and all form secondary spores.

The spores of the Homobasidiomycetes (figs. 99–143) are also illustrated. The spores of the family Corticiaceae (figs. 99–124) are no less heterogeneous than those of the Heterobasidiomycetes. All those figured are hyaline. Spores are also illustrated from the families Thelephoraceae (figs. 125–130), Coniophoraceae (figs. 100, 131–132), the spores of which are subhyaline, yellowish to yellowish-brown, and smooth; and Hydnaceae (fig. 133), in which the hymenium covers spines and teeth. Of the family Clavariaceae, the spores of *Kavinia bourdotii* (Bres.) Pilát (fig. 134) are yellow, and those of the family Polyporaceae (figs. 135–143) are hyaline.

FIGURES 91–143.
Spores of resupinate Basidiomycetes. Fig. 91, *Helicogloea lagerheimii.* Fig. 92, *Platygloea vestita.* Fig. 93, *Sebacina incrustans.* Fig. 94, *S. effusa.* Fig. 95, *Tulasnella albida.* Fig. 96, *T. rutilans.* Fig. 97, *Gloeotulasnella helicospora.* Fig. 98, *G. calospora.* Fig. 99, *Uthatobasidium fusisporum.* Fig. 100, *Jaapia argillacea.* Fig. 101, *Botryobasidium botryosum.* Fig. 102, *Cristella confinis.* Fig. 103, *C. trigonospora.* Fig. 104, *C. stellulata.* Fig. 105, *C. christiansenii.* Fig. 106, *Cristella* sp. Fig. 107, *C. variecolor.* Fig. 108, *Xenasma tulasnelloideum.* Fig. 109, *Galzinia pedicellata.* Fig. 110, *Gloeocystidiellum citrinum.* Fig. 111, *G. leucoxanthum.* Fig. 112, *Tubulicrinis glebulosus.* Fig. 113, *T. vermifer.* Fig. 114, *Athelia epiphylla.* Fig. 115, *Byssocorticium atrovirens.* Fig. 116, *Tylospora fibrillosa.* Fig. 117, *T. asterophora.* Fig. 118, *Peniophora quercina.* Fig. 119, *Hyphoderma polonense.* Fig. 120, *H. roseocremeum.* Fig. 121, *Hypochnicium punctulatum.* Fig. 122, *H. bombycina.* Fig. 123, *Epithele typhae.* Fig. 124, *Subulicystidium longispora.* Fig. 125, *Tomentella mucidula.* Fig. 126, *T. macrospora.* Fig. 127, *T. luteomarginata.* Fig. 128, *T. microspora.* Fig. 129, *T. atromentaria.* Fig. 130, *T. bresadolae.* Fig. 131, *Coniophora arida.* Fig. 132, *C. bourdotii.* Fig. 133, *Steccherinum fimbriatum.* Fig. 134, *Kavinia bourdotii.* Fig. 135, *Meruliporia taxicola.* Fig. 136, *Lindtneria trachyspora.* Fig. 137, *Xylodon versiporus.* Fig. 138, *Podoporia sanguinolenta.* Fig. 139, *Ceriporia viridans.* Fig. 140, *Poria subincarnata.* Fig. 141, *Chaetoporus euporus.* Fig. 142, *Coriolellus colliculosus.* Fig. 143, *Antrodia mollis.*

Figures 91–143.

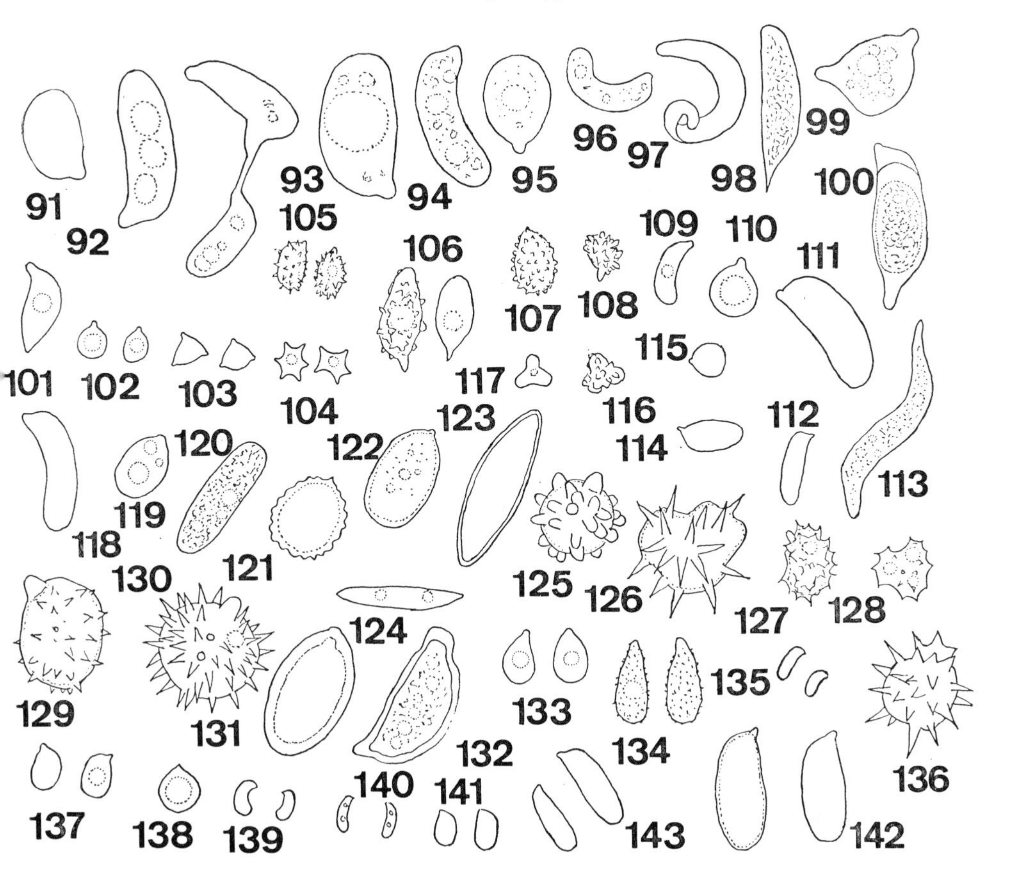

## References

BOURDOT, H. AND A. GALZIN. 1927 (1928). Hyménomycètes de France. Sceaux, 764 p.

DONK, M. A. 1956. Notes on resupinate Hymenomycetes. III. Fungus 26: 3–24.

———. 1957. Notes on resupinate Hymenomycetes. IV. Fungus 27: 1–29.

ERIKSSON, J. 1950. *Peniophora* Cke. sect. *Coloratae* Bourd. & Galz. Symb. Bot. Upsal. 10: 1–76.

———. 1958. Studies in the Heterobasidiomycetes and Homobasidiomycetes Aphyllophorales of Muddus National Park in North Sweden. Symb. Bot. Upsal. 16: 1–172.

JACKSON, H. S. 1948. Studies of Canadian Thelephoraceae. I. Some new species of *Peniophora.* Canad. J. Res. C 26: 128–139.

## Discussion

DUBLIN: Figure 92 showed what appeared to be a dividing or conjugating spore. Could you explain this?

CHRISTIANSEN: It was a secondary spore being produced by a basidiospore of a heterobasidiomycete.

SMITH: A comment was made about drying fungi with heat. In recent years we have had great success with drying with silica gel when large quantities of water were not involved. Silica gel might be ideal for the resupinates. It is not very good for boletes where you have to get rid of a pint of water. I will admit heat is best for these, but probably not for resupinates and small fungi. We found that activated silica gel gave the best results short of carrying a portable deep-freeze drier in the field.

OLEXIA: Do you consider secondary spore production a character of equal weight with basidial septation in separating the Heterobasidiomycetes from the Homobasidiomycetes?

CHRISTIANSEN: The key character here is secondary spore production and this is the character on which I would place an organism in the Heterobasidiomycetes.

ROBERT L. GILBERTSON

*Professor of Plant Pathology*
*University of Arizona, Tucson, Arizona*

# PHYLOGENETIC RELATIONSHIPS OF HYMENOMYCETES WITH RESUPINATE, HYDNACEOUS BASIDIOCARPS[1]

In accepting this topic I am fully aware of the difficulty of discussing the phylogeny of a heterogeneous group that touches virtually all natural groups of the Aphyllophorales. In fact, it is probably impossible to do so adequately on the basis of our current knowledge. At best, I can summarize current thinking on the phylogeny of these fungi and perhaps contribute a few new suggestions regarding the natural relationships of these Hymenomycetes with resupinate, hydnaceous basidiocarps.

The development of a natural system of classification of the Aphyllophorales must be securely based on detailed knowledge at the species level. This has not yet been attained for many species and much still needs to be done to elucidate the regional flora in many areas of the world, particularly the tropics. However, our level of knowledge has advanced to the point where a better understanding of phylogenetic relationships is emerging and the taxonomy of the Aphyllophorales is now in a state of flux after more than one hundred years of domination by the classical Friesian system.

In the Friesian system of classification, the Hymenomycetes with resupinate, hydnaceous basidiocarps were placed in several genera of the family Hydnaceae. The practicality of the Friesian system enabled it to become so firmly established that it has been used with minor modification for major taxonomic works on resupinate hydnaceous Hymenomycetes to this day. These include those of Velenovsky (1920–1922), Cejp (1928), Miller and Boyle (1943), and Nikolajeva (1961). In these works the "resupinate Hydnaceae" are for the most

[1] Since the presentation of this paper an important publication by Parmasto has become available. This publication is pertinent to much of the subject matter in the present paper, particularly to the genera *Echinodontium*, *Cystostereum*, *Steccherinum*, *Vararia*, *Scytinostroma*, *Phanerochaete*, *Basidioradulum*, and *Hyphodontia*. See Parmasto, E. 1968. Conspectus systematis corticiacearum. Inst. Zool. et Bot. Acad. Sci. R.P.S.S. Estonicae. 261 p.

part placed in the genera *Grandinia* Fr., *Odontia* Fr., *Mycoacia* Donk, *Radulum* Fr., *Steccherinum* S. F. Gray, *Asterodon* Pat., *Caldesiella* Sacc., and *Mucronella* Fr., or their nomenclatural equivalents, based primarily on the configuration of the "teeth" or other macroscopic features. The base for this discussion will be essentially the North American fungi placed in these genera by past workers. A number of authors (Miller & Boyle, 1943; Cejp, 1928; Nikolajeva, 1961) have reviewed the history of the taxonomy of this group and it will not be repeated here.

Revision of the Friesian system was notably initiated by Patouillard (1900), and his system was largely followed by Bourdot and Galzin (1928). Even in Patouillard's modification, however, the genera of hydnaceous fungi, although realigned in several subtribes and series of the Porohydnés, remained essentially the same. Since the time of Patouillard, the intensified study of micromorphology of basidiocarps by his contemporaries and later workers such as Romell, Bresadola, von Hoehnel, Litschaeur, and Bourdot provided the detailed data necessary for further steps in the development of a natural system. Bourdot recognized several natural groupings in sections within genera such as *Corticium* and *Peniophora*, some of which have since been raised to generic rank. This intensified study of micromorphology of basidiocarps, combined with research on cultural characteristics and sexuality of these organisms, has indicated relationships that appear more natural and which have stimulated the proposal of new genera and families and realignment of families of Aphyllophorales. Recent workers who have been particularly active in this regard include H. S. Jackson, D. P. Rogers, M. A. Donk, John Eriksson, M. P. Christiansen, J. Boidin, E. Parmasto, F. Kotlaba, and Z. Pouzar. Discussing the phylogenetic position of resupinate hydnaceous hymenomycetes is necessarily based to a large degree on the contributions of these mycologists.

In discussing the phylogeny of the Basidiomycetes, Savile (1955) stated that "convergent evolution is probable if not inevitable in the achievement of important ends if there are few methods of obtaining them. Characters of great functional value—such as those aiding dispersal in plants . . . develop very readily and may appear repeatedly in different groups with bewildering similarity." The hydnaceous hymenophore is presumably such a character. Savile further stated that "it is not merely possible but almost inevitable that superficially similar basidiomycetous fruit bodies have been derived from distinct evolutionary lines arising from the primitive Hymenomycetes." It certainly seems much more reasonable to assume that the hydnaceous hymenophore arose independently in a number of series than to assume that a complex of characters such as those typical of the tomentelloid fungi or the "xanthochroic series" arose independently in various families distinguished by configuration of the hymenophore. It is now generally accepted by most students of the Aphyllophorales that the old Friesian family Hydnaceae was an artificial assemblage of species in which convergent evolution had established a common character—the hydnaceous hymenophore. This paper proceeds on the premise that this is the case.

In terms of numbers of species, the hydnaceous resupinates are a relatively small group. Bourdot and Galzin (1928) recognized about 60 species from France, Miller and Boyle (1943) 43 from Iowa, Cejp (1928) 41 from Czechoslovakia, Nikolajeva (1961) 85 from the Soviet Union, and Eriksson (1958b) about 43 from Muddus National Park in Sweden. My records from North America indicate approximately 90 reasonably well-characterized species described to date. These are modest figures compared to the estimated numbers of other corticioid fungi, agarics, or polypores. However, the development of a natural system of classification including these fungi is part of the larger problem concerning the Hymenomycetes and particularly the Aphyllophorales, as a whole. Consequently, the taxonomist striving to incorporate these hydnaceous fungi in a natural system of classification becomes involved in a study of the entire group of Aphyllophorales and must develop a detailed knowledge at the species level throughout the group. The difficulties involved in developing such a background and the time required have undoubtedly been responsible in large part for the perpetuation of the Friesian system of classification up to recent times. It has been more practical to specialize in Friesian families or genera. However, progress in development of a phylogenetic system will owe much to specialists in the old Friesian families and genera for supplying much of the detailed morphological data at the species level necessary for the development of a natural classification.

The taxonomy of the Aphyllophorales has developed almost entirely on the morphology, at first macroscopic and later microscopic, of the basidiocarp. Indeed, there are many instances throughout the literature where the terms "basidiocarp" and "species" are apparently considered to be synonymous (viz., resupinate species, stipitate species, brown species, etc.). Morphology of the basidiocarp will continue to be important in the taxonomy of the Aphyllophorales, but a truly phylogenetic system will probably develop eventually on the basis of data obtained from the "total organism." Contemporary taxonomic research is disclosing important sources of information from other phases, both structural and functional, in the life cycles of the Homobasidiomycetes. Nobles (1958, 1965), working primarily with polypores, has already convincingly demonstrated the role of cultural characters in a classification system. One of the obstacles in development of a natural system of classification of the Aphyllophorales has been a dearth of data on cultural morphology, physiology, and sexuality of many species because of the slowness in obtaining in culture the fungi with resupinate fruiting bodies. Many of them have thin, fragile basidiocarps that are not amenable to tissue culture methods and are difficult to handle in setting up spore prints. Some have spores that do not germinate readily and some grow slowly or not at all on standard culture media. For example, there is literally no information available on cultural characters or sexuality in tomentelloid fungi because specific metabolic requirements for spore germination and growth have not been worked out (Larsen, 1967a). The development of cultural methods for these fungi will greatly facilitate solving some of the current problems in the taxonomy of the group. Fortunately, more workers are becoming concerned with cultural studies of the resupinate hymenomycetes and the number of species in culture is rapidly increasing.

Data from the total organism would then include macroscopic and microscopic morphology of basidiocarps, morphology of the vegetative mycelium both in nature and in culture on a defined medium, type of mating system, biochemical characteristics, enzyme reactions, nuclear behavior in vegetative hyphae, basidia, and basidiospores, and perhaps factors such as phenology, substratum relationships, and geographic distribution. There are probably other sources of relevant evidence on phylogeny of higher Basidiomycetes that should also be considered. For example, recent contributions show definite relationships between certain groups of fungus-inhabiting insects and species or species complexes of hymenomycetes, particularly polypores (Lawrence, 1967; Graves, 1960; Paviour-Smith, 1960). This relationship is presumably an expression of biochemical or physical properties of the fruiting bodies and therefore may hold valuable clues in the search for a phylogenetic classification of the fungi.

Since these resupinate hydnaceous fungi are a heterogeneous group, it is difficult to make any meaningful generalizations on them. However, a few observations seem worthy of mention.

Although the literature on these fungi in culture is far from complete, it does seem that a positive oxidase reaction is characteristic of virtually all of them. My own field observations on rot associated with about 60 species indicate that they are (with the exception of *Dacryobolus sudans*) associated with a white rot. This is a rather unusual situation, as other Friesian families containing wood-rotting fungi (viz., Polyporaceae, Thelephoraceae, and Agaricaceae) are composed of some species associated with white rots and some with brown rots. Nobles (1958) has concluded that since the ability to utilize cellulose is common to all wood-rotting fungi it is a primitive character. The ability to degrade lignin, on the other hand, has been acquired by fewer species and is therefore considered to be an advanced character.

When the substratum relationships of this oxidase-positive group are examined, we find that there is a definite indication of a preference for the wood of angiosperms. About 60% of 80 species surveyed from North America are found on angiosperms, about 22% are found on both conifer and angiosperm wood, and about 18% are found on conifer wood. These figures are similar to those published by Nobles (1958) for the oxidase-positive polypores. She found 70% to be restricted to broad-leaved trees or to occur only rarely on conifers, 17% to occur exclusively or usually on conifer hosts, and 12% to show no preference for either substratum. Therefore the resupinate hydnaceous fungi, like the oxidase-positive polypores, show preference for the wood of angiosperms, considered to be a more advanced group than the conifers.

All three types of mating systems known in the homobasidiomycetes are found in the hydnaceous resupinates. Of 23 species for which type of mating system has been determined, 11 (or 47%) are heterothallic and tetrapolar, 6 (or 26%) are heterothallic and bipolar, and 6 (or 26%) are homothallic. These percentages agree closely with those given by Boidin and Lanquetin (1965), who summarized reports on sexuality of 150 species from 52 genera, mostly resupinate hymenomycetes. They found 75 (or 50%) heterothallic and tetrapolar

FIGURES 1–4.

Hyphal systems in some resupinate hydnaceous fungi. Fig. 1, Dimitic hyphal system of *Steccherinum ciliolatum* (FP 100447) with nodose-septate generative hyphae (1a) and skeletal hyphae (1b). Fig. 2, Monomitic hyphal system of *Hyphodontia floccosa* (RLG 3513) with nodose-septate hyphae. Fig. 3, Monomitic hyphal system of *Odontia romellii* (RLG 298) with septate hyphae. Fig. 4, Dimitic hyphal system of *O. subcrinalis* (RLG 6322) with septate (very rarely nodose-septate) generative hyphae (4a) and skeletal hyphae (4b).

FIGURES 1–4.

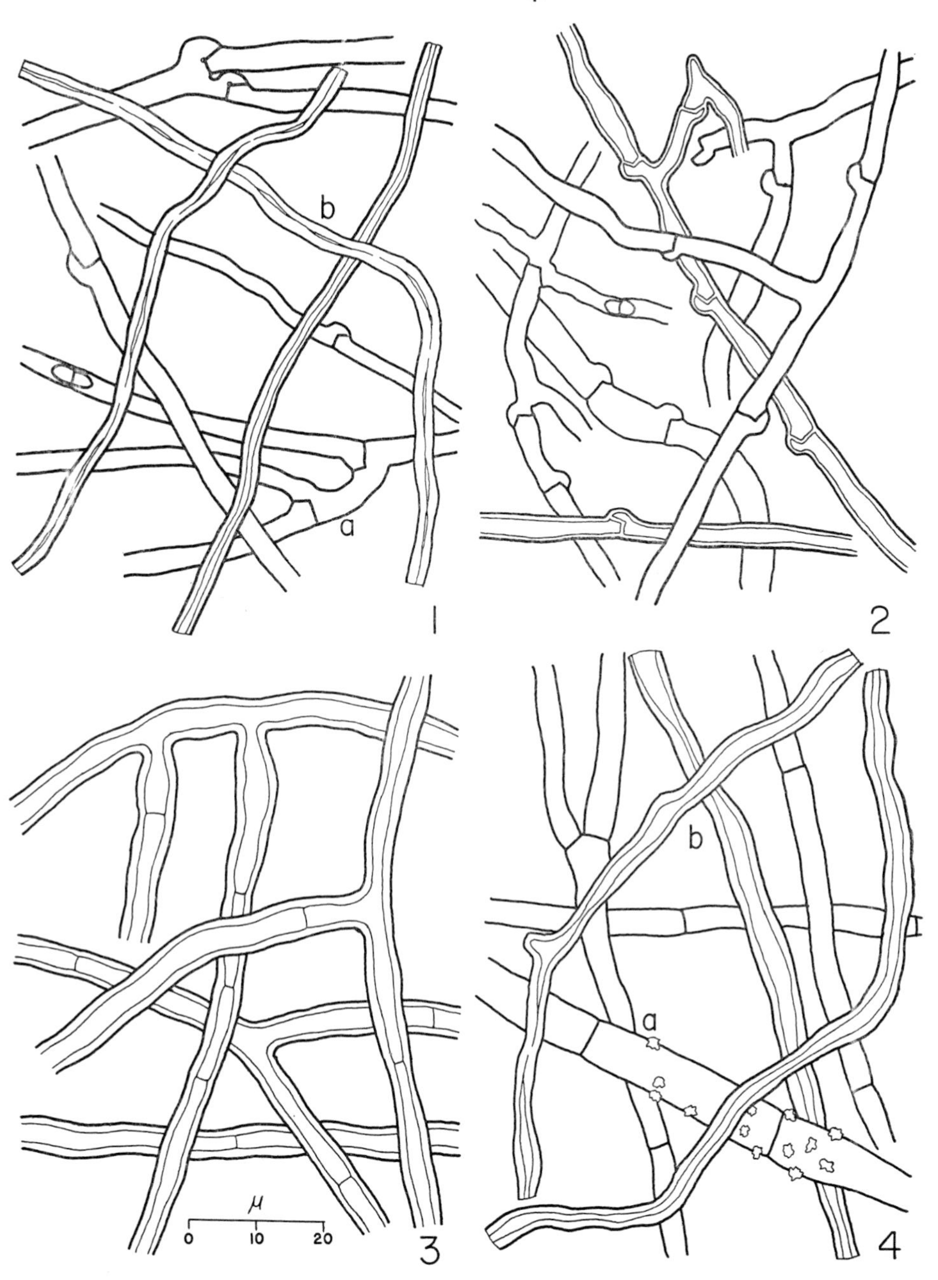

FIGURES 5–13.
Some sterile hymenial elements in resupinate hydnaceous fungi. Fig. 5, Cystidium of *Hyphodontia floccosa* (RLG 3413). Fig. 6, Dichohyphidium of *Vararia granulosa* (RLG 3376). Fig. 7, Cystidioles of *Odontia bicolor* (RLG 4430). Fig. 8, Gloeocystidia of *Dentipellis separans* (JLL 11683). Fig. 9, Cystidia of *H. arguta* (RLG 7429). Fig. 10, Cystidia of *Steccherinum fimbriatum* (RLG 4921). Fig. 11, Cystidia of *O. furfurella* (JLL 4417). Fig. 12, Setae of *Asterodon ferruginosus* (RLG 2669). Fig. 13, Cystidium of *Hyphoderma setigera* (RLG 6933).

FIGURES 5–13.

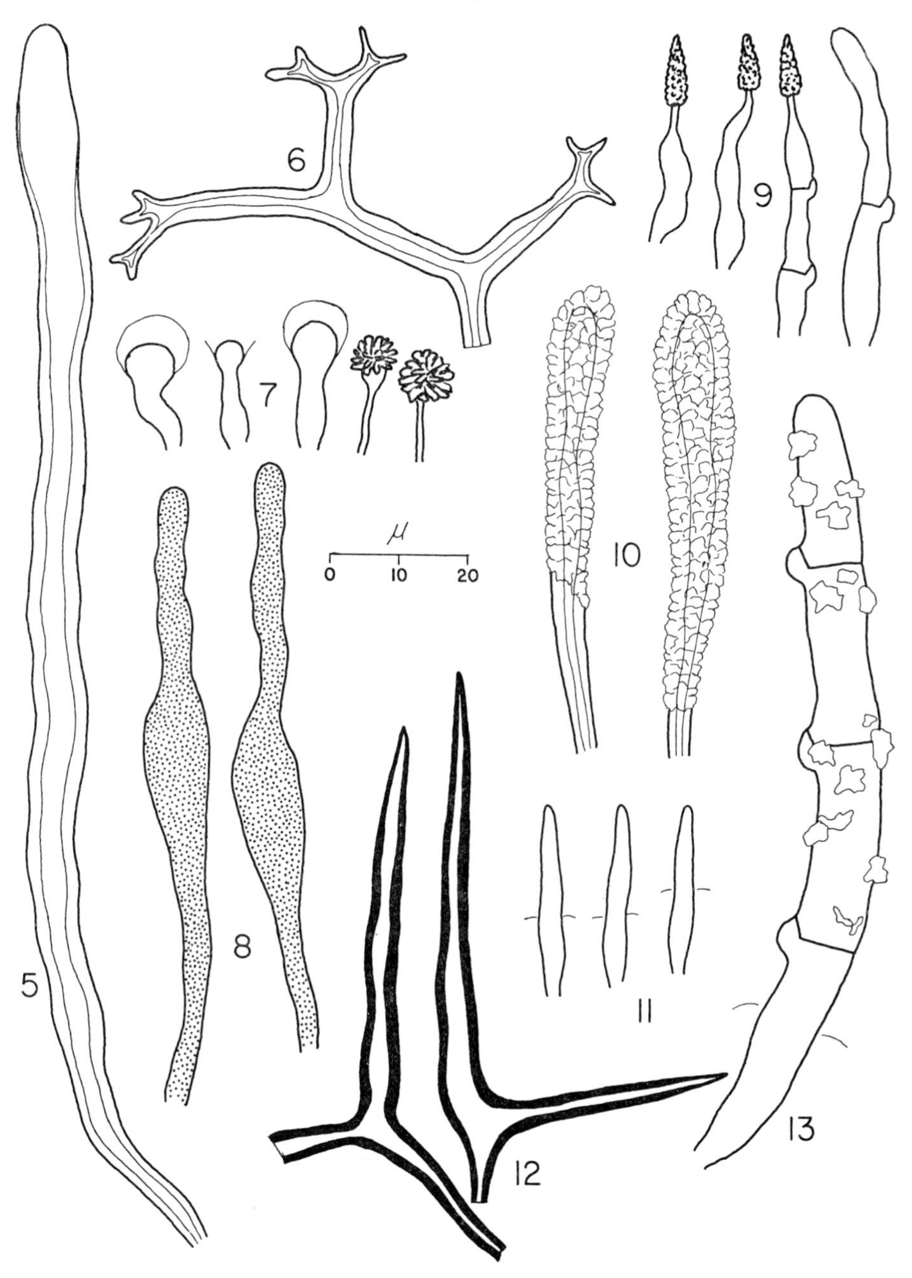

species, 36 (or 24%) heterothallic and bipolar species, and 39 (or 26%) homothallic species.

This suggests that the distribution of types of mating systems is about the same in the resupinate hydnaceous group as it is in the resupinate hymenomycetes as a whole, and that no general correlation exists between the hydnaceous hymenophore and type of mating system. However, some genera that form reasonably natural groupings seem to be rather uniform as far as type of sexuality is concerned. *Hyphodontia* and *Steccherinum* species for which the mating system has been determined are all tetrapolar. Bipolar heterothallism occurs in some species of *Hyphoderma* and homothallism occurs in species of *Mycoacia* and the *Phanerochaete* group.

Most of the fungi considered here have a monomitic hyphal system with clamp connections (fig. 2). A dimitic structure with nodose-septate generative hyphae (fig. 1) is found primarily in the genera *Steccherinum* and *Cystostereum* and the family Auriscalpiaceae. A small group of species characterized by a monomitic hyphal system with clamps absent or rare also exists. These may eventually find a natural position in *Phanerochaete* when the limits of that genus are determined. Some of these have rare opposite or whorled clamps on the hyphae of the basidiocarp and in culture, a character associated with homothallism by Boidin and Lanquetin (1965).

Hymenial cystidia are found in many species and many different types are represented in the group (figs. 5, 7, 9, 10, 11, 13). They are generally significant taxonomically at the species level. One genus, *Steccherinum*, does have very similar thick-walled, heavily incrusted cystidia (fig. 10) in all species but these are associated with skeletal hyphae of a dimitic hyphal system. The species of *Hyphodontia* typically have hymenial cystidia, but they show great variation from one species to another (figs. 5, 9) and are not important as a generic character.

Gloeocystidia are found in several genera. In *Gloeodontia, Gloiodon, Dentipellis* (fig. 8), and some species of *Vararia* and *Scytinostroma* they are associated with amyloid spores. In *Cystostereum* and *Basidioradulum* they are associated with nonamyloid spores.

Our knowledge of temperate zone species indicates that most of them have a cosmopolitan distribution. The keys of Bourdot and Galzin (1928) or Nikolajeva (1961) serve reasonably well for the identification of North American specimens. However, a few species (such as *Radulum concentricum, Odontia furfurella,* and *Scytino-*

*stroma arachnoideum*) apparently have an extremely limited or widely disjunct distribution. Since large areas of the world are still relatively unexplored mycologically, the picture of distribution that we now have may be deceptive. It is difficult to evaluate the phylogenetic significance of distribution patterns and to determine the centers of origin of various taxa, particularly in the virtual absence of a fossil record.

The relatively small number of species with the hydnaceous hymenophore is of interest from a phylogenetic standpoint. Perhaps the hydnaceous hymenophore is not as efficacious in spore dispersal as are the smooth, lamellate, or tubular hymenophores or at least has not had as much survival value. It may also indicate that they represent recently evolved fungi in each line of development.

The objective of contemporary taxonomic research is to try to place the species of "resupinate hydnums" into genera that appear to represent natural groupings. When the genera and families listed by Donk (1964), supplemented with a few other taxa, are used as a basis, about 23 genera in 9 families are found to be involved. Approximately three-quarters of the resupinate hydnaceous fungi can be accommodated reasonably well in these 23 genera. The remainder present a difficult problem.

A few other fungi that may have resupinate fruiting bodies with hymenophores that appear hydnaceous are not included in this analysis because they have not generally been placed in the Hydnaceae by past workers. For example, *Veluticeps berkeleyi* (Berk. & Curt.) M. C. Cooke is a widely distributed fungus associated with decay of pine in North America (Gilbertson, Lombard, and Hinds, 1968). It has basidiocarps that appear hydnaceous because of abundant fascicles of sterile hyphae that penetrate the hymenium and project beyond it. *Veluticeps* has been placed in the Stereaceae (Donk, 1964).

The following section attempts to provide a key to the genera of Aphyllophorales containing species with resupinate hydnaceous basidiocarps. A short discussion of each follows.

### KEY TO GENERA OF APHYLLOPHORALES CONTAINING SPECIES WITH RESUPINATE HYDNACEOUS BASIDIOCARPS

1. Teeth usually arising directly from the substratum, subiculum absent or scanty (Clavariaceae Chev.) . . . . . . . . . . . . . . . . . . 2

1. Teeth arising from a distinct subiculum .................. 3
   2. Tissue monomitic; gloeoplerous hyphae present or absent; spores weakly amyloid ................ *Mucronella* Fries
   2. Tissue dimitic, gloeoplerous hyphae absent; spores not amyloid ........................... *Deflexula* Corner
3. Setae present; tissue permanently darkening (xanthochroic) in KOH solution .............. (Hymenochaetaceae Donk)–4
3. Setae absent; tissue not permanently darkening in KOH solution ........................................ 5
   4. Asterosetae abundant throughout the subiculum; hymenial setae also present .......................... *Asterodon* Pat.
   4. Asterosetae absent; hymenial setae present ........................... *Hydnochaete* Bres.
5. Hyphae pigmented, loosely arranged; basidiospores pigmented, echinulate to verrucose or strongly warted (Thelephoraceae Chev.) .................. *Tomentella* Pers ex Pat.
5. Hyphae and spores hyaline with few exceptions; spores smooth or echinulate .................................. 6
   6. Hyphal system dimitic, dextrinoid hyphidia absent ...... 7
   6. Hyphal systems monomitic or with dextrinoid hyphidia ... 11
7. Basidiospores amyloid, echinulate, or minutely verrucose .... 8
7. Basidiospores not amyloid, smooth ...................... 10
   8. Basidiocarps perennial; gloeocystidia absent (Echinodontiaceae Donk) ... *Echinodontium* Ell. & Ev.
   8. Basidiocarps annual; gloeocystidia present (Auriscalpiaceae Maas G.) ...................................... 9
9. Tissue brown, coarsely tomentose; gloeocystidia present ............................... *Gloiodon* Karst.
9. Tissue pale colored; gloeocystidia and incrusted cystidia present ............................. *Gloeodontia* Boid.
   10. Gloeocystidial vesicles present; tissue compact, not fibrous (Stereaceae Pilát) ................ *Cystostereum* Pouz.
   10. Gloeocystidial vesicles not present; thick-walled, incrusted skeletocystidia present (residual genus of Hydnaceae *s.l.*) ....................... *Steccherinum* S. F. Gray
11. Spores cyanophilous in cotton blue; tissue green in ferric sulphate solution (Gomphaceae Donk) ......... *Kavinia* Pilát
11. Spores not cyanophilus in cotton blue; tissue not green in ferric sulphate solution ................................ 12

12. Spores subglobose, thick-walled, amyloid; gloeocystidia present (Hericiaceae Donk) .......... *Dentipellis* Donk
12. Spores globose to cylindric, thin-walled, amyloid or not amyloid, gloeocystidia present or absent (the remaining genera have been placed in Corticiaceae Herter) ...... 13
13. Dextrinoid hyphidia present; hymenium of the catahymenial type ........................................................ 14
13. Dextrinoid hyphidia not present; hymenium of the euhymenial type ........................................................ 15
14. Dextrinoid hyphidia with conspicuous dichotomous branching ........................ *Vararia* P. Karst.
14. Dextrinoid hyphidia not with conspicuous dichotomous branching ...................... *Scytinostroma* Donk
15. Basidia urniform, usually 6–8 sterigmate .... *Sistotrema* Fries
15. Basidia clavate, with or without a median constriction, 4-sterigmate ........................................ 16
16. Teeth capped by an amber droplet .... *Dacryobolus* Fries
16. Teeth not capped by an amber droplet ................ 17
17. Hyphae with clamps absent or rare, when present often opposite or whorled .... *Phanerochaete* P. Karst. emend. Donk
17. Hyphae with abundant simple clamps .................. 18
18. Hymenophore raduloid ........................ 19
18. Hymenophore hydnoid or odontioid .................. 20
19. Moniliform hyphal apices and gloeocystidia present ........................ *Basidioradulum* Nobles
19. Moniliform hyphal apices and gloeocystidia absent ........................ *Radulomyces* P. Chris.
20. Cystidia present ........................................ 21
20. Cystidia absent ........................................ 22
21. Basidia and spores large ............... *Hyphoderma* Wall.
21. Basidia and spores small ............ *Hyphodontia* J. Erikss.
22. Basidiocarps fragile, hyphae swollen at septa, spores of most species echinulate .......... *Trechispora* P. Karst.
22. Basidiocarps firm, hyphae uniform in diameter, spores smooth ........................... *Mycoacia* Donk

## Remarks on Genera

*Mucronella* Fr. (Hym. Eur. p. 629. 1874) is a small genus characterized by small, simple or branched, pendent, tooth-like basidiocarps usually not united by a common subiculum. Patouillard (1900) mentioned the affinity of *Mucronella* to "les Clavariés inferieurs" and Corner (1950) and Donk (1964) consider the individual teeth to be clavarioid fruiting bodies. Corner placed *Mucronella* in his *Clavariadelphus*-series and Donk placed it in the family Clavariaceae. Hyphal structure is monomitic with clamp connections present. Basidiospores of some species are weakly to distinctly amyloid, smooth, and may vary considerably in size on the same basidiocarp. Some of my collections of *M. aggregata* Fr. have a thin, arachnoid, sterile subiculum connecting the individual teeth. The species described from North America are poorly defined and a detailed study of the genus is badly needed to clarify specific entities. Some North American collections have conspicuous gloeoplerous hyphae which are similar to those found in *Dentipellis*. The combination of this character and the amyloid spores found in the genus suggests a possible relationship to the Hericiaceae.

*Deflexula* Corner (Clavaria and Allied Genera, p. 394. 1950) is included in this survey because it contains one species, *Deflexula ulmi* (Peck) Corner, originally described from North America as a *Mucronella*. Corner (1950) placed *Deflexula* in his Pteruloid-series and Donk (1964) placed it in the subfamily Pteruloideae of the family Clavariaceae. The hyphal system of *D. ulmi* is dimitic with clamp connections on the generative hyphae. The spores are nonamyloid, ellipsoid, conspicuously apiculate, and 8–12 x 12–16 μ. The other species of *Deflexula* recognized by Corner are from tropical regions of the world.

*Asterodon* Pat. (Bull. Soc. Mycol. France 10:130. 1894) is a monotypic genus characterized by resupinate, brown, strongly hydnaceous basidiocarps with soft, cottony subiculum. Microscopically the abundant asterosetae and hymenial setae (fig. 12) are distinctive. The tissue gives a typical xanthochroic reaction in KOH solution. *Asterodon ferruginosus* Pat. is a cosmopolitan species. Patouillard (1900) placed *Asterodon* in his *Série des Astérostromes* with the genus *Asterostroma* Massee. Bourdot and Galzin (1928) followed this arrangement but mentioned that the hymenial setae were similar to those of *Hymenochaete*, *Phellinus*, and *Xanthochrous*. Donk (1964) placed *Asterodon* in the subfamily Hymenochaetoideae of the family Hymenochaetaceae. *Asterdon ferruginosus* is associated with a white rot as are other members of this family, and in culture gives a positive oxidase reaction (F. F. Lombard, personal communication). The

Asterosetae are similar in shape to those of *Asterostroma*, but those of *Asterostroma* are not xanthochroic. Other characters, particularly the small, hyaline, and smooth spores (fig. 31), and the hymenial setae indicate a relationship to *Hymenochaete* Lev.

*Hydnochaete* Bres. (Hedwigia, 25:287. 1896) is characterized by thin, resupinate to reflexed basidiocarps with a strongly hydnaceous hymenophore that results from the early splitting of tubes. The tissue is brown and xanthochroic and hymenial setae are abundant. This genus is also well placed in the Hymenochaetaceae by Donk (1964). *Hydnochaete olivaceum* (Schw.) Banker is the only species in North America. It is perhaps most closely related to some of the species of poroid fungi such as *Polyporus iodinus* Mont., with thin, setulose basidiocarps. Patouillard (1900) considered it to represent an intermediate between *Phellinus* and *Hymenochaete*. Many species of polypores have tubes that become thin and deeply lacerated with age, simulating the hydnaceous hymenophore. *Hydnochaete olivaceum* is a species in which this occurs very early in the development of the basidiocarp. Banker (1914) gave an account of the taxonomy and morphology of *H. olivaceum*. It is associated with a white rot and gives a positive oxidase reaction (F. F. Lombard, personal communication).

*Tomentella* Pers. ex Pat. (Hym. Eur., p. 154. 1887) is a large genus characterized by pigmented hyphae, pigmented and ornamented spores (fig. 30), and loose, cottony texture. A number of species have distinctly to slightly hydnaceous basidiocarps and some of these have been placed in the genus *Caldesiella* Sacc. and included in the Hydnaceae *s.l.* Basidiocarps of numerous other tomentelloid fungi not previously placed in *Caldesiella* also have a tendency to be hydnaceous. The relationship of these hydnaceous fungi to the tomentelloid fungi with a smooth hymenophore was recognized by Patouillard, who placed *Caldesiella* and *Tomentella* in his *Série des Phylactéries*. Rogers (1934) stated, "not by the wildest exercise of the imagination can *Caldesiella* be held to be anything other than a *Tomentella* with hymenophore sculpturing in somewhat higher relief." Larsen (1967b) has recently transferred the species formerly in *Caldesiella* to *Tomentella*. Species with a hydnaceous hymenophore include *Tomentella crinalis* (Fr.) M. J. Larsen, *T. calcicola* (Bourd. & Galz.) M. J. Larsen, *T. viridis* (Berk.) H. G. Cunningh., *T. subcalcicola* M. J. Larsen, and *T. italica* (Sacc.) M. J. Larsen.

The presence of the hydnaceous hymenophore in this group is

FIGURES 14–45.
Basidia and basidiospores of some resupinate hydnaceous fungi. Fig. 14, Basidium of *Vararia granulosa* (RLG 3376). Fig. 15, Basidium of *Hyphoderma setigera* (RLG 6933). Fig. 16, Basidium of *Dentipellis separans* (RLG 3795). Fig. 17, Basidium of *Tomentella crinalis* (RLG 6556). Fig. 18, Basidium of *Sistotrema raduloides* (RLG 4344). Fig. 19, Basidium of *Odontia pruni* (JLL 10685). Fig. 20, Basidium of *Dacryobolus sudans* (RLG 4957). Fig. 21, Cluster of basidia of *Hyphodontia abieticola* (RLG 3682). Fig. 22, Cluster of basidia of *O. romellii* (RLG 289). Fig. 23, Spores of *Mucronella aggregata* (RLG 3651). Fig. 24, Spores of *Hyphodontia floccosa* (RLG 3513). Fig. 25, Spores of *O. pruni* (JLL 10685). Fig. 26, Spores of *O. subcrinalis* (RLG 6322). Fig. 27, Spores of *Hyphoderma setigera* (RLG 6933). Fig. 28, Spores of *Steccherinum ciliolatum* (JLL 6074). Fig. 29, Spores of *Hyphodontia barba-jovis* (RLG 3659). Fig. 30, Spores of *T. crinalis* (RLG 6556). Fig. 31, Spores of *Asterodon ferruginosus* (RLG 2669). Fig. 32, Spores of *Mycoacia alboviride* (RLG 4138). Fig. 33, Spores of *Kavinia himantia* (JLL 10115). Fig. 34, Spores of *Radulum casaerium* (RLG 6350). Fig. 35, Spores of *Radulum concentricum* (RLG 221). Fig. 36, Spores of *Sarcodontia crocea* (E. Ross). Fig. 37, Spores of *Gloeodontia discolor* (JLL 4136). Fig. 38, Spores of *Dentipellis separans* (RLG 3795). Fig. 39, Spores of *Dacryobolus sudans* (JLL 4417). Fig. 40, Spores of *Trechispora farinacea* (RLG 3639). Fig. 41, Spores of *Radulum magnoliae* (RLG 26). Fig. 42, Spores of *Hyphodontia arguta* (RLG 6299). Fig. 43, Spores of *Cystostereum pini-canadensis* (JLL 10615). Fig. 44, Spores of *H. alutacea* (RLG 7572). Fig. 45, Spores of *H. abieticola* (RLG 7209).

Figures 14–45.

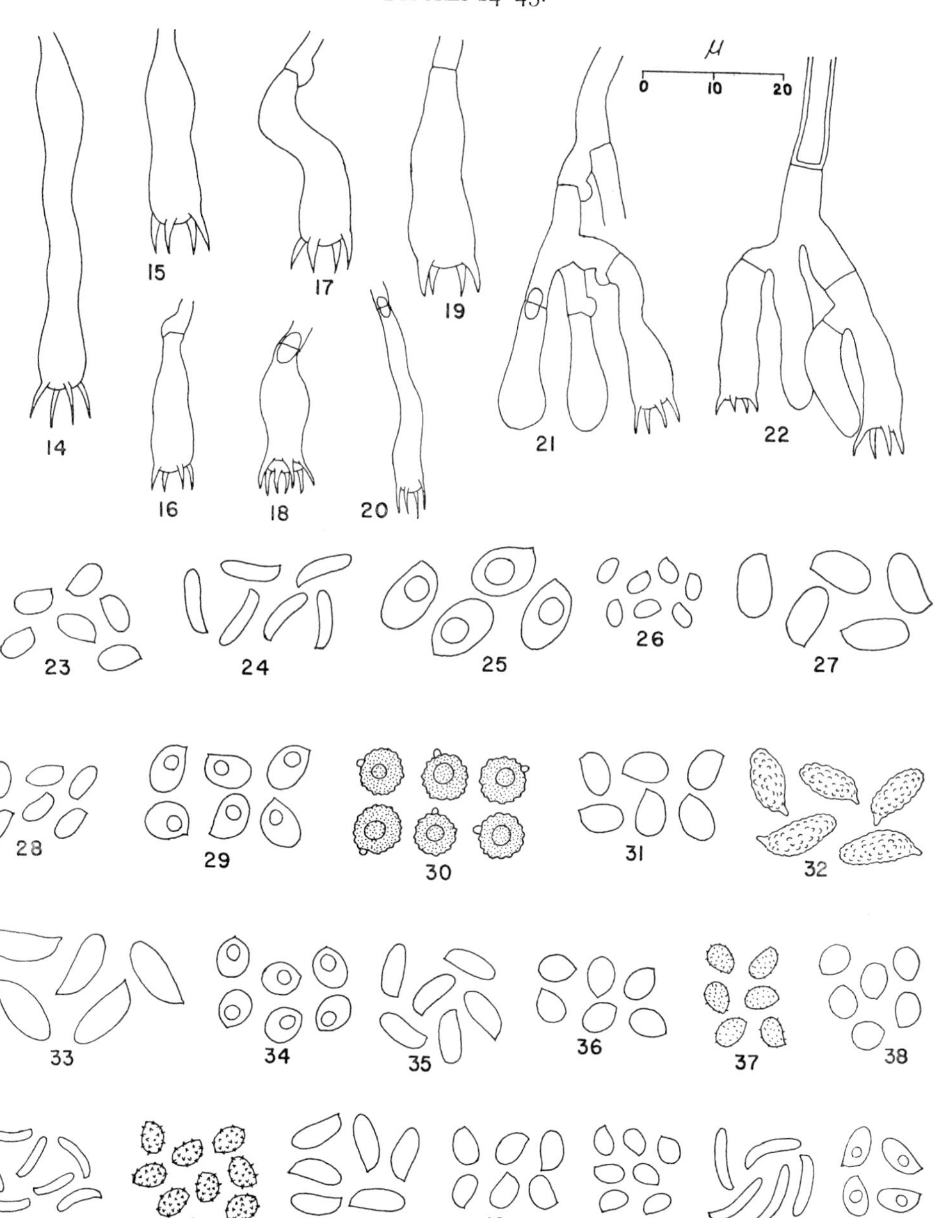

an excellent example of convergent evolution of this character. The tomentelloid Hymenomycetes (*Tomentella*, *Pseudotomentella* Svrček, and *Knieffiella* P. Karst.) form a natural group that are perhaps related to some of the stipitate hydnaceous fungi through resupinate to reflexed or substipitate forms such as *Thelephora terrestris* Fr. (Larsen, 1967a). There is virtually no information on cultural characters or sexuality of these fungi as methods of culturing them have not been discovered.

*Echinodontium* Ell. & Everh. (Torrey Bot. Club. Bull. 27:41. 1900) was based on *E. tinctorium* (Ell. & Everh.) Ell. & Everh., a species with sessile, ungulate basidiocarps. Some species subsequently placed in this genus have basidiocarps that may be resupinate. These include *E. japonicum* Imazeki, *E. sulcatum* (Burt.) Gross, and probably *E. ballouii* (Banker) Gross, and *E. taxodii* (Lentz & McKay) Gross. The genus is distinguished by its perennial, woody to leathery basidiocarps with smooth to warted or strongly hydnaceous hymenophore, stratified subhymenium, nodose-septate hyphae with secondary simple septa, thick-walled, incrusted cystidia, and hyaline, amyloid, echinulate basidiospores. Donk (1961) proposed the family Echinodontiaceae to include the genus *Echinodontium*. Gross (1964) has provided a detailed monograph of the Echinodontiaceae, which he considered to be a monotypic family consisting of six species of *Echinodontium*. This group of six species forms a strikingly distinct and closely related group which shows no close relationships to any other fungi. The wide geographic distribution of *E. sulcatum*, a common saprophyte, and the relatively restricted distribution of the other five species, mostly associated with heartrots of living trees, is of interest. These fungi may have evolved from an *E. sulcatum*-like ancestor with a wide geographical distribution. Host-specificity may have been the isolating mechanism that resulted in the development of the other more geographically restricted species, along with the appearance of the more strongly hydnaceous hymenophore.

*Gloiodon* P. Karst. (Medd. Soc. Fauna Fl. Fenn. 5:42. 1879) is a monotypic genus based on *G. strigosus* (Schw. ex Fr.) P. Karst., a cosmopolitan species. Basidiocarps of this fungus are resupinate or reflexed and are composed of a tough, coarse, but loosely constructed tomentum bearing spines on the lower surface. This tough tomentum is composed of strands of generative hyphae with clamp connections and skeletal hyphae. Gloeocystidial hyphae extend into the hymenium and the basidiospores are subglobose, amyloid, and slightly verrucose. *G. strigosus* is reported to be tetrapolar (Fries, 1941) and to give a positive oxidase reaction (F. F. Lombard, personal communication). Nikolajeva (1961) placed *Gloiodon* (as *Sclerodon* Karst.) with *Auriscalpium* P. Karst. in the tribe Auriscalpieae under the family Hydnaceae. Maas Geesteranus (1963) proposed the new family Auriscalpiaceae and included both genera in it. In this same publication he also placed the agaricaceous genus *Lentinellus* in the same family. Singer (1962) and Romagnesi had previously pointed out the similarity in micromorphology of *Auriscalpium* and *Lentinellus*. Donk (1964) also included the three genera in the Auriscalpiaceae. Boidin (1966) has since added the genus *Gloeodontia* to this family.

*Gloeodontia* Boid. (Cahiers de la Maboke 4:22. 1966) is characterized by a dimitic hyphal system, nodose-septate generative hyphae, incrusted skeleto-cystidia, sulfobenzaldehyde-positive gloeocystidial hyphae, and minutely verrucose, amyloid basidiospores (fig. 37). Boidin (1966) transferred one species, *Irpex discolor* Berk. & Curt., to this genus, which he placed in the family Auriscalpiaceae. *G. discolor* has been found in the southeastern United States (Gilbertson, 1965a) and in Africa (Boidin, 1966). A very similar fungus with hydnaceous basidiocarps occurs throughout the northwestern United States and in western Canada but I am not satisfied that it is conspecific with *G. discolor*.

Eriksson (1958a) described a new species, *Gloeocystidiellum nannfeldtii* John Erikss., and discussed the similar species *G. heterogeneum* (Bourd. & Galz.) Donk. Both of these have amyloid, asperulate spores, gloeocystidia, and a dimitic hyphal system with nodose-septate hyphae. These two species, by virtue of their similar micromorphology, appear closely related to, if not congeneric with, *Gloeodontia discolor*. Perhaps they represent a corticioid element of the Auriscalpiaceae.

*Cystostereum* Pouzar (Česká Mykol. 13:18. 1959) is based on *C. murraii* (Berk. & Curt.) Pouz., which Pouzar (1959) stated as the only species in the genus. One other fungus, however, *Radulum pini-canadensis* Schw., may be placed in *Cystostereum*. It is distinguished by a smooth to strongly hydnaceous hymenophore, dimitic hyphal system, imbedded gloeocystidial vesicles with yellowish contents, and small, hyaline, nonamyloid spores (fig. 43). As pointed out by Gilbertson and Larson (1965), *Radulum pini-canadensis* is close to *Stereum murraii* (Berk. & Curt.) Burt (=*C. murraii*). I believe the two species are congeneric and the combination **Cystostereum pini-canadensis** (Schw.) comb. nov. is proposed (basionym = *Radulum pini-canadensis* Schw., Am. Phil. Soc. Trans. N. S. 4:164. 1832). *Cystostereum murraii* also shows considerable variation in configuration of the hymenophore and many specimens approach a hydnaceous condition.

*Steccherinum* S. F. Gray (Nat. Arr. Brit. Plants 1:597. 1821) is characterized by thin, fibrous, resupinate, or reflexed basidiocarps with distinctive micromorphology. The hyphal system is dimitic (fig. 1), with thick-walled, aseptate skeletal hyphae that give rise to conspicuous, often heavily incrusted skeleto-cystidia (fig. 10) in the hymenial region. Generative hyphae are thin-walled and nodose-septate. Basidiospores are small, hyaline, smooth, nonamyloid, and cylindric to ovoid in shape (fig. 28). The basidiocarps tend to be pinkish in color. Some species of hydnaceous fungi that belong here are *Steccherinum ochraceum* (Fr.) S. F. Gray, *S. robustior* (J. Erikss. & Lundell) J. Erikss., **S. ciliolatum** (Berk. & Curt.) Gilbertson, comb. nov. (basionym = *Hydnum ciliolatum* Berk. & Curt., Hooker's Jour. Bot. and Kew Gard. Misc. 1:235. 1849), and **S. fuscoatrum** (Fr.) Gilbertson, comb. nov. (basionym = *Hydnum fuscoatrum* Fr., Syst. Myc. 1:416. 1821), and *S. laeticolor* (Berk. & Curt.) Banker.

*Steccherinum ochraceum* is heterothallic and tetrapolar (Kimura, 1954) and gives a strong oxidase reaction on gallic and tannic acid media. *Steccherinum fimbriatum* also gives a strong oxidase reaction (Boidin, 1958), as does *S. ciliolatum*, *S. fuscoatrum*, and *S. setulosum* (F. F. Lombard, personal communication). Information on cultural characters of the other species is not available.

The similarity of this group of hydnaceous fungi to some species in the Polyporaceae has been noted by Eriksson (1958b), who commented that *Poria eupora* (Karst.) Cooke could be referred to it. Other polypores with the same pattern of micromorphology are *Poria rixosa* Karst., *P. luteoalba* (Karst.) Sacc., *P. fimbriatella* (Peck) Sacc., and *P. radula* Pers. ex Fr. These species have been placed in the genus *Chaetoporus* P. Karst. emend. Bond. by contemporary authors. Donk (1967) has further discussed similarities between species of *Steccherinum* and *Chaetoporus*. Three of the species (*P. eupora*, *P. rixosa*, and *P. luteoalba*) studied by Lombard and Gilbertson (1966) had similar cultural characters and gave strong oxidase reactions on tannic and gallic acid agar media. *Poria eupora* is reported as tetrapolar by Nobles, Macrae, and Tomlin (1957).

The available evidence on morphology of basidiocarps, cultural characteristics, and mating systems strongly indicates that the species of *Steccherinum* and *Chaetoporus* listed form a natural group. It is perhaps significant that the species of polypores mentioned have tubes that tend to become lacerate and deeply split, approaching a hydnaceous condition. Their basidiocarps also tend to be pinkish in color. Some additional species of resupinate polypores that do not fit into this natural group have been placed in the genus *Chaetoporus*.

One other interesting species, *Odontia subcrinalis* (Peck) Gilbertson, could logically be placed in *Steccherinum*. It has the dimitic hyphal system (fig. 4), incrusted cystidia, small spores (fig. 26), and pinkish basidiocarp typical of the *Steccherinum* species but has only rare clamp connections on the generative hyphae. It does not seem to have affinities with any of the other species with clamps lacking or extremely rare. It has a wide distribution in North America but has not been reported elsewhere (Gilbertson, 1963). A similar species is known in Europe and is placed in *Mycoleptodon*, an equivalent of *Steccherinum*, by Nikolajeva (1961). This is *Mycoleptodon kavinae* Pilát, which, judging from a specimen sent by Dr. Nikolajeva, is very similar to *O. subcrinalis*. *Steccherinum setulosum* (Berk. & Curt.) Miller does not have as well-developed and conspicuous skeletals as the rest of the species listed above but is considered a true species of *Steccherinum* by Maas Geesteranus (1962).

*Kavinia* Pilát (Stud. Bot. Cech. 1:3. 1938) includes two common cosmopolitan species of hydnaceous resupinates with cyanophilous spores and a positive reaction (green) in ferric sulphate solution. These are *K. himantia* (Schw.) J. Erikss. and *K. sajanensis* Pilát. Donk (1956) reported that *K. sajanensis* was the same fungus as *Hydnum alboviride* Morgan [ = *Mycoacia alboviride* (Morg.) Miller & Boyle, 1943]. My observations on *K. himantia* show the spores to be smooth to verrucose (fig. 33) and moderately cyanophilous. *Mycoacia alboviride* has verrucose (fig. 32), strongly cyanophilous spores. Basidiocarp tissue of both gives a green reaction in ferric sulphate solution. Cultures of both species give a positive oxidase reaction (F. F. Lombard, personal communication). Eriksson (1954) discussed the relationship of these (as *Clavaria himantia* and *C. bourdotii*) to species of other clavariaceous fungi and later (Eriksson, 1958b) stated that the correct place for these two species was in the Clavariaceae. Donk (1964) placed *Kavinia* in the family Gomphaceae along with some clavarioid and cantharelloid fungi.

*Dentipellis* Donk (Persoonia 2:232. 1962) was erected to accommodate some species with resupinate to rarely effuso-reflexed basidiocarps bearing densely crowded, fragile teeth on a soft, membranous subiculum. Spores are amyloid, smooth (fig. 38), and gloeocystidia (fig. 8) are present. The type species is *Dentipellis fragilis* (Pers. per Fr.) Donk and *Dentipellis separans* (Peck) Donk is given as a second member of the genus. The latter was considered by Donk to be the correct name for the fungus known to North American authors as *Mycoacia macrodon* (Fr.) Miller & Boyle. Donk (1964) placed *Dentipellis* in the family Hericiaceae but emphasized the close relationship to the Corticiaceae through *Gloeocystidiellum.* The subsequent description of the genus *Gloeodontia* by Boidin (1966) provided another related taxon. That genus is characterized by a dimitic hyphal system, gloeocystidia, and amyloid, echinulate spores. The thick-walled refractive hyphae in *Dentipellis* appear more like oleaginous or gloeocystidial hyphae than typical thick-walled skeletal hyphae. It seems likely that *Dentipellis* may represent a connection between the corticioid fungi and *Hericium* while *Gloeodontia* represents a connection between the corticioid fungi and other Auriscalpiaceae, both representing lines of development from *Gloeocystidiellum*-like forms.

*Vararia* P. Karst. (Bidr. Kanned. Finl. Nat. Och. Folk 62:96. 1903) is characterized by a catahymenium and dichohyphidia that are dextrinoid in Melzer's reagent (fig. 6). Some species have gloeocystidia and some have ornamented, amyloid spores. The only member of this genus that has been placed in the Hydnaceae *s. l.* is *Vararia granulosa* (Fries) M. Laur. (= *Grandinia granulosa* Fries), which has a grandinioid or papillose hymenial surface. It has been reported to be heterothallic and bipolar and to have a positive oxidase reaction (Maxwell, 1954). Donk (1964) included the genus *Vararia* in the Hymenochaetaceae. In Donk's discussion of the Hymenochaetaceae he emphasized the absence of clamp connections as a family character. Clamp connections on the generative hyphae of the basidiocarp are characteristic of most of the nine species of *Vararia* known from temperate North America (Gilbertson, 1965b) and are found in most cultures of species of *Vararia* deposited with the Forest Disease Laboratory, U. S. Forest Service at Beltsville, Maryland (F. F. Lombard, personal communication). The occurrence of the catahymenial structure, clamp connections, amyloid, ornamented spores, the dextrinoid hyphidia, and gloeocystidia seem to argue more strongly for a position elsewhere with the closely related genus *Scytinostroma* Donk. This conclusion has also been reached by Jackson (1948), Lentz and McKay (1966), Welden (1965), and Eriksson (1958b). Perhaps the best disposition of this genus will be to recognize a separate family to include *Vararia* and *Scytinostroma.*

*Scytinostroma* Donk (Fungus 26:19. 1956) seems to be very close to *Vararia* for reasons detailed in the discussion of that genus. Like *Vararia,* it has a catahymenial structure and includes some species with amyloid, ornamented spores and some with gloeocystidia. It is also characterized by dextrinoid hyphidia but these are not distinctly dichotomously branched as in *Vararia.* Some species have clamp connections and some apparently do not. One species, *Scytinostroma arachnoideum* (Peck) Gilbertson [ = S. *quaesitum* (Jacks.) Donk] has a strongly hydnaceous hymenophore. It has a wide distribution in North America (Gilbertson, 1962) but has not been reported elsewhere. S. *arachnoideum* has amyloid, echinulate spores and large hymenial cystidia. Basidiocarps of S. *galactinum* (Fries) Donk occasionally show a grandinioid or papillose hymenial surface but are usually smooth. Donk (1964) suggests that *Scytinostroma* may be more closely related to genera such as *Gloeocystidiellum* than to *Vararia.* Donk also mentions that *Scytinostroma* differs from *Vararia* in that some of the species of *Scytinostroma* have clamps. Clamps are typical of most species of *Vararia,* as pointed out in the discussion under that genus. *Scytinostroma* includes one species, S. *portentosum* (Berk. & Curt.) Donk, with hyphidia that approach the typical dichotomously branched hyphidia of species of *Vararia.* In fact, Cunningham (1953) placed this species in *Vararia.*

*Sistotrema* Fries (Syst. Mycol. 1:426. 1821) includes species with urniform basidia, often with 6–8 sterigmata (fig. 18). Species previously placed in the Hydnaceae are S. *raduloides* (Karst.) Donk and S. *muscicola* (Pers.) Donk, both of which have strongly to slightly hydnaceous basidiocarps, and S. *brinkmannii* (Bres.) J. Erikss., which frequently has a grandinioid hymenial surface. Other species included in this group by Rogers (1944, under the generic name *Trechispora*) have a poroid or smooth hymenophore. This group comprises essentially the "*Urnigera*" section recognized by Bourdot and Galzin in *Corticium* and *Gloeocystidium,* with several species from other genera of hymenomycetes including *Grandinia* and *Poria.*

*Dacryobolus* Fr. ex Fr. (Summa Veg. Scan. 2:404. 1849) is a monotypic genus based on *Hydnum sudans* Alb. & Schw. ex Fries [ = *Odontia sudans* (Alb. & Schw. ex. Fr.) Bres.]. This distinctive fungus is characterized by a slightly hydnaceous hymenophore with a droplet of amber-colored material secreted at the apex of each tooth. The hyphal system is monomitic with distinctive, thick-walled, nodose-septate hyphae. The basidia are narrowly clavate, only 3–3.5 μ wide (fig. 20), and the spores are hyaline, smooth, narrowly allantoid, and 4–5.5 x 0.7–1 μ (fig. 39). *Dacryobolus sudans* was reported to be heterothallic and tetrapolar by Biggs (1938) and to have a negative oxidase reaction by Boidin (1958). My specimens are apparently associated with a brown rot with the same distinctive mycelium in small patches in the decayed wood. The affinities of *D. sudans* among the other corticioid fungi are obscure at this time. The negative oxidase reaction and brown rot are very unusual among the hydnaceous resupinates, especially in combination with tetrapolar heterothallism.

*Phanerochaete* Karst. em. Donk. (Persoonia 2:223. 1962) is a broadly conceived genus the limits of which have not yet been established. It would correspond with *Peniophora* sect. *Membranacea* group A of Bourdot and Galzin (1928) with the addition of some additional species of *Peniophora*, *Corticium*, and *Odontia.* Eriksson (1958b) recognized this grouping under the tentative generic designation "*Membranicium ad int.*" Boidin (1958) recognized essentially the same group as "*Corticium* sect. *Subeffibulata*" and also mentioned that *Odontia corrugata* could be included. Donk (1964) listed several groups of species that might be referred to *Phanerochaete* when more information became available. One of these includes those with toothed hymenophore and cystidia lacking or inconspicuous. *Oxydontia chrysorhizon* (Torr.) Rogers & Martin is in this group. Another includes those with toothed hymenophore and conspicuous cystidia. *Odontia hydnoides* (Cooke & Mass.) v. Hoehn. and *Odontia corrugata* (Fr.) Bres. are species listed in this group by Donk. *Odontia laxa* Miller, *Odontia pruni* Lasch., and *Radulum magnoliae* Berk. & Curt. might also fit here. Most of the fungi tentatively placed in this genus have rare clamp connections on hyphae of the basidiocarp and in culture show some clamps that are double or whorled. Boidin and Lanquetin (1965) stated that all of the species studied by them with clamps opposite or whorled were homothallic. Cultural data on many species that may fall in this *Phanerochaete* group are not yet available and so the significance of this relationship as a generic character cannot be fully evaluated. *Odontia hydnoides* and *Oxydontia chrysorhizon* have been reported to be homothallic (Brown, 1935). This is a very good example of a group for which cultural information is an essential adjunct to basidiocarp morphology in the determination of generic limits.

*Basidioradulum* Nobles (Mycologia 59:192. 1967). Nobles proposed this new genus as the old name *Radulum* Fries has been shown to be based on an Ascomycete and therefore not available for a genus of Basidiomycetes (Donk, 1956). *Basidioradulum radula* (Fr.) Nobles (= *Radulum orbiculare* Fr.) was given as the type of this genus and no additional transfers were proposed (Nobles, 1967). The moniliform terminal cells and gloeocystidia of this species together with the large nonamyloid, cylindric spores form a highly distinctive combination of characters not found in other hydnaceous fungi. *B. radula* is tetrapolar and gives a positive oxidase reaction. *Radulum* Fr. is an artificial genus and none of the other species previously placed in it show any close relationship to *Basidioradulum radula*. It seems preferable to consider *Basidioradulum* as a monotypic genus, at least for the present.

*Radulomyces* M. P. Christiansen (Dansk. Bot. Arkiv. 19[2]:230. 1960) is a genus without very distinctive characters, erected to accommodate some species previously placed in *Corticium*, *Radulum*, and *Irpex*. Those with a hydnaceous hymenophore include *Radulum molare* Chaill. in Fr. and *Irpex pachyodon* (Pers.) Quel. Christiansen described *Radulomyces* as having basidiocarps with waxy and somewhat gelatinous or coriaceous consistency, hymenophore smooth to warted, raduloid or irpicioid when fresh, often smooth on drying, hyphae thin-walled, nodose-septate, basidia clavate, somewhat undulate, 2–4 spored, and spores hyaline, smooth, thin-walled, nonamyloid, and ellipsoid to subglobose. The type species is *Thelephora confluens* Fr. [ = *Corticium confluens* (Fr.) Fr.] Oberwinkler (1965) considered *R. confluens* to belong in a series of primitive Basidiomycetes with "plastischer basidie" and to be closely related to the family Xenasmataceae, described by him in the same publication. The other species included in *Radulomyces* by Christiansen presumably have a different basidial type, the typical clavate homobasidium, and would be doubtfully congeneric with *R. confluens*.

*Hyphodontia* J. Erikss. (Symb. Bot. Upsalienses 16 [1]:101. 1958) was published with a very brief and rather vague description as follows: "Genus *Hymenomycetum* corticioidum et hydnoidum, generi

*Hyphodermati* affine sed differt basidiis minoribus, fructificationibus fibrosis et hyphis angustioribus." Eriksson stated that its "characteristics are, however, not very striking and it is therefore difficult to give it a description clear also to the inexperienced mycologist." As I understand this genus, it is characterized by a smooth to hydnaceous hymenophore; fibrous texture; monomitic hyphal system; nodose-septate hyphae that are much branched at the septa, some rather thick-walled (fig. 2); cystidia variable, ranging from small, often apically swollen structures to large thick-walled tubulae (figs. 5,9); basidia small, about 5–6 μ in diameter, urniform with a median constriction, 4 sterigmate, developing in candelabrum-like clusters (fig. 21); basidiospores small, ovoid to cylindric, smooth, nonamyloid (figs. 29, 42, 44, 45). All of the species of *Hyphodontia* for which the type of mating system has been determined are tetrapolar. All species I have collected are associated with a white rot and those studied in culture give a positive oxidase reaction. As thus defined this genus would include a number of species previously placed in the artificial genus *Odontia*, as well as some species from *Corticium* and *Peniophora*. Donk (1967) also suggested that a polyporoid element then placed in *Poria versipora* (Pers.) Lloyd was also closely related. In addition to those transferred to *Hyphodontia* by Eriksson (1958b), the following species would belong in this group: **Hyphodontia spathulata** (Schrad. ex Fr.) comb. nov. (basionym = *Hydnum spathulatum* Schrad. ex Fr., Syst. Myc. 1:423. 1821); **Hyphodontia stipata** (Fr.) comb. nov. (basionym=*Hydnum stipatum* Fr., Syst. Myc. 1:425.1821); **Hyphodontia rimosissima** (Peck) comb. nov. (basionym = *Odontia rimosissima* Peck, N. Y. State Mus. Rep. 50:114. 1897); and **Hyphodontia burtii** (Peck) comb. nov. (basionym = *Odontia burtii* Peck, N. Y. State Mus. Rep. 53:847. 1900). *Hyphodontia* was placed in the Corticiaceae by Eriksson (1958c) in the subfamily Hyphodermoideae with the closely related genera *Hyphoderma*, *Hypochnicium*, *Amphenema*, and *Fibricium*.

*Hyphoderma* Wallr. em. Donk (Fungus 27:14. 1957) includes fungi with both smooth and hydnaceous hymenophores. Gloeocystidia are present in some species and others have smooth or incrusted cystidia. The hyphal system is monomitic with clamp connections on the hyphae, which are generally thin-walled. Basidia are large (about 6–8 x 20–40 μ) and urniform with a median constriction (fig. 15). Basidiospores are smooth, nonamyloid, thin-walled, ovoid to cylindric, and relatively large (fig. 27). Species with hydnaceous basidiocarps include *H. setigerum* (Fr.) Donk, the type species, and *H. mutatum* (Peck) Donk. Both of these species have been reported to be bipolar. *Hydnum radula* Fr. ex Fr. [ = *Radulum orbiculare* (Fr. ex Fr.) Fr.] has also been placed in *Hyphoderma* but has been subsequently designated as the type of a new genus *Basidioradulum* Nobles. *Basidioradulum* has been discussed previously. *Hyphoderma* was placed in the Corticiaceae by Donk (1964), Eriksson (1958c), and Christiansen (1960). It is apparently closely related to *Hyphodontia* J. Erikss.

*Trechispora* Karst. (Hedwigia 29:147. 1890) contains several fungi with hydnaceous basidiocarps. It is characterized by short basidia with long sterigmata, nodose-septate hyphae with swellings at the septa, and small spores that in most of the species are echinulate (fig. 40). The name *Cristella* Pat. has been applied to this genus but Liberta (1966) believed this name was impriorable and that *Trechispora* Karst. was the nomenclaturally correct name. *T. alnicola* (Bourd. & Galz.) Liberta, *T. farinacea* (Pers. ex Fr.) Liberta, and *T. fastidiosa* (Pers. ex Fr.) Liberta are some species with hydnaceous basidiocarps. This genus has been placed in the Corticiaceae (as *Cristella* Pat.) by Eriksson (1958c) and Donk (1964). The type species, *T. onusta* Karst., is polyporoid and the other species placed in *Trechispora* are corticioid.

*Mycoacia* Donk (Meded. Nederl. Mycol. Ver. 18–20:150. 1931) is characterized by fleshy to waxy basidiocarps with more-or-less well-developed teeth and cystidia absent or represented by inconspicuous cystidioles. This is an artificial taxon in which species such as *M. stenodon* (Pers.) Nikol., *M. setosa*, and *M. uda* have been placed. Bipolar heterothallism has been reported for *M. setosa* and *M. uda*, and Boidin found them to have the "astatocènocytique" type of nuclear behavior. This combination of characters is typical of the group he designated as true *Phlebia* (Boidin, 1964) along with several other species formerly placed in *Phlebia*, *Merulius*, and *Corticium*. *Odontia hydnoides*, tentatively assigned to *Phlebia* previously, was found by Boidin to be homothallic and to have the "holocènocytique" form of nuclear behavior. It is now placed in the *Phanèrochaete* group.

From this review of groups in which the resupinate hydnaceous fungi may be placed it is evident that a great deal of work remains to be done before they will be accommodated in a generally accepted system. The fact that most of the species are assigned to genera tentatively placed in the large heterogeneous family Corticiaceae indicates that the problem of family relationships of this group still has difficult unsolved aspects.

## Remarks on some Residual Species

The following are some residual species that do not seem to fit into any of the preceding generic groupings.

*Odontia bicolor* (Fries) Bres. has small capitate cystidia with a thin halo-like cap (fig. 7). It also has abundant spherical clusters of crystals at the apices of cystidial hyphae. The hyphae are nodose-septate but become densely aggregated and gelatinized with individual hyphae difficult to discern. Except for the incrusted cystidia, it is quite close to *Corticium furfuraceum* Bres. Eriksson (1958c) considered both species to be near the genus *Phlebia* Fr. em. Donk. *C. furfuraceum* was reported to have an *Oedocephalum* stage in culture (Maxwell, 1954), a character not shared by *Odontia bicolor* (Nobles, 1953).

*Radulum casaerium* (Morg.) Lloyd, a common species on *Populus* in North America, has a dimitic hyphal system, and small, hyaline, nonamyloid spores (fig. 34). Although it lacks the gloeocystidial vesicles typical of *Cystotereum murraii* and *C. pini-canadensis*, it can logically be placed near these two species. Basidiocarps of *Radulum casaerium* also have a consistency and general aspect similar to those of the *Cystostereum* species. *R. casaerium* is tetrapolar (Nobles, Macrae, & Tomlin, 1957) and is associated with a white rot. It does not find a natural place in any other group.

*Radulum concentricum* Cooke & Ellis is a very distinctive species known only from western North America (Gilbertson, 1964). The hymenium and subhymenial tissue are a pinkish-buff to salmon color and the tissue of the interior of the coarse, flattened teeth and the subiculum are a chocolate brown. Clamp connections are abundant on hyphae in both layers. Basidiospores are hyaline, cylindric, and nonamyloid (fig. 35). It has no apparent affinities with any of the genera or species considered here. No information on cultural characters of *R. concentricum* is available.

*Odontia furfurella* Bres. has basidia and a monomitic hyphal system with clamps typical of *Hyphodontia* Erikss. and perhaps could be placed in that genus. This interesting fungus, known only from North America, has very unusual branches or extensions of the hyphal walls terminated by globose structures (Gilbertson, 1963). Both the branch and the globose structure appear to be solid and their function is unknown.

*Odontia lateritia* Berk. & Curt. is commonly found on dead American chestnut in eastern North America. The reddish-orange basidiocarps are strongly hydnaceous with short crowded and branched teeth. Hyphae are nodose-septate and some are thick-walled and heavily incrusted. The latter may project in clusters at the apex of the teeth. *O. lateritia* is associated with a white rot. The relationships of this fungus are obscure at this time but it could most logically be placed in *Steccherinum*.

*Odontia romellii* Lund. ap. Erikss. is characterized by rather thick-walled, septate hyphae (fig. 3), cystidia which approach the form of those in *Hyphodontia barba-jovis* and related species, and small allantoid spores. The basidia are small and borne in candelabrums (fig. 22). Except for the absence of clamp connections it fits well in *Hyphodontia* and perhaps should be placed in that genus.

*Hydnum sulphurellum* Peck has a pale yellow grandinioid hymenial surface. Microscopically it is characterized by small dendrohyphidia, cylindrical, curved, smooth, nonamyloid spores, nodose-septate generative hyphae, and extremely fine and much branched, aseptate hyphae. The hymenium is of the euhymenial type. Although I previously proposed a transfer to *Laeticorticium* (Gilbertson, 1962), I now question its relationship to that genus. It has some characters (dendrohyphidia, cylindrical, curved spores, the euhymenium) similar to those of *Peniophora polygonia* but lacks the gloeocystidia found in that species.

## Acknowledgment

The author's research on resupinate hymenomycetes has been supported in part by grants G-24382 and GB-7120 from the National Science Foundation.

## References

Banker, H. J. 1914. Type studies in the Hydnaceae—VII. The genera *Asterodon* and *Hydnochaete*. Mycologia 6: 231–234.

Biggs, R. 1938. Cultural studies in the Thelephoraceae and related fungi. Mycologia 30: 64–78.

BOIDIN, J. 1958. Essai biotaxonomique sur les Hydnés résupinés et les Corticiés. Rev. Mycol. Memoire hors-série No. 6. 387 p.

———. 1964. Valeur des caractères culturaux et cytologiques pour la taxinomie des Thelephoraceae résupinés et étalés-réfléchis (Basidiomycètes). Bull. Soc. Bot. France 111: 309–315.

———. 1966. Basidiomycètes Auriscalpiaceae de la République centrafricaine. Cahiers de la Maboké, Tome IV, Fasc. 1, pp. 18–25.

BOIDIN, J., AND P. LANQUETIN, 1965. Hétérobasidiomycétes saprophytes et Homobasidiomycètes résupinés X.–nouvelles données sur la polarité dite sexuelle. Rev. Mycol. 30: 3–16.

BOURDOT, H., AND A. GALZIN. 1928. Hyménomycètes de France. Hetérobasidiés-Homobasidiés gymnocarpes. Paul Lechevalier, Paris. 761 p.

BROWN, CLAIR A. 1935. Morphology and biology of some species of *Odontia*. Bot. Gaz. 96: 640–675.

CEJP, K. 1928. Monographie des Hydnacées de la republique Tchécoslovaque. Bull. Int. Acad. Sci. Boheme. 102 p., 2 plates.

CHRISTIANSEN, M. P. 1960. Danish resupinate fungi. II. Homobasidiomycetes. Dansk. Bot. Arkiv. 19(2): 61–388.

CORNER, E. J. H. 1950. A monograph of *Clavaria* and allied genera. Oxford Univ. Press, London. 740 p.

CUNNINGHAM, H. G. 1953. Revision of Australian and New Zealand species of Thelephoraceae and Hydnaceae in the herbarium of the Royal Botanic Gardens, Kew. Proc. Linn. Soc. N. S. W. 77: 275–299.

DONK, M. A. 1956. The generic names proposed for Hymenomycetes–V. "Hydnaceae." Taxon 5: 69–80, 95–115.

———. 1961. Four new families of Hymenomycetes. Persoonia 1: 405–407.

———. 1964. A conspectus of the families of Aphyllophorales. Persoonia 3(2): 199–324.

———. 1967. Notes on European polypores–II. Persoonia 5: 47–130.

ERIKSSON, JOHN. 1954. *Ramaricium* n. gen., a corticioid member of the *Ramaria* group. Svensk. Bot. Tidskrift. 48: 188–198.

———. 1958a. Studies in Corticiaceae (*Botryohypochnus* Donk, *Botryobasidium* Donk and *Gloeocystidiellum* Donk). Svensk Bot. Tidskrift 52: 1–17.

———. 1958b. Studies in the Heterobasidiomycetes and Homobasidiomycetes-Aphyllophorales of Muddus National Park in North Sweden. Symb. Bot. Upsalienses XVI: I. 172 p., 24 plates.

———. 1958c. Studies of the Swedish Heterobasidiomycetes and Aphyllophorales with special regard to the family Corticiaceae. Inaugural Dissertation, University of Uppsala. 26 p.

FRIES, NILS. 1941. Uber die Sexualitat einiger Hydnaceen. Bot. Notiser 285–300.

GILBERTSON, R. L. 1962. Resupinate hydnaceous fungi of North America I. Type studies of species described by Peck. Mycologia 54: 658–677.

———. 1963. Resupinate hydnaceous fungi of North America II. Type studies of species described by Bresadola, Overholts, and Lloyd. Papers Mich. Acad. Sci. Arts Lett. 48: 137–149.

———. 1964. Resupinate hydnaceous fungi of North America III. Additional type studies. Papers Mich. Acad. Sci. Arts Lett. 49: 15–25.

———. 1965a. Resupinate hydnaceous fungi of North America V. Type studies of species described by Berkeley and Curtis. Mycologia 57: 845–871.

———. 1965b. Some species of *Vararia* from temperate North America. Paper Mich. Acad. Sci. Arts Lett. 50: 162–184.

GILBERTSON, R. L., AND M. J. LARSEN. 1965. Resupinate hydnaceous fungi of North America IV. Some western species of special interest. Bull. Torrey Bot. Club 92: 51–61.

GILBERTSON, R. L., FRANCES F. LOMBARD, AND T. E. HINDS. 1968. *Veluticeps berkeleyi* and its decay of pine in North America. Mycologia 60: 29–41.

GRAVES, R. C. 1960. Ecological observations on the insects and other inhabitants of woody shelf fungi (Basidiomycetes:Polyporaceae) in the Chicago area. Ann. Ent. Soc. Amer. 53: 61–78.

GROSS, HENRY L. 1964. The Echinodontiaceae. Mycopath. et Mycol. Applicata 24: 1–26.

JACKSON, H. S. 1948. Studies of Canadian Thelephoraceae II. Some new species of *Corticium*. Canad. J. Res. C. 26: 143–157.

KIMURA, K. 1954. On the sex of some wood-destroying fungi II. Bot. Mag. Tokyo 67: 34–35.

LARSEN, MICHAEL J. 1967a. The genera *Pseudotomentella* and *Tomentella* in North America. Ph.D. Thesis, State Univ. Coll. Forestry, Syracuse, N. Y. 278 p.

———. 1967b. *Tomentella* and related genera in North America IV. Taxonomy and nomenclature of *Caldesiella*. Taxon 16: 510–511.

LAWRENCE, JOHN F. 1967. Delimitation of the genus *Ceracis* (Coleoptera: Ciidae) with a revision of North American species. Bull. Mus. Comp. Zool. Harvard Univ. 136: 91–144.

LENTZ, P. L., AND H. H. MCKAY. 1966. Delineations of forest fungi: Morphology and relationships of *Vararia*. Mycopath. et Mycol. Applicata. 29: 1–25.

LIBERTA, A. 1966. On *Trechispora*. Taxon 15: 317–319.

LOMBARD, FRANCES F., AND R. L. GILBERTSON. 1966. *Poria luteoalba* and some related species in North America. Mycologia 58: 827–845.

MAAS GEESTERANUS, R. A. 1962. Hyphal structures in hydnums. Persoonia 2: 377–405.

———. 1963. Hyphal structure in hydnums. II. Proc. Ned. Akad. Wet. (C) 66: 426–436.

MAXWELL, M. B. 1954. Studies of Canadian Thelephoraceae XI. Conidium production in the Thelephoraceae. Canad. J. Bot. 32: 259–280.

MILLER, L. W., AND J. S. BOYLE. 1943. The Hydnaceae of Iowa. Univ. Iowa Studies Nat. Hist. 18(2): 1–92.

NIKOLAJEVA, T. L. 1961. Flora plantarum cryptogamarum URSS VI. Fungi (2). Familia Hydnaceae. Leningrad. 339 p., 78 plates.

NOBLES, MILDRED K. 1953. Studies in wood-inhabiting Hymenomycetes I. *Odontia bicolor*. Canad. J. Bot. 31: 745–749.

———. 1958. Cultural characters as a guide to the taxonomy and phylogeny of the Polyporaceae. Canad. J. Bot. 36: 883–926.

———. 1965. Identification of cultures of wood-inhabiting Hymenomycetes. Canad. J. Bot. 43: 1097–1139.

———. 1967. Conspecificity of *Basidioradulum* (*Radulum*) *radula* and *Corticium hydnans*. Mycologia 59: 192–211.

NOBLES, MILDRED K., RUTH MACRAE, AND BARBARA P. TOMLIN. 1957. Results of interfertility tests on some species of Hymenomycetes. Canad. J. Bot. 35: 377–387.

OBERWINKLER, F. 1965. Primitive Basidiomyceten. Revision eineger Formenkreise von Basidienpilzen mit plastischer Basidie. Ann. Mycol. ser. II. 19(1–3): 1–72, with 21 plates.

PATOUILLARD, N. 1900. Essai taxonomique sur les familles et les genres des Hyménomycètes. Lons-le Saunier. 184 p.

PAVIOUR-SMITH, K. 1960. The fruiting bodies of macro-fungi as habitats for beetles of the family Ciidae (Coleoptera). Oikos 11: 43–71.

POUZAR, Z. 1959. [Czech title.] New genera of higher fungi III. Česká Mykol. 13: 10–19.

ROGERS, D. P. 1934. The basidium. Univ. Iowa Studies Nat. Hist. 16 (2): 160–182.

———. 1944. The genera *Trechispora* and *Galzinia* (Thelephoraceae). Mycologia 36: 70–103.

SAVILE, D. B. O. 1955. A Phylogeny of the Basidiomycetes. Canad. J. Bot. 33: 60–104.

SINGER, R. 1962. The Agaricales in modern taxonomy. 2nd ed. J. Cramer, Weinheim. 915 p. and 72 plates.

VELENOVSKY, J. 1920–1922. Česke Houby. Prague. 950 p.

WELDEN, A. L. 1965. West Indian species of *Vararia* with notes on extralimital species. Mycologia 57: 502–520.

## DISCUSSION

DONK: You have spoken about *Vararia* and *Scytinostroma* that I now tentatively place in another family, but that had already been done by Reid who erected the family Lachnocladiaceae. It is associated with clavarioid genera.

PETERSEN: I might ask Dr. Donk if he agrees with Dr. Reid's segregation of Lachnocladiaceae, including some of the genera that you put in the Hymenochaetaceae.

DONK: Yes. The genera *Vararia* and *Scytinostroma.* I put a special subfamily there because I didn't know where to place them. A few months later the paper by Dr. Reid appeared and he was daring enough to take them out.

NOBLES: When you mentioned what we call *Radulum casaerium,* you said it did not seem to be related to any of the hydnaceous things. It belongs to my group 2–4, whose cultures have positive oxidase reactions, simple septa in the peripheral zone and clamps elsewhere, and the astatocenocytic nuclear situation. It seems to me that there are many species that don't seem to be related to others whose fruit bodies are similar in appearance.

ALBERT PILÁT
*Head, Mycology Department*
*Narodni Museum, Prague, Czechoslovakia*

# DIVERSITY AND PHYLOGENETIC POSITION OF THE THELEPHORACEAE

The Thelephoraceae seems to be a well-arranged family in the system of Fries and other earlier authors and were it not for the great number of species, it would not be difficult to make oneself familiar with them, at least from the standpoint of the older mycologists.

At the beginning of this century, with more attention being paid to the study of these primitive Basidiomycetes, it was found that this was a very heterogeneous group and that neither the conception of the family nor the definition of individual genera by earlier authors was tenable, because they combined heterogeneous elements which did not belong to one another from the viewpoint of phylogeny, and because they were united on the basis of superficial features which were often morphologically striking but phylogenetically of little significance. The greatest credit for the investigation of the Thelephoraceae *s. ampl.* and its systematics built upon the basis of phylogenetic relations should be given in particular to P. A. Karsten, V. Fayod, N. Patouillard (1900), V. Litschauer, E. M. Wakefield (1914), E. A. Burt (1919), S. Lundell, H. Bourdot, G. H. Cunningham (1952, 1963), M. A. Donk (1931, 1933, 1954, 1956, 1958a, 1958b, 1959), D. P. Rogers (1935), H. S. Jackson (see Rogers & Jackson, 1943), J. Boidin (1958), G. W. Martin (1938), L. S. Olive (1957b), P. H. B. Talbot (1954), P. L. Lentz (1957), W. B. Cooke (1961), J. Eriksson (1958), A. S. Welden, D. A. Reid (1963), E. Parmasto, M. Svrček (1960), A. E. Liberta (1960, 1962), and M. P. Christiansen (1959, 1960), in addition to many other investigators.

In this "family" we find, without doubt, the beginnings of the evolutionary branches of almost all Eubasidiomycetes as well as the meeting point of various branches of the Auriculariales and Tremellales. All this makes the situation complicated, so that it will take a considerable amount of time before a system can be worked out which will be fully acceptable from the viewpoint of phylogeny. For the time being we have not succeeded in finding two original systematic studies which use the same system. I think that this creative chaos will last for a long time.

What is the cause of these obscurities? There are many reasons. In my opinion the most important are as follows:

1. As has already been said, the evolutionary branches of almost all Basidiomycetes meet here; hence, it is a very heterogeneous group of fungi, even though the species and genera belonging to it seem to be similar because of their simple appearance. The reason for this similarity is primitivism in conjunction with a relatively small morphological differentiation. Macromorphological features do not suffice for the determination and definition of genera; moreover, they may often lead us astray. For the characterization of genera mainly anatomical, cytological, and physiological features may be taken into account, and occasionally also others, but no single one of these is reliable in all circumstances since we often do not know which of these features, the microscopical, chemical, physiological, or ecological, is reliable enough to serve as a basis for the combination of species into genera that would be acceptable from the viewpoint of phylogeny.

2. It also has been mentioned that morphological features are very scarce in these primitive fungi and insofar as they exist, they are often unreliable from the point of view of phylogeny. It is generally known that the morphological configuration of the hymenophore is a very unreliable feature. Many hymenial configurations occur in species that are often unrelated, such as the forming of tubes, or of a hymenophore imperfectly poroid (meruloid), either rugose or warted, spiny to coralloid, positively or negatively geotropic.

It is probable that the evolutionary possibilities of living matter are limited and that the gene combinations are, with regard to morphological development, restricted only to certain plans of construction. These, however, may be repeated in various developmental branches so that phenomena occur which are usually designated as parallel evolution. If we consider these apparently similar features as characters of one developmental line, we obtain entirely incorrect results, for they are only superficially similar without great phylogenetic importance. These features are the ones on which the system of the earlier authors was built.

3. E. Gäumann (1964) favored a third concept, which certainly

was also a hypothetical one. He claimed that it was impossible to speculate about the monophyletic evolution of the Basidiomycetes, but only about a polyphyletic one. He assumed that the origin of the exospores occurred independently several times throughout the long development in a past geological time and that a series of phylogenetic branches took their origin in the Ascomycetes, which developed convergently, independently of one another, and that particularly in primitive types there developed species and genera which appeared to be similar but were entirely different from the viewpoint of phylogeny. I, too, accept this concept, because I think it is the only way to explain the apparent resemblance and at the same time the great variety of numerous primitive types of Basidiomycetes assigned to the Thelephoraceae in the widest sense of meaning.

Of course, to demonstrate evolution as it factually proceeded is an extraordinarily difficult task and that is why the views of individual authors are considerably different. R. Heim, for instance, considers the Tulasnellaceae to be primitive. L. S. Olive (1957a), on the contrary, accepted the family as a definite link in a progressive evolutionary line of development of the Basidiomycetes, arising from the *Tremella*-type. Similarly, different opinions are held by authors with respect to the development of the Gastromycetes. R. Heim regarded them as being reduced Agaricales. R. Singer (1962), however, thought they were the original type that gave rise to the Agaricales. In my opinion, and as many other authors report, the Gastromycetes certainly came into existence polyphyletically. It is impossible to discuss this question in greater detail for the lack of time.

## The Evolution of Basidiomycetes

Having thought over the origin of the Thelephoraceae, we must at the same time look into the origin of the Basidiomycetes in general, because the Thelephoraceae is the most primitive family of this class of fungi. The view generally held is that the basidium is the transformed ascus where the spores do not form inside (endogenously) but seemingly outside (exogenously), and after meiosis the nuclei enter the finger-like excrescences of the wall of the ascus, forming usually 4 exogenous spores at the ends, but also 2 or 1, and rarely more than 4.

However, there exist various types of basidia; in addition to holobasidia there are also various types of phragmobasidia. In studying their development we meet with various views, for some authors consider the original type to be the holobasidium which later gave rise to phragmobasidia, others believing that the opposite is true and supporting their views by various arguments.

Savile (1955) considered phragmobasidia to be the most primitive, assuming that the Auriculariales, which have a transversely divided phragmobasidium, developed from the hypothetical ascomycete "Protaphrina" and that their evolution reaches the highest levels in the Tremellales with a longitudinally divided phragmobasidium, and on the other hand types with holobasidia as we may see in the Corticiaceae (in a wide sense of meaning) or the Clavariaceae, and elsewhere. R. Heim regarded the Basidiomycetes of the *Tulasnella* type as the connecting link between Ascomycetes which developed into two branches, one of which led to the Clavariaceae, while the other led to the Tremellales continuing in the Auriculariales with the transversely divided basidium. From the viewpoint of evolution the most important organ of the sporophore of the Basidiomycetes is the basidium, where karyogamy takes place. The important types of basidia are one-celled basidia, holobasidia, found in the majority of the Basidiomycetes assigned as Homobasidiomycetes, and on the other hand, basidia divided transversely or longitudinally by walls usually into four independent parts. This is the so-called phragmobasidium, which is taxonomically important in the Heterobasidiomycetes to which the Auriculariales and the Tremellales belong. The latter has basidia that are longitudinally or obliquely septate, with the secondary septa at right angles to the primary septa. A special basidial type is formed by the Tulasnellales with its sporelike epibasidia separated from the hypobasidium by septa, and by the Ceratobasidiaceae—commonly inserted herein—with the epibasidia not separated from the hypobasidium by septa. This family forms the transition to the Homobasidiomycetes and the related species were formerly placed in that subclan.

Heterobasidiomycetes usually form spores which do not germinate into filaments but often form secondary spores or produce conidia (Pilát, 1957a). In this group also belong the obligately parasitic Uredinales and Ustilaginales with a very complicated formation of spores. These undoubtedly are very old organisms (at least 200–250 millions of years), for their ancestors had already been parasitic on carboniferous ferns.

The question still remains: which is more primitive, the holobasidium or phragmobasidium? The majority of authors consider the phragmobasidium to be the more primitive, whereas others, for instance Gäumann (1964), believe the opposite—namely, that phragmobasidia developed from holobasidia. This problem has not been satisfactorily solved up to the present time, and therefore it is impossible to maintain with certainty that the species whose holobasidia resemble heterobasidia are older than types with a stereotype holobasidium. The argument that holobasidiomycetes are primitive because they have the most species with resupinate sporophores is not convincing, since these may be reduced types, similar to the Uredinales, which owe their reduced form to a parasitic way of life and the very complicated formation of spores.

As for the Tulasnellaceae, whose basidia are formed differently, the authors' views disagree on the explanation of the family's significance. Some authors consider the epibasidia, which are in groups of 4–7, as constituting a divided basidium. Others, for instance E. Gäumann, regard the epibasidia as spores which germinate by a sterigma-like outgrowth on which a fungal spore is borne. Basidia are usually placed terminally, rarely in a chainlike row. Inside the basidium, karyogamy takes place as well as the reduction division of the nucleus (meiosis); each nucleus from the tetrad passes into a globose excrescence at the apical end of the basidium. These excrescences, subsequently separated from the basidium by walls, are the places where postmeiotic nuclear division usually proceeds. This basidiospore usually forms a sterigmatoid outgrowth while still sitting on the basidium. At the end of the outgrowth two dikaryotic conidia, or sometimes secondary basidiospores (Rogers, 1932, 1933), are formed.

The Ceratobasidiaceae are noted for basidia which are somewhat similar to those of the Tremellales, being most often compared with the basidia of the genus *Sebacina* but having undeveloped septa. So they considerably resemble holobasidia. In some genera, the overlap is almost complete, for instance, in *Uthatobasidium* Donk, which is connected by numerous intermediate stages with some forms of the family Corticiaceae *s. ampl.* These species of the Corticiaceae are usually considered to be primitive. A good survey of them has been given by F. Oberwinkler (1965). They are above all the species of the Corticiaceae which bear more than four spores on their mostly urniform basidia, such as *Botryobasidium* Donk and *Sistotrema* Donk in Rogers, *Sistotremastrum* J. Eriksson, *Paullicorticium* J. Eriksson, and the closely related genera *Botryohypochnus* Donk and *Waitea* Warcup & Talbot, which bear only four spores on the basidium, similar to the Ceratobasidiaceae, and which certainly are closely related to the genus *Botrybasidium* Donk.

I do not regard it as confirmed that these genera are in fact primitive and not derived, since although we may assume that they are primitive yet we may equally consider them to be reduced from more progressive types. An example in the genus *Sistotrema* Fr. is *S. confluens* (Pers.) ex Fr., which is sometimes placed among the Hydnaceae, or sometimes among the Polyporaceae, because it forms carpophores with caps 1–3 cm in diameter with short tangled lamellae attached to the underside of the caps. To this species, M. A. Donk and J. Eriksson connect *Odontia brinkmannii* Bres. [ = *Grandinia brinkmannii* (Bres.) Bourd. et Galz.] and similar fungi which are corticioid, with a warted or smooth hymenophore and with similar basidia. Even if the basidia were the proof of close phylogenetic relations, *G. brinkmannii* could represent a form derived from a more progressive one.

Interesting basidia, the so-called pleurobasidia, are formed by the Xenasmataceae Oberwinkler, where the author places the genera *Xenasmatella* Oberw., *Xenasma* Donk, *Litschauerella* Oberw., *Xenosperma* Oberw., and *Acanthobasidium* Oberw., which form laterally placed basidia on the more-or-less horizontally proceeding ends of the hyphae. The basidia bear 2–7 sterigmata. Cystidia are either present or lacking. The carpophores are resupinate, usually gelatinous or waxy-gelatinous, since most of the hyphae of the carpophore gelatinize, similar to the Tremellales.

A very interesting species is *Acanthobasidium delicatum* (Wakef.) Oberw., which was described as *Aleurodiscus* by E. M. Wakefield. It has more-or-less uniform pleurobasidia, which are covered with elongated warts on the lower half resembling the acanthophyses of several species of the genera *Aleurodiscus*, *Stereum s. ampl.*, *Mycena*, and others (Lemke, 1963, 1964). The spores are echinulate with amyloid warts.

When the pleurobasidium elongates, there originates the podobasidium, which differs from the holobasidium by a mere protuberance on the basal part which, however, is not always distinctly discernible. According to Oberwinkler, it occurs partially in the genus *Athelia* and in several species of the genera *Corticium*, *Hyphoderma*, and *Peniophora*.

Similar to these podobasidia originate the holobasidia of other Corticiaceae also occurring in substantial numbers. Club-shaped excrescences are formed on one cell at the end of the horizontal hypha. These first extend clublike vertically and then terminally divide their own basidia by septa. Furthermore, basidia may occur in two ways, one of which is lateral probasidium production below the terminal basidium. This is a very widespread way, found in the Poriales (i.e., Aphyllophorales) and Agaricales, as well is in numerous representatives of the Corticiaceae. Noteworthy is the second way, which is similar but distinctive. The genus *Repetobasidium* J. Eriksson may serve as an example, since its later basidia do not occur laterally below the terminal basidium. After the maturing of the terminal basidium and its subsequent collapse, a new basidium is formed directly below it and the process is repeated, so that the remnants of the membrane of the older, withered basidia envelop the basal part of the next youngest basidium.

In addition to normal basidia, there often occur differently shaped types. They are short and broadly cylindrical basidia, slightly narrowed in the middle, usually bearing more than 4 sterigmata, which have already been mentioned here. Also narrowed in the middle are the utriform basidia, which are not short but, on the contrary, often very long and regularly bear 4 sterigmata. Such basidia may be found, for instance, in the genera *Coniophora*, *Galzinia*, *Aleurodiscus*, *Laeticorticium*, *Vararia*, and others. In most cases there occurs an almost round to ellipsoid probasidial vesicle. The probasidium represents a tubelike apical elongation, the so-called metabasidium, on which sterigmata are found. Donk assumed—and I think he was right—that it was impossible to regard the utriform basidium as a diagnostic feature of primitive forms, which would point to the close relations of Heterobasidiomycetes, but that the reason for its development was the fact that the basidium was rooted deep inside the catahymenium or in the conspicuously thickened euhymenium.

A thorough study of the most primitive Basidiomycetes, as has already been mentioned, was practically initiated in the twentieth century. Formerly it had been carried out only on the basis of macroscopic morphological features, and because mostly simple sporophores were treated which furnished only few morphological characters, individual species were distinguished very imperfectly. The systematic use of the microscope enabled study of these fungi to be pursued in greater detail, disclosed their large number of species, and helped to outline their phylogeny on the basis of anatomical features. But it is necessary to bear in mind that an overemphasis on some anatomic microscopic features may be misleading, similar to the overemphasis of some macroscopic characters.

In microscopic features, too, there is often a question of mere apparent resemblance but not of close phylogenetic relations. It is certainly very difficult to decide in advance which microscopic feature may or may not be regarded as phylogenetically important. That is why the systematics of these fungi has been imperfect up to now in spite of its continuous development and evident progress. It is often difficult to assess generic features and it is still more difficult to determine higher taxa. In the majority of cases it is possible to say that the fewer characters a species has, the more difficult it is to define it. Scientific progress, of course, always gives us increasing possibilities to define larger numbers of characters by means of the latest methods, which enable a more precise definition of individual species and genera. However, in using these new methods it is necessary to proceed critically and avoid any undue estimation of their significance in advance, since the newly disclosed characters may often be misleading.

The configuration of the hymenophore is commonly not a deciding feature from the viewpoint of phylogeny, for it repeatedly occurs in parallel series in very similar forms. This applies not only to the polyporoid or sublamelloid up to the lamelloid hymenophore, which often occurs in the family Polyporaceae, but also to the hydnoid hymenophore. Between the hydnoid and polyporoid hymenophore there are many intermediate stages, whose series end with primitive or seemingly primitive types with a smooth hymenophore on a resupinate sporophore.

Thus, for instance, *Peniophora hydnoides* Cooke & Massee has a smooth hymenium, whereas the same species called *Odontia hydnoides* von Höhn. et Litsch. (= *O. conspersa* Bres.) has a hydnoid hymenophore. Closely related are *Peniophora incarnata* (Pers.) Bourd. et Galz. and *Radulum laetum* Fr., assigned as a variety of *P. incarnata* by Bourdot and Galzin. In this case, however, in my opinion it is a question of two species which are very near to each other but nevertheless different. *Corticium confluens* Fr. and *R. membranaceum* (Bull. ex Fr.) Bres. represent another such class. But these also are very closely related species, not mere ecomorphs.

Donk established the family Gomphaceae for the genera *Ramaricium* J. Erikss., *Kavinia* Pilát, and *Ramaria* (Fr.) Bonord. em. Donk, which are very closely related as evidenced by their shape and color of the spores, by their spore membranes and ornamentation and, to a considerable extent, absorption of cotton blue. Very near to them is *Gomphus* Pers. per S. F. Gray. *Ramaricium* was placed among the Corticiaceae, *Kavinia* among the Hydnaceae, and *Ramaria* among the Clavariaceae. Another hydnoid type, whose pileus and stem are centrally to eccentrically placed, is represented by the genus *Beenakia* Reid, described from Australia. In my survey of middle-European Clavariaceae, I place all these species among the Clavariaceae. The establishment of the new family Gomphaceae may be considered as acceptable, even if all four genera placed therein are diametrically different in their macroscopic appearance.

Bondarzew and Singer place the genera *Fistulina* and *Porothelium* in the Cyphellaceae; later Singer (1945) also added the genera *Campanella*, *Favolaschia*, and *Leptotus* (the family Leptotaceae of R. Maire). In my opinion, this is in some cases a question of parallel series not of close affinity. This concerns mainly the genus *Fistulina*, which represents a considerably different type and approaches, I think, much more the family Polyporaceae *s. ampl.* than the Cyphellaceae, as the earlier authors assumed.

It is well known that several of the Tremellales have a hydnoid hymenophore, as for example *Pseudohydnum gelatinosum* (Fr.) Karst. or *Protodontia* von Höhn. Between these hydnoid and tremelloid types there is certainly a great affinity. Phragmobasidia, somewhat related to the genus *Tremella*, were found by Teixeira and Rogers (1955) in *Poria canescens* Karst., which today is called *Aporpium caryae* (Schw.) Teix. et Rog. These authors regarded this fungus as belonging to the order Tremellales, some of which have a polyporoid hymenophore. However, in its other characters it resembles numerous species of *Poria*. In addition to this, its phragmobasidia are not perfect in all cases, since the longitudinal septa in the basidia may not divide the whole basidium and are not always perfectly developed (Macrae, 1955). I think that this fungus is not closely related to the genus *Tremella*, but that it is a species of the family Polyporaceae, which in regard to the basidia is an example of parallel evolution.

The blue coloration of spores caused by iodine need not be at all times an important character, since this reaction may be positive or negative in the species of one genus. This is especially significant where there is no doubt about the phylogenetic origin of the genus, as for example in *Amanita*. This reaction, however, may undisputably draw our attention to a phylogenetic connection with morphologically fairly different types. The amyloid reaction is not always equally intense, sometimes being strong, and at other times weak. In the genus *Lentinellus* Karst. the hyphae in some species are strongly amyloid. Several authors overemphasize this reaction and join genera which are not related. I do not consider it right to put the genera *Hericium*, *Dentipellis*, and *Creolophus* together with *Bondarzewia montana* in the family Bondarzewiaceae and I agree with Donk (1964) that of the above-mentioned genera only the genus *Bondarzewia* may be placed in this family. H. Romagnesi (1964) drew attention to the fact that identical microchemical reactions, sometimes even two different ones at the same time, may often be found in various groups of fungi, so that it is difficult to say whether they point to a phylogenetic affinity or whether they occur in various evolutionary branches at the same time. There must be some question about the significance of the amyloidity of spores—also often of amyloid warts on them, or of lactiferous ducts and macrocystidia with lipid content, which also respond to sulphoaldehydic substances, as for example in the families Russulaceae and Hydnangiaceae and also the genera *Lentinellus*, *Auriscalpium*, *Hericium*, and often *Gloeocystidiellum* and *Clavicorona pyxidata*.

The direction of the dividing nuclear spindles in the basidia is regarded as an important character for the estimate of phylogenetic relations. There are chiefly two cases, the so-called stichobasidial position, where the first division occurs in the central part of the basidium and is directed to a great extent longitudinally. The second-division spindles of the nuclei are separated from one another in this case; they never occur together in the upper part and often proceed in two planes, the direction of which is random.

In the chiastobasidial situation the first and second divisions are within the terminal part of the basidium and are mostly transverse. The second divisions proceed in the same plane.

N. Penancier (1961) found in numerous species of Basidiomycetes that nuclear division in the basidium proceeds in most cases chiastobasidially, except for the genera *Cantharellus s. str.*, *Clavulina cin-*

*erea*, and *C. cristata*, *Hydnum repandum*, and *Clavulicium*, where it proceeds stichobasidially.

Whether the species of *Cantharellus*, *Clavulina*, and *Hydnum repandum* are very closely related remains a question (Corner, 1957). What great phylogenetic significance the stichobasidium has as compared with chiastobasidium is unclear up to now. Some authors regard the manner of nuclear division as very important, but I believe that its importance is being much overemphasized (Pilát, 1958). In the family Clavariaceae, it is only *Clavulina* which has stichobasidia and it is well separated from the other genera of the Clavariaceae with regard to other distinctive features, such as the character of the spores. The significance of the stichobasidia is evident in this case. However, the genera *Cantharellus* Fr., *Craterellus* Pers., and *Pseudocraterellus* Corner show close relations to the genus *Clavariadelphus*, so that some species almost form an evolutionary line which connects these genera, at least with respect to their morphological features. The division of the nucleus in the first three genera is stichobasidial, similar to the genus *Clavulina*, whereas *Clavariadelphus* has a chiastobasidial division of its basidial nucleus.

Juel and Maire considered the stichobasidial type to be a primitive one because it regularly occurred in Ascomycetes. In Basidiomycetes, however, it is found in so many various species that it is difficult to make many conclusions as to its phylogenetic development. In the genus *Exobasidium* there are different positions of the dividing spindle, not only in individual species but also in one and the same species.

As to the definition of the species and their relationships, much may be explained by the criterion of interfertility in diploidization and "sexual" polarity. In most of the cases, along with other criteria, they allow for a more precise definition of relationships. Among interesting results obtained in this respect by a number of authors are those reported by J. Boidin and M. des Pomeys. From their study it is evident that the species of the genus *Stereum s. str.* are homothallic and differ considerably from the other species of this genus *s. ampl.*, which nowadays are placed in other genera for good reasons—for example, *Laxitextum* (*bicolor*), *Lopharia* (*cinerescens*), *Stereum murraii*, which are tetrapolar, as are many other species of the genera *Peniophora*, *Hyphodontia*, *Radulomyces*, and others.

Homothallism and heterothallism do not inform us much about

the developmental relations of individual types. J. Boidin found that of 85 species of resupinate Basidiomycetes, 61 were heterothallic and 24 (28.2%) homothallic. Of 40 species, 22 were tetrapolar, 18 (45%) bipolar. In the Agaricaceae, of 145 species, 36 (24.8%) were bipolar, and in the Polyporaceae, studied by Quintanilha and Pinto-Lopes (1950), 3 were homothallic, 12 tetrapolar, and 15 bipolar.

The presence of clamps on the septa of the hyphae is of significance in a few cases only. The diplonts of several species are, according to J. Boidin, sensitive to seration and may pass from the dikaryotic stage with clamps to a cenocytic stage without clamps and vice versa. The diplonts in 42% of the species studied inconsistently had clamps, or had none at all.

## Hyphal Systems

Great progress in the study of the Aphyllophorales has resulted from investigation of hyphal systems. It is often possible to infer from them the phylogenetic affinity of various genera. The hyphal system is said to be "monomitic" when all the hyphae of the basidiocarp are of the generative series; when the generative and skeletal series are present, the system is "dimitic"; when generative skeletal, and binding series are present, then the system is "trimitic" (Corner, 1948).

Some genera are characterized by the presence of special hyphal structures such as gloeocystidia (*Bourdotia* Bres., *Gloeotulasnella* von Höhn. & Litsch., *Gloeocystidium* Karst., *Dryodon* Quel. [= *Hericium* Pers. ex S. F. Gray], and *Parapterulicium* Corner). Gloeocystidia are also present in species of *Stereum*, *Aleurodiscus*, and *Merulius*. Singer (1945) found that gloeocystidia of the corticioid type were present in *Favolaschia*. Their shape is very variable. Singer also reported that the pseudo-cystidia which were present as the termination of the laticiferous hyphae in *Russula* did not change their color after the use of cresyl blue, whereas, in *Laschia* they were strongly metachromatic; their content changed to blue when stained, while the walls became lilac.

Very important also are vascular hyphae which are tubelike. That is, they are nonseptate or sparingly septate, being also conspicuous by their large diameter or distinctive contents or both. Similar are the sanguinolentous hyphae (or tannic acid hyphae)—thick-walled, vascular hyphae which contain a liquid or solid reddish-brown matter but which are homologous in structure with the liquid-containing hyphae (P. L. Lentz, 1954).

Of importance in the systematics of the Basidiomycetes in general and particularly for the systematics of the corticioid fungi is the termination of some hyphae on the surface of the sporophore, especially in the hymenium, assigned altogether as cystidia and paraphyses. As their origin and shape are different, they are given different names.

Setae, too, are very important; they are brown, spinelike, sterile projections, appearing in various parts of xanthochroic basidiocarps. They are never markedly incrusted but are thick-walled, and they darken conspicuously when moistened with a potassium hydroxide solution. Here also belong the hymenial setae, the embedded setae, and the pileosetae and stellate setae. They may be found in the hymenium of the Basidiomycetes of both the stereoid (*Hymenochaete*) and the polyporoid type (*Phellinus*, *Inonotus*), and rarely of the clavarioid type (*Clavariachaete*). They are formed not only in the hymenium but also under the hymenial surface projecting into the trama, sometimes with hooks at their ends like recurved apices. They originate from the hyphae of the trama in the genus *Hymenochaete*, from colorless, thin-walled, septate hyphae of the context (Brown, 1915). Similarly, in *Inonotus cuticularis*, the brown color of the seta extends throughout the whole length of this hyphal system including its base, usually as far as the septum, which is separated from the other cells of the hypha. The other cells commonly do not differ from the hyphae of the trama.

H. L. Gross (1964) placed *Stereum sulcatum* in the genus *Echinodontium* and in the family Echinodontiaceae Donk emend. Gross, together with *E. tinctorium* (Ell. et Ever.) ("Indian Paint Fungus"). Ellis and Everhardt, who described *Echinodontium*, included it in the family Hydnaceae. In the same family Gross placed fungi with dimidiate, conchate to effuso-reflexed sporophores of woody consistency and with colored hyphae. The hymenium becomes thick and contains colored cystidia which resemble setae. The spores have smooth, amyloid membranes. The type of the genus is *E. tinctorium*, whose crushed sporophores were used by Indians for makeup and painting of the body. The cystidia in this fungus closely resemble setae, but they are not true setae. There exists a number of transient forms between metuloid cystidia and setae. For instance, *Peniophora laevigata* (Fr.) Karst. has thick-walled cystidia of a dirty yellow to brown.

## The Thelephoraceae s. str.

The Thelephoraceae *s.str.*, namely, the Phylacteriaceae *sensu* Bourdot and Galzin, is a very natural family (Istvanffi, 1896), in which belong primitive types like the genera *Tomentella* Pat. (= *Hypochnus* Fr. emend. Karst.), *Tomentellina* von Höhn. & Litsch. with tramal cystidia, *Caldesiella* Sacc. with resupinate sporophores and hydnoid hymenophore, and *Lindtneria* Pilát, whose hymenophore is polyporoid. Of types that are more progressive from the viewpoint of evolution, the following are included: *Thelephora* [= *Phylacteria* (Pers.) Pat] with dimidiate (fan-shaped) or clavarioid sporophores, and other genera of pileate and stipitate types with a hydnoid hymenophore, *Hydnellum* with brown and spiny spores as well as *Phellodon* and *Bankera* with colorless and aculeolate spores, which Donk placed in the family Bankeraceae. Donk for the first time related the genus *Boletopsis* Fayod, to which *Polyporus leucomelas* with colorless, gibbose spores belongs, in an affinity with *Thelephora.* I do not consider this classification to be reasonable, because I think that the affinity here is apparent only with regard to both the spores and the color of the sporophores.

Noteworthy is the genus *Polyozellus*, whose shape resembles the genus *Cantharellus*, but whose spores are reminiscent of species of the genus *Thelephora.* G. Malençon and R. Bertault described a fungus under the name *Lenzitopsis oxycedri*, which grows in Morocco in the Rif Mountains on the bark of *Juniperus oxycedrus.* The fungus resembled *Gloeophyllum* but had a monomitic structure and brown, warted spores; that is why the authors thought that it belonged among the Thelephoraceae *s. str.* In this case, too, it is a question of a parallel evolutionary branch.

## Characters which Enable the Classification of Species into Groups

1. The structure of monomitic, dimitic, occasionally trimitic hyphae.

2. Growth, either apical or by hyphal inflation. The first case is

characterized by the formation of rather thinner hyphae, which when adult do not inflate but sometimes become thick-walled. In the second case (inflationary growth) adult hyphae, too, are thick-walled, but they inflate considerably, increasing substantially the dimensions of the sporophore. This often occurs in primitive types as for instance in big, fleshy fruit bodies of the family Clavariaceae.

3. The presence of special paraphysoid or cheilocystidioid systems in the hymenium, or of hyphal apices which emerge from the trama of the sporophores into the hymenium, where various cystidioid formations occur and are often very conspicuous.

4. Shape and character of basidia.

5. Shape and character of sterigmata.

6. Shape, character, color, and stainability of the spores.

7. Positively or negatively geotropic position of the hymenium.

8. Clamps on hyphal septa. The significance of this character is very variable. Sometimes it is possible to regard it as being fairly important, but in other genera it is entirely negligible from the viewpoint of phylogenesis.

9. In addition to progressive evolution, there is also a regressive one, where from more perfect and more complicated types some forms arise which appear to be more simple and more primitive. All these are often fairly marked by morphological features, which, however, may lead us astray. As a more detailed study may show, they may occur simultaneously in fungi which anatomically are very different and surely belong to evolutionary groups considerably distant from one another. For instance, the Polyporaceae have a polyporoid hymenium, similar to *Lindtneria* Pilát, which conversely is smooth or nearly smooth. *Corticirama* Pilát (1957b) is phylogenetically very near *Tomentella*. *Caldesiella* Sacc. has spines similar to the Hydnaceae but is very closely related to *Tomentella*, whose hymenium is smooth or nearly smooth. *Corticirama* Pilát (1957b) forms ramarioid carpophores, but it is clearly related to *Corticium s. l. Ramaricium* Eriksson resembles, by its appearance, the species of *Corticium*, but its spores are similar to those of *Ramaria* S. F. Gray. The fruit body of *Kavinia* Pilát looks like those of some species of the Hydnaceae, but *Kavinia* is much nearer to the family Clavariaceae. Setae in the hymenium are found in the genus *Hymenochaete*, as well as in the family Polyporaceae. They also occur simultaneously in *Clavariachaete* Corner of the family Clavariaceae, and so on.

A very interesting fungus of the stipitate type belonging to the family Hydnaceae is *Beenakia dacostae* Reid from Australia and New Zealand, as was found by Maas Geesteranus (1963), with the stipe arising from a monomitic subiculum as in *Ramaricium occultum*. Moreover, the stipe and pileus of this fungus are monomitic and the spores are highly reminiscent of some species of *Ramaria* because of their shape, color, verrucose ornamentation, and the long oblique and blunt apiculus. Furthermore, the spores stain dark blue in cotton blue. The very slender basidia also resemble those in several species of *Ramaria*, as do the fragile and slightly inflating hyphae of the context.

Finally, I should like to point out that in polymorphic groups of fungi such as the Thelephoraceae *s. ampl.*, there occur both related types and others which are only superficially or apparently similar, and which in reality belong to other evolutionary branches. We must also take into account that some features which nowadays are suggested to be phylogenetically important are in reality unimportant and that in these cases it is only parallel evolution, not one and the same phyletic branch.

## References


Boidin, J. 1958. Essai biotaxonomique sur les Hydnés résupinés et les Corticiés. Etude spéciale du comportement nucléaire et des myceliums. Rev. Mycol. Mem. Hors–sér. No. 6. 388 p.

Brown, H. P. 1915. A timber rot accompanying *Hymenochaete rubiginosa* (Schrad.) Lev. Mycologia 7:1–20.

Burt, E. A. 1919–1926. The Thelephoraceae of North America. Pts. 1–15. Ann. Missouri Bot. Gard. 1–13.

Christiansen, M. P. 1959. Danish resupinate fungi. Part I. Dansk Bot. Archiv 19: 1–55.

———. 1960. Danish resupinate fungi. Part II. Dansk Bot. Archiv 19: 56–388.

Cooke, W. B. 1961. The Cyphellaceous fungi. A study in the Porotheliaceae. Beih. Sydowia 4: 1–144.

Corner, E. J. H. 1948. *Asterodon*, a clue to the morphology of fungus fruit-bodies: with notes on *Asterostroma* and *Asterostromella*. Trans. Brit. Mycol. Soc. 31: 234–245.

———. 1957. *Craterellus* Pers., *Cantharellus* Fr., and *Pseudocraterellus* gen. nov. Sydowia, Festschr. Petrak, pp. 266–276.

CUNNINGHAM, G. H. 1952. Revision of Australian and New Zealand species of Thelephoraceae and Hydnaceae in the herbarium of the Royal Botanic Garden, Kew. Proc. Linnaean Soc. N.S.W. 77: 275–299.

———. 1963. The Thelephoraceae of Australia and New Zealand. New Zealand D.S.I.R. Bull. 145. 359 p.

DONK, M. A. 1931. Revise van de Nederländse Heterobasidiomycetae en Homobasidiomycetae-Aphyllophoraceae. Wageningen. 200 p.

———. 1933. Revision der Niederländischen Homobasidiomycetae-Aphyllophoraceae. II. Mededeel. Bot. Mus. Univ. Utrecht No. 9. 278 p.

———. 1954. A note on sterigmata in general. Bothalia 6: 301–302.

———. 1956. Notes on resupinate Hymenomycetes. Reinwardtia 3: 363–379.

———. 1958a. Notes on the basidium. I. Blumea Supple. 4: 96–105.

———. 1958b. Notes on the basidium. II. Persoonia 2: 211–216.

———. 1959. Notes on "Cyphellaceae"—I. Persoonia 1: 25–110.

———. 1964. A conspectus of the families of the Aphyllophorales. Persoonia 3: 199–324.

ERIKSSON, J. 1958. Studies of the Swedish Heterobasidiomycetes and Aphyllophorales with special regard to the family Corticiaceae. Uppsala. 26 p.

GÄUMANN, E. 1964. Die Pilze. 2. Aufl. Basel & Stuttgart. 541 p.

GROSS, H. L. 1964. The Echinodontiaceae. Mycopath. & Mycol. Appl. 24: 1–26.

ISTVANFFI, G. DE 1896. Untersuchungen über die physiologische Anatomie der Pilze mit besonderer Berücksichtigung des Leitungssystems bei den Hydnei, Thelephorei und *Tomentella*. Jahrb. Wiss. Bot. 29: 391–440.

LEMKE, P. A. 1963. The genus *Aleurodiscus* (*sensu stricto*) in North America. Canad. J. Bot. 42: 213–282.

———. 1964. The genus *Aleurodiscus* (*sensu lato*) in North America. Canad. J. Bot. 42: 723–768.

LENTZ, P. L. 1954. Modified hyphae of Hymenomycetes. Bot. Rev. 20: 135–199.

———. 1957. Studies in *Coniophora* I. The basidium. Mycologia 49: 534–544.

LIBERTA, A. E. 1960. A taxonomic analysis of section *Athele* of the genus *Corticium*. I. Genus *Xenasma*. Mycologia 52: 884–914.

———. 1962. The genus *Paullicorticium* (Thelephoraceae). Brittonia 14: 219–223.

Maas Geesteranus, R. A. 1963. Hyphal structures in hydnums. III. Proc. Koninkl. Ned. Akad. Wet., Ser. C 66: 437–446.

Macrae, R. 1955. Cultural and interfertility studies in *Aporpium caryae.* Mycologia 47: 812–820.

Martin, G. W. 1938. The morphology of the basidium. Amer. J. Bot. 25: 682–685.

Oberwinkler, F. 1965. Primitive Basidiomycetes. Sydowia 19: 2–72.

Olive, L. S. 1957a. Two new genera of the Ceratobasidiaceae and their phylogenetic significance. Amer. J. Bot. 44: 429–435.

———. 1957b. The Tulasnellaceae of Tahiti. A revision of the family. Mycologia 49: 663–679.

Patouillard, N. 1900. Essai taxonomiques sur les familles et les genres des Hyménomycètes. Lons le Saunier. 184 p.

Penancier, N. 1961. Recherches sur l'orientation des fuseaux mitotiques dans la basidie des Aphyllophorales. Trav. Labor de "La Jaysinia," pp. 57–71.

Pilát, A. 1957a. Übersicht der europäischen Auriculariales und Tremellales mit besonderer Berücksichtigung der tschechoslowakischen Arten. Acta Mus. Nat. Pragae 13B: 115–210.

———. 1957b. *Corticirama petrakii* gen. et sp. n. Clavariacearum jugoslavica. Sydowia, Festschr. Petrak, pp. 128–131.

———. 1958. Übersicht der europäischen Clavariaceen unter besonderer Berücksichtigung der tschechoslowakischen Arten. Acta Mus. Nat. Pragae 14B: 129–255.

Quintanilha, A., and L. Pinto-Lopes. 1950. Aperçu sur l'état actuel de nos connaissances concernant la "conduit sexuelle" des espèces d'Hyménomycètes. I. Bol. Soc. Broteria 24 (ser. 2): 115–290.

Reid, D. A. 1963. Notes on some fungi from Michigan. I "Cyphellaceae." Persoonia 3: 97–154.

Rogers, D. P. 1932. A cytological study of *Tulasnella.* Bot. Gaz. 94: 86–105.

———. 1933. A taxonomic review of the Tulasnellaceae. Ann. Mycol. 31: 181–203.

———. 1935. Notes on the lower Basidiomycetes. Univ. Iowa Stud. Nat. Hist. 17: 3–43.

Rogers, D. P., and H. S. Jackson. 1943. Notes on the synonymy of some North American Thelephoraceae and other resupinates. Farlowia 1: 263–328.

Romagnesi, H. 1964. Sur deux réactions microchimiques associées chez certains Basidiomycètes supérieurs. Rev. Mycol. 20: 93–100.

Savile, D. B. O. 1955. A phylogeny of the Basidiomycetes. Canad. J. Bot. 33: 60–104.

Singer, R. 1945. The *Laschia*-complex (Basidiomycetes). Lloydia 8: 170–230.

———. 1962. The Agaricales in modern taxonomy. 2nd ed. Weinheim. 915 p.

Svrček, M. 1960. Tomentelloideae Čechoslovakiae (Genera resupinata familiae Thelephoraceae s. str.). Sydowia 14: 170–245.

Talbot, P. H. B. 1954. Micromorphology of the lower Hymenomycetes. Bothalia 6: 249–299.

Teixeira, A. R., and D. P. Rogers. 1955. *Aporpium*, a polyporoid genus of the Tremellaceae. Mycologia 47: 408–415.

Wakefield, E. M. 1914. Some notes on the genera of the Thelephoraceae. Trans. Brit. Mycol. Soc. 4: 301–307.

## Discussion

Singer: I have been cited from statements I made about thirty years ago regarding the Cyphellaceae and the presumably related groups of Fistulinaceae, Leptotaceae, etc. May I say that I have since changed my mind and published these changes in various papers. Right now I find myself in full agreement with the feelings voiced by Dr. Pilát and other mycologists in this regard.

I might also elaborate on the consideration of the cystidia of the Russulaceae and the gloeocystidia of the Aphyllophorales (*Favolaschia*). The cystidia of the Russulaceae are referred to in Romagnesi's terminology as macrocystidia and the others as gloeocystidia. In the Russulaceae I have studied the cresyl blue reaction, which does not normally cause the cystidial contents to become dark blue. However, about ten years ago, I investigated some groups of *Russula* in this regard, and in the *R. foetens* group (*R. foetens* and related species) there is a blue discoloration of the cystidial contents just as in gloeocystidia, and these *Russula* cystidia at the same time also react with the aldehydes. According to tradition in these cases they would be at the same time gloeocystidia and macrocystidia.

Pilát: In the case of these cystidia we should know more about the chemical origin of these phenomena in order to interpret them. The amyloid reaction itself may well depend on different chemical phenomena. For instance, in some cases substances that are adherent to the hyphal surface are amyloid whereas in other cases it is the membrane itself. In *Lentinus* and related groups I have often found that the iodine reaction is different—weaker or stronger—in young specimens versus old specimens so that there is also a question of individual development in the species.

SMITH: Since we were talking about chrysocystidia perhaps it is worthy of note that Hesler and I ran into all sorts of variation in these so-called chrysocystidia in *Pholiota*. In some you get a dark red-brown reaction in the cystidial contents in iodine which is called dextrinoid, and in KOH you get a dark yellow content in some. We found all kinds of modifications in morphology—simple basidioles with a chrysocystidial content to large leptocystidia with yellow contents which do not fit into any of these groups. The chemical content of cystidia is variable and if we use just the statement, "chrysocystidia present or absent," we have not described them completely enough. For instance, Thiers and I have chrysocystidia in one variety of *Tylopilus felleus*. In *Tylopilus* one of the characters which is going to help in the definition of the genus is the fact that the pleurocystidia, as we call them without wanting to designate them as to type, have a strong dextrinoid reaction. Somebody must do some microchemical work to determine just how these cystidia vary chemically, and then we might find that the fungi that have them are quite unrelated on this basis.

PETERSEN: When you talk about chrysocystidia of *Pholiota*, are these analagous at all to the structures one finds as colored oleiferous structures apparent in the hymenium of the species of *Gymnopilus*? In staining capriciously three or four species of *Gymnopilus*, I found that some of these chrysoid bodies do not stain at all in cotton blue, some stain bright blue, and some stain bright grass green. This varies from species to species.

SMITH: No, these are different structures. These are chrysocystidia in the sense of what you find in *Hericium* or *Clavicorona*.

PILÁT: I agree with what Dr. Smith has said about chrysocystidia in *Pholiota*. I have made an almost monographic study of *Pholiota* in Czechoslovakia years ago. I also observed very different and interesting types of cystidia, and also variations between young and old specimens, and I think that here is a development where an initially liquid substance may in the end become solid (resinous) and therefore have different optical properties.

BOIDIN: You mentioned in the 8-spored species like *Sistotrema* that the basidia is utriform and that utriform basidia are an advanced character. In such a basidium, the nuclear condition is stichic. Do you think that this character is primitive or advanced?

PILÁT: I have not made any cytological investigations on this group, and I took my data from the literature. I am not sure whether this is a progressive or a regressive movement. I must reserve my answer. I believe that even the Uredinales may have been derived from more complicated forms.

DEREK A. REID
*Mycologist*
*Royal Botanic Gardens, Kew, England*

# INTERMEDIATE GENERIC COMPLEXES BETWEEN THE THELEPHORACEAE AND OTHER FAMILIES

One of the greatest difficulties facing the taxonomist attempting to classify fungi is the circumscription of families. Starting with species and progressing through genera to family level, diversification of form is greater and relationships become increasingly obscured both on this account and also because of the reverse process of convergent evolution.

The Thelephoraceae present many problems of classification. However, it must first be realized that this family can be accepted in a broad, traditional sense or with a modern restricted circumscription.

Fungi referred to the Thelephoraceae in the restricted sense (*s. str.*) may have fruit bodies with almost any growth-form ranging from resupinate, effuso-reflexed, dimidiate, or stipitate. In those species the fruit bodies of which are resupinate the hymenium may be smooth (*Tomentella*), warted (*Kneiffiella*), or toothed (*Caldesiella*). The effuso-reflexed fruit bodies have the fertile surface smooth or inconspicuously warted (*Thelephora*) but in the dimidiate forms it may be imperfectly lamellate (*Lenzitopsis*). There are also merismatoid fruit bodies in which the hymenium is reticulate-venose or cantharelloid (*Polyozellus*) and stipitate fungi in which the pileus may be divided into erect flattened clavarioid segments (*Thelephora*) or the stipe may be central and the pileus either funnel-shaped (*Thelephora*), plane, or convex and either thick and fleshy or thin and coriaceous. In the mesopodal fungi the hymenium is either toothed (*Hydnellum* and *Sarcodon*) or poroid (*Boletopsis*).

Thus, there is enormous variation in external morphology among fungi assigned to the Thelephoraceae *s. str.*—so much diversity that the genera involved are scattered through most of the major families, with the exception of the Agaricaceae, in the Friesian system of classification. Nevertheless they all have a similar microstructure and a characteristic type of spore. The latter are basically globular or ellipsoid, although frequently modified by development of conspicuous tuberculate swellings. Further, they are usually warted or spinous and either dark brown or at least faintly tinted. The hyphae are usually thin-walled and may or may not possess clamp connections.

In this restricted sense the family is very natural, and evidence for outside relationships is weak. There are superficial resemblances to members of the Gomphaceae where species of *Ramaria* have long, narrow, elliptical brown spores ornamented with warts and spines, but these are strongly cyanophilous, unlike the spores of members of the Thelephoraceae *s. str.* Hence, a close relationship between these two families is unlikely. Another possibility is of a link between the hydnoid members of the Thelephoraceae and the Bankeraceae, for without question there is a very close external similarity between *Bankera* and *Phellodon* in the Bankeraceae and *Sarcodon* and *Hydnellum* in the Thelephoraceae. However, the two genera of the Bankeraceae have round, hyaline, spiny spores. Nevertheless, if one admits *Scytinopogon* to the Thelephoraceae—a genus in which the species have hyaline, elliptical, spiny spores, a case could be argued for the inclusion in the Thelephoraceae of those fungi currently assigned to the Bankeraceae. With this possible exception, however, there appear to be few connections between the Thelephoraceae *s. str.* and other families.

An entirely different situation exists with regard to the Thelephoraceae taken in a wide sense (*s.l.*), for included are those Basidiomycetes in which the species have a smooth hymenium. As such the family comprises an extremely heterogeneous assortment of fungi which by modern standards have little in common and is useless to the taxonomist as a guide to natural relationships of the genera so classified; generic complexes exist which can be linked to almost any of the "modern" families. Many of the Friesian genera assigned to the Thelephoraceae *s.l.* are themselves heterogeneous and comprise groups of unrelated species. Hence, what has been said of the family is also applicable at generic level. Since it is impracticable to consider the relationships of the Thelephoraceae *s.l.* as a whole, one can study a single Friesian genus, *Stereum s.l.* The situation as it affects this taxon is similar to that affecting the entire family, although on a much smaller scale.

Like many traditional genera, *Stereum* has grown rapidly—from 5 species in 1821 when the name was validly republished by S. F.

Gray, to 80 in 1836, 200 in 1888 and now includes between 300–400 species. These have been placed in the genus merely because they had a leathery texture and smooth hymenium, irrespective of whether they were resupinate, effuso-reflexed, or stipitate. Virtually nothing was known of their anatomical structure until Burt published his paper on the genus *Stereum* in North America. This clearly showed the genus to consist of groups of species which although superficially similar had quite different microcharacters.

Before progress could be made in restricting the genus to a more natural group of species it was necessary that it should be typified. In selecting S. *hirsutum* (Willd. ex Fr.) S. F. Gray as the type species from among the original five fungi included by S. F. Gray, the genus was circumscribed to cover the biggest single group of species in an otherwise heterogeneous assemblage. Henceforth, all members of the genus *Stereum s. str.* had to conform closely in both external morphology and in microcharacters to those of S. *hirsutum*.

S. *hirsutum* produces its fructifications on woody substrates as thin, leathery, effuso-reflexed, imbricate brackets, of which the upper surface is covered by a greyish or ochraceous tomentum while the hymenium is smooth and egg-yellow (this type of fruit body is subsequently referred to as stereoid). The flesh is dimitic and consists of (*a*) solid, unbranched, highly refractive skeletal hyphae and (*b*) thin-walled, branched generative hyphae, which, in the fruit body, lack clamp connections. Toward the surface of the pileus there is a golden-brown cuticular zone formed of skeletal hyphae bound together by modified generative hyphae which are coralloid and have brown walls. From this cuticular zone the surface hairs arise as elongated, unbranched, cylindrical, thick-walled hyphae. These are not terminations of skeletal hyphae, although they could be interpreted as modified skeletals of limited growth. In the lower portion of the flesh some skeletals do curve down into the hymenium and terminate at the same level as the basidia without protruding beyond the current hymenial surface. The tips of these skeletals, usually slightly enlarged and thin-walled, are modified to form secretory organs known as pseudocystidia. In addition to the basidia and pseudocystidia there are other sterile elements or basidioles in the hymenium. These arise from the same hyphae which give rise to the basidia as very narrow, thin-walled organs with obtuse or pointed apices. The spores which are white in mass are hyaline, elliptical, and strongly amyloid.

When restricted in this way to species with similar form and structure to S. *hirsutum,* there are about 14 American taxa referable to *Stereum s. str.* These form by far the largest single group within the genus as treated by Burt. However, even in the restricted sense *Stereum* is often divided into the sections *Stereum, Cruentata,* and *Phellina.*

Species belonging to the section *Stereum* are very like S. *hirsutum,* although some possess basidioles which bear a few minute spines at the apex (e.g., S. *ostrea*). The North American representatives are S. *hirsutum,* S. *sulphuratum,* S. *rameale* (of American authors), S. *striatum* (= S. *sericeum*), S. *ochraceoflavum,* S. *subtomentosum,* and S. *ostrea.*

Members of the section *Cruentata* are very similar to those of the previous group except that the pseudocystidia are filled with colored sap and as a result the hymenium "bleeds" when bruised. In addition, the distinction between skeletal and generative hyphae is less obvious, and most species have aculeate-tipped basidioles. American species belonging in this section are S. *sanguinolentum,* S. *gausapatum,* S. *rugosum,* S. *australe,* and S. *styracifluum.* Since this section is so close to the previous one, however, it is doubtful if any useful purpose is served by its recognition. It was segregated at generic level by Pouzar under the name *Haematostereum,* owing to the mistaken belief that the species had a monomitic structure.

Species of the section *Phellina* have a brownish flesh, conspicuously spiny-tipped basidioles (acanthophyses), small spores, and pseudocystidia which are often brownish and sometimes also bear aculeate processes. American representatives are S. *frustulosum* and S. *subpileatum.* This section is recognized at generic level by Boidin under the name *Xylobolus* Karst.

In the modern system of classification *Stereum s. str.* has been excluded from the Thelephoraceae *s. str.* and has been made the type genus of a separate family—the Stereaceae Pilát.

Having discussed the species of *Stereum s. str.,* we must consider the other North American fungi which Burt placed in the genus, starting with the effuso-reflexed species as represented by S. *chailletii.* This produces stereoid fructifications with a dimitic hyphal structure. The skeletal hyphae are brownish and the generatives hyaline with clamp connections. Some of the skeletals curve into the first-formed hymenium to terminate as brown, apically incrusted cystidia. Unlike the pseudocystidia in species of *Stereum s. str.,* these organs are of limited growth and become buried as the hymenium thickens, although similar bodies arise in large numbers at all levels in the new hymenia to take their place. Finally, the spores are amyloid. Thus, although the cystidia are not quite comparable with the pseudocystidia of the true *Stereum* spp., clearly S. *chailletii* shows a close relationship with these fungi and has to be included in the Stereaceae. Largely on account of the distinctive cystidia, however, it has been made the type species of the genus *Amylostereum* Boidin.

Another species which has been split from *Stereum* and made the type of the genus *Laurilia* is S. *sulcatum*, and to this genus has been added the recently described American species *L. taxodii*. S. *sulcatum* produces its fructifications, which may be entirely effused or narrowly effuso-reflexed, on conifers. The flesh is trimitic, with hyaline skeletal, branched binding, and clamp-bearing generative hyphae, and in the hymenium there are conspicuous incrusted metuloids. These organs are not very clearly differentiated since they tend to be very long, cylindrical or somewhat lanceolate bodies. It is the apical incrustation which causes them to appear more highly differentiated than is really the case. They are of limited growth and become buried as the hymenium thickens, but others arise at various levels to replace them. The first-formed metuloids arise deep in the context. Nevertheless, they are not terminations of skeletal hyphae as are the pseudocystidia of the true *Stereum* spp., but could be interpreted as highly modified and very short skeletals. The species of *Laurilia*, which also have amyloid, although globose and minutely echinulate spores, still show an obvious relationship with true *Stereum*. It is possible, however, to detect an increasing divergence from these fungi with regard to the type of sterile hymenial structure and spore characters. Gross has recently transferred the genus *Laurilia* to the Echinodontiaceae as emended by him, but he stated that the family had a phylogenetic position near *Stereum*. Personally, I prefer to retain *Laurilia* in the Stereaceae.

With S. *abietinum* and S. *ambiguum*, which are now assigned to the genus *Columnocystis*, a major link with *Stereum s. str.* is broken, for whereas the foregoing genera represent species with amyloid spores, these two species have nonamyloid spores. They produce stereoid brackets with a brown, dimitic context consisting of brown skeletal hyphae and clamp-bearing, hyaline, generative hyphae. The hymenium is strongly thickened and there are very conspicuous, elongated, brown, cylindrical, finger-like cystidia which protrude beyond the basidia. Despite the nonamyloid, elliptical spores the genus *Columnocystis* is still included in the Stereaceae.

The next group of species for consideration includes S. *cinerascens*, S. *papyrinum*, S. *umbrinum*, and S. *heterosporum*. This is a difficult group, for whereas they have been separated into the genus *Lopharia*, there is some doubt whether they are all congeneric. The type species of the genus is *L. lirosella*, which is thought to be a tropical form of S. *cinerascens*, with the hymenium ornamented with ridges and plates of tissue. The fruit bodies of S. *cinerascens* are commonly resupinate or narrowly pileate and have a dimitic structure consisting of hyaline skeletal hyphae and clamp-bearing generatives. The hymenium thickens and there are conspicuous, heavily incrusted conical metuloids. The first-formed metuloids arise deep in the context but it is not clear whether they are modified terminations of skeletal hyphae or completely modified skeletals of limited growth. These organs, which may project beyond the current hymenial layer, are peculiar in that the nonincrusted basal portion is distinctly brown whereas the apical region is hyaline. The spores are elliptical and nonamyloid.

The remaining species are similar but have brown or pale brown flesh. S. *umbrinum* is dimitic, but to distinguish between the pale brown skeletals and the generative hyphae, which lack clamps, is not always easy. Here the metuloids appear to be terminations of skeletal hyphae. In S. *papyrinum* the metuloids, like those of S. *cinerascens*, have a brown base and a hyaline apical region, and again the distinction between generative and skeletal hyphae is not always well marked.

Evidently with this group of fungi there is partial breakage of another link with true *Stereum* for the dimitic hyphal construction is often rudimentary or wanting.

S. *heterosporum* is of uncertain position. I have included it in *Lopharia*, as the fruit bodies, although normally resupinate, may form narrowly reflexed pilei. The flesh is brown and consists predominantly of clamp-bearing generatives with pale walls but there are other, much darker hyphae with thicker walls, which are unbranched. However, when of considerable length these hyphae may bear an occasional clamp connection. They curve down through the flesh and terminate with little or no modification in the lowest portion of the hymenium and must surely be regarded as poorly differentiated skeletals. Likewise, the flesh must be interpreted as rudimentarily dimitic. The hyphae from which the basidia arise grow up between the skeletals but they also produce sterile elements that are virtually skeletal hyphae of very limited growth which are frequently secondarily septate at the apex. In addition, this fungus has conspicuous lanceolate metuloids which project beyond the hymenium. These organs are brown in the basal portion but hyaline beneath the apical incrustation. The first-formed metuloids have a very deep-seated origin and appear to be terminal modifications of skeletal hyphae. The spores, which are pink in mass, are elliptical and nonamyloid.

The color of the spore print has led Boidin to conclude that *S. heterosporum* would be better placed in the genus *Peniophora s. str.* in the Corticiaceae. Although I would not accept this I am willing to concede that it might well be included in the genus *Duportella*, which contains species with similar rudimentarily dimitic structure and spores which are pink in mass. The type species, *D. velutina*, however, lacks metuloids but possesses gloeocystidia. This genus is also assigned to the Corticiaceae by Donk. Both Cunningham and Pouzar are prepared to accept *Duportella* with an enlarged circumscription to include *S. umbrinum* and others of these brown-fleshed *Lopharia* spp. Before adopting this view, however, it is necessary to know the color of the spore print and the reaction of the sterile elements in the hymenium to sulphoaldehyde reagents in these brown-fleshed species. Clearly, the genus *Lopharia* presents a problem to the taxonomist.

*Stereum purpureum*, well known as the cause of "Silver-Leaf Disease," is another species which can no longer be retained in *Stereum s. str.*, despite the fact that it produces well-developed stereoid fructifications. These have an entirely different structure from that of true *Stereum* and as a result the fungus has been segregated into the genus *Chondrostereum*. The flesh of the brackets is monomitic, consisting of clamp-bearing generative hyphae with distinctly thickened walls. Hence, the dimitic character as a link with *Stereum s. str.* is finally lost (Pouzar maintains that the structure of the flesh is dimitic—a feature which I am unable to confirm). Furthermore, the spores are nonamyloid and the only sterile elements in the hymenium are inconstant, thin-walled, protruding hairlike organs. The presence of these elements has been responsible for the unwarranted recognition of *S. rugosiusculum*. In addition to these characters, *S. purpureum* develops large saclike vesicles in a zone above the hymenium. This structure is so different from true *Stereum* that it is possible the affinities of *Chondrostereum* will be found to lie with such genera as *Merulius s. str.* in the Corticiaceae. Pouzar has even postulated that its relationships may be with such genera as *Skeletocutis* and *Hirschioporus* in the Polyporaceae, although I rather doubt this suggestion.

*Stereum murraii*, which produces widely effused and occasionally very narrowly pileate fructifications with a whitish tuberculate and often deeply cracked hymenium, is rather similar to *Chondrostereum purpureum* in certain microscopic features. Thus, it produces abundant vesicular gloeocystidia in zones throughout the thickened hymenium. According to Pouzar this species is dimitic with very scarce, light-colored skeletal hyphae. Working with dried material I am unable to confirm this feature, but examination is difficult because of the extremely thin flesh in comparison with the strongly thickened hymenium. Pouzar has segregated *S. murraii*, however, into the monotypic genus *Cystostereum* largely on account of its texture. Gilbertson has drawn attention to the fact that this species is probably closely related to *Radulum pini-canadensis*.

Of those species of *Stereum s.l.* which produce effuso-reflexed, stereoid fructifications, there remains *S. bicolor*, which has been made the type species of the genus *Laxitextum* Lentz. This fungus has a soft, spongy, monomitic flesh formed of loosely arranged, thin-walled, clamp-bearing, brown hyphae. The hymenium contains numerous, elongated, narrow, undulating gloeocystidia with highly refractive contents which tend to fracture transversely. These organs scarcely project beyond the hymenium. In this species, however, the small, broadly elliptical or ovate spores are minutely echinulate and strongly amyloid. Despite the amyloid spores the type of sterile element in the hymenium and the hyphal construction is so different from that of true *Stereum* that *Laxitextum bicolor* is thought to belong in an entirely different family—the Hericiaceae. The latter is composed, in the main, of genera with spinous hymenophores.

With *S. rufum*, *S. pini*, *S. erumpens*, *S. albobadium*, and *S. versiforme* another link with true *Stereum* is broken, for the ability to produce a truly pileate fruit body is finally lost. These fungi form greyish, lilac, or brownish resupinate, pustulate or discoid fructifications with an adherent or raised margin. They have a monomitic context in which at least some of the hyphae nearest the substrate are usually brown. The generative hyphae bear clamps and there are conspicuous granule-incrusted metuloids which project above the current hymenial layer but subsequently become buried as thickening occurs. All these species have elliptical or suballantoid spores which are pink in mass, and with the exception of *S. rufum* and

S. *pini* possess brown dendrophyses in the hymenial layers. Both S. *pini* and S. *rufum* produce conspicuous gloeocystidia in the hymenial and subhymenial region, however. Whereas these organs are oboval in S. *pini* they are elongated and either pointed or obtuse with very strongly thickened and gelatinized walls in S. *rufum*. At one time S. *pini* was placed in the monotypic genus *Sterellum* Karst., whereas S. *rufum* was assigned to *Cryptochaete* Karst., but these two fungi together with the other above-mentioned species are currently referred to the genus *Peniophora s. str.* in the Corticiaceae.

Having investigated the effuso-reflexed and resupinate fungi which were included by Burt in *Stereum s.l.*, we must examine the few remaining stipitate species. These lack even a morphological resemblance to true *Stereum* and likewise their microstructure is different. The stipitate species are now distributed among five genera of which *Podoscypha* Pat. and *Cymatoderma* Jungh. are undoubtedly very closely related. Members of the former genus produce medium-sized pleuropodal, mesopodal, or merismatoid fruit bodies with a smooth hymenium, and a dark brown or golden-brown upper surface which is glabrous or minutely pruinose with thick-walled, obtuse, cylindrical, or clavate pileocystidia. However, in some species the surface is tomentose but then the tomentum is formed of hyphae lacking clamp connections. The flesh is dimitic or rudimentarily trimitic, consisting of hyaline skeletal hyphae and clamp-bearing generatives, and the hymenium contains long, undulating gloeocystidia of unlimited growth which often traverse the entire width of the thickened hymenium. Several species produce conspicuous caulocystidia which are much like the cystidia on the pileus but longer. The spores are usually small to very small, broadly elliptical or ovate, and nonamyloid. This genus is represented in North America by S. *ravenelii*, S. *cristatum*, and S. *aculeatum*.

Species of *Cymatoderma*, of which only *C. caperatum* and *C. fuscum* var. *floridanum* occur in North America, restricted to the southern states, are very similar in structure to the members of the genus *Podoscypha*. They have similar dimitic or trimitic hyphal structure and an identical type of gloeocystidium in the thickened hymenium. However, unlike *Podoscypha* spp. they produce large fructifications which are usually covered by a very thick, pallid tomentum formed of clamp-bearing generative hyphae and never develop caulo- or pileocystidia. Their spores vary from very small and ovate to large and elliptical. Clearly, there is a close relationship between the genera

*Podoscypha* and *Cymatoderma* and consequently they have been separated into a distinct family—the Podoscyphaceae.

Now included in the Podoscyphaceae are a number of other stipitate fungi which had previously been placed in *Stereum s.l.* There are *S. diaphanum*, *S. pannosum*, and *S. undulatum*, which are now assigned to the genus *Cotylidia* Karst. These fungi produce pale-colored fruit bodies which may be pleuropodal, mesopodal, or merismatoid. They are always monomitic and consist of generative hyphae, lacking clamp connections. The hymenium is strongly thickened and contains conspicuous sterile elements of unlimited growth which project markedly beyond the current hymenial layer as elongated, thin-walled, cylindrical, finger-like organs. In *C. undulata* there are also similar sterile bodies on the surface of the pileus. The spores are elliptical and nonamyloid.

The genus *Stereopsis* Reid, represented in North America by *S. burtianum* and *S. hiscens*, may or may not prove to be a natural entity, but it contains species very like those I have assigned to *Cotylidia* except that they lack the characteristic protruding cystidia and have broadly elliptical or subglobose spores. The degree of relationship existing between *Podoscypha* and *Cymatoderma* on the one hand and *Stereopsis* and *Cotylidia* on the other has still to be evaluated, but for the present they are all retained in the Podoscyphaceae as they have nothing in common with *Stereum s. str.*

There remains one small muscigenous species which has soft, white flabelliform pilei with a monomitic structure. The narrow generative hyphae lack clamps and the spores are small, elliptical or broadly elliptical and nonamyloid. This species, *Cantharellus laevis* (= *C. pogonati*), is now assigned to the genus *Cyphellostereum* Reid and is not related to the species of *Cotylidia* or *Stereopsis* but is a reduced agaric. It is thought to be closely related to the genus *Leptoglossum* in the Tricholomataceae.

Having surveyed the genus *Stereum* in North America as circumscribed by Burt we find that the 42 species he included are now scattered through about 15 genera distributed among 4 distinct families. The stipitate species show little if any relationship to true *Stereum* and have been segregated into the Podoscyphaceae, while *Cyphellostereum* has been assigned to the Tricholomataceae in the Agaricales. The resupinate species have been placed in the genus *Peniophora* in the Corticiaceae, while the effuso-reflexed species *S. bicolor*, which belongs in the genus *Laxitextum*, has been transferred to the Hericiaceae. Of the remaining segregate genera with effuso-reflexed fruit bodies, only the species of *Amylostereum* and *Laurilia* show close relationships with *Stereum s. str.* The residue of genera, although currently grouped in the Stereaceae, are often so classified largely on their gross morphology and may eventually have to be removed from this family and be redisposed elsewhere. Conversely, some of the amyloid-spored species of *Aleurodiscus* appear to show affinities with true *Stereum* and may have to be moved into the Stereaceae from the Corticiaceae where they are currently assigned.

Having seen the position as it affects a single genus over part of its geographical range, we must conclude that the situation is obviously far more complex for *Stereum* on a world basis. A similar state of affairs exists in most of the old traditional genera of the Thelephoraceae, indicating a measure of the problem at family level.

The taxonomist's aim is to produce a system of classification based on homogeneous taxa, whereby species assigned to a given genus show close natural relationship. Only when this ideal has come nearer to fruition can it be used with any confidence in predicting lines of possible evolution within the fungi.

## Discussion

Singer: What do you think of the genus *Caripia*, which is such a common fungus in the neotropics? Years ago I referred it to the Stereaceae in the larger sense, when the family was developed. I did that because it certainly is not an agaric, although it had these lamellar ends at the margin. Do you think now that *Caripia* would fit anywhere in the Stereaceae?

Reid: I have to confess that I haven't studied this. But it is a very woody thing. One would think that it would have more the type of structure of the Stereaceae of the Aphyllophorales, rather than that of an agaric. I too would have thought it a member of the Aphyllophorales. But I don't know whether it fits into the Stereaceae.

Petersen: Corner, of course, first included *Caripia* as a genus of the clavarioid fungi, but then he repeated the genus in his *Cantharellus* book.

Boidin: *Lopharia crassa*, which I have recently received from Gabon, has white spore prints.

REID: This would tend to put it with *L. cinerascens*, which also has a white spore print. I'm very interested to hear that.

BOIDIN: I will have to figure out nuclear behavior. *Duportella tristicula* and *Stereum heterosporum* have pink spores and for me are to be grouped into *Peniophora s. str.* or at least so close by as to put *Duportella* at the subgenus level. I think that "*Stereum*-Cooke" is convergent, and perhaps the family Stereaceae is not as homogeneous as you think. No more than the Corticiaceae, at least. Perhaps in the future only two or three genera will stand.

REID: This is really what I said too. I am in agreement with you. I think that a lot of these genera will have to be taken out and put in other families.

BOIDIN: As you have pointed out, for *Aleurodiscus* the limit is also very difficult to find. Discussion should be made to point out the differences between *Aleurodiscus*, *Xylobolus*, and *Stereum.*

DONK: In connection with the remark by Dr. Boidin, I wish to mention that when I spoke of *Cyphella* I fleetingly suggested that in the future this genus would perhaps become the crystallization point for a new family and that I thought it possible that the amyloid–spored species of *Stereum*, *Aleurodiscus*, and the like will have to be transferred to such a family that could be called the Cyphellaceae.

REID: Seeing some of those tiny *Stereum* spp. from North America reminds one of *Cyphella* anyway.

PILÁT: What you think about *Stereum murraii* in Europe and in North America? This species in Czechoslovakia is distributed in the Carpathian Mountains, and only on *Abies alba.* All specimens which I have seen from America grew on oaks.

REID: I collected this fungus in both countries—in Czechoslovakia, as you know, in the Tatra, and North America. They certainly look very similar. But it may be a case, as with *Stereum pini*, of one of these North American equivalents, where there is perhaps some sort of genetic disjunction. I suppose one ought to investigate cultural characters of these fungi from the two continents and see if they are interfertile.

PILÁT: I made many microscopic preparations from both—that is, from America and from Europe. They appear under the microscope the same, but biologically the fungi are different.

REID: This happens many, many times in mycology. When one goes to another continent, the behavior and the biology of the fungus are so different. In Australasia there are fungi which in Europe and North America are almost invariably associated with conifers but in the southern hemisphere they occur on *Eucalyptus*.

BOIDIN: Interfertility tests should be made to try out the relationship.

ROGERS: I collect *Stereum murraii* very commonly on *Abies*.

POMERLEAU: For us it is always on birch.

PETERSEN: It seems we are dealing with a ubiquitous species.

REID: In North America you have another species in the *S. ostrea* complex which may be passing under the name *S. fasciatum*. A new species was described from Europe by Pouzar—*S. subtomentosum*—and this to the naked eye looks like *S. fasciatum*. When I was up in Michigan in 1961, before this was described, I noticed that the so-called *S. fasciatum* there bled with a yellow milk if bruised. This was the very character noted by Pouzar in this new species from Europe. Undoubtedly, *S. subtomentosum* does occur in North America and had passed under the name *S. fasciatum* in Michigan. Just how widely this is distributed here I don't know, but I think you might be interested to look for this character of the yellowing in so-called *S. fasciatum* to decide just how widely this other species is distributed.

WELDEN: Burt recognized *S. styraciftuum* as bleeding yellow, and I have found specimens or varieties of *Stereum ostrea* which bleed yellow; then they will slowly turn red. Others which are younger will bleed yellow and remain yellow. I think this is a variable character.

REID: In order to identify *S. subtomentosum* you have to look at the basidioles, which is not an altogether easy matter, to see whether they are aculeate at the tips or not. In *S. subtomentosum* they are not. In some of the other species like *S. ostrea* they are.

PILÁT: A similar biological problem as in *S. murraii* is also in *Polyporus montanus*, or *P. berkeleyi*. *P. montanus* in Europe grows only on *Abies alba*, and *P. berkeleyi* on oak in America. The difference between these two species, macroscopic and microscopic, is small.

SINGER: *P. montanum* has no latex in it.

DONK: Not much at least. I collected it fresh six months ago, but at least it didn't ooze out.

NOBLES: A very small point, but it is perhaps parallel. Some years ago I published on *S. pini*. I thought what we had in Canada was the same as *S. pini* from Europe, but when Dr. Boidin gave us single-spore cultures of the European species, they were not interfertile with ours. When we looked at the fruit bodies of our *S. pini*, we then found differences, and Weresub and Gibson described a new species, *Peniophora subpini*. It is very similar to the European *P. pini*, but not identical. There may be other pairs of species which look very similar, but may not be exactly the same.

BOIDIN: The genus *Stereum s. str.* is very difficult to delimit specifically, because they are all fixed homothallic lines and there is no crossing between them.

RONALD H. PETERSEN
*Associate Professor of Botany*
*University of Tennessee, Knoxville, Tennessee*

# INTERFAMILIAL RELATIONSHIPS IN THE CLAVARIOID AND CANTHARELLOID FUNGI

The clavarioid and cantharelloid fungi seem to occupy a unique and important position in the phylogeny of the higher Basidiomycetes. That position is not clear, however, and what is more, it could be at either end of the total spectrum of the class. If the primitive fruiting body of the Basidiomycetes is a simple club, the clavarioid fungi occupy the most primitive positions. If the most primitive fruiting body is resupinate, then the simple club may represent one of the most reduced forms—gymnocarpic with the absence of a differentiated hymenopodium. But in any case several lines pass through these fungi, making them conspicuous as intermediates. I might anticipate the body of my paper in this regard, and tell you that I have little answer for the direction of evolution outside these groups of fungi, so that the exact direction of basidiomycete evolution will not be concisely elucidated in this paper.

The Friesian (1821, 1874, etc.) system of classification of the clavarioid fungi was based on preceding systems, especially that of Persoon (1797, 1801, etc.). It grouped the "Clavati" regardless of their mode of spore production, including *Geoglossum*, *Calocera*, and the "true" clavarias, in a single large assemblage. As was the custom, gross morphology of the fruiting body was used by Fries (1821) in his earliest system, dividing the genus *Clavaria* into "tribes" by the shape of the fruiting body—simple to fasciculate, branched with narrow stipe, and branched with thick stipe. Later, Fries (1838) began to use spore color as a basic character, and this was adopted by Karsten (1879, etc.) in his description of *Clavariella*. The early English workers, including Cooke, Berkeley, Massee, and others, used the Friesian system almost exclusively, bringing few new ideas to the classification scheme, even though it was eventually an Englishman who revolutionized the taxonomy of the group. The French workers Boudier, Patouillard, Bourdot, and others used more taxa, especially at the species level, but did not utilize all of the segregate genera already available to them. Finally, Donk (1933, etc.) resurrected and recognized additional supraspecific taxa, and Coker (1923, 1947) informally divided the genus *Clavaria* into groups roughly corresponding to the more natural generic taxa adopted later, setting the stage for Doty's (1948) and Corner's (1950) keys to the genera of clavarioid fungi.

The traditional system using *Cantharellus* and *Craterellus* for all of the basidiomycetes with thick lamellar folds inferior to the pileus extension went without revision for many years. Within the last few years several additional cantharelloid genera have been described, although acknowledged not to be always closely related to *Cantharellus s. str. Pseudocraterellus* Corner (1958) and *Podoserpula* Reid (1963) are examples. The resurrection of *Gomphus* by Donk (1933) —taking priority over *Neurophyllum*—and the description of *Polyozellus* might be added to other examples of progress toward establishment of more natural systematics in the group. The treatment of the cantharelloid fungi from the northwestern United States by Smith and Morse (1947) was not critical, and the recent treatment of the Michigan species by Smith (1968) is still disappointingly nomenclaturally and taxonomically inexact.

Even though Maire (1902, 1914) early related the cantharelloid and clavarioid fungi, it was not until much later that other mycologists began to recognize such a relationship, and then the pendulum swung perhaps too far, with some authors grouping the chanterelles and clavarias together as though each were a homogeneous group that could be juxtaposed as a unit. Corner's investigations, and, I hope, my own work have dispelled both dreams. Instead, several rather natural groups of species and/or genera have emerged as concepts, to be tested in further investigations.

The vast heterogeneity of the clavarioid fungi is quite easily demonstrated with examples like *Tremellodendron* with cruciately septate basidia, *Paraphelaria* Corner with phragmobasidia, and *Aphelaria* with semi-septate basidium, all doubtfully related or outright unrelated to the homobasidiomycetous clavarioids. More subtly perhaps, one could point out *Pterula* (Pteruloideae of the Clavariaceae *sensu* Donk—1964) and *Lachnocladium*, the latter placed in the Hymenochaetaceae by Donk (1964) and made the type genus of the Lachnocladiaceae by Reid (1965), as examples of genera representing separate lines of evolution within the Homobasidiomycetes, although unrelated to *Clavaria* or *Ramaria*, the generally accepted stereotypes of the group. In the same way, *Scytinopogon* Singer may be thought of (*à la* Corner, 1950) as a somewhat prodigal thelephoroid with the superficial appearance of a clavaria. None of these is very closely related to the generic complex including *Clavulinopsis*, *Clavaria*, *Multiclavula*, *Pistillaria*, and others, and none of these latter genera is closely related to the *Ramaria* species assemblage. In the cantharelloid fungi, *Cantharellus* seems most closely related to *Pseudocraterellus* and *Craterellus*, but *Gomphus* and *Dicantharellus* seem to represent separate and distinct evolutionary lines. Somewhat less obvious, and still hard for some to accept completely, is the probable inclusion of the clavarioid *Clavicorona* with the hydnoid *Hericium* and resupinates *Gleocystidiellum* and *Laxitextum* to compose the family Hericiaceae Donk. Just as interesting is the statement by Donk (1964) that *Hydnum* is closely related to the Cantharellaceae, to which I concur completely. But if anything is clear, it is that gross morphology is deceiving, and that additional characters must be relied on just as heavily in determining probable relationships between groups of organisms.

Two general groups of organisms have attracted my attention during the past several years, and my view of their interrelationships has changed almost annually. To be perfectly clear, the remarks made here are the latest in this oscillating pathway, and may—nay, will—change and refine again and again within the next years.

From time to time I have been convinced that just as *Gomphus* and *Ramaria* were related, although of different growth form, *Clavulinopsis* and *Cantharellus* were also so related. I experimented with an additional idea, which hesitatingly explored the concept that *Cantharellus* and its relatives *Craterellus* and *Pseudocraterellus* were not directly related to *Clavulinopsis-Clavaria* alliance, but might be related to *Clavariadelphus*, which in turn might also be related to the Clavariaceae *sensu* Donk (1964). Moreover, if one looks at the spores, basidia, gill structure, pileus trama, and other such features of several species of *Hygrophorus* or *Clitocybe* it is not impossible to imagine either the precursor to the reduced cantharelloid form, or the result of sophistication of that same form. My conclusions, however tentative, are summarized in figure 1.

Even in my limited experience, I have seen intermediate forms between the fleshy fruiting bodies of *Cantharellus cibarius*, *C. subalbidus*, *C. odoratus*, and *C. formosus* and the waxy, slender, often perforate species like *C. infundibuliformis*, *C. tubaeformis*, *C. minor*, and *Pseudocraterellus sinuosus*. The various fruiting body forms in the perforate series of species, especially in the species without clamp connections (generally thought of as either *Pseudocraterellus* if secondary septation is present, or *Craterellus*, if without secondary sep-

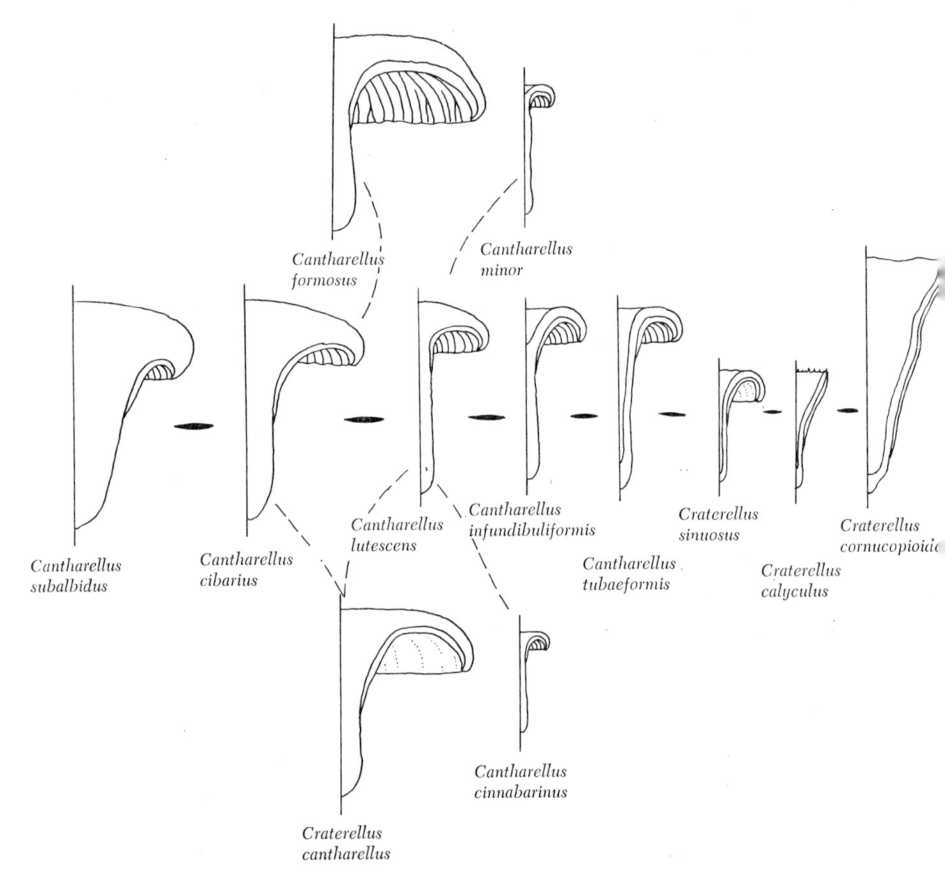


FIGURE 1. Proposed evolutionary pathway in the cantharelloid fungi.

ta) show a tendency toward the widening of the perforation into an obconical form, quite probably having led to the classic form of the dark-colored *Craterellus* species, exemplified by *C. cornucopioides* in Europe and *C. fallax* in North America. In this regard, Corner's (1958) separation of *Pseudocraterellus* on the basis of secondary septation was questionable for it cut across what seemed to be relatively valid evolutionary lines to link a number of the species of *Cantharellus*, as well as some species placed in *Craterellus* by many authors including Corner himself. Preliminary indications are that several species would have to be transferred to *Pseudocraterellus* on this single character, but that this would distort other character associations. Sections of pileus surface and pileus trama tissues show wide variation in the abundance of secondary septation in various collections of one species (*Cantharellus sinuosus*), and in young versus old fruiting bodies in the same collection. Secondary septation is most obvious in inflated hyphae, and the amount of inflation often seems to be a function of the age of the hyphae themselves.

The artificiality of the present systematics of this group is reflected in the following brief notes on North American species. (*a*) Although *Craterellus fallax* should exemplify the genus characters of *Craterellus*, nevertheless common secondary septation is seen and would, on that character alone, force placement of this species in *Pseudocraterellus*. (*b*) *Craterellus carolinensis* (Petersen, 1969) has the microscopic and anatomical morphology of *Cantharellus hystrix* Corner, including the preponderance of 5-sterigmate basidia, but is without clamps and bears relatively abundant secondary septa. If placed in *Pseudocraterellus*, it would be without immediate relatives and so has been described as a *Craterellus*. (*c*) The shape of the mature fruiting body in *C. carolinensis*, as well as others best placed in *Craterellus*—namely *P. calyculus* and *P. subundulatus*—is far from cornucopioid. *C. carolinensis* possesses a relatively long and distinct stipe, and the pileus extension is relatively small and flaring, although often tardily perforate. It would seem that its ontogenetic development is between *Cantharellus* and *Craterellus*, although its dark colors are similar to other *Craterellus* species. Generic parameters are difficult to draw in this smooth-spored portion of the cantharelloid fungi.

But even if the *Cantharellus* and *Craterellus* species compose a relatively natural series, are the smooth-spored cantharelloid fungi related directly to the clavarioid fungi? Corner (1966) stated that *Pseudocraterellus* is to *Cantharellus* what *Clavaria* is to *Clavulinopsis*. This may well be true, but are *Pseudocraterellus* and *Cantharellus* directly related to *Clavaria* and *Clavulinopsis*? I think not—at least directly. Presumably, the clamped genera *Craterellus* and *Clavulinopsis* would be closest, evolving their clampless analogs *Craterellus*,

*Pseudocraterellus*, and *Clavaria*. First, there are no forms known in *Clavulinopsis* which form single or gregarious fruiting bodies together with turbinate or very broadly clavate form, as found in *Clavariadelphus*. Second, few forms are seen which bear any indication of rudimentary lamellar fold formation, although Heim (1948) has reported such forms from Africa and Madagascar. Third, if nuclear orientation in the incipient basidia can be given weight in classification, the species of *Cantharellus* are reported (Corner, 1966) as stichic, whereas those of *Clavulinopsis* and its allies have been stated to be chiastic (Donk, 1964, and others).

As usual, however, there is conflicting evidence. First, the pigmentation of the *Cantharellus* species generally eligible for close relationships with *Clavulinopsis* on morphological grounds is almost exclusively carotenoid in composition. The work summarized by Fiasson *et al.* (1967, 1968) has been supported by some investigations carried on here on the pigments of *Cantharellus* and related taxa. The French workers, led by Arpin *et al.* (1966) have shown that a group of *Cantharellus* species (*C. lutescens*—the identity of which may still be in doubt—and *C. tubaeformis*) exists which accumulates neurosporene, a carotenoid pigment with relatively limited distribution in the fungi. In addition, two species complexes apparently occur, one accumulating beta-carotene, the other anthraquinonic-appearing compounds, although the latter identification appears tentative. We have found that *Cantharellus cibarius* (both the typical form—two other forms not yet published—and var. *pallidofolius* Smith) apparently produce large amounts of beta-carotene, as well as several other carotenoid compounds in smaller quantities. This matches Fiasson's summary. We also have found that some species of *Clavulinopsis* (*C. miniata* and *C. aurantio-cinnabarina*) produce almost all their pigmentation in the form of carotenoids. Heim (1949) had previously reported carotenoid crystals in *C. cardinalis*, which Petersen (1968a) has placed in synonymy under *C. aurantio-cinnabarina.* However, *C. fusiformis* and *C. corniculata* produced pigments which, although soluble in acetone, were not partitionable in petroleum ether, or having been suspended in methanol, were not partitionable in petroleum ether. In neither case was the pigment characterized as carotenoid. *Clavariadelphus pistillaris* gave identical results. Arpin (private communication) had clearly indicated the presence of anthraquinones in the clavarias. All of this would seem to indicate that, although there might be some supportive evidence for a tie between the weakly apiculate-spored species of *Clavulinopsis* and the *Cantharellus cibarius* complex, an equally good case might be made for a relationship between the strongly apiculate-spored species of *Clavulinopsis* and the "*Cantharellus griseus*" group of species, if the evidence were simply chemical. But if fruiting body morphology is also taken into account, the latter chemical data can be called nothing more than coincidence, for the species of *Cantharellus* which seem to produce anthraquinonic compounds are also those with a perforate or funnel-shaped fruiting body, presumably not close to the solid, simple club of *Clavulinopsis*.

It would appear that from the few species tested, the chemical evidence of carotenoid versus noncarotenoid pigment production, the genus *Clavulinopsis* might well be broken into two subdivisions. Corner (1950) and Petersen (1968a) already indicated such a division of the genus, but along strictly morphological lines, identifying a group whose spores were weakly apiculate (and which now also are found to apparently produce carotenoid pigments) and a group whose spores were strongly apiculate (and which produce noncarotenoid pigments). The chemical evidence in this case supports the more classical morphological information.

Table 1 summarizes macrochemical reactions in the white-spored clavarioid fungi and is offered as conflicting evidence in the possible links of *Cantharellus* and the clavarias. In this table it is possible to compare several species and genera at a glance. Note that there is a similar macrochemical $FeSO_4$ reaction with several species of *Clavulinopsis* and *Clavariadelphus*, although there is a negative reaction with all the species of *Clavaria* and the species of *Clavulinopsis* with weakly apiculate spores. Again, this reinforces the concept that perhaps *Clavariadelphus* and *Clavulinopsis* are related. If the $FeSO_4$ is followed by a drop of ethyl alcohol, the reaction is much stronger, with the color deepening to green-black, usually very rapidly in all species shown here to have a positive test.

To these data might be added the reactions on *Cantharellus cibarius*, where $FeSO_4$ produces a watery grey-green to grey reaction, *C. subalbidus* a grey reaction, and *C. tubaeformis*, no reaction. A species close to *C. infundibuliformis* gives a strong vinaceous red color on the stem flesh with $FeSO_4$, as does one form of *C. cibarius* in this area. Again, with positive reactions, the *C. cibarius* complex would seem most closely related to the strongly apiculated species of *Clavulinopsis*, but this in opposition to the pattern displayed in pigment analysis.

Most authors, however, including Corner (1966), have related *Cantharellus* with the clavarioid fungi through *Clavariadelphus*.

TABLE 1. MACROCHEMICAL REACTIONS ON CLAVARIOID FUNGI.

| *NAME* | $FeSO_4$ | *KOH* |
|---|---|---|
| *Clavariadelphus pistillaris* | greenish yellow | orangy |
| *Clavariadelphus truncatus* | dull green | pale cherry red |
| *Clavariadelphus ligulus* | deep green where bruised | deep vinescent where bruised |
| *Clavulinopsis fusiformis* | dull green | watery orange |
| *Clavulinopsis corniculata* | grey-green | orange |
| *Clavulinopsis laeticolor* | grey-green | yellow-green |
| *Clavulinopsis miniata* | o | o |
| *Clavulinopsis aurantio-cinnabarina* | o | o |
| *Clavaria inaequalis* | o | coppery |
| *Clavaria rosea* | o | watery green |
| *Clavaria vermicularis* | o | o |
| *Clavaria zollingeri* | o | o |

Such an association is based on monomitic hyphal construction, long basidia, smooth spores, and the tendency of *Clavariadelphus* toward truncate or turbinate fruiting body form, and the common presence in these same species of rudimentary longitudinal hymenial folds. However, *Clavariadelphus* has been reported to bear chiastic basidia, whereas *Cantharellus* has stichic. With the new species of Wells and Kempton (1968), especially *C. borealis*, such an evolutionary relationship between the cantharelloid and clavarioid fungi would seem likely. If so, then it would be the *C. cibarius* group which would be most eligible, and it would share the positive iron salts reaction with the truncated species of section *Clavariadelphus* (Wells & Kempton, 1968) even to the obscure slate green color of the reaction. Even more interesting has been the discovery of a species of *Cantharellus* (*C. cuticulatus* Corner) with a hymeniiform pileus surface (Corner, 1966) and the details of this surface in the truncate species of *Clava-*

*riadelphus* (Wells & Kempton, 1968). The upper surface of *Clavariadelphus truncatus* (fig. 2) compares very favorably with that of *Cantharellus cuticulatus* as drawn by Corner (1966), although clamp connections are missing from the *Cantharellus*. There is immediately the suspicion that what Corner dealt with was really a degenerate *Clavariadelphus* without clamps, for he shows no illustration of the fruiting body, but the description including gill folds 3–5 mm in height, and basidia with 4–5–6-sterigmata suppresses such suspicion. Nomenclatural vagaries do not cloud the issue that *Clavariadelphus truncatus* and its ilk may be related to this organism, although presently *Cantharellus cuticulatus* is the sole representative of section *Cutirellus*, subgenus *Cantharellus* of *Cantharellus sensu* Corner. Moreover, *Clavariadelphus truncatus*, sometimes *C. unicolor*, and usually *C. borealis* form rugose to rather regularly longitudinal folds of the hymenium as well.

Again, though, there is conflicting evidence. There seems to be an almost universal tendency in the *Cantharellus cibarius* group to stain on handling or bruising. *C. cibarius* (all forms collected) and *C. subalbidus* stain rusty orange where cut or bruised, and *C. odoratus* and *Canth. lateritius* (= *Craterellus cantharellus*), which I retain as separate species, stain pale yellow-green on handling, *then* dull rust color. The species of *Clavariadelphus* which stain under the same conditions, do so into the vinescent or brunnescent shades. In addition, the pigment-analysis data already presented also conflict. We have obtained a negative carotenoid analysis for *Clavariadelphus pistillaris*. Just the opposite data have been obtained from the *Cantharellus cibarius* group, including *C. odoratus*. What becomes imperative are tests on other species of *Clavariadelphus*, especially the *C. ligulus* group, placed in a separate subgenus by Wells and Kempton (1968) and presenting an often more distinct yellowish coloration. The question of the links between *Cantharellus* and *Clavariadelphus*, then, are open to question.

With smooth, usually white, multiguttulate spores and a monomitic hyphal construction, it would seem relatively easy to pass from *Clavariadelphus* into *Clavulinopsis* and *Clavaria*, and thence to the reduced *Multiclavula-Pistillaria* group. With the thickening hymenium as an added shared character, *Clavulinopsis* shares clamp connections with *Clavariadelphus* and, within the genus, passes from gregarious but single, to fascicled to branched fruiting body form. It is in *Clavulinopsis* that one may observe the derivation of the branched

FIGURE 2. Trichodermoid pileocystidia of the apex of the fruiting body of *Clavariadelphus truncatus*.

fruiting form from the unbranched by the division of the meristematic region into several points and concomitant growth. Fasciculation is attained when several individuals arise from very closely juxtaposed primordia, with some forms like the fastigiate form of *C. corniculata* apparently performing both modes of development at the same time. Corner (1950) assessed the relationship between *Clavulinopsis* and *Clavaria* but kept the two genera separate on the presence or absence of clamp connections. Leathers (1955) disputed the use of this character, but I have been unable to support his observations of clamps in species of *Clavaria s. str.*

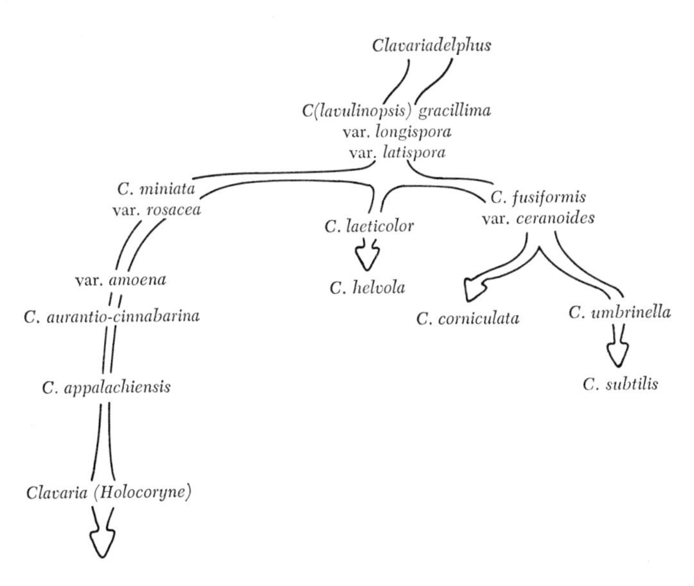


FIGURE 3. Proposed (abbreviated) evolutionary pathways in *Clavulinopsis.*

Figure 3 is distilled from the more detailed treatment of *Clavulinopsis* by Petersen (1968a). It indicates the two major subdivisions of *Clavulinopsis* as secondary, presumably derived from species which are neither weakly nor strongly apiculate—in this country, notably, *C. gracillima,* a synonym of which is *Clavaria luteo-alba* Rea. This species also gives equivocal tests for carotenoid pigments. The genus *Clavaria* may be derived from the species of *Clavulinopsis* whose spores are weakly apiculate, with *Clavulinopsis appalachiensis* also lacking clamps at many septa. Any *Clavariadelphus* ancestor postulated is subject to the questions raised formerly. Van Overeem (1923) described *Clavulinopsis* on the variable sterigmata, both in number and placement, and this may represent what Smith has conjectured as a primitive feature.

Petersen and Olexia (1969) have indicated that the two genera might better be placed together again as subgenera of *Clavaria* because of some intermediate species bearing combinations of the generic characters. These would include *Clavulinopsis corallino-rosacea*, which bears clamps but has the hyphal construction of *Clavaria*; *Clavaria longispora*, which lacks clamps but has the hyphal configuration of *Clavulinopsis* and bears numerous 2-sterigmate basidia; and *Clavulinopsis appalachiensis*, which bears both clamped and unclamped septa on the contextual hyphae. The discovery of intense yellow coloration both in *Clavaria inaequalis* (over which there has been no small nomenclatural controversy) and an undescribed species of *Clavaria* from the southern Appalachian Mountains has strengthened this thesis. An interesting study might be an investigation of the spore ornamentation in the echinulate-spored species of *Clavaria* (Petersen & Olexia, 1969), namely, *Clavaria atrofusca*, *C. asperulospora*, and *C. californica*, and its possible similarity to that of *Clavulinopsis asterospora* and *C. helvola*.

At this point it might be wise to insert a thought or two on the genus *Ramariopsis*. As Corner surmised, the genus is probably derived from the stock of *Clavulinopsis-Clavaria*, contrary to the opinion expressed by me earlier (Petersen, 1966). Hyphal construction is monomitic, and the septa bear clamps. The spores, however, are usually small (an exception is *R. lentofragilis*, with spores up to 7 μ long). Most fruiting bodies are branched, and some have a small basal mycelial pad or mat from which the fruiting body arises. The anatomical features and spores are remarkably similar to *Cristella* species, as pointed out by Petersen (1966), especially species like *C. farinacea* and *C. alnicola*. This might be mentioned as the first of several lines of clavarioid groups to reduce to resupinate fruiting forms.

In partial summary (fig. 4), then, there seems to be both positive and negative evidence for a somewhat close relationship between *Clavariadelphus* and the smooth-spored cantharelloid fungi on the one hand, and the Clavariaceae *sensu* Donk (1964) on the other, but little doubt that *Clavulinopsis, Clavaria, Ramariopsis,* and the rest of the Clavariaceae *sensu* Donk make up a homogeneous group.

Corner (1966), although elaborating on the probable relationship of *Clavariadelphus* to *Cantharellus,* was loath to accept the thesis of Maire and later Heim (1954) that *Clavariadelphus* was related to

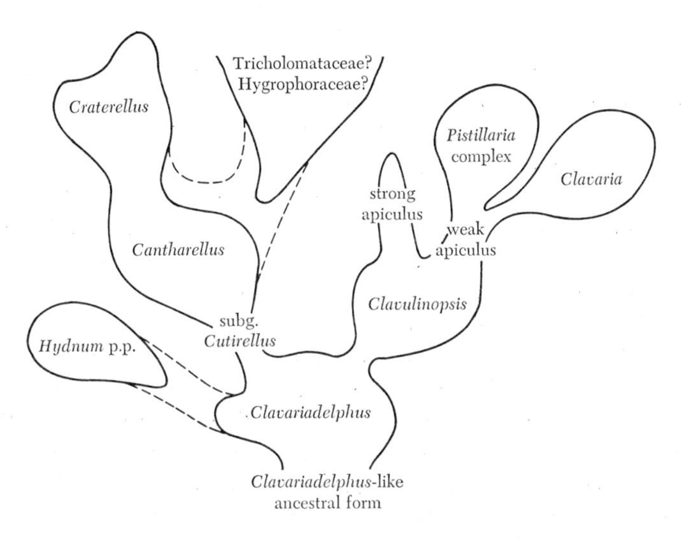


FIGURE 4. Proposed phylogeny of the Clavariaceae and Cantharellaceae.

*Gomphus.* I have suspected such a relationship for some time, in fact long rejecting the probable relationship just introduced between *Clavariadelphus* and the smooth-spored cantharelloid fungi. Again, the detailed study of Wells and Kempton (1968) has shed light on the possibility of such a relationship. Several characters may be used. The truncate or turbinate growth form found in *Clavariadelphus* might match that found in *Gomphus* subgenus *Gomphus,* including *G. guadelupensis, G. viride,* and perhaps more important, *G. clavatus. G. clavatus,* and the entire subgenus *Excavatus* show positive iron salts reactions, and the other species of *Gomphus* may be equally positive, but I have seen them only in the dried state. Wells and Kempton (1968) described a unique rosy-red or rosy-pink KOH reaction of the fruiting body of *Clavariadelphus truncatus* and observed a cherry red coloration in other species of the subgenus *Clavariadelphus* (as I have), other than *C. unicolor.* I have obtained precisely the same reaction on the fruiting bodies of *Gomphus clavatus*. Moreover, the hymenial surface of some species of *Clavariadelphus* has been described as pink to lilac or purplish, perhaps matching the colors found in *G. clavatus*. Both the truncate species of *Clavariadelphus* and the *Gomphus floccosus* complex tend to become naturally vinescent and to bruise brown. Microscopically, the basidia of both genera are very long and attenuate, and the hymenium in both genera thickens. In *Gomphus* (all species examined) and in many species of *Ramaria* the clamp connections themselves are inflated and often thick-walled. With the hyphal diameter of perhaps 3–8 μ, the inflated clamps are often 15–20 μ in thickness, including the parent hypha. Although the wall never thickens so as to obscure the cell lumen, nevertheless under phase contrast the wall is thick enough to be distinctly more refringent than the wall of the parent hypha. In some species of *Ramaria*, specifically the species which would be included in Brinkmann's genus *Phaeoclavulina*, in *Lentaria*, and in the *Ramaria stricta-R. apiculata* complex, this is the extent of the modified structure, the walls of the inflated clamps being internally and externally smooth, although often thickened. Wells and Kempton report such features in all truncate species of *Clavariadelphus*. But in all species examined in *Gomphus*, and in the largest subdivision of *Ramaria*—I have called it the *R. formosa* complex—the walls of many of the inflated clamps extend inward in very delicate processes (Fig. 5), either simple or occasionally branched or forked, not unlike stalagtites in caves. Rather frequently, these inflated areas are not restricted to clamps but occur as clavate hyphal tips or as inflated areas in the midst of otherwise uninflated hyphae. Although the tendency for inflated clamps is surely not restricted to the groups mentioned, the extent and constancy of the inflated clamp are unique, I think. Moreover, the delicate internal ornamentation of the wall in so consistent a fashion in *Gomphus* and several species of *Ramaria* would seem to link the two genera. In both genera, distribution seems not to influence the structure's presence—I have seen identical structures on *G. clavatus* from Fries's herbarium at Uppsala, from France, and from all over the distribution through North America. I have seen the same feature on the neotype specimen of *G. floccosus* in Schweinitz's herbarium at Philadelphia, and in all of the various forms of the species as they occur from eastern Canada to the Pacific coast. They are present in *Ramaria sinapicolor* and *R. subbotrytis* which are clampless, and *R. botrytoides* and *R. brunnea* which bear clamps. In short, they are a constant feature of all species of *Gomphus* (which include all species reported from North America, with representatives from European herbaria, and some extralimital species), and several species of *Ramaria* (see Petersen & Olexia, 1967), most apparently those with large, fleshy fruit bodies and rugose

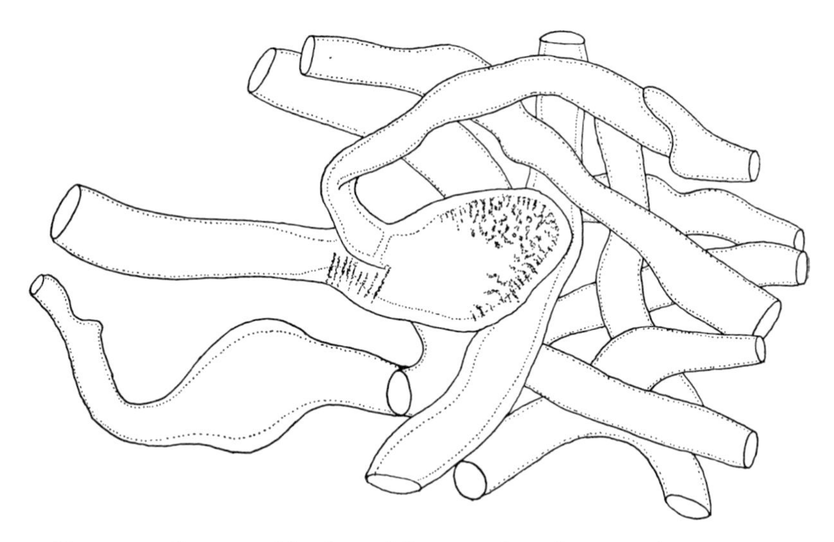

FIGURE 5. Contextual hyphae of *Ramaria-Gomphus* complex, showing punctate, inflated clamp connection.

spores and their relatives. These structures have been conspicuous by their total absence in *Cantharellus*, *Craterellus*, and the species of the Clavariaceae *sensu* Donk.

I think that the inflated clamps, especially with the thickened wall, may closely link *Clavariadelphus* and *Gomphus* or *Ramaria*. The non-ornamented inflated clamp mentioned above for certain groups within *Ramaria* is quite similar to that found in *Clavariadelphus*.

All this would seem to attest to links between *Gomphus* and *Ramaria*—links which are not missing by any means, but what the relationships are between these various subgenera of *Ramaria*, *Gomphus*, and *Clavariadelphus* only more exhaustive investigation will tell.

Wells and Kempton (1968) ascribe an ochre spore print to *Clavariadelphus truncatus*, although second-hand from the notes of Leathers on a single specimen. The growth forms of *C. subfastigiata* and

*C. cokeri* are clustered to connate, with the latter (known only from the type specimen) inclined toward antler-like branching. I have observed such branching but have identified the specimens as *C. pistillaris*, although not critically so.

There would seem no reason in this day to belabor the probable relationships between *Gomphus* and *Ramaria*, or the separation of *Gomphus* and *Cantharellus*, although some investigators are still hesitant to accept either postulate. Petersen (1967a) has summarized the case for both concepts, supporting the familial arrangement by Donk (1961, 1964). The ochraceous to creamy, ornamented spores of both *Gomphus* and *Ramaria*, the cyanophilous spore ornamentation in a characteristic pattern, the presence not only of inflated, thick-walled clamp connections, but of punctate ornamentation on these inflated clamps, and, much less important, the shared ecological preference of the two genera all suggest strongly that they are closely related. Corner (1966), writing of the possible connection between *Gomphus* and *Ramaria*, stated, "There is no fungus known to connect them as *Clavariadelphus truncatus* connects *Clavariadelphus* and *Cantharellus*." I would dispute the spirit of this concept, although not its letter. Several species of *Ramaria*, including *R. obtusissima* which was largely described on such a character, become pistillariform apically in age. These truncate branch apices are common in older material, and with color fading, often render the true identity of the fungus much more difficult to ascertain. Moreover, species of the subgenus *Gomphus* are rarely even depressed on the surface, but possess a solid stipe and pileus unlike other members of the genus which are funnel- or trumpet-shaped. A number of species and infraspecific taxa produce branched or cespitose fruiting bodies—namely *G. bonarii*, *G. floccosus*, *G. rainieriensis*, and *G. guadelupensis*.

As stated by Petersen (1968b), the spores most heavily ornamented in both genera are those of *Gomphus guadelupensis* and *G. viride*. Both these species have fruiting bodies which are cespitose or branched and have an undifferentiated pileus surface of generally erect, sometimes fascicled, narrow hyphal tips. This same type of surface is found in most truncate fruiting bodies of *Clavariadelphus*, according to Wells and Kempton (1968), whose conclusions are supported by my own observations. Whether the extent of spore ornamentation is the result of progressive excessive secretion of cyanophilous material, or the most primitive state from which the reduced ornamentation of the *Ramaria-Gomphus* spore was derived, is open to question. If *Clavariadelphus*, with its smooth, white spores, is considered primitive, then ornamentation and color would seem to be additive, but additive genes have always left me skeptical.

Corner (1966) conjectured that *Gomphus* had progressed from a merismatoid ancestral form, to become pileate more recently. On the contrary, if the establishment of the branched growth habit could be accomplished by the breakup of the meristematic region into small points or patches on the primordium, then the merismatoid growth habit could be the result of a similar phenomenon, in which the meristematic regions result in the production of petaloid or pleuropodal processes instead of cylindrical. In the merismatoid species, the hymenium is inferior just as in the pileate species and not amphigenous as it is in *Sparassis*. The type of *G. guadelupensis* which is included in the subgenus *Gomphorellus* by Corner is really branched, not merismatoid, with each of the subpilei having the hymenium extending completely around the stipe portion. *G. pallidus*, on the other hand, is merismatoid, but it would seem a reduced form, with its small size, small spores, and off-white colors. In passing, Corner has described the spores of *G. pallidus* as smooth, but examination of the type has proven the spores to be rough and quite typically cyanophilous, as in the rest of the genus. Corner's *G. pseudoclavatus* turns out not to be a *Gomphus* at all, but a large, unique *Pseudocraterellus*.

In order to better conceptualize the relationships between the species of *Gomphus* and the possible relationship of the whole genus to other taxa, two ground plans have been drawn up. It may seem strange to draw up two ground plans for the same genus, but I felt that when the related groups were not clear, and therefore character states could not be easily determined, then it was best to start with both the possible ancestral forms and make plans in both ways. In doing so, I have assumed that *Gomphus* was either the product of degeneration from a former agaric ancestral form or the product of sophistication of a *Clavariadelphus*-like ancestral form. Again, to anticipate, I will not be able to predict which plan is correct, so you will not want to hold your breath.

In both plans I started with fifteen characters, ranging from clamps versus no clamps, smooth pileus surface versus scaly pileus surface, presence or absence of coscinoids, presence or absence of mediostratum, and extent of spore ornamentation (rough versus very rough, which may sound arbitrary, but with knowledge of the species, is really not so). These characters are summarized in table 2.

TABLE 2. CHARACTERS USED IN CONSTRUCTION OF GROUND PLANS IN *Gomphus*.

| | |
|---|---|
| 1. Presence or absence of clamps | 9. Spores rough versus very rough |
| 2. Smooth pileus surface versus scaly | 10. Fruiting bodies single versus cespitose |
| 3. Cystidia versus no cystidia | 11. Scales small versus large |
| 4. Excavate versus solid | 12. Lamellar folds versus none |
| 5. Laticiferous hyphal system versus none | 13. Mediostratum versus none |
| | 14. Coscinocystidius versus none |
| 6. Punctate clamps versus none | 15. Mesopodal versus pleuropodal fruiting bodies |
| 7. Coscinoids versus none | |
| 8. Basidia 1–4 sterigmate versus basidia 4 sterigmate | |

In the first plan (fig. 6), using an agaric ancestral type, the genus may be divided into two major lines. As might be expected, the species and forms resembling *G. floccosus* are tightly grouped. *G. clavatus*, *G. viridis*, and *G. guadelupensis* are grouped, all with smooth

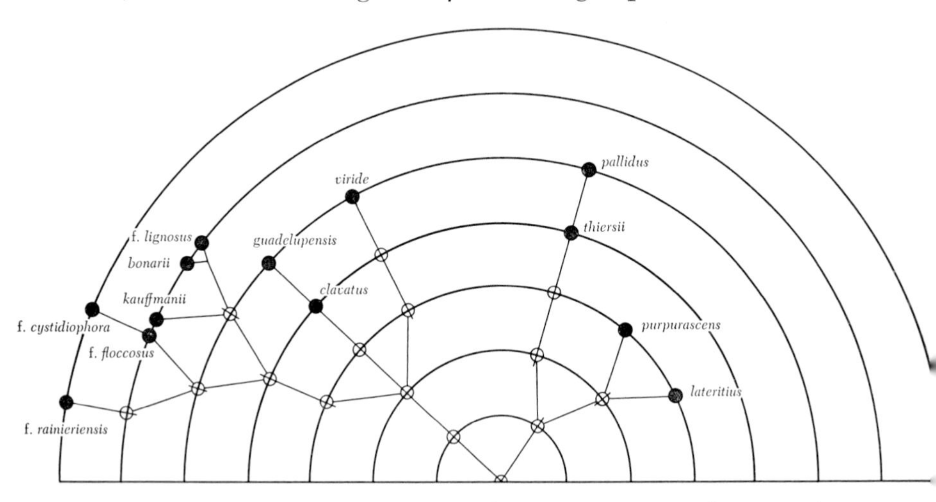


FIGURE 6. Ground plan for *Gomphus* species—agaricoid ancestor.[1]

[1] The reader is cautioned that f. *lignosus*, originally intended for proposal herein, has since been found to be *Cantharellus floccosus* f. *wilsonii* Smith *in* Smith & Morse. It is treated in a forthcoming paper in Nova Hedwigia.

pileus surface. *G. viridis* bears cystidia whereas *G. guadelupensis* does not, although they both produce highly ornamented spores. The major distinction between the two large subdivisions is the absence of coscinoidal elements in the group just mentioned, and their presence in the remaining four species. Petersen (1968b) has already commented on the very close similarity of *G. purpurascens* and *G. lateritius*, which are really much too far apart on this chart. *G. thiersii*[2] and *G. pallidus*, the former new and named in honor of Dr. Harry Thiers, its collector, bear no coscinocystidia as do

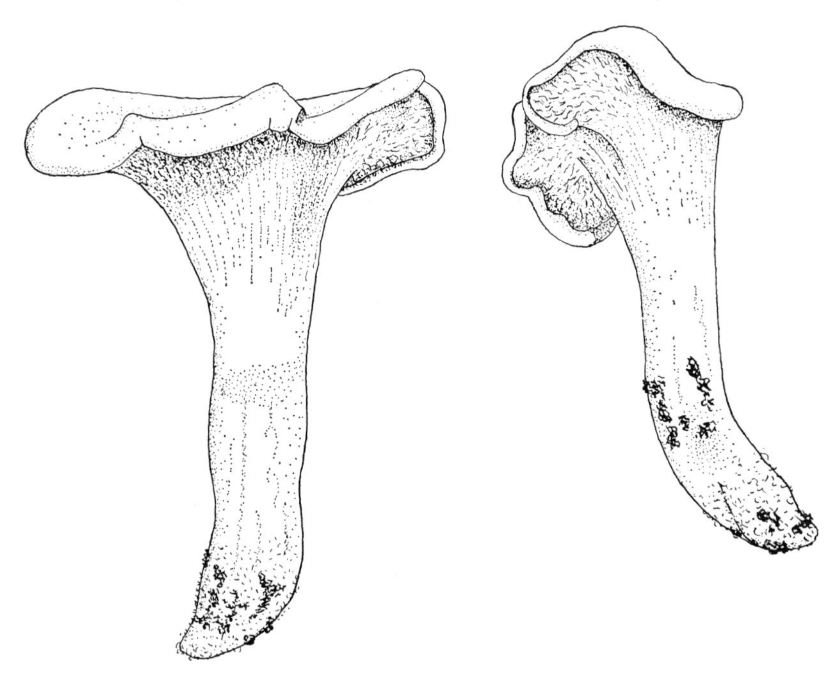

FIGURE 7. *Gomphus thiersii*, fruiting bodies (reconstructed from dried specimens).

[2] *Gomphus thiersii* sp. nov. Fructificationes –10 cm altae, –7.5 cm latae, solitariae vel gregariae (fig. 7); pileus solidae, planae, laevus ad subglaber, "ochraceous buff" ad "apricot buff" (Ridgway, 1912) (siccus); stipes laevus; plicae humilae, decurrentes, venulis transversaliter connexae. Sporae in cumulo pallide ochraceae, 10.6–13.5 × 5.3–6.3 μ leniter rugulosae, cyanophilae. Hyphae contextualae tenuitunicatae, saepe inflatae, hyalinae, sine fibulis. Hyphae oleiferae copiae, uninflatae, "coscinoidal," flavae ad fulvo-flavae; septa rara. Typus: Herb. Thiers; San Francisco State College, s.n. Calavaras Big Tree State Park, Oct. 15, 1967.

*G. purpurascens* and *G. lateritius*. For these reasons, even though I would personally rather retain this complex as a subgenus of *Gomphus*, I suppose that these species might make up the nucleus of the genus *Gloeocantharellus* Singer, which was described based on *Cantharellus purpurascens*. *Linderomyces*, based on *Paxillus lateritius* Petch, must be placed in synonymy under *Gloeocantharellus*, at whatever the level the taxon is admitted. Coscinoids and coscinocystidia as found in *G. purpurascens* and *G. lateritius* have been illustrated by Petersen (1968b, and fig. 8).

If one constructs a ground plan using a *Clavariadelphus*-like ancestral form (fig. 9), some of the character states change. For instance, using an agaric ancestor, lamellar ridges are more primitive than a rugose hymenium, and a mediostratum more primitive than the lack of it. Using a *Clavariadelphus*-like ancestral form, however, a rugose hymenium is more primitive, as is the lack of a mediostratum. When these and other character states are modified (although none eliminated), the ground plan changes. In this case, there seem to be four distinct lines within the genus. *G. clavatus* stands alone, understandable in its merismatoid growth form, purplish hymenium, and other characters. *G. guadelupensis* and *G. viridis* share the highly ornamented spores without having *G. clavatus* in the evolutionary line. Finally, the same two major complexes again occur—those with coscinoidal elements, and all the various taxa around *G. floccosus*, *G. kauffmanii*, and *G. floccosus* f. *rainieriensis* could just as easily be turned around, perhaps preferable in light of coloration. I see some advantages in this plan, especially in the separation of *G. clavatus* from the *G. guadelupensis* and *G. viridis* line, but not really enough to make a clear-cut choice of one over the other.

I might digress for just a moment to examine the occurrence of oleiferous hyphae, commonly called laticiferous hyphae, in the group in question. In several members of *Gomphus* and *Ramaria* such oily hyphae occur, sometimes exhibiting a metachromatic reaction with cresyl blue, sometimes not, but always refringent under phase contrast so that they almost shine in squash mounts of contextual tissue. They are most common in the stem tissue, although often present in the pileus context as well. In most species, this laticiferous system is represented by hyphae with relatively homogeneous contents, even though very different from the normal generative contextual hyphae. In some species, however, these hyphae are more than nominally yellow under bright field, and in some fewer species the contents are

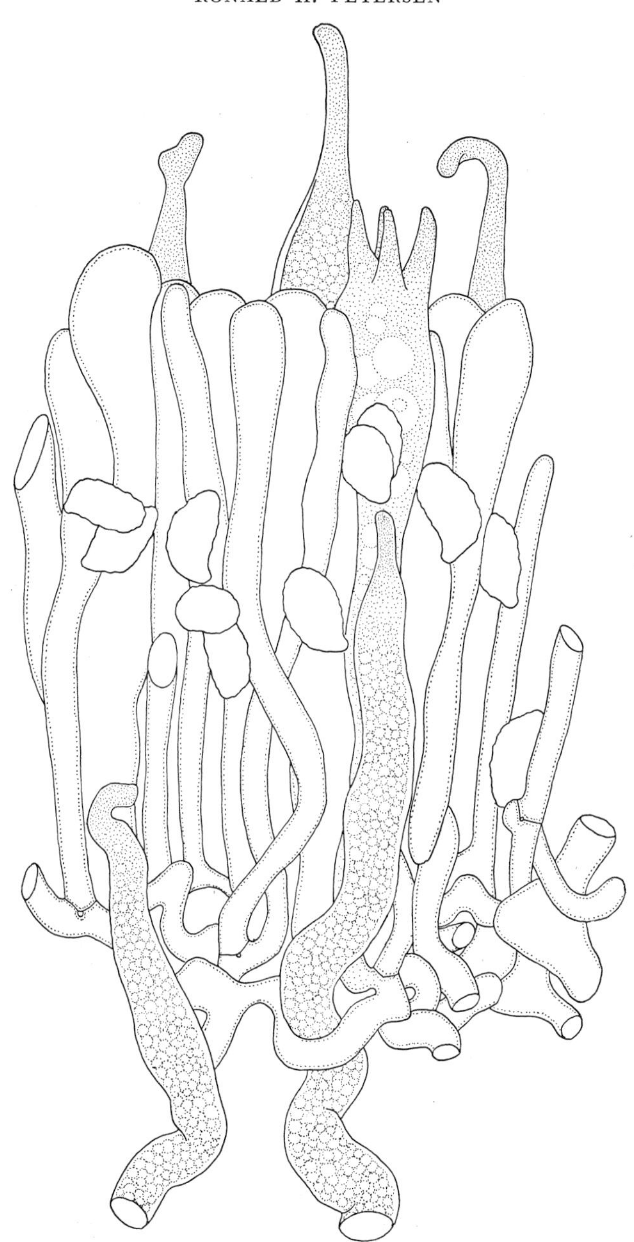

FIGURE 8. *Gomphus purpurascens.* Hymenium with coscinocystidia and basidioles.

heterogeneous—almost obscurely foamy. This appearance is finally exemplified by the conspicuously foamy content of the coscinoidal elements of the *G. purpurascens–G. thiersii* complex. Whether or not the coscinoids and coscinocystidia of *G. purpurascens* and *G. lateritius* are related to the laticiferous hyphal systems in the rest of *Gomphus* and *Ramaria* is again open to question, but intermediate

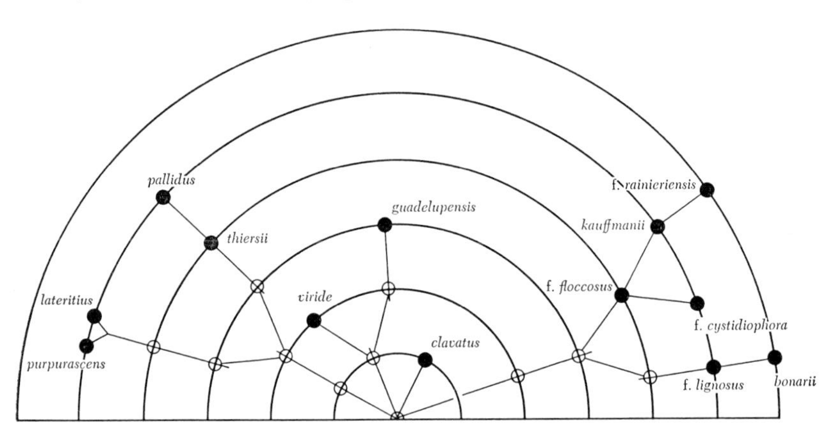


FIGURE 9. Ground plan for *Gomphus* species—*Clavariadelphus* ancestor.

states of these hyphal systems occur frequently, and, I suspect, not without significance.

The genus *Ramaria* has been informally subdivided both by Corner (1950) and Petersen (1967a), although rather differently. I tentatively identify four relatively well-defined subgeneric groups of species as follows: (*a*) *Phaeoclavulina* group (falling within that relatively little-known genus by Brinkmann), including *R. zippellii*, *R. grandis*, *R. invallii*, *R. myceliosa*, *R. murrillii*, and other species with echinulate, usually rusty brown spores; (*b*) *Ramaria* group (falling within the strict genus *Ramaria* if typified by *R. botrytis*), including *R. botrytis*, *R. crassipes*, *R. australiana*, *R. xanthosperma*, and other species with large, fleshy fruiting bodies and striate spores; (*c*) *Clavariella* group (in the strict sense, typified by *R. apiculata* following Doty), including *R. stricta*, *R. apiculata*, *R. acris* with rough spores, and *R. pinicola* with ochraceous smooth spores, and other species with small spores and thick-walled contextual hyphae. It is in this group that Corner (1961) had described dimitic and trimitic fruiting bodies, surely not related to the dimitic fruiting bodies of

*Pterula.* This group might well be accepted at genus rank, with *Lentaria* as a pale-spored subgenus. (*d*) Formosa group (with no former genus name applicable to my knowledge, even though the group includes perhaps more species than all the other groups combined), including *R. formosa, R. conjunctipes, R. subbotrytis, R. botrytoides, R. fennica,* and others with fleshy fruiting bodies and roughened, ellipsoid, ovoid, or cylindrical spores. I have seen specimens of a species called *R. flavo-brunnescens* in the western United States, but not really that species, in which binding hyphae seem to occur with generative hyphae, but without the presence of skeletals.

The relationships between *Gomphus* and *Ramaria* cannot be drawn very specifically, since the intermediates are now probably extinct, for the most part. Some small specific similarities still exist, however. *Gomphus clavatus, Ramaria fennica, R. testaceoviridis,* and two undescribed species of *Ramaria* from the western United States all share a violaceous color of the young fruiting body, turning various colors in age or from spore deposits. One of these undescribed species and *R. fennica* react with KOH to produce a coral to rosy-red pigment, the former on the hymenium, the latter just at the cut surface. *G. guadelupensis* and the *R. grandis–R. longicaulis* complex share rich brown coloration, rusty spore color, very coarse spore ornamentation, and agglutinated basidia. But here even the indirect close similarities stop. Granted, the same general coloration is shared in the *R. formosa* complex and the subgenus *Excavatus* of *Gomphus,* but this is hardly valid considering the very large discrepancy in growth form. In fact, it would appear to be microscopic characters which hold the groups together. In this way, I would have to partially agree with Corner when he states, "While still considering *Gomphus* and *Ramaria* as distantly related, I think that they offer a different and more remote problem from the close affinity of *Clavariadelphus* and *Cantharellus.*"

The origin of the *Clavariella* group would seem to be within *Ramaria s.l.,* perhaps through some of the complexes surrounding *R. formosa.* From the *Clavariella* group there seems to be a natural series through *R. pinicola* var. *robusta* and *R. pinicola* into *Lentaria* (Petersen, 1967b) with very pale to white smooth spores, and even more emphasis on thick-walled generative hyphae. Perhaps here the ramarioid tendency ends. On the other hand, a reduction series could be envisioned from the *Clavariella* group through *Kavinia* Pilát, although the spores of this genus do not appear totally similar to those of the other members of the series, to *Ramaricium* Eriksson, a resupinate genus. In this regard, Corner's genus *Lentaria* was admittedly heterogeneous when described. Petersen (1967c) has split out the phycophilous species into the genus *Multiclavula* (a name which may now be replaced by the prior, although until recently obscure, *Stichoclavaria* Ulbrich). The remaining homogeneous group in *Lentaria*, characterized by white, smooth spores and rather thick-walled generative hyphae, although still ramarioid form, is typified by *L. epichnoa* (Fr.) Corner. The representative specimen of Karsten's of this species (and Fries referred to Karsten as having the right idea about the species) not only bears basidia on the ramarioid portion, but on the fertile corticioid mat on the wood substrate, bridging this genus with a very possible resupinate group yet to be delineated, or at least unknown to me. An interesting conjecture would have *Clavariadelphus* as a white-spored, reduced form of *Gomphus*, as *Lentaria* is of *Clavariella*. Such a thought would necessitate the agaric ancestral model for the ground plan.

I would summarize the line generally thought of as the Gomphaceae Donk (1961) (Fig. 10) as consisting of *Clavariadelphus* with its

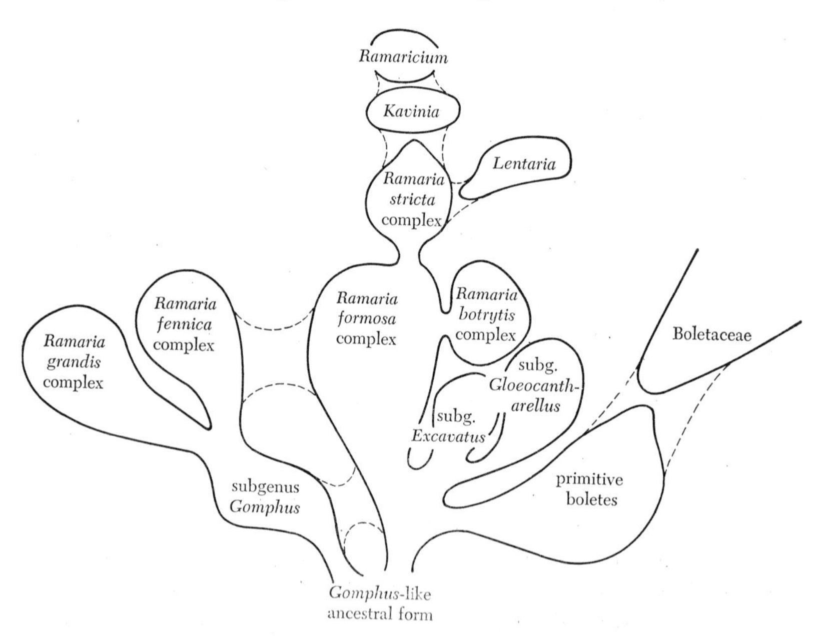


FIGURE 10. Proposed phylogeny of the Gomphaceae.

simple to rudimentarily branched species, *Gomphus*, showing fruiting bodies ranging from simple and solitary to cespitose or branched, and solid to funnel-shaped, and *Ramaria*, all the species of which are branched. In passing, Corner has stated that no species of *Ramaria* has been found which forms simple fruiting bodies. I would submit that if such a species were found, it would be forced into *Gomphus* or *Clavariadelphus* anyway, for of such forms are these genera made.

It should have become obvious that I have included *Clavariadelphus* in all the evolutionary lines of the Clavariaceae *s. str.*, Cantharellaceae *s. str.*, and Gomphaceae. Petersen (1967a) disputed Donk's (1964) similar disposition of this genus, but I have it upon good authority, that he (Petersen) is living to regret that printed doubt. *Clavariadelphus* indeed possesses characters from all these lines. I can remember a long discussion in the lab of Alex Smith, in which he tried to convince me of the very arguments I present herein—with vigorous opposition on my part. I offer my apologies to both Smith and Donk, hoping that I have added some scientific fuel to the fire of their arguments.

## References


Arpin, N., P. LeBreton, and J. Fiasson. 1966. Recherches chimotaxonomiques sur les champignons II. Les caroténoïdes de *Peniophora aurantiaca* (Bres.) (Basidiomycetes). Bull. Soc. Mycol. France 82: 450–459.

Coker, W. C. 1923. The clavarias of the United States and Canada. Univ. North Carolina Press. 209 p.

———. 1947. Further notes on clavarias, with several new species. J. Elisha Mitchell Sci. Soc. 63: 43–67.

Corner, E. J. H. 1950. A monograph of *Clavaria* and allied genera. Ann. Bot. Mem. 1: 1–740.

———. 1958. *Craterellus*, *Cantharellus* and *Pseudocraterellus*. Beih. Sydowia 1: 266–276.

———. 1961. Dimitic species of *Ramaria* (Clavariaceae). Trans. Brit. Mycol. Soc. 44: 233–238.

———. 1966. A monograph of the cantharelloid fungi. Ann. Bot. Mem. 2: 1–255.

Donk, M. A. 1933. Revision der Niederländischen Homobasidiomycetae-Aphyllophoraceae II. Mededeel. Bot. Mus. Herb. Utrecht 9: 1–278.

———. 1961. Four new families of Hymenomycetes. Persoonia 1: 405–407.

———. 1964. A conspectus of the families of the Aphyllophorales. Persoonia 3: 199–324.

DOTY, M. S. 1948. A preliminary key to the genera of clavarioid fungi. Bull. Chicago Acad. Sci. 8: 173–178.

FIASSON, J., AND N. ARPIN. 1967. Recherches chimotaxonomiques sur les champignons. V. Sur les caroténoïdes mineurs de *Cantharellus tubaeformis* Fr. Bull. Soc. Chim. Biol. France 49: 537–542.

FIASSON, J., P. LEBRETON, AND N. ARPIN. 1968. Les caroténoïdes des champignons. Bull. Soc. Nat. Arch. l'Ain 82: 47–67.

FRIES, E. M. 1821. Systema mycologicum. Vol. 1. Reprint, 1952. 520 p.

———. 1838. Epicrisis systematis mycologici. Stockholm. 610 p.

———. 1874. Hymenomycetes Europaei sive Epicriseos systematis mycologici. 2nd ed., Stockholm. 755 p.

HEIM, ROGER. 1948. Trois clavariées de Madagascar. Vol. jub. Rene Maire. 9 p.

———. 1949. Une clavaire cantharelloides australienne a pigment carotinien cristallise. Rev. Mycol. 14: 113–120.

———. 1954. A propos de trois chanterelles americaines. Rev. Mycol. 19: 47–56.

KARSTEN, P. A. 1879. Rysslands. Finlands och den Skaninaviska halfons hattsvampar. Bidr. Kann. Findlands Natur Folk 37: 1–257.

LEATHERS, C. R. 1955. The genus *Clavaria* Fries in Michigan. Thesis, Univ. of Michigan.

MAIRE, RENE. 1902. Recherches cytologiques et taxonomiques sur les Basidiomycètes. Bull. Soc. Mycol. France Supple. 18, Fasc. 2: 1–209.

———. 1914. La flore mycologiques des forets de Cedres de l'Atlas. Bull. Soc. Mycol. France 30: 199–220.

OVEREEM, C. VAN. 1923. Beiträge zur Pilzflora von Niederländischen Indien. Bull. Jard. Bot. Buitenzorg III. 5: 247–293.

PERSOON, C. H. 1797. Commentatio de fungis claviformibus. *In* Holmskjold, Coryphaei clavarias ramariasque. Lipsaei, 239 p.

———. 1801. Synopsis methodica fungorum. Gottingae. 706 p.

PETERSEN, R. H. 1966. Notes on clavarioid fungi. V. Emendation and additions to *Ramariopsis*. Mycologia 58: 201–207.

———. 1967a. Evidence on the interrelationships of the families of clavarioid fungi. Trans. Brit. Mycol. Soc. 50: 641–648.

———. 1967b. Notes on clavarioid fungi. VIII. *Clavaria pinicola* and *C. stillingeri.* Bull. Torrey Bot. Club 94: 417–422.

———. 1967c. Notes on clavarioid fungi. VII. Redefinition of the *Clavaria vernalis–C. mucida* complex. Amer. Midl. Nat. 77: 205–221.

———. 1968a. The genus *Clavulinopsis* in North America. Mycologia Mem. 2:1–39.

———. 1968b. Notes on cantharelloid fungi. I. *Gomphus* S. F. Gray and some clues to the origin of the ramarioid fungi. J. Elisha Mitchell Sci. Soc. 84: 373–381.

———. 1969. Notes on cantharelloid fungi. II. Some new taxa, and notes on *Pseudocraterellus.* Persoonia 5: 211–223.

PETERSEN, R. H., AND P. D. OLEXIA. 1967. Type studies in the clavarioid fungi. I. The taxa described by Charles Horton Peck. Mycologia 59: 767–802.

———. 1969. Notes on clavarioid fungi. XI. Miscellaneous notes on *Clavaria.* Canad. J. Bot. 47: 1133–1142.

REID, DEREK. 1963. Fungi Venezuelani: VI. New or interesting records of Australasian Basidiomycetes: IV. Kew Bull. 16: 437–445.

———. 1965. A monograph of the stipitate stereoid fungi. Beih. Nova Hedwigia 18: 1–382.

RIDGWAY, R. 1912. Color standards and color nomenclature. Published privately, Washington, D. C. 43 p. + 53 pl.

SMITH, A. H. 1968. The Cantharellaceae of Michigan. Mich. Bot. 7: 143–183.

SMITH, A. H., AND E. E. MORSE. 1947. The genus *Cantharellus* in the western United States. Mycologia 39: 497–534.

WELLS, V. L., AND P. E. KEMPTON. 1968. A preliminary study of *Clavariadelphus* in North America. Mich. Bot. 7: 35–57.

## DISCUSSION

DONK: May I draw your attention to *Typhula erythropus,* which has been found by Romagnesi to have amyloid spores. The construction of the setiform stalk is precisely like that in some tropical species of *Marasmius.* Is it possible that we are confronted in this case with an example of a reduced *Marasmius*? Other clavariaceous genera may theoretically also be derived directly from agarics.

PETERSEN: I hope that I indicated in the beginning that the two groups discussed before —that is, the Clavariaceae and the Gomphaceae—I considered to be two lines of evolution which we would consider, but that the *Clavicorona* group, the *Pterula* group, *Lachnocladium*, and a number of other genera of clavarioids were not included in this discussion, yet represented separate lines. I would not doubt that we might find that other clavarioids have arisen from agarics in other positions.

PILÁT: I observe that the fleshy clavariaceae in America generally are very different from those of Europe, whereas the nonfleshy representatives are much more homogeneous to the continent. I think there might be some significance in this. Would you comment on this?

PETERSEN: The distribution of these things is very poorly known—I collected a species last year in the Pacific northwest that I cannot separate from the European *Ramaria mairei*, and another that I cannot separate from the Australian *R. sinapicolor*. I have seen type material of *R. sinapicolor*, and I still cannot separate what I have from it. I have not seen material of *R. mairei*, so I cannot tell whether we are dealing with precisely the same thing. I might say this: of the nonfleshy species (and I assume by nonfleshy you may be talking about what I called the smooth-spored series, the *Clavulinopsis-Clavaria* group) *Clavulinopsis corniculata* and *C. fusiformis* occur in Europe, but I do not think you get *C. miniata* or *C. aurantio-cinnabarina*. I think, in fact, that our species are slightly different. I spent a summer in Nova Scotia to find out where the two floras met. A significant species there was *C. helvola*, which appeared a number of times and which does not get south of northern New England apparently, and yet occurs quite freely in Europe. We do not get it in the southern United States at all. I have a half-dozen species now whose distribution goes through Asia, one or two collections in Michigan and New York, and then I find them in the southern Appalachians in great abundance. *C. miniata* is one of these.

A species' distribution appears different when you begin to ferret out the synonyms. There may be one name from Malaya, but a different name for the same species from Michigan. *Clavaria luteotenerrima* var. *borealis* from Michigan equals *C. inaequalis* from Scandinavia and perhaps *C. amoenoides* from India. If one begins to lump synonymy, distribution becomes a little clearer.

OLEXIA: You said *Ramariopsis* may get into resupinate forms, but you did not elaborate on its possible origins.

PETERSEN: Two or three years ago I published that I thought *Ramariopsis* had no relationship to the *Clavulinopsis-Clavaria* line—and that I thought it arose from a resupinate ancestor. At that time I knew of no echinulate-spored species of *Clavaria.* I now know about three. At that time I knew of only one echinulate-spored species of *Clavulinopsis.* I now know two. So I am much more able to envision a reduction into *Ramariopsis* and degeneration into a resupinate form, rather than an origin from a resupinate. I also labored with the idea that if *Ramariopsis* arose from a resupinate form, I had to supply an origin for the resupinate form. And I could not.

DONK: There is still another example of a clavarioid fungus with a fertile corticioid mat, viz., *Sparassis simplex*, described by Dr. Reid. It shows a considerable resupinate layer from which the branched fruiting body arises. (Compare also *Pterulicium* mentioned in my first paper.)

PETERSEN: When we culture S. *crispa* we get no fruiting body at all, but just fertile hymenium across the agar. By the way, this makes a beautiful experimental organism for cytological work because the basidia form progressively from the center inoculation block, right out across the plate. So you can cut a slice a half a millimeter or a millimeter across the agar and get all stages from the incipient to the mature basidia.

DONK: You included *Cantharellus* subgen. *Cutirellus* Corner in your discussion. Have you seen specimens? I have not seen any, but the basidia as drawn by Corner are so different from those of the *Cantharellus* group that I doubt that the two taxa have any close connection with each other. This subgenus seems to me more closely related to the *Leptoglossum* group, judging from the evidence published by Corner.

PETERSEN: I have not seen specimens. I raised the point that there could be suspicion that Corner was dealing with an aberrant form, but I certainly could not say it was *Clavariadelphus*-like because it has very well formed gill folds.

SINGER: You mentioned the relationship between the Agaricales and the *Gomphus* group. You also indicated your opinion of the genus *Linderomyces* as overlapping *Gomphus.* If this were actually so, you would have a very strong case in linking the two groups. I must, however, take exception to this, because we have found a second, very closely related species of *Linderomyces* in South America. I might add that this was collected by Corner, and I called it *Linderomyces corneri.* He collected it as a paxillaceous fungus, and Petch described the type of *Linderomyces* as *Paxillus lateritius.*

There are also anatomical differences that would separate the two genera rather strongly. The structure of the hymenophore of *Linderomyces* is a typically boletinoid bilateral structure and not merely divergent hyphae as you find in many cantharelloid fungi. The mediostratum is parallel as you say, but then you have the lateral stratum that is well differentiated from the mediostratum and is not only immediately divergent but goes almost at right angles outward as you have it in the Boletaceae.

PETERSEN: I find a series between the genera. I cannot separate *G. purpurascens* and *G. lateritius* out of the same complex. I find them very close, but then I find *G. thiersii,* which has no coscinocystidia but which has coscinoids, but it has reduced gills. Both *G. purpurascens* and *G. lateritius* have good gill folds.

SINGER: *G. purpurascens* is perfectly all right in *Gomphus.* If you remember, the forerunner of the family Gomphaceae was my subfamily Gomphidioideae, and I put *Gloeocantharellus* in that when I published it. I take exception only to the identification of the genus *Linderomyces* with *Gloeocantharellus.*

PETERSEN: I have seen the type specimen of *P. lateritius,* and I have collected and examined fresh material of *Cantharellus purpurascens* from the type locality and have seen its type specimen, and I still hold to their similarity, but perhaps further investigation will serve to clarify the situation.

KENNETH A. HARRISON

*Research Associate*
*University of Michigan Herbarium, Ann Arbor, Michigan*

# THE EVOLUTIONARY LINES IN THE FUNGI WITH SPINES SUPPORTING THE HYMENIUM[1]

The fungi that increase the hymenium-bearing surface through the development of spines or warts have been placed in a number of genera in different families. Close scrutiny makes it obvious that they are specialized forms that have found rather specialized ecological niches, and it seems logical to assume that they are remnants of very early evolutionary lines—the survivors that found a special habitat and were able to hold it against all competitors. Very few of them are common enough in any region for mycologists to become familiar with more than a dozen or so in a lifetime, and the specialist often must use herbarium specimens accumulated over long periods of time.

The presence of spines, or processes resembling spines, was the character used by Fries (1821) to distinguish a species as a member of the genus *Hydnum*. By 1874, Fries distinguished *Hericium*, *Radulum*, *Grandinia*, *Kneiffia*, *Odontia*, and *Mucronella*. At the same time he also recognized *Irpex*, *Sistotrema*, *Phlebia*, and *Tremellodon*, spined forms since removed to other groups by different workers. Quélet (1886) (Cooke & Quélet, 1878) and Karsten (1879) independently decided that there were still other genera and divided the Hydnaceae along what now appear to be more natural lines, Karsten being the first to publish his genera in accordance with the rules of nomenclature. Quélet's genera were not validly published until Karsten (1881) decided to use Quélet's names instead of his own, and in turn validly published them. The taxonomy would be simple if this were the whole story, but S. F. Gray (1821) published a different

[1] This paper was prepared during the tenure of NSF grant GB 4853 awarded for a study of the stipitate hydnums of North America under Professor A. H. Smith, Director of the University of Michigan Herbarium, and my thanks are extended to him and various associates in the Herbarium for their encouragement and suggestions during the development of the ideas presented here.

classification, which was not recognized as having any significance until Fries (1821) was established as the starting date for nomenclature. Banker (1906) and Coker and Beers (1951) must be considered the major North American contributors.

Miller and Boyle (1943) called attention to the fact that Gray introduced *Dentinum* for *H. repandum* Fries, and used *Hydnum* for *H. imbricatum* Fries. Gray was also the validating publisher for *Auriscalpium*, *Hericium*, and *Steccherinum*. An argument has been advanced against using *Dentinum* instead of *Hydnum* for the smooth-spored *H. repandum* of Linnaeus (Donk, 1956). My argument is that *Dentinum* was typified by Gray when he included both *D. repandum* (Fries) S. F. Gray and *D. rufescens* (Schaeff. ex Fries) S. F. Gray. The latter is doubtfully distinct, being often placed as a variety of *D. repandum* as it is little more than a darker color variant. Fries considered *H. rufescens* Schaeffer as synonymous with *H. repandum*, so I cannot see that *Dentinum* has any other possible type than *H. repandum* (L.) Fr.

The following is a very abbreviated key used for the genera of stipitate hydnums referred to in this paper.

KEY TO HYDNACEAE

| | |
|---|---|
| A. Sporophore lignicolous | |
| a. Spores hyaline, smooth | |
| 1. Flesh woody, spores amyloid | 1. *Echinodontium* |
| 2. Flesh tough | 2. *Steccherinum* |
| b. Spores hyaline, echinulate | |
| 3. Flesh and spores amyloid | 3. *Hericium* |
| 4. Spores amyloid | 4. *Gloiodon* |
| 5. Spores amyloid, on pine cones | 5. *Auriscalpium* |
| c. Spores colored, echinulate | 6. *Beenakia* |
| B. Sporophore terricolous | |
| a. Spores hyaline, smooth | 7. *Dentinum* |
| b. Spores hyaline, echinulate | |
| 1. Growth indeterminate, flesh tough | 8. *Phellodon* |
| 2. Growth determinate, flesh fragile | 9. *Bankera* |
| c. Spores colored, tuberculate | |
| 3. Growth indeterminate, flesh tough | 10. *Hydnellum* |
| 4. Growth determinate, flesh fragile | 11. *Hydnum* |

The genus *Steccherinum* S. F. Gray *s.l.* has been divided into a number of very small genera which appear to be as diverse a group as the Polyporaceae. At the moment the relationships are so obscure, and many species are so imperfectly known, that it is much more practical to refer to them under one genus—which certainly speeds up the recognition of taxa. They cannot be traced back to a common ancestral source or a prototype that we can recognize. They are too diverse morphologically to be anything but the ends of long lines of development in their particular habitats and can easily be offshoots from the same ancestors as the resupinate Polyporaceae.

There are surprisingly few morphological characters that can be used in tracing the relationships of the hydnaceous fungi listed above. Some of these characters are so easily influenced by environment that recognition of species from different regions often presents a problem. The research worker can only search for more morphological characters common to the variables from all areas, to indicate what changes are ecologically induced and which are genetic.

The more obvious morphological characters present are: spines—varying in size, shape, and spacing; habit of growth—either determinate or indeterminate; average size of fruit body; pigments—yellow, orange, olive browns, grays, black; surface characters—glabrous, matted, tomentose, scaly (I have not found any gelatinizing hyphae producing a viscid layer in any species, although the scales can be washed off *H. imbricatum* Fries in age); the organization of the context in the basidiocarp—simple or duplex; and the manner of development of the very young basidiocarps. Microscopic characters are: the types of hyphae present; the presence or absence of clamps on the hyphae and basidia; whether hyphae are incrusted or smooth, and with or without cell inclusions; and the presence of intercellular debris; hymenial structures such as cystidia are usually lacking, but of course basidioles are abundant. Some species in *Steccherinum* have complex structures, and *Hericium* Fries has gloeocystidia. Spores vary in color, size, and ornamentation; some are difficult to record so that they are useful as distinguishing characters. Chemical data remain largely ungathered. Differences in odor and taste are present in many genera, while KOH and $FeSO_4$ reactions occur occasionally and may eventually be found to have some taxonomic value. Melzer's reagent is also useful when examining some species, as there is a very peculiar gray reaction to inclusions in the tissues and hyphae: this has been described as "apparent amyloid" by Harrison (1964). The odor of slippery elm that is present in all species of *Phellodon* Karsten and *Bankera* Coker & Beers ex Pouzar when dried, is an indication of a chemical common to both genera and seems to indicate that they evolved from a common ancestor. It may be wishful thinking, but it is possible that answers to relationships can be secured through the chemistry of the various compounds—first between species and then between genera.

The impression grows that these are primitive groups of fungi and that each has not advanced very far since it became established in the particular restricted ecological niche now occupied. *Hydnellum*, *Hydnum*, *Phellodon*, and *Bankera* are all genera whose species are suspected of being mycorrhizal with coniferous trees. They have not been cultured. I have tried a number. A few short strands of mycelium have developed from isolates, but then growth ceased. The four genera produce basidiocarps in a gregarious pattern, and it is not unusual to find five or six species in one area. Twelve different kinds were once found in a three-acre block of mixed conifers.

Among the other genera listed, *Auriscalpium* S. F. Gray has one species of worldwide distribution (Harvey, 1958), and it is confined to the cones of conifers, especially *Pinus* spp. This is an outstanding example of a fungus that has a very special ecological niche. *Gloiodon* Karsten, with one species in North America, is confined to very wet rotting wood, especially ash. *Dentinum* S. F. Gray has a worldwide distribution on soil in mixed and coniferous woods. It is outwardly variable and, according to the temperament of the individual mycologist, is either disposed of as one species with numerous forms or varieties, several subspecies, or even regarded as autonomous species. There are a few variations with distinctively different-sized spores, but all the other characters intergrade. Size, shape, and color of sporocarps vary from tiny umbilicate forms in cedar swamps or high mountains to the large irregular, pallid autumnal forms of the Pacific Coast regions. Usually there are distinctive white forms (albinos?) in most regions. This is the one species of stipitate hydnum that is common and well known, in some form or another, over large parts of the world, and is highly regarded as an esculent.

*Echinodontium* E. & E. is lignicolous, perennial, and an important tree parasite—also a real relic of the past. It has recently been placed in a separate family because it is certainly neither a polypore nor a hydnum. It appears to be a survivor of an ancient primitive evolutionary selection that developed sufficient staying power to persist as long as its coniferous hosts survive. This is the only species that has developed a perennial fruit body, a distinct advantage in the struggle for survival.

*Beenakia* Reid (see Maas Geesteranus, 1963) is a rarity, mentioned here because it inhabits tree ferns south of the equator. This is an outstanding example of a fungus surviving in a special habitat.

Another interesting genus is *Hericium* Pers. ex S. F. Gray, with a number of widely distributed species, although occasionally common in restricted areas. One species causes a heart rot of oaks, a second of conifers, while a third is common on rotting aspen. Its amyloid reaction of both fruit body trama and spores is unusual in the fungi (Maas Geesteranus, 1959, 1962). It is a primitive genus of relatively few species. Spore sizes are diagnostic for each species, but fortunately are associated with other morphological characters that permit identification in the field. The different species exhibit an interesting parallelism by the formation of a unique growth form called "alpestre." This form, named by Bresadola, occurs in *H. coralloides* (Scop. ex Fries) S. F. Gray, *H. ramosum* (Bull. ex Mérat) Letellier, and in *H. abietis* (Weir ex Hubert) K. Harrison. In all three, occasional fruit bodies are found growing erect and branched with the teeth coming out in all directions, but with the spores typical of the species. This is one of the variations that has led to the confusion of species in the genus. It is now possible to map the distribution of the species of *Hericium* in North America and to obtain correct geographic ranges for each. *H. coralloides* has been reported as being common in the Pacific northwest, but I have yet to find a single collection. It is probably there but has not been collected to date. *H. ramosum* is abundant on deciduous wood, and *H. abietis* on coniferous.

I will deal with *Phellodon*, *Bankera*, *Hydnum*, and *Hydnellum* together. They all thrive in the same type of habitat, and the various species possess a greater range of characters than in the other genera of hydnums. The individual genera are separated first on spore characters into two groups, and then further on flesh characters. They require plenty of moisture and the pilei grow so close to the ground that the spines are almost in the duff. *Hydnellum* and *Phellodon* have fruit bodies that develop for one to three months under normal weather conditions, and some are so tough that they persist over winter after being killed by frost. They grow most luxuriantly when rainfall is frequent and humidities are high. Some species can be found year after year in the same spots under the same trees (thirty years in one block of *P. resinosa* Ait. for *H. pineticola* K. Harrison).

*Phellodon* and *Hydnellum* have a type of growth that I have termed indeterminate (Harrison, 1961). It starts as a hyphal column that grows upward from a mycelial pad and expands outward at the apex, thickening and spreading indefinitely during the summer until growth ceases with cold weather in the fall. Pilei can grow to quite impressive sizes under optimum conditions. The margin expands continuously, and new teeth develop as long as moisture and temperature are favorable. It is this feature of continuous growth that creates some of our most troublesome problems of identification of species in these genera in the late summer and fall. Dry weather can kill the center, or heavy rains can drown the central part by saturating the tissue with water, and then when conditions improve, new growth may start again on the margin, and outwardly exhibit a new appearance. Slugs can eat the surface, and the new growth will differ from the original. The fruit body of *H. mirabile* (Fries) Karst. will be white and tomentose instead of brown and strigose, as originally developed. Growth can be renewed in a most interesting fashion in practically every species I have observed in nature.

*Hydnum* and *Bankera* are more agarics in their manner of growth—that is, determinate. The very young fruit bodies that I have found formed very early a recognizable pileus and stipe which enlarge simultaneously. This has not been confirmed for the solitary species but has been observed a number of times with cespitose forms. Even when young, the injured parts are rarely renewed by new hyphal growth, and there is a definite limit to the size to which pilei can develop after the basidiocarp has started to grow. These are basic differences in development. At the same time it presents a case of parallelism between the white-spored echinulate-spored forms and the brown-spored, tuberculate-spored taxa. However, the basic differences between *Hydnum* and *Hydnellum* are bridged by *Hydnum stereosarcinon* (Wehmeyer) K. Harrison. It closely resembles *Hydnellum humidum* Banker, which is partially fleshy but definitely indeterminate in growth habit. The latter does not have inflated cells in the context of the pileus, a *Hydnum* character, but these are present in *Hydnum stereosarcinon*, though of an unusual type. The spines of both groups begin to develop as soon as there is a slight overhang of the pileus. The tips of the spines remain sterile and grow throughout their lifetime. This explains the great variation that can exist in size, shape, and length of spines.

## Speculations on Evolution in Hydnums

The next part of this paper is speculative, using examples from the hydnums to support various ideas. These are not new, but it is interesting to note the evolutionary steps that exist in the different genera.

The first fungus bearing an organized fruiting surface was in all probability a mat—a pad of mycelium producing spores on structures extending upward (the typical culture of many fungi in a Petri dish). Selection would favor the most prolific spore producers and those whose spores were best adapted for the widest dispersal. Forms were eliminated which lacked aggressiveness in finding suitable habitats for survival or could not resist competition for available food supplies. Ecological niches have been occupied by variants, and the extant species surviving in any niche are not necessarily the original occupant, for there must have been a long succession of species that could utilize every niche available.

Of the spine-bearing fungi, *Auriscalpium vulgare* S. F. Gray must have special characteristics to be associated with cones of conifers. It has probably retained this niche since very ancient times. However, *Gloiodon strigosus* (Fries) Karst. is a primitive type that is struggling to survive on wet soggy wood and is a rare species that conceivably could become extinct because of competition from more efficient forms.

The first natural selection pressures were those that favored variants that could effectively utilize substrates with unusual food supplies. Then as suitable substrates were occupied, competition favored fertility improvements and protective antibiotic devices. The organisms survived that produced sufficient numbers of spores to insure attainment of suitable habitats before other competitors arrived. This placed emphasis on developments that increased the spore-bearing surface of the hyphal mat, and on spores with adaptions that favored wider distribution. The development of the mat surface to make available all the area possible undoubtedly led to the early appearance of the resupinate hymenium facing downward. The first modifications of the hymenium were probably either warts, spines, or wrinkles—all very simple structures—that increased the hymenial area for spore production. Spines were the most efficient of these, but seem rather low in efficiency when compared with pores or lamellae. The hymenium on the spines is on the lower portion of a slender cone, an extension of the hymenium that appears first on the pileus. The tip of the spine grows actively and basidia are produced amphigenously, after the ends of the outer layer of actively growing hyphae turn outward. These hyphal ends in turn become fascicles of basidia and basidioles that crowd together in a very complex manner, their arrangement being extremely difficult to study but very orderly in any one species. The evolution of the regulatory system to control the overall development is not perfect in many of the hydnums. It takes very little extra growth to crowd the spines together, and then tramal hyphae appear, forcing their way out between the basidia. They become interwoven with other hyphae and the bases of the spines fuse. Changes in weather, such as lower temperatures, cause some of these fusions to occur while the young spines are developing and the resulting pilei will show bands that resemble the fruiting surfaces of polypores more closely than those of hydnums. I have seen twelve species in five genera that display polyporoid modification of the spines, and possibly it will be found in all species when sufficient collections can be obtained.

Another evolutionary pathway involves the modification of spores to favor distribution. The echinulations on the spores of *Phellodon* and *Bankera*, and the tubercles on those of *Hydnum* and *Hydnellum* all increase the spore surface and therefore their buoyancy. As wind became the essential factor for successful distribution, modifications that elevated the spore-bearing surface became more important. The development of elevated hymenial-bearing structures undoubtedly was a long tedious process with many early trends being eliminated. The first elevated forms could have been twisted ropelike strands of hyphae that bore the hymenium on the surface. Similar structures appear in cultures of fungi. These expanded at the apex and became simple primitive forms that could be modified into any number of elevated shapes. The clavate clubs could lead to inverted cones and turbinate pilei, all represented by various *Clavaria*, *Ramaria*, *Thelephora*, *Craterellus*, *Polyozellus*, *Cantharellus*, and finally the complex pilei of the stipitate hydnums, stipitate polypores, and agarics. The most difficult steps of evolution were undoubtedly the genetic controls necessary for the orderly development of the pilei of higher fungi. The genetic control system necessary for the orderly growth of a *Hydnum* or a *Hydnellum* appears to be quite different from that needed for the development of a *Clavaria* or a *Thelephora.*

The evolutionary steps necessary to control the positively geotropic spines on the lower surface of a pileate structure are too complicated to consider here. *Hydnum* and *Hydnellum* differ very little from the stipitate polypores, and *Boletopsis* Fayod and *Polyozellus* Murrill are almost identical except for having pores in the first, and a wrinkled hymenium in the latter.

Several directions of evolution are exemplified in various genera of the hydnums. In *Auriscalpium* the basidiocarp begins in a mat of mycelium as a small stalk with an active growing point developing upward, and after reaching a certain point it suddenly turns at right angles. Next, the growing point expands laterally to form the spoon-like pileus. This growth habit produces small eccentric pilei or much larger and almost centrally stipitate structures. If the growing tip is killed by unfavorable conditions, another growing point will develop lower down on the injured stipe and will elongate until a fruiting structure is successfully formed. Development is in a series of stages in which the hyphae change to other forms of growth in sudden steps. The genetic control is apparently induced by influences from the environment during the growth of the fungus. Growth is negatively geotrophic until the stipe is formed, then there is a right-angled turn in the growing point, and this is followed immediately by expansion into a pileus bearing the positively geotrophic spines.

In *Gloiodon* there is the same basic strand formation, usually without the primary development of a stipe. In the pileus, whether imbricate or stipitate, there is a branching of the strands within a mat of thick-walled tomentum, and the spines grow downward from the lower surface. This could be considered the formation of a fruit body in a tomentose mat of hyphae. The most complex development of this kind, but not necessarily the most advanced, is in *Hericium* and *Mucronella*—both genera primitive—with the fruit bodies ranging from single spines to tubercles of fused strands with spines growing in all directions on branches, along the lower surface of branches, or in tufts from the ends of branches. Finally, the massive bearded tubercle of *Hericium erinaceus* (Bull. ex Fries) Pers. with its long spinelike teeth is nothing more than a hymenium borne amphigenously on twisted strands of hyphae hanging downward. The relationships in this group are to be found in the method of the development of the fruit body, a very primitive type that evolved very early. There was an apparent trend from single spines to a complex fused tubercle in which the strands ultimately separated and produced spines. This in itself only indicates that evolution or natural selection had very little to select from for the development of a superior structure in this group of fungi. They survived because of their wood-rotting ability and the abundance of easily transportable spores.

*Echinodontium* is another archaic form that found a specific niche and survived because of the vast number of spores produced and its parasitic habitat on a number of conifers. It is also strongly favored by forming perennial fruiting structures which remain active for a great number of years. The mycelium can survive one hundred years or more of unfavorable conditions as a heart rot in its host.

I am going to mention the steccherinums only as a group. They are as diverse as the Polyporaceae with all the evidence pointing to their having evolved under similar conditions as for that group, and coming off second-best in competition for many favorable habitats. This could be our best example of the greater efficiency of pores versus spines. Perhaps they became too highly specialized and fixed too early in evolutionary history. Probably after the evolutionary trends in the Polyporaceae have been resolved, it will be profitable to work out the pattern in the steccherinums.

The most numerous spine-bearing fungi are *Phellodon* and *Bankera*; *Hydnum* and *Hydnellum*. They form parallel lines of evolution with the spores being remarkably consistent within each series—small, hyaline, echinulate spores in the first two, and brown, tuberculate spores in the latter. They exhibit many similarities in appearance, habits of growth, and hymenial organization. The same sites favor *Boletopsis*, *Polyozellus*, and various stipitate polypores. The similarities between these groups and the two series of hydnums are too numerous to be overlooked.

The species in these genera can be abundant in limited areas, and conditions that favor one will favor many others. When one species of hydnum is found, it is almost a certainty that others will be found in the same limited area. The same intergrading series of variations exist in each genus. The rarity of some forms makes it all but impossible to draw an exact line between all species. For example, *Phellodon melaleucus* (Fries) Karst. is brownish-black whereas *P. atratus* K. Harrison is black. When large, the former can grade into *P. graveolens* (Pers.) Karst., and the latter into *P. niger* (Fries) Karst., whereas *P. niger* var. *alboniger* (Peck) K. Harrison often can only be separated from *P. niger* when young and actively growing. *P. confluens* (Pers.) Pouz. is difficult to separate from *P. niger* var. *alboniger* or *P. graveolens* when conditions for growth are unfavorable. The type of *P. putidus* (Atkinson) Banker certainly appears to be distinct from *P. confluens*, but all collections that have been identified since the original are closer to *P. confluens*. The type may only have been a luxuriant form that was found when young and actively developing. The same situation is evident in other genera. Species are distinct under favorable—or optimum—growing conditions, but when either extremely favorable or unfavorable conditions prevail, specimens of some of the rarer species are difficult to identify. Hydnum genera have been linked to the Thelephoraceae primarily because thelephoric acid is common to both groups. I wish to point out, however, that their relationships to the stipitate polypores are even more convincing. The number of hydnum species with a counterpart in the polypores is too high not to be significant.

Fries (1821, p. 398), in commenting on the genus *Hydnum*, mentions the resemblance between six polypores and six hydnums. *Polyporus perennis* was listed with *Hydnum cyathiforme* (= *Phellodon tomentosus*); *P. subsquamosus* with *H. subsquamosum*; *P. ovinus* with *H. repandum* (= *Dentinum repandum*); *P. carbonarius* with *H. melaleucum* (= *Phellodon melaleucus*); *P. tomentosus* with *H. velutinum* (= *Hydnellum velutinum*); and *P. abietinus* with *H. ochraceum* (= *Steccherinum ochraceum*).

The outward similarity between polypores and hydnums is very striking. It is not unusual to find *Polyporus tomentosus* Fries in herbaria, mixed with the hydnums because of its similarity to the *Hydnellum velutinum* (Fries) Karst. complex. When collecting in North America it is necessary to turn over *Hydnum crassum* K. Harrison before being certain that it is not *P. confluens* Alb. & Schw. ex Fries. *Hydnellum frondosum* K. Harrison was so named because of its resemblance to *P. frondosus* Dicks. ex Fries. Similarities in appearance exist between *Hydnum cristatus* Bresadola in Atkinson and *P. cristatus* Pers. ex Fries; *H. stereosarcinon* and *P. peckianus* Cooke; *H. subfelleum* K. Harrison and *Boletopsis leucomelas* (Pers. ex Fries) Fayod; and *H. languinosum* K. Harrison and *P. hirtus* Quél. The various species fall into stirps or sections, and only improved methods of chemical classification hold any hope of our being able to basically understand the complex of forms and varieties in each. In any one state or province it is possible to recognize the species that exist. However, when the various species occurring in North America are brought together, it is evident that intergradation is present on the continental level between many of them.

## Summary

In summary, spined taxa are present in a number of genera that are not even closely related, and spines can best be regarded as primitive. It appears highly probable that spined forms were among the earlier evolutionary developments—but so early that there is little hope of tracing lines of development from our present-day morphological studies. However, the developing knowledge of chemical pathways may give us new angles of approach. No relationships are apparent between such genera as *Echinodontium*, *Hericium*, *Auriscalpium*, *Gloiodon*, and *Beenakia*, and the various species (or genera) in *Steccherinum* are so variable that it is not possible to consider them as having a common ancestor. They could just as easily be modifications that originated from the same ancestral sources as resupinate Polyporaceae and Thelephoraceae. *Dentinum* is definitely not related to other spined genera and is a highly successful fungus in spite of its primitive characters. None of the variants seen in *Dentinum* so far has established itself as a competitor of *D. repandum*.

The four genera with the greatest genetic variability (number of species) with evolutionary possibilities are the echinulate, white-spored genera *Phellodon* and *Bankera*, and the tuberculate, brown-spored genera *Hydnum* and *Hydnellum*. These two groups have a striking parallelism in development of species. Evolution is probably very slow, when fungi can persist for hundreds of years associated with the roots of forest trees—probably two or three hundred times slower than in fungi with one generation a year. The habitat and spore characteristics indicate affinities between *Hydnum*, *Bankera*, *Polyozellus*, and *Boletopsis*, and through them to the stipitate polypores and/or Thelephoraceae. We have no method of weighing the evolutionary importance of characters which are most difficult to change genetically. Possibly spines, colors, and shapes are the more easily changed, with hymenium, habit of growth, and spore types perhaps much more strongly fixed. *Phellodon* and *Hydnellum* develop in a primitive manner from a tomentose mat of mycelium, a character that is also present in *Auriscalpium* and *Gloiodon.* The indeterminate habit of growth is another character that indicates a common ancestral source, but the differences in spores and habitats are so distinctive that any relationship appears very distant. It can be argued that the hydnum genera are related in other ways by other characters, but until these can be proved as primitive or recent, it is only an exercise in imagination.

## References


Banker, H. J. 1906. A contribution to a revision of the North American Hydnaceae. Mem. Torrey Bot. Club 12: 99–194.

Coker, W. C., and A. H. Beers. 1951. The stipitate hydnums of the eastern United States. Univ. N. C. Press, Chapel Hill. 211 p.

Cooke, M. C., and L. Quélet. 1878. Clavis synoptica hymenomycetum Europaeorum. London. pp. 195–198.

Donk, M. A. 1933. Revision der niederländischen Homobasidiomycetae—Aphyllophoraceae. II. Utrecht. 278 p.

———. 1956. The generic names proposed for Hymenomycetes. V. Taxon 5: 69–80, 95–115.

Fries, E. M. 1821. Systema Mycologicum. Vol. 1. Holmiae. 520 p.

———. 1874. Hymenomycetes Europei sive Epicriseos systematis Mycologici. 2nd ed. Uppsala. 755 p.

Gray, S. F. 1821. Natural arangement of British plants. London. 624 p.

Harrison, K. A. 1961. The stipitate hydnums of Nova Scotia. Publ. 1099: 1–60. Queen's Printer, Ottawa.

———. 1964. New or little known stipitate hydnums Canad. Jour. Bot. 42: 1205–1233.

Harvey, R. 1958. Sporophore development and proliferation in *Hydnum auriscalpium* Fries. Trans. Brit. Mycol. Soc. 41: 325–334.

Karsten, P. A. 1879. Symbolae ad mycologiam Fennicum. Medd. Soc. Fauna Fl. Fenn. 5: 40–42.

———. 1881. Enumeratio Hydnearum Fr. Fennicarum systemata novo dispositarum. Revue Mycologique 3: 19–21.

Maas Geesteranus, R. A. 1959. The stipitate hydnums of the Netherlands. IV. *Auriscalpium* S. F. Gray, *Hericium* Pers. ex S. F. Gray, *Hydnum* L. ex Fr. and *Sistotrema* Fr. em. Donk. Persoonia 1: 115–147.

———. 1962. Hyphal structures in hydnums. Persoonia 2: 377–405.

———. 1963. Hyphal structures in hydnums. II, III, IV. Proc. Koninkl. Ned. Akad. Wetenschap. Ser. C 66: 426–457.

MILLER, L. W., AND J. S. BOYLE. 1943. The Hydnaceae of Iowa. Univ. Iowa St. Nat. His. 18: 1–92.

QUÉLET, L. 1886. Enchiridion fungorum in Europa media, 188–193. Lutetiae.

## DISCUSSION

LUTTRELL: I think you brought out in *Echinodontium* survival by means of a compensatory mechanism. The mycelium could survive indefinitely in the heart wood of a tree. A mechanism has evolved which compensates for perhaps an inefficient method of spore dispersal. What is the peculiarity in the habitat of those that occur in the soil which compensates for the same inefficiency? Are there conceivable factors in microclimate or something that might make this toothed form superior, say, to gilled or pored forms?

HARRISON: I have referred to that sort of thing as a niche—some habitat that has a special condition in which a particular species can have priority of occupation. *Echinodontium* establishes itself as a heart rot of these conifers, is able to resist all competition, and can complete its life history. But if the lumbermen have a free hand, there won't be any trees left and a habitat will be eliminated. In soil you are dealing with other things such as all the antibiotic substances that the normal soil inhabitants put out. I did not go into great detail, but some of our stipitate hydnums are probably mycorrhizal. I haven't any proof of this, but they are almost in constant association with certain species of trees. If I want to find *Hydnellum diabolus* or *H. stereosarcinon* in Nova Scotia I look under spruce trees, even though there may be other things that they are associated with. This is where I will be sure I can find them. If I want to find these hydnellums on the west coast I go to the Sitka spruce forests of northern California.

LUTTRELL: Is there a possibility that in very humid conditions teeth might be superior to gills—that gills would be too wet?

HARRISON: I do not know of any group of fungi that is more sensitive to humidity and the amount of water supply. Every development that I can see in the *Hydnum-Hydnellum* group seems to be based on a very fine control of humidity or the amount of water available for their growth. This is definitely the factor that makes certain habitats in this country unusually favorable for this group of *Hydnum*, *Hydnellum*, *Phellodon*, *Bankera*, *Boletopsis*, and *Polyozellus*. They all have the same requirement (together with the stipitate polypores) and they have a broad common niche in that they are usually associated with conifers.

Mathematically, the amount of hymenium in a toothed form is less than a pore form. The space available for it is not used up as efficiently with teeth as it is with pores. I have done some mathematics of the gill and there I think many more spores are being liberated from the same area of basidiocarp.

GRAND: Along these same lines have you noticed any kind of spatial arrangement of the teeth? I have collected what I have called *Steccherinum adustum* in great abundance this spring in the Piedmont region of North Carolina. The thing that struck me about the collections was the closeness of the teeth. From a distance, say arm's length, it appears like a polypore, but upon closer inspection it is obvious that the hymenium is on spines.

HARRISON: I see each species as having a system of spacing rather common to that species. There are always small teeth which start to develop between the large ones. They can be seen only with a hand lens or under the binocular microscope. They are suppressed. Moreover, under usual conditions these small spines are completely sterile, but occasionally they grow up so that the hymenium starts developing down their sides. I think there is a spacing mechanism under some sort of genetic control, but it can be broken down by the environment. I was interested in the *Panus* that Miller mentioned because it showed exactly the same development as *Auriscalpium*.

LUTTRELL: There is danger in putting so much emphasis on spore dispersal. I would think it was quite possible that there might be some circumstances under which certain forms of hymenium are not the most efficient.

PETERSEN: Quite possible. But suppose with this inefficient system of spore dispersal, the somatic mycelium produced such a strong toxin that within three feet of it no other organism could exist. It would essentially clear space in the ground for itself and could throw up one spore per 100 years and still survive, just because it is competing somatically.

BURDSALL: You refer to the "tooth" in two ways, one as primitive and one as a derived process. Did evolution go from the tooth to polypore and then back? I ask this because I know how many polypores become dentoid when old.

HARRISON: I refer to the tooth as derived only in that it is derived from a flat surface that grows up into a lump. I didn't intend to imply that the tooth is an advanced structure. To me it is always primitive. It has one basic way of happening.

There is a separating mechanism that leads to apparent teeth which takes place in some polypores after the pores have developed. In the hydnums there is some mechanism that permits only so many spines to a certain area of surface. This breaks down under moisture conditions or some other control, so that occasionally the spines fuse into a superficially polyporoid structure. Something breaks up the normal genetic response to an outside influence and out comes the tramal tissue between these spines. It is not just a straight development of the pore as in the polypores, or ridges and reticulations like *Daedalea.*

PETERSEN: Perhaps we are dealing with some sort of auxinic control that we do not understand. Perhaps the auxinic production or flow may be modified by the configuration of the tramal hyphae. We do not know whether it streams laterally back from the margin, or whether it goes straight down to form a tooth.

MCLAUGHLIN: You said that some species of hydnums produce very little growth in culture. Were you referring only to the stipitate hydnums?

HARRISON: Only to the stipitate hydnums. *Steccherinum* and a lot of wood-rotters grow nicely in culture. I have heard of one case of *Bankera* in culture, but this has not been confirmed.

NOBLES: In some cultures of polypores, we get formation of fruit bodies. Some species of Polyporaceae form fruit bodies with flat hymenial surfaces, appearing like a thelephoraceous fungus. In other cases we get distinct pores formed which look quite normal. It seems to me I can see two definite ways in which they develop. In one case there seems to be a hyphal layer that has the potentiality of fruiting all over, and it develops as though it were having holes or pores dug out of it. In the other case little folds or teeth are produced on the young developing fruit body. There must be several ways of developing all of these kinds of hymenophores, but we can't lump them all together since they may have arisen in different ways.

OLEXIA: When looking at a specimen—possibly a dried specimen—how do you determine whether it grows determinately or indeterminately?

HARRISON: I have to see it growing. It is pretty well established that *Hydnum* fruiting bodies do not spread indeterminately. There is a borderline case in *Hydnum stereosarcinon*, which is very close to being indeterminate. There are additional separating characters, though. The *Hydnums* all have inflated tramal cells in the pileus so you can determine quite easily that it is a *Hydnum* that way. There is supposedly more thelephoric acid in the *Hydnums* than in any others. That is why they were placed with *Boletopsis* in the Thelephoraceae. I am very anxious for someone to look at specimens of *Boletopsis* to see if there is a not tremendous amount of thelephoric acid in them. When the hyphae of *Boletopsis* are mounted in KOH, there is a characteristic blue-green reaction.

BRADY: Let me summarize some of the details of chemotaxonomy in *Hydnellum*. Perhaps the greatest point I could make is that chemotaxonomy up to this point isn't as helpful in the hydnellums as it is elsewhere. We can find a number of factors present that have some potential but most of the differences we are finding are on chemical indicators that we consider to have fairly low chemotaxonomic potential. Thelephoric acid is one of the most useful compounds, but it appears to be common to a fairly large number of taxa in the Hydnaceae. It is interesting, for example, that we can find no evidence of it in *Dentinum*. In fact, *Dentinum* in our chemical survey is just about as far removed from the family as you can imagine, so I was very interested to hear Dr. Harrison's comments in that regard. There are a few indications in the *Hydnellum scrobiculatum-zonatum* complex that chemotaxonomic separation may clarify resolution of specific entities. We have looked only at material from western Washington and we are fully aware that this is a most variable complex. It was Dennis Hall who has worked on this, and he plans to continue with it.

DONK: I assumed that no thelephoric acid was present in the fruiting body of *Bankera* when I segregated the family Bankeraceae. There are of course other characters to support the family: none of its species (of the two genera, *Bankera* and *Phellodon*) have clamp connections; they all have a peculiar odor which is absent in the brown-spored genera (*Hydnellum* and *Sarcodon*) of the Thelephoraceae. The spores are also different. Did Dr. Brady find thelephoric acid in *Bankera*?

BRADY: *Bankera* does not grow in the Pacific northwest so we have not examined it. All the species of *Phellodon* that we looked at contain thelephoric acid.

DONK: You called the fruiting body of *Hydnellum* "indeterminate" and that of *Sarcodon* less indeterminate. I think you can express this also in terms of hyphal structure. The hyphae of the fruiting body of a typical *Hydnellum* are noninflating and appear to have dormant stages in the zones of the flesh. They resume growth if environmental conditions become favorable. However, in *Sarcodon* the flesh is not zonate and the hyphae inflate as in fleshy agarics; this condition makes them ill-adapted for rejuvenation and the formation of new hyphae. This is what seems to be at the heart of your terms. I leave out of consideration, here, such examples as *Hydnum stereosarcinon*, but in other cases separation on the basis of hyphal structure and hyphal behavior is not difficult. In Europe *H. mirabile* was often taken to be a species of *Sarcodon*, but microscopical examination of the hyphae leaves little doubt about its real position.

HARRISON: *H. humidum* is a rare species with very soft, watery fruit bodies, but definitely indeterminate. I have never seen it in the fresh condition. *H. piperatum* is the same way.

GILBERTSON: In what sense are you using the name *Steccherinum*? Are the species with large, sessile, fleshy fruiting bodies included, like *S. pulcherrimum* and *S. septentrionale*?

HARRISON: Yes, they are all included. I use the genus in the broadest sense possible. In 1964, I published a species from Mexico, and actually I think it is the very simplest form of *Steccherinum* possible.

M. A. DONK

*Senior Mycologist*
*Rijksherbarium, Leiden, Holland*

# MULTIPLE CONVERGENCE IN THE POLYPORACEOUS FUNGI

## Restriction of the Polypores

To arrive at a definition of what is to be understood by "polypores" during this discussion it seems advisable to start from Fries's latest conception (1874) of the "Polyporei." To him this group was made up of all hymenomycetes with a more-or-less tubular hymenophore. His definition ran: "Hymenio effigurato poroso: hymenophore tubular." That is all. The tubes ranged from shallow pits to long, narrow pipes open at one end.

Let us first discard the merulioid hymenium configuration, which produces folds that come into being through excessive intercalary growth of the initially smooth hymenium. It is easy to understand why these folds should be fertile at the edge. This removes *Serpula*, *Merulius*, and similar genera. Next come the tubes that are mutually free from one another. This eliminates such genera as *Solenia* (correct name, *Henningsomyces*), *Porotheleum* (correct name, *Stromatoscypha*), and *Fistulina*. Finally come the tubes that are fertile only at the bottom and which are formed by sterile tissue protruding beyond the hymenium as dissepiments. This excludes *Porogramme* and a few similar groups.

What thus remains are the species in which the dissepiments of the tubes have sterile edges, while the hymenium lines the inside. Just the same, it is necessary to go on with the elimination. Out goes *Boletus* in Fries's conception, now forming one or two distinct families of Agaricales. Out go a number of species and genera such as *Mycenoporella* (synonym, *Filoboletus* sensu Singer), *Favolaschia*, and the like, now sometimes also considered to be agarics, but not *Polyporus* itself. Out go also the genera with more-or-less typical tremellaceous basidia, *Aporpium*, and apparently *Protodaedalea* as well.

This brief enumeration of groups with tubular hymenophore, but now excluded from the more restricted version of the polypores, is by itself telling evidence that this type of hymenophoral configuration has been recognized for a long time as being of multiple origin.

Adding by inference a few groups with nontubular hymenophore, like *Lenzites* and *Daedalea*, to what remains, we arrive at the polypores of this lecture. Some mycologists, like Overholts (1953) and Cunningham (1965), call these polypores the Polyporaceae, implying a natural family; others consider them to be a grade, an artificial assemblage accepted only for the sake of convenience. It is our problem to try to decide between the two views. In any case it is the group of which Corner stated: "I believe that the natural classification of the polypores will be the hardest problem in the systematics of Basidiomycetes."

## Radicals versus Conservatives

How widely current views often diverge is demonstrated in the case of the genus *Ganoderma*. While some authors include it in a special family together with the related genus *Amauroderma*, others claim that the characters emphasized this way are not really important enough to remove *Ganoderma* (inclusive of *Amauroderma*) from the neighborhood of *Fomes*, and therefore keep it in the Polyporaceae. Still others (and these I would call the conservatives) do not even think the recognition of the genus justifiable, particularly, they say (and here I quote Overholts), because that segregate is based largely on spore characters. This third category refers the perennial species to *Fomes* and the annual species to *Polyporus*, both these genera being taken in an extremely wide sense.

True conservatives are not only content to accept unwieldy genera that are admittedly heterogeneous, but they also act as though the demand for a natural classification cannot be met, or, at least, is of no importance. Lloyd was a conservative in this sense of the word. The terror he spread by ridiculing everything that reminded him of taxonomic innovation in the systematics of fungi, and of the bigger Aphyllophorales in particular, has undoubtedly been a contributing factor to the wide cleft existing between the taxonomic thinking of European and North American mycologists. Another such factor was Murrill's failure to further a natural classification by excessive multiplication of artificial genera. This was a warning.

What Murrill did was to continue along the line that other splitters had followed before him: that is, to use characters by which Fries had subdivided genera. What was really needed were "new" characters, such as some of those used by Patouillard, coupled with noteworthy disregard of the importance previously attached to some of the old ones. His system of the polypores became better known throughout Europe through the work of Bourdot & Galzin (1928). Although debatable in many respects, this contained enough attractive suggestions to inspire a generation of young mycologists in Europe. Murrill, Lloyd, and Overholts, each in his own way, conditioned North American mycologists to disregard this rejuvenation of interest in the taxonomy of the polypores. To them the determination of specimens remained more important than the development of a natural classification.

The European authors have gone too fast. They act as though a new classification for polypores can be made by taking into account little more than the European species. This of course is a weak premise. Not only is Europe a continent poor in species, but any advanced system must also absorb the large number of species found only in other parts of the world, especially the tropics. Around 1942 the European polypores had been distributed over many genera, of which a good number were patently artificial, even though a considerable array of microscopical features had been used in defining them at the specific level, thanks to the pioneer work by Bourdot & Galzin. There can be no doubt now that in about 1942 the effort had reached an impasse.

It was time for new inspiration. This came from careful study of the structure of the fruit body, the importance of which, however, can be understood only by reflection on the function of the fruit body.

## The Function of the Fruit Body

The fruit body of the true Hymenomycetes serves a single, all-important purpose: it produces and efficiently liberates forcibly discharged basidiospores—so-called ballistospores. Certain types of polypore fruit bodies are more complicated and more efficient, if we reckon by the number of spores produced. From the outstanding studies by Buller, supplemented by reports from Buchwald (1938), Ingold (1957), and other authors we now know that *Ganoderma applanatum* and *Fomes fomentarius* are among the most highly adapted species. The yearly formation of a new hymenophore makes it possible for the fruit body to function for many years. Throughout its growth the developing fruit body is sufficiently responsive to gravity and also perhaps too light to place the hymenophore in a horizontal position. The context, effectively supported by a hard crust, is sufficiently strong to assist in keeping it in that position while the hymenophore itself forms and keeps the very narrow, long tubes in a surprisingly accurately vertical position, so that the liberated spores can fall unhindered and reach the air currents underneath the fruit body without becoming stranded along the hymenium. It has been suggested that the tubular hymenophore in these species is more efficient in using the area below the cap than the system of gills of the agarics.

In spite of the complicated construction of the fruit body of such wood-inhabiting polypores as *G. applanatum* and *F. fomentarius*, it is nevertheless striking that their fruit bodies are so much more variable in shape (and often also in other characters) than those of the terrestrial agarics. This is one of the principal reasons that make it so difficult to construe a sensible classification of the polypores on the basis of external characters.

This versatility appears strongly correlated with habitat. The new station of the fruit body that migrated away from the soil to other substrata called for a wide range of improvisation, if you will allow me to express it this way. The variation within certain wood-inhabiting species of polypores under normal conditions may range from an elaborately developed to a strongly reduced fruit body that nevertheless still produces basidiospores, a kind of variation that will suggest to the evolutionist a phylogenetic line of development.

## Hyphal Analysis

Perhaps the most important single stimulus that has affected the course of the classification of the polypores in modern times has been Corner's publication (1932a-c) of three meticulous studies on the hyphal construction of the fruit bodies of a few tropical polypores. He boldly foresaw (Corner, 1932a: 71) that "the hyphal system of the fruit body must be considered foremost in the morphology of polypores as it will provide the key to a natural classification." His studies were to lie dormant until, after World War II, several authors began to take notice of them. Except for breaking new ground, Corner himself published only comparatively few analyses of polypore fruit bodies. However, the number of more really thoroughly analyzed species is now rapidly increasing and it can be predicted that after a few more years the hyphal construction of the European polypores will be comparatively well known. If Europe were the only continent in the world, the classification of the polypores in this respect would by then have been brought up to date. If Cunningham's work (1965) on the Australian and New Zealand polypores had been done more carefully and with fewer errors, the same could have been said about these regions. Without a thorough knowledge of the huge number of tropical species, however, a regional classification like the one taking shape in Europe can be accepted as only tentative; it will undoubtedly be radically improved in the future.

Hyphal analysis has by now become so commonplace that I can safely take it for granted that you know what the terms generative, skeletal, and binding hyphae stand for, and that these hyphal systems enter into the formation of mono-, di-, and trimitic tissues. It should not be overlooked that the generative hyphae may become modified secondarily, not seldom even very characteristically. According to Corner's conception, where skeletal hyphae occur these are initiated along growing margins such as the margin of the cap and of the hymenophoral gills and dissepiments of tubes, and the tips of teeth. The binding hyphae are formed at a later stage, in the context itself.

It becomes more and more evident from recent publications that the terms mono-, di-, and trimitic have acquired almost magical status. Often authors do not go beyond merely finding out the bare "miticity" and then give an oversimplified description of the hyphal structure. There is a tendency to neglect complications and epigenic alterations.

The increasing number of hyphal analyses has shown the existing terminology to be inadequate. In the growing margins, an imposing array of hyphal types may be initiated. Of these the skeletal hyphae are only one. To mention some of the others: the arboriform hyphae of *Ganoderma*, the vermiform hyphae of *Poria* [*Melanoporia*] *niger*, the macrosetae in many Hymenochaetaceae, the gloeoplerous hyphae in the Hericiaceae, and so on. In the Auriscalpiaceae skeletal hyphae and gloeoplerous hyphae occur together and are formed simultaneously, making the context "tetramitic," but not in the sense of Corner. I hope to publish an emended terminology in the near future.

## Clamp Connections

Corner's hyphal studies have also taught us where to look for clamp connections; when they occur at all, they are found on the generative hyphae only. Awareness of this has increased the importance of clamp connections as a taxonomic feature. They now play a well-earned role in the systematics of the polypores. It is, therefore, not surprising that contradictory views have been expressed about their taxonomic value. Let us start by ruling out such preconceived ideas as that microscopical characters should be ignored when defining genera, as Overholts ignored them for spore characters. Even so, it appears that present views range widely. Some authors go so far as to claim that a species with and one without clamps automatically belong to different phyletic groups, and they are inclined to divide the polypores into two groups on this basis.

On the other hand Nobles (1958: 892) recorded the repeated isolation of monokaryotic (which implies clampless) mycelia of various species from decays in trees as proof that such mycelia were common in nature. She also reported that in the laboratory single-spore cultures of many species produced fruit bodies which appeared normal. (From this I conclude that she was suggesting that since they were produced in culture, monokaryotic fruit bodies may be expected in nature.) Hence, she continued, it appeared possible that some species with hyphae lacking clamps were monokaryotic counterparts of heterothallic species. Nobles also thought it likely that heterothallic species without clamps were reduced forms of ancestors with clamp-bearing hyphae and were comparable to modern, clamp-bearing, heterothallic species. "If this is true," she concluded, "only species with [clamp-bearing] hyphae should be considered in establishing a system of classification for the Polyporaceae, and the species with hyphae consistently or mainly [without clamps] should be classified with corresponding [clamp-bearing] species. In this interpretation, the presence or absence of clamp connections has no taxonomic significance beyond the species level . . . ."

Nobles' theoretical conclusions would be devastating to our faith in those taxa that have been defined by relying heavily on the presence or absence of clamps. In my opinion the ultimate guide in estimating the value of taxonomic characters must be experience: the simple registration of facts unbiased by preconceived ideas. The golden rule in taxonomy is to judge each character on its own merits, and for each group separately, not allowing conclusions drawn from one taxon to dictate in advance conclusions for others, which is the way artificial systems come into being. Of course, this rule is easier to formulate than to follow, and, I would add, opinions derived from other groups naturally act as sources of inspiration.

Let me mention some conclusions as they are based on presently known facts. For instance, not a single clamp has ever been reported for a big family or order like that of the Septobasidiaceae (over 200 species). As far as our knowledge goes in this special case there is nothing against according the absence of clamps a very high mark in the definition of genus, family, and order.

When Donk (1948: 474) published the family name Hymenochaetaceae he made the absence of clamps a leading character. A few years later, Kühner (1950a, b), who studied more than thirty species in culture and investigated their nuclear behavior, concluded that "the absence of clamps is certainly a character of the highest importance in the series of the Igniaires [= Hymenochaetaceae]." Corner (1950: 19) was of the same opinion. "The absence of clamps so distinguishes the xanthochroic series . . . that their presence is proof that a species does not belong," he said. Boidin, in his recent studies, did not mention clamps either.

Nobles (1958: 917) studied in culture the characters of about 40 polyporoid species of Hymenochaetaceae (25 in her publication of 1965) and two species of *Hymenochaete*, a genus that has a smooth hymenophore. With the exception of *Polyporus* [*Phaeolus*] *schweinitzii*, these were all placed in a single group of oxidase-positive species, with the remark that in all these species the hyphae lacked clamps, except for the very rare clamps in a number of them. From a later publication (Nobles, 1965) I would conclude that these species with rare clamps are *Polyporus* [*Inonotus*] *cuticularis* and *P.* [*Phellinus*] *gilvus*, of which she then wrote, "hyphae of the advancing zone all simple-septate or with very rare single clamp connections, usually difficult to find." No further details were given about this very important observation.

Recently, Imazeki and Kobayasi (1966: 44) reported having seen a few clamp connections in a single fruit body of *Polyporus* [*Col-*

*tricia*] *montagnei* var. *greenei*, but neither in other specimens of the same variety nor in the other Japanese species of *Coltricia* was it possible to repeat this observation.

These exceptions to the rule that the Hymenochaetaceae lack clamps become very important because of their rareness. They are recommended for careful study, as this may throw some light on the fundamental reason for their almost complete absence in this family.

At the moment European authors mostly occupy an intermediate position as to the taxonomic value of clamps in polypores other than the polyporoid Hymenochaetaceae. They tend to consider the absence of clamps to be of generic value and this supports their recognition of such genera as *Bondarzewia*, *Laetiporus*, *Oxyporus*, *Rigidoporus*, and *Ceriporia.* In all these examples the absence of clamps was first established from fruit bodies collected in nature. Cultures have since shown that in a number of instances clamps do occur, after all, in the mycelial mat. Leaving *Fomes annosus* aside for the moment, this calls for a sharper differentiation of these polypores into at least two groups: genera (or species) lacking clamps completely, and genera (or species) practically lacking clamps in the fruit body, but producing clamps in cultures. We now know that clamps have never been reported (not even from cultures) in *Polyporus sulphureus*, *P. montanus*, and *P. fibrillosus*, the type species of the small genera *Laetiporus*, *Bondarzewia*, and *Pycnoporellus*, as well as in a few other species that have not been segregated on this basis from the genera in which they are now placed (for instance, *Polyporus* [*Tyromyces*] *mollis*).

Of the other genera mentioned above as lacking clamps in the fruit body we now know that at least in one or a few of the species of each genus clamps have been reported from cultures. They are *Oxyporus* and *Rigidiporus*, and *Ceriporia* in a narrow sense. Nevertheless the European taxonomist is still prepared to maintain these genera.

In connection with clamps I would like to mention as another example the genus *Irpex sensu strictissimo*, embracing the *Irpex lacteus* complex and *Polyporus* [*Irpex*] *tulipiferae*. These two species are microscopically so much alike that I find it impossible to suggest reasons for generic separation, although *I. lacteus* has clamps throughout the fruit body and *P. tulipiferae* lacks clamps altogether, even in cultures.

As for the final example, *Fomes annosus*, recently Pouzar (1966: 363), in redefining the genus *Heterobasidion* to which this is now referred as the sole species, assumed that clamps were lacking. Both Roll-Hansen (1940) and Nobles (1948: 319), however, reported isolations showing clamps among a huge majority in which no clamps could be found. It is not likely that misidentifications were involved since in view of its remarkable *Oedocephalum* state this species is highly characteristic in culture.

This is not the occasion to inquire into the reasons why clamps may be lacking. Suffice it to state that there are at least several reasons. Some of the above-mentioned examples, especially *Irpex*, will undoubtedly form interesting objects for a closer study.

## Cultural Characters—I. Oxidase Tests

Possibly second in importance among the new tools in the arsenal of the taxonomist working with polypores is the study of cultures. The fact that many fungi cause serious decays in trees and timber has stimulated cultivation of these fungi under laboratory conditions in order to facilitate their determination. Most of the isolates appeared to be Hymenomycetes, the great majority of them, polypores. In practice these isolations are usually made without knowing about the fruit body. Nobles (1958, 1965) has twice summarized the results of the very extensive studies she and a few other workers in this field have made of species that nearly all grow in temperate North America, mainly Canada.

The intensive study of cultures of polypores caused students of this group to drift apart to some extent, so much so that it now becomes possible to speak of culture-mat taxonomists and fruit body taxonomists. In this connection it should be kept in mind that the primary aim of the cultural studies is to provide the possibility of naming the species to which these isolates belong and this resulted in the construction of keys that lead to species-groups of which all members have the same characters. Such species-groups may or may not be homogeneous. It should also be kept in mind that the species of natural genera need not share all characteristics used in key codes and hence may key out in different species-groups. A rigid adherence to key codes could very well result in overlooking this last category of taxa or, in contrast, in considering unrelated species as forming natural taxa.

Nobles has placed the results of oxidase tests at the head of the sets of characters by which she divides the polypores into species-groups. On this basis two main divisions are obtained, the larger of which contains the species that have given positive results in tests for extracellular oxidase, and the smaller, negative results. The species whose cultures lack extracellular oxidase are usually associated with brown rots, whereas the species of the other group are associated with white rots. According to her these two groups represent natural subdivisions of the family Polyporaceae. And by "subdivisions" she means "primary subdivisions." Moreover, she claims that those species that do not produce extracellular oxidase constitute a primitive group whereas those that do produce extracellular oxidase are more advanced.

These suggestions are interesting enough to be subjected to careful scrutiny. The first objection I would raise is that without knowledge of the fruit body it is impossible by cultural methods to delimit the polypores from the other Hymenomycetes; the family Polyporaceae as such is accepted *a priori*. Another objection is that about 10% of the species give erratic results in tests for extracellular oxidase, which means that they are not definitely assignable to either of the two recognized groups and in fact constitute a third group. That this group is significant follows by comparing it with the oxidase-negative group which makes up nearly 24% of the total. That nearly one in ten species is not directly assignable is a serious flaw in an attempt to distribute species over two main divisions.

What is at the least also debatable is the assumption that the polypores form a single taxon divisible into two subdivisions. I would oppose to this the main thesis developed in this lecture, which is that the polypores are a grade formed by multiple convergence of several not closely related lines.

## Cultural Characters——II.

Nobles not only believes that in the results of the tests for extracellular oxidase she has found a set of characters that is all-important, to the exclusion of all the others hitherto used, but she also considers her arrangement based on cultural characters as coming close to a taxonomic arrangement of the polypores. It should be added at once that she is clearly aware, as she emphatically states, that "at this time, while a correlated study of the sporophores remains to be done, the system can be only provisional. Hence no changes in nomenclature are suggested and no names are proposed for the designation of groups of species of suprageneric or generic rank" (1958: 895).

At the time of this statement Nobles included among the cultural characters one that she has dropped in her latest survey (1965), viz., characters derived from the basidiospores. As to some of the other sets of characters, it is also a matter of argument whether they are really and specifically cultural. For instance, the color of the mycelial mat will call to mind that of the context of the fruit body. The presence or absence of clamps on thin-walled hyphae in culture no doubt corresponds in the main with the presence or absence of clamps on the generative hyphae of the fruit body, although it may be true that it will sometimes be easier to derive more precise information on this subject from cultures rather than from old and poorly preserved herbarium specimens.

The hyphal analysis of a polypore fruit body is often a time-consuming task that requires a great deal of technical skill. If it could be replaced by a study of the mycelial mat developing in culture this would be a very welcome simplification. Therefore, it is desirable to give some preliminary thought to this matter. The first questions that arise are these. Do the several kinds of hyphal systems that may occur in a normally developed fruit body reappear in the mycelial mat? And if, or where, this is the case do they retain their features more-or-less unimpaired? No very definite answer can be given at present simply because Pinto-Lopes (1952), Nobles, and other authors describe the hyphal composition of the cultural mat in languages quite different from that which we use at present for the hyphal structure of a fruit body. It would even seem as though Nobles deliberately avoided using Corner's terminology, which might indicate that she couldn't use it. It also appears that generally the careful methods of analysis on which Corner insists have not been complied with. Only small tracts of hyphae are depicted and these do not tell much about the hyphal unit as a whole. What we want to know about the construction of the cultural mat are more precise facts comparable to what has been stated about the corresponding fruit body. If there are skeletal hyphae in the mat, we want to know enough about them so we can decide whether or not there are significant differences from the skeletal hyphae of the fruit body.

The evaluation of what Nobles calls "fiber hyphae" as a taxonomic character is not made easy. According to her definition fiber hyphae have thick, refractive walls, either hyaline or brown, with the lumina reduced or apparently lacking. They arise as elongated terminal cells and are without septa; they are rarely branched in some species, frequently branched in others. In a few instances I cannot escape the conclusion that thick-walled generative hyphae with septa and clamps have also been included with the fiber hyphae.

Although Pinto-Lopes has answered the above questions about the identity of the various kinds of hyphae in fruit body and cultural mat in the affirmative, I tend to disagree. The following may be an extreme example, but it speaks for itself. In *Polyporus* [*Laricifomes*] *officinalis* the context of the fruit body is mainly built up of short skeletal-like hyphae that have been termed "acicular hyphae," as well as of complexes of strongly branched hyphae, the binding hyphae (Teixeira, 1958). Both kinds would presumably have been called fiber hyphae if they had occurred in the mycelial mat, but according to Nobles (1958: 328; 1965: 1155), and barring misidentification, no fiber hyphae at all are found in culture.

It is now well known that considerable differences in hyphal makeup may occur in the several parts of a single fruit body. The parts are the flesh of the cap and the dissepiments and the context of the crust. The fruit body is an infinitely more complicated body than the mycelial mat. The morphogenetic forces that come into play during its development conjure up particularities of hyphal construction that cannot, and should not, even be suspected from cultures. Accordingly, homologous organs of both may be underdeveloped or less characteristically developed—if developed at all—in the mat.

In short, cultural mycelial mats are mere ghosts of natural fruit bodies—if they are comparable at all. Until now they have provided few characters above the generic level that cannot be matched by corresponding ones of the fruit body.

This is not to say that cultural characters are unimportant. Far from that! The work of Drs. Miller, Boidin, and Nobles has convincingly demonstrated this. Nobles, for instance, has pointed out that certain groups of the species she has segregated coincide with established genera, such as *Ganoderma* and *Polyporus s. str.* This implies that she has found supporting characters for them. In other cases it is quite possible that cultural characters will prove decisive in an eventual sharp delimitation of genera that had previously been inadequately defined, like *Coriolellus* [= *Antrodia*]. One of the finest examples of a supporting character furnished by Nobles is found through the oxidase tests. All the species of the Hymenochaetaceae she studied—that is, all polyporoid species investigated and two species of *Hymenochaete*—were all oxidase-positive, with only one species (*Polyporus* [*Phaeolus*] *schweinitzii*) "undecided." All five species of *Ganoderma* listed by her were also oxidase-positive. But the reverse thesis that all oxidase-positive are *per se* related I find unacceptable.

## Multiple Convergence

Nobody now doubts that Fries's Polyporei of 1874 is a heterogeneous group tied together only by the tubular hymenophore, an abstraction. If in his time contemporary mycologists had suggested that *Merulius*, *Porotheleum*, *Fistulina*, *Boletus*, *Laschia* Mont., and such species as *Protodaedalea hispida* and *Poria* [*Aporpium*] *caryae* did not belong to the Polyporei, even if they had disclosed their reasons, others would have had no more to do with them. I suppose that at this moment none of you would be prepared to defend Fries's circumscription of the polypores. When at a later stage Patouillard had suggested and performed most of the elimination that is now accepted as self-evident, he was made game of by such die-hards as Lloyd. Now that after many years even the obstinate echo of Lloyd's penetrating voice has died down, it is safe to assume that nobody, even in Lloyd's own country, would again extend the limits of the polypores beyond those that were accepted by Overholts (1953), which is the self-same group that is now under discussion. I would now put the rhetorical question, Why should this restricted conception of the Polyporaceae (bear in mind the capital *P* and the Latin form) be final?

Many mycologists, mostly European, refuse to stop where Overholts stopped; they do not speak of "Polyporaceae" but of the "polypores" (without a capital), thus making clear that they consider the much restricted conception of the polypores as still a grade, an artifically conceived, although convenient, group: to use a different wording, a group formed by representatives of more than one line of convergent evolution.

And even here, within this strongly reduced conception of the polypores, we encounter from the start the genius of Patouillard. Among his most daring innovations that have stood the test of time was his conception of the *Séries des Igniaires*, which has now grown out to the family Hymenochaetaceae, perhaps the Hymenochaetales of the future. The original *série* consisted of one genus with smooth hymenophore, *Hymenochaete*, one with an *Irpex*-like hymenophore, two widely inclusive genera of polypores, and one with cyclomycetoid gills, somewhat reminiscent of the agarics, viz., *Cyclomyces*. The modern version of these "Igniaires" (from which I now exclude, following Reid, the Lachnocladiaceae) comprises in addition a genus with toothed hymenophore, *Asterodon*, and the coralloidly branched genus *Clavariachaete* derived from the Clavariaceae *s.l.*

The guiding principle was at first based on those characteristic organs that are now called the setae. They had already been known for a long time and in 1809 Link made them one of the leading features of his emendation of the genus *Stereum*, the same taxon that is now called *Hymenochaete*. However, setae are not always present and this has drawn attention to other features: in combination these form an imposing support for the family. I have already discussed some of these characters, the lack of clamps (with perhaps very few exceptions) and the tests for extracellular oxidase, which are positive. Add to this a chemical test, usually called the xanthochroic, and you have defined a long series of genera that are closely bound together. They were taken from the Friesian Clavariei, Thelephorei, Hydnei, Polyporei, and if you wish, Agaricini, as far as *Cyclomyces* is concerned. This is a microcosmos within the Aphyllophorales, but with the same span! It will certainly make the evolutionist's mouth water. I have little hesitation in cutting off the Hymenochaetaceae from the residual polypores.

About the next episode in this process of dismantling I will be brief, but it is telling. In 1889 Fayod published the genus *Boletopsis* for a fleshy, terrestrial species of polypores. It has irregularly tuberculate spores that suggest those of the Thelephoraceae, modern version. This did not escape Bourdot & Galzin, who remarked, "[*Polyporus leucomelas*] seems to be a lengthening of the 'Phylacteriés' among the polypores; when old it shows a rather strong likeness to a *Sarcodon*." When I studied the species (Donk, 1933) now many years ago and compared it carefully with *Sarcodon* I was so struck by the similarity between the two, even in details, that I resurrected Fayod's genus and transferred it to the Phylacteriaceae, an incorrect name for the modern version of the Thelephoraceae.

At that time I had not yet completely lost my respect for the tubular hymenial configuration and made of *Boletopsis* a special tribe of this family, but if one of these days someone were to make it a mere section or even species of *Sarcodon*, I would hardly be surprised. The step from the aberrant, flattened teeth that may occur incidentally in several species of *Sarcodon* to the rather irregular tubes of *Boletopsis* is so short, and the inner picture of the two genera so completely alike, that in my opinion there should be no hesitation in transferring *Boletopsis* to the Thelephoraceae. Lundell (1946: No. 1309) objected that the spores were hyaline under the microscope, and the spore powder white, but this discrepancy is in my opinion not only slight (it perhaps contributes to the generic character of *Boletopsis*) but also not quite correct; in the experience of other mycologists the spores are faintly colored in a print. Chemical studies are much needed in this case.

As was intimated in my previous lecture, the Thelephoraceae are another strongly coherent microcosmos within the Aphyllophorales. The family is derived from Fries's Clavariei, Thelephorei, Hydnei, Polyporei (viz., *Boletopsis*), Agaricini (*Lenzitopsis*), and, if you wish to keep these apart from the agarics, also the Cantharellaceae (*Polyozellus*).

Still another minor blood-letting consists in removing *Bondarzewia* from the polypores. It is a genus with a few members with big fruit bodies, among which is a North American species, *B. berkeleyi*, that may ooze out a milky juice when broken, which will explain its synonym *Polyporus lactifluus*. Not only in its system of gloeoplerous hyphae but also in other respects *Bondarzewia* suggests the Russulaceae. Thus, the spores closely resemble the spores of this family in shape and in the ornamentation with strongly amyloid spines and crests. After I had recommended that this taxon be made a distinct genus (Donk, 1933: 121), Singer (1940: 4) found that its spore ornamentation was amyloid and he compared it with that of *Russula* and *Lactarius*. The combination of these features in *Bondarzewia* has been responsible for its being placed in a separate family that belongs to the largely uncharted division where the Russulales, the clavarioid genus *Amylaria*, and perhaps also the Hericiaceae, Auriscalpiaceae, and still other groups belong.

These preceding amputations removed from the polypores some groups that showed little relationship with them. *Boletopsis* and

*Bondarzewia* revealed affinities with other existing taxa that easily absorbed them. The status of the Ganodermataceae is a problem of a different nature.

In many respects *Ganoderma* resembles *Fomes fomentarius*, and in my opinion the following question is legitimate. If *Ganoderma* had colorless, thin-walled, smooth, inamyloid spores like *F. fomentarius*, would it be included in the same group? I believe it would, although I have little doubt that because of certain peculiarities of the hyphal makeup it would be kept distinct from *Fomes*. This raises the next question. Are the spores really so remarkable that they warrant the separation of *Ganoderma* into a family of its own? This is very much a matter of personal opinion. On the whole, characters of the spores of the Hymenomycetes have proven to be among the most important and in several families they do play a decisive, or at least a strongly supporting, role. As a matter of fact, the spores of *Ganoderma* (a taxon that, I believe, will eventually be divided in a few genera) and of another currently accepted segregate, *Amauroderma*, are distinctly uniform in certain details. Their wall is a very complicated structure that as far as I am aware has no match anywhere among the Hymenomycetes (Furtado, 1962; Heim, 1963). (The truncation of the apex of the spores in *Ganoderma* is omitted from this discussion because it is typical of only a part of the group.)

This has led to different views. Some consider the main characters of the spores so much alike throughout the group that on the one hand they wish to admit only a single genus, and on the other hand do not find the spore characters sufficiently important to separate the inclusive genus *Ganoderma* from *Fomes* except at the generic level.

Personally, I favor the view that the construction of spore wall is remarkable enough to earn for the group the status of a family. What would be very welcome are supporting characters. These are now gradually being added: the Ganodermataceae are, as far as is known, oxidase-positive and in culture form a peculiar type of so-called cuticular hyphae; the fruit body is always pileate, stalked to adpressed with a narrow attachment, while eventually some hyphal features are very likely to be of importance. One difficulty is that as a family the Ganodermataceae must primarily be contrasted with the residual polypores, but these are still a chaotic mass of which nobody knows the exact delimitation or the family character.

When I suggest that at this stage we can cease to eliminate and temporarily call the residual mass the Polyporaceae (with a capital), I hope that I have not given the impression that this is done for any other reason than that this remainder still includes the type genus. I cannot follow Singer (1962: 156), who excluded the genus *Polyporus* as it is now often emended and *Pseudofavolus* in order to transfer them to the Agaricales where they are associated with *Mycobonia* and with a mixed assortment of agarics (for instance, with *Schizophyllum*) to form a new family Polyporaceae. This, I believe, is not advisable for various reasons, the nomenclatural of which I shall not mention. First, it is questionable whether or not the polyporoid genera that were removed formed a homogeneous entity; one of the modern requirements for a well-founded opinion is precise knowledge of the hyphal structure, and this requirement has as yet not been satisfied for either the transferred polyporoid genera or the agaric elements that acted as the magnet. Some recent studies on these agarics would suggest that another solution might have been to segregate certain agaric genera and transfer them to the Polyporaceae. Secondly, with our present knowledge it is not at all clear that the cleft between the Agaricales and *Polyporus s. str.* is less significant than that between *Polyporus* and the rest of the Polyporaceae. As long as these questions have not been studied by up-to-date methods, to deprive the Polyporaceae of the genus *Polyporus* was a debatable decision. What I do not wish to criticize is another suggestion, viz., that the limits between the Agaricales and Aphyllophorales are often arbitrary and capricious and of scarcely any scientific value.

It will not be possible to do much about the restricted family Polyporaceae before its genera have been clearly defined on a worldwide scale. The fragmentation into smaller genera is going on, but it is a pity that this is still done in Europe without a reasonable knowledge of the structure of the tropical and other non-European polypores. During the peeling out of some of the modern genera that have been taken to be more or less "natural" it appeared once more that the tubular hymenophore was a poor guide. Some of these genera contain in addition to polyporoid species also elements with nontubular hymenophore or else they appear to be closely related to "nontubular" species or groups. To mention a few examples: the effused genus *Cristella*, besides a few porias, now also contains species with hydnoid or corticioid hymenophore. A similar remark can be made for *Sistotrema*, which has urniform basidia. The restricted genus *Irpex* consists of a distinctly polyporoid species, *Polyporus* [*Irpex*] *tulipiferae*, and an irpicioid element, *Irpex lacteus*. As a genus it is, to say the least, doubtfully distinct from the hydnaceous genus *Steccherinum*. Another series that seems to be a natural one, although it crosses the borderlines between the corticiums, hydnums, and polypores, consists of the genera *Mycorrhaphium* (hydnaceous), *Hyphodontia* (corticioid and hydnaceous), and *Schizopora* (irpicioid and poroid). I could mention still more examples of this kind. All this shows that during this period of transition in the classification of the Hymenomycetes it becomes more rather than less difficult to define strictly artificial families!

It will be the task of the evolutionist to digest the evidence gathered by the taxonomist and to make up his mind about what is primitive and what is advanced or reduced. Until now he has usually been inclined to look upon the poria fruit body as primitive. However, the tide is now turning and several mycologists who wish to express evolutionistic views are inclined to appraise the poria state as reduced. However, such terms acquire more sense when they are based on taxonomic series. Isolated taxa are more difficult to place in the reduced-advanced scale.

## References


Bourdot, H., and A. Galzin. 1928. Hyménomycètes de France. Sceaux. 761 p. "1927."

Buchwald, N. F. 1938. Om Sporeproduktionens Størrelse hos Tøndersvampen *Polyporus fomentarius* (L.) Fr. [On the size of the spore-production of the Tinder Fungus, *Polyporus fomentarius* (L.) Fr.] Friesia 2: 42–69.

Corner, E. J. H. 1932a. The fruit-body of *Polystictus xanthopus* Fr. Ann. Bot. 46: 71–111.

———. 1932b. A *Fomes* with two systems of hyphae. Trans. Brit. Mycol. Soc. 17: 51–81.

———. 1932c. The identification of the brown-root fungus. Gdns' Bull., Straits Settl. 5: 317–350.

———. 1950. A monograph of *Clavaria* and allied genera. Ann. Bot. Mem. 1: 1–740.

———. 1953. The construction of polypores—1. Introduction: *Polyporus sulphureus, P. squamosus, P. betulinus* and *Polystictus microcyclus.* Phytomorphology 3: 152–167.

CUNNINGHAM, G. 1965. Polyporaceae of New Zealand. Bull. New Zeal. Dep. Scient. Ind. Res. No. 164. 359 p.

DONK, M. A. 1933. Revision der niederländischen Homobasidiomycetae-Aphyllophoraceae II. Utrecht. 278 p.

———. 1948. Notes on Malesian Fungi. I. Bull. Bot. Gdns. Buitenz. III 17: 473–482.

FAYOD, V. 1889. Sopra un nuovo genere di Imenomiceti. Malpighia 3: 69–73.

FRIES, E. M. 1874. Hymenomycetes europaei . . . . Upsaliae. 755 p.

FURTADO, J. S. 1962. Structure of the spore of the Ganodermoideae Donk. Rickia 1: 227–241.

HEIM, R. 1963. L'organisation architecturale des spores de *Ganoderma.* Revue Mycol. 27: 199–211.

IMAZEKI, R., AND Y. KOBAYASI. 1966. Notes on the genus *Coltricia* S. F. Gray. Trans. Mycol. Soc. Japan 7: 42–44.

INGOLD, C. T. 1957. Spore liberation in higher fungi. Endeavour 16: 78–83.

KÜHNER, R. 1950a. Absence de boucles chez les Basidiomycètes de la série des Igniaires et comportement nucléaire dans le mycelium des *Hymenochaete* Lév. C.r. Acad. Sci., Paris 230: 1606–1608.

———. 1950b. Comportement nucléaire dans le mycélium des Polypores de la série des Igniaires. C.r. Acad. Sci., Paris 230: 1687–1689.

LUNDELL, S. 1946. 1309. *Polyporus leucomelas . . . . In* Lundell and Nannfelt, Fungi exs. suecici Fasc., 27–28.

NOBLES, M. K. 1948. Studies in forest pathology. VI. Identification of cultures of wood-rotting fungi. Canad. J. Res. C 26: 281–431.

———. 1958. Cultural characters as a guide to the taxonomy and phylogeny of the Polyporaceae. Canad. J. Bot. 36: 883–926.

———. 1965. Identification of cultures of wood-inhabiting Hymenomycetes. Canad. J. Bot. 43: 1097–1139.

OVERHOLTS, L. O. 1953. The Polyporaceae of the United States, Alaska and Canada. . . . Univ. Michigan Stud., Sci. Ser. 19, Ann Arbor. 466 p.

PINTO-LOPES, J. 1952. "Polyporaceae." Contribuição para a sua biotaxonomica. Mem. Soc. Broteriana 8: 1–215.

POUZAR, Z. 1966. Studies in the taxonomy of the polypores II. Folia geobot. phytotax. 1: 356–375.

ROLL-HANSEN, F. 1940. Undersøkelser over *Polyporus annosus* Fr., saerlig med henblikk på dens forekomst i det sønnefjelske Norge. Meddr. norske Skogfors Ves. 7 (1): 1–100.

SINGER, R. 1940. Notes sur quelques Basidiomycètes. VIe. Série. [1. Un genre nouveau de Polyporaceae]. Revue Mycol. 5: 3–4.

———. 1962. The Agaricales in modern taxonomy. 2nd ed. Weinheim. 915 p.

TEIXEIRA, R. A. 1958. Studies on microstructure of *Laricifomes officinalis*. Mycologia 50: 671–676.

## DISCUSSION

DUBUVOY: I understand from what I have heard and seen that skeletal hyphae originate from generative hyphae and that the skeletal hyphae are clampless. How do you explain the absence of clamp connections in skeletal hyphae, which actually originate from clamped generative hyphae? The beginning part of the hypha will be clamped but then it will become clampless because skeletal hyphae are clampless—but their origin is from the generative hyphae.

BOIDIN: The problem is to know what type of septum one observes. It may separate living parts on both sides, or living from dead parts. The clampless parts are skeletalized.

DONK. I agree with what Dr. Boidin said. You must be very careful and from the start "outlaw" one kind of septum altogether as false or adventitious. These are different from the septa that usually are associated with clamp connections and which are true septa; these septa should have dolipores. I assume that the former have no dolipores and are merely *cloisons de retrait*, septa formed in connection with the retreat of the bulk of protoplasmic contents.

NOBLES: In my experience, the skeletal hyphae having clamp connections are formed as the ends of hyphae. There is a clamp connection at the base of that segment which may be difficult to see in highly differentiated tissue. It is a skeletalized end-cell which becomes very, very long and has either no septa or, as Dr. Boidin has said, retreat septa, where the cytoplasm has apparently just withdrawn and where a false wall or something is formed. There are no true septa in that differentiated part of the hypha.

PETERSEN: According to Corner, this end portion, which is a skeletal hypha or whatever you call it, is one cell, but may extend from one end of the mature fruiting body to the other.

DUBUVOY: But do you see the clamp connection where the hypha originates in the fruiting body?

DONK: Let me return to one of the slides I used (Corner, 1953, fig. 3 on p. 154), which illustrates what Dr. Nobles said. In it, you see a tissue of monomitic hyphal structure and without clamp connections on the hyphae which are all of one type and called generative hyphae. However, more often than not you will find that in generative hyphae the septa are associated with clamp connections. In the next slide (Corner, 1953 fig. 2) you see that skeletal hyphae are formed in the growing margin of the fruiting body. Here, again, it is common to find clamp connections on the generative hyphae and also at the bases of the skeletal hyphae. Sometimes septa are formed in the youngest (terminal) portion of a skeletal hypha. These have no clamps. The margin of the fruiting body may grow on for quite a long time and the already formed skeletal hyphae often keep pace over considerable distances; they have only a basal clamp.

The hyphal systems that originate in the growing margin might be called primogenic (produced first, so to speak). Later, inside the fruiting body other systems of hyphae may be formed, such as the system of binding hyphae. It binds together the generative and the skeletal hyphae. Finally, a third kind of hyphae may come into being: although the generative hyphae inside the fruiting body usually deteriorate quickly, they may remain "active" and undergo transformation, by becoming thick-walled for instance, or by acquiring other peculiarities. In this way we are not only faced with primogenic hyphal systems, but also with neo- or allogenic hyphal systems (such as the binding hyphae) and epigenically transformed generative hyphae. I tried to avoid these aspects of the various systems because mentioning them in this manner would have made the paper too long.

PILÁT: The term "skeletal hyphae" is an unfortunate one because they are not really skeletals, but are what you would call trichomes in the phanerogams. In phanerogams, the trichomes occur only on the surface of the organs which are plectenchymatous, whereas in the fungi they also occur inside. I am not certain (as in *Phellinus*, for example) that the skeletal hyphae always have very much taxonomical importance. I agree with Dr. Donk that the skeletal hyphae are not indefinite in growth but definite, and that the generative hyphae have normal septa and the skeletals do not.

DONK: I believe that on the whole skeletal hyphae do play a part in supporting the fruiting body. This can be inferred from the dimitic fruiting body (lacking binding hyphae) of *Heterobasidion annosum*, where the generative hyphae become neither firmly glued together nor inflated (and the like) to such a degree that the firm context of the fruiting body is explained. In most cases typical skeletal hyphae do form a kind of skeleton or supporting framework.

PILÁT: The comparison of skeletals with trichomes was metaphorical.

DONK: Skeletal hyphae grow indefinitely. One of the characteristics of skeletal hyphae is that they originate in the growing margin. The indefinite growth is one of the factors that make hyphal analysis often very difficult, because isolating a deep-seated single skeletal of many millimeters or even centimeters length from its base to its apex is no easy task. If you have performed such a manipulation successfully only a few times you will be satisfied and will not look for longer ones in the same object.

Dr. Pilát said that in his opinion skeletals could have arisen in different groups of fungi; he mentioned the Hymenochaetaceae as an example. This is precisely what I also suggested in my first lecture. In it I asked a rhetorical question about what would happen to a terrestrial agaric when it migrated onto the side of a stem and then to the underside of a log. It would be monomitic as a terrestrial agaric; but it might become dimitic and even trimitic when the cap is formed high in the air on the side of a tree-stem; and then on further migration onto the underside of a log as a thin corticioid structure it might become monomitic again because as a thin layer it no longer has need of skeletal and binding hyphae. If in addition it is accepted that certain groups of Aphyllophorales are derived from as many groups of agarics, then this theory would be an answer to the supposed multiple origin of the "same" hyphal system.

I shall try to answer still another question. It should be possible by using chemical tests to establish the (taxonomic) series of Hymenomycetes to which a given system of skeletal hyphae belongs. Congo red, Giemsa *lent* (the use of which has been encouraged by Kühner), cresyl blue, cotton blue, and other aniline dyes already have proven their importance as tools in the classification of the Aphyllophorales. This shows once more that chemical tests should also occupy a very important role in the study of polypores that aims at fixing relationship. For instance, the series of *Mycorrhaphium*, *Poria versipora*, and *Hyphodontia* is an example in which the use of chemical tests has assisted in the recognition of a certain type of skeletal. I suspect that the skeletals of *Heterobasidion*, for instance, will appear to react differently when adequately tested.

BOIDIN: I will dwell on the first case. We must distinguish the true septa between living parts from retreat septa (*cloison de retrait*) between living and dead parts. It is very important.

Do you hope to publish an emended hyphal terminology in the future? I hope that you will publish a terminology for hyphae in culture at the same time.

DONK: Dr. Lentz has spoken about terminology of hyphae in his lecture. Therefore I shall not say much about it. The terminology I use is perhaps already overloaded; all the same I would rather list characters first and think of terms later. Taxonomy must not be encumbered by too many new terms.

BOIDIN: One must know the nuclear condition to judge intelligently the value of presence or absence of clamps in taxonomy. For instance, last September in Maboké, I found some *Podoscypha* specimens without clamps. They were only parthenogenetic, uninucleate forms so this does not change the definition of *Podoscypha* that I elaborated and which was further delineated by Reid. I agree, however, that the presence of clamps is an important item in the classification of these fungi.

DONK: You have given a clear example yourself. *Irpex lacteus* has clamps and the hyphae are "normal" with dicaryons, whereas *Polyporus tulipiferae*, which looks not very different as a species, has coenocytic cells without clamps.

Something of this kind is found in *Hygrophorus*. The genus *God-*

*frinia* was published by R. Maire on the basis of two-spored basidia. He was dealing with haplo-parthenogenesis so that no clamps were to be expected. Even if all species of *Hygrophorus* had clamps (in their synkaryotic stage), I would not exclude *Godfrinia* from *Hygrophorus* solely for the reason that clamps were lacking. This example shows that it is often necessary to know why no clamps are formed before we use their absence as a taxonomic character. In most cases the absence of clamps is accepted by direct observation in a strictly empirical manner. We have to wait until someone explains why clamps are absent; not until then will the time come for a critical reevaluation of the taxonomic importance of the character. This opinion will explain why I mentioned at some length the reports on the very few clamps that have been found in the Hymenochaetaceae. This paucity may be due to various causes. At any rate, because of their unexpected presence such exceptions should be carefully studied; the results are likely to furnish indications of relationship with other groups. Exceptions are really very important cases. In my lecture I also mentioned as examples a number of genera defined (*inter alia*) by the lack of clamps in the fruiting body, but at the same time I pointed out that *Polyporus* [*Tyromyces*] *mollis* is still kept in the genus *Tyromyces* although it has no clamps, not even in culture. Since we do not know why this is the case we should be careful in founding a new genus for this one species on this single character. I hope that in the future, fruit-body mycologists, who know little or nothing about the nuclear situation in their material, and culture-mat mycologists will go arm-in-arm. They should keep close contact because neither of the two can proceed without the help of the other, which is a main thesis of my lecture.

DUBUVOY: I'd like to know more about the nature of the septa in the skeletal hyphae with clamps at their bases.

DONK: When no septa are formed, there will be no clamps. Therefore, typical skeletal hyphae have no clamps; when in spite of this a few septa are formed these are not true septa but false ones, which are abundantly different. The same appears to be the case in the metabasidia of the tremellaceous fungi. In some species a septum separates the upper portion of the basidium from an anucleate stalk. It is a false septum and may be expected to have no dolipore. In the same basidium another type of septum is formed in conjunction with nuclear division; these are true septa. In *Tulasnella* the young metabasidium has no septa, but later on septa are formed across the bases of the epibasidia or (as I call them) the sterigmata. These false septa are *cloisons de retrait*. In my opinion such septa should be left out of account in defining holo- and phragmobasidia.

It should be kept in mind that in nature a lot of things are not always clear-cut. Skeletal hyphae that are in the process of being formed often "cannot make up their mind," which perhaps is due to their formation somewhere between two morphogenetic regions or near another such region. In *Hyphodontia*, in some species of *Poria*, and in *Chaetoporus* the most "perfect" skeletals originate in the hymenophore just beyond the border between the subiculum and the hymenophore. Others may originate somewhat "too early" and during their growth have to cross that border. These often show signs of "hesitation" or indefiniteness. Corner's medial hyphae are an example of this kind. Skeletals bending into the hymenium and terminating as cystidia are another example of the result of successive morphogenetic forces of two different regions.

PETERSEN: In this context, do you consider gloeoplerous hyphae as a fourth basic hyphal type?

DONK: Yes. I would regard macrosetae (and so-called setiform or setal hyphae) as a further example; also what are called skeletal hyphae in *Ganoderma*, which usually become branched in their apical portion, and many other kinds of specialized hyphae. This seems preferable to calling them all skeletal hyphae as is now sometimes done, because all of them originate in the growing margin, I think. With this interpretation it is possible to come across fruiting bodies with generative hyphae, binding hyphae, and more than one type of hyphal system formed in the growing margin. There is little justification in an expanded terminology that hangs on the terms mono-, di-, and trimitic by trying to press all the possible combinations of hyphal systems into these three classes. "Dimitic" already means two different situations: dimitic with binding hyphae and dimitic with skeletal hyphae.

SINGER: I do not think it is absolutely necessary to introduce a more and more complicated terminology for other types. Would you not agree with me that in the Agaricales, the best that could be done now would be to describe what one sees, instead of forcing on it a terminology that originated in other groups?

DONK: The Agaricales, excepting certain genera like *Lentinellus*, *Lentinus*, and *Panus*.

SINGER: *Lentinellus* is in the Aphyllophorales.

DONK: In case such genera are excluded. The term "monomitic" has little meaning in the agarics. It is preferable to sketch in details; for instance, presence or absence of inflated hyphae and so-called fundamental hyphae, whether or not the hyphae are cemented together, or whether or not thickening of walls occurs. In one of the Hymenochaetaceae which I mentioned (*Cyclomyces* sp.; cf. Corner, 1953, fig. 12 on p. 165; as *Polystictus microcyclus*) the hyphal system is monomitic but nevertheless the fruiting body is leathery because the non-inflated, thick-walled hyphae become firmly glued together, a very characteristic condition. In certain fleshy fruiting bodies interweaving hyphae may occur which look like binding hyphae but differ in being thin-walled and septate with clamps. We need a few extra terms but it is often more important to add an adequate description of what is actually seen; this last requirement is often neglected.

PETERSEN: We have come up with the very thing you are describing in the clavarias. Descriptions have stated "monomitic" but we find agglutinated or coherent hyphae, inflation, secondary septation, and whatever else. These must all be described again more accurately in separate groups.

SINGER: [to Donk] Do you agree with the interpretation I put forward after the paper by Dr. Pilát on the setae of the Agaricales? In *Marasmius*, for instance, we have several species where you can really find no differences in the setae themselves. I heard you say now that wherever in the Aphyllophorales the setae have major importance, they arise from hyphae which are not clamped, but in the Agaricales I think they arise from clamped hyphae.

DONK: There are no clamps to be found on the setiferous hyphae (of the Hymenochaetaceae) in the Aphyllophorales. I suspect that there are chemical differences of the walls in the two groups. This needs investigation. The number of Hymenochaetaceae that have been studied in cultures by Kühner, Boidin, and Miss Nobles is as high as about 65, and all appeared to be oxidase positive (with perhaps one exception).

SINGER: This has not been tabulated for *Marasmius*.

DONK: Such chemical data will have to be worked out before it will be possible to correlate or dissociate two types of structures that have been called setae and occur in widely different groups. I know a number of these tropical species of *Marasmius*: their "setae" are pliable, which is not typical of the setae in Hymenochaetaceae.

SINGER: The ones we have we cannot bend. But as long as the definition of setae exists, I cannot see why we should introduce new terms.

KIMBROUGH: In examining some agaric monographs for tissue terminology, I'm a little discouraged to see that descriptions of supposedly new tissues exist, but for tissues that have been described in other fungi, a new term is often chosen just because it exists in an agaric. For example, in the monomitic system of generative hyphae perhaps two or three sets of tissues described by Starback[1] are created by the arrangement or development of the generative hyphae. I think there should be some approaches to standardizing terminology of tissues in fungi, incorporating such principles, for example, as Remsberg[2] did in *Typhula* for sclerotia. Tissues common to agarics, polypores, discomycetes, and other fleshy Ascomycetes should be described in a single terminology. I am also reminded that the terminology used by Orson Miller (see that paper) and taken from Richard Korf applied perfectly to agaric cultures, because in looking at the cells and tissues, certain of them are very similar, for instance, in a sclerotium of a discomycete or a basidiomycete. And it is the same way with some of the other tissues.

GILBERTSON: I do not think I can agree with the way Dr. Donk interpreted the significance of the oxidase reaction. What is basically involved here, and what Dr. Nobles is interested in, is whether these organisms have the ability to degrade lignin or whether they do not. I think this is a very basic metabolic characteristic, and one that is of very large taxonomic and phylogenetic significance. The majority of the organisms that have the ability to degrade lignin have an extra-

[1] Starback, K. 1895. Discomyceten-Studien. Bihang Kongl. Svenska Vet.-Akad. Handl. 21, 3: 1–42.

[2] Remsberg, R. E. 1940. Studies in the genus *Typhula*. Mycologia 32: 52–96.

cellular oxidase system which can be detected by the usual tests—the use of tannic and gallic acid in the agar medium, or a solution of gum guaiac. However, some of them are apparently chemically different enough that their mechanisms cannot be detected by these convenient techniques. That does not have any bearing at all, however, on this basic characteristic of lignin degradation. I do not think anyone who has studied *Poria aneirina*, for instance, in nature could ever question that it has the ability to degrade lignin, yet it does not give a positive oxidase reaction with these tests. I will try to mention some reasonably natural phylogenetic groups I believe you would agree with me on. The Hymenochaetaceae, for example, every member of which has the ability to degrade lignin, with no exceptions. The genus *Coriolus* is another good natural group where there is no exception to the ability to degrade lignin. We could mention the *Coriolellus* group as studied by Sarkar (1959),[3] every species of which lacks the ability to degrade lignin, with no exceptions. Every species of the genus *Chaetoporus* has the ability to degrade lignin. And as I go down and look at various groupings that seem to be good natural taxonomic groups there does not seem to be any variation from this pattern. To me this characteristic seems to be much more significant and important that you would indicate by your presentation. But I think it is not the oxidase reaction itself that is basically important here, but whether or not these organisms have the enzyme system present to degrade lignin. The oxidase test is simply a convenient method of detecting some of these enzyme systems. In a small percentage of cases it doesn't work out, however.

DONK: I agree completely with Dr. Gilbertson. I may add that if you put agarics to the oxidase test a number of them will appear oxidase-positive and others oxidase-negative. When this difference is held to be a primitive character the resulting groups will show very mixed assortments. I cannot believe in advance that these groups consist of related components. That is the point. In the case of the Hymenochaetaceae I was very happy to enlist "oxidase-positive" as a sustaining character, because it also appeared to be consistent.

GILBERTSON: I agree with you on that.

[3] Sarkar, Anjoli. 1959. Studies in wood-inhabiting Hymenomycetes. IV. The genus *Coriolellus* Murr. Canad. J. Bot. 37: 1251–70.

AMBURGEY: Dr. Gilbertson talked about extracellular oxidase reactions. This reaction is based on perhaps a single enzyme difference in the breakdown of lignin, and on that ground one might argue that there is not that much difference between those organisms which can break down lignin, and those that cannot. But the utilization of lignin once it is broken down requires a whole new series of enzymes. This is quite a difference.

DONK: That is about what Dr. Gilbertson said. It is not simple.

NOBLES: In 1948 I published a key for the identification of 125 species of wood-inhabiting hymenomycetes. At that time I said that it was impossible on the basis of cultural characters to decide to which certain family or genus a fungus belonged. This was my idea in 1948. As I worked to identify some 3,000 isolates per year, it became obvious that some of these species were so similar in their cultural characters that you could not separate them out at the specific level without using interfertility tests. It seemed obvious to even the most reluctant taxonomist that those groups of species must be related. I did not know anything about the taxonomy of polypores at that time, but when I started studying, I found that those species which were alike in their oxidase reaction, and their association with a brown rot or white rot, and their hyphal characters, etc., also belonged to a cultural group. By 1958 I was so persuaded about this that I wrote a paper in which I suggested that cultural characters could be used to revise or assess the classical taxonomic system. I am not saying that you can base a system solely on cultural characters, but that you can use these as additional characters to be evaluated always along with the fruiting body characters.

A fungus has a certain genetic makeup and it must express it. It may express it in different ways, but it is the same genetic makeup whether it is growing in culture, or in a tree, or forming a fruit body, and any character which can be applied to it in one stage can be used in correlation with characters of another stage.

You may notice that I tend to put Polyporaceae in quotes. It is a very artificial group as I have used it. I might mention one of my most recent problems, in which I have been trying to identify a fungus that has been isolated repeatedly from recently felled trees in British Columbia. I did not know whether it was *Ischnoderma resinosum* or *Pleurotus serotinus*. It happens to be *Poria zonata*, but in this case you can't tell a polypore from an agaric, their cultural characters are so close. All three belong to a very homogeneous group based on their cultural characters, and I think it is time to start looking at their fruit body characters to see if they have something in common.

One example that Dr. Donk mentioned was that of *Fomes* and *Ganoderma*. In my key, they come out right together. Sometimes they cannot be separated in culture, but I separate them on spore characters.

ROGERS: In about 1945, Corner made the statement that careful study of a basidiomycete meant study in terms of hyphal systems. This is, of course, a dogma. I think it is completely unsupported by experience and unwarranted in principle. There are many groups of Basidiomycetes in which "hyphal analysis" (that is the term he used) is completely irrelevant. As has been said twice recently, the important thing is not hyphal analysis, not description in terms of Corner's or anyone else's terminology, but description of the structures as they exist.

SINGER: I am happy to hear what Dr. Donk said about *Bondarzewia*, because it coincides with observations we have made in South America. We have a third species, *Bondarzewia quaitecasensis*, which is very important because it is an extremely serious parasite on and destroyer of *Nothofagus*. This species also produces latex and has an almost indistinguishable relative in the astrogastraceous series of the gastromycetes—*Hybogaster*. It's the only multipileate form of the Bondarzewiaceae. One cannot tell from a few yards away whether it is *Bondarzewia* or *Hybogaster*, which is a new genus of gastromycetes. We obtained cultures of the species and ran them through Miss Nobles' keys and came out exactly where she now has put the American and European species. The species has no clamp connections in culture and conforms to this homogeneous group from all points of view. I would say that this is the only group of Aphyllophorales so far that we can link with any gastromycetous group. That puts them right at the same level with the Russulaceae, and therefore I would agree perfectly with Dr. Donk that this removes the Bondarzewiaceae in some sense from the rest of the Polyporaceae.

HARRY D. THIERS

*Professor of Biology*
*San Francisco State College, San Francisco, California*

# SOME IDEAS CONCERNING THE PHYLOGENY AND EVOLUTION OF THE BOLETES

❦

The title of this paper is somewhat frightening to me, and I am sure that it is not necessary to remind the reader of the numerous pitfalls and uncertainties to be contended with in promulgating theories concerning the origin, phylogeny, and evolution of any group of fleshy fungi. This is true even though serious study has been given to the fleshy fungi for over a century. There is still a severe paucity of data available which relate or contribute to the formulation of such theories. Only recently Dr. Corner has indicated and emphasized one of the major contributing factors to the lack of progress in this field, estimating that perhaps as few as only one-fourth of the existing fungi are known at the present time. Thus, the first major basis for reluctance in discussing this topic is the poor state of knowledge of the present flora. Other factors, particularly the almost complete lack of fossil evidence, the difficulty in culturing and especially of obtaining bolete fruit bodies in culture, the difficulties inherent in our inability to demonstrate conclusive hybridization, and the lack of substantiating cytological investigations contribute materially to the difficulties enumerated above. Therefore, with these handicaps in mind and the consequent severe limitations they impose, I hope you will understand that there is really little more than circumstantial evidence to support any ideas that I may develop during the course of this paper.

The group of Homobasidiomycetes which I will discuss are commonly referred to as the boletes or fleshy pore fungi. They are variously placed in one or more families in either their own order, the Boletales, or together with the gill fungi in the Agaricales. Because of their distinctive hymenophore they constitute an easily recognized group of fleshy fungi. Also, because of the poroid hymenophore many early workers classified these fungi with the polypores because of the superficial similarity of the configuration of the hymenium-bearing tissues. Few present-day mycologists subscribe to this concept, however, and almost without exception, these fungi are considered to have close affinities with the agarics or gill fungi. Generally, we can describe the boletes as having fleshy, putrescent basidiocarps of which the hymenophore is poroid and the basidiospores forcibly discharged. With the exception of only a few taxa, the fruit bodies are large, often brightly colored, and are assumed to be almost exclusively mycorrhiza formers. The microscopic anatomy of these fungi very closely parallels that of the gilled fungi, particularly in the structural elaborations of the hymenium and trama.

As is true with practically all groups of fungi, there is something less than general agreement as to generic, and, to a lesser extent, familial concepts within the group. Presently one or two families are recognized; if two are recognized, the basis for separation is most commonly found in the nature of the spore wall. Species in which the epispore is smooth are placed in the Boletaceae and those with roughened spores in the Strobilomycetaceae. The more commonly accepted genera are as follows.

Strobilomycetaceae

*Strobilomyces*: Basidiocarp typically dark gray to black; epicutis dry, typically strongly squamulose; hymenophore gray to black at maturity; spores globose to subglobose, rough-walled, black in deposit.

*Boletellus*: Basidiocarp variously colored; hymenophore variously colored, but typically yellowish to vinaceous at maturity; spores variously roughened, sometimes obscurely so, some shade of olive or brown in deposit.

*Phylloboletellus*: Hymenophore lamellose; spores roughened; not seen by the author.

Boletaceae

*Boletus*: Basidiocarp variously colored, dry to viscid, glabrous to squamulose, often massive; hymenophore yellow to red or reddish, often white to pallid when young; stipe often reticulate; spores smooth, typically some shade of brown in mass.

*Pulveroboletus*: As in *Boletus*, but with a pronounced, floccose yellow veil.

*Gyroporus*: Basidiocarp typically pallid, relatively small, dry; hymenophore typically depressed, pallid, and usually not highly pigmented; spores smooth, pallid to hyaline in deposit.

*Tylopilus*: Basidiocarp variously colored, often some shade of brown; glabrous to tomentose to at least subscaly, typically dry; hymenophore pallid to white at first, becoming vinaceous to flesh-colored to dark reddish vinaceous at maturity.

*Leccinum*: Basidiocarp various in color, often orange to orange-red to brown; glabrous to fibrillose to fibrillose-scaly, dry to less frequently viscid, often massive; hymenophore often white to pallid when young, becoming yellowish to brownish at maturity; stipe characteristically covered with numerous small, often dark-colored scales; spores typically large, smooth, some shade of brown in deposit.

*Suillus*: Basidiocarps variously colored, frequently some shade of brown or yellow; cuticle viscid or dry; hymenophore typically some shade of yellow; pores small to large, irregular to almost lamellose; surface of stipe often glandulose; veil or false veil present or absent; fascicled cystidia usually present which stain dark brown in KOH; spores smooth, yellow-brown to olive brown in deposit. This emended concept includes such genera as *Boletinus*, *Paragyrodon*, and *Gyrodon*.

*Fuscoboletinus*: As in *Suillus* but with the spores colored vinaceous-brown to purple-brown in deposit.

*Phylloporus*: Hymenophore lamellose, although frequently intervenose or poroid; spores and hymenophoral trama typically boletoid; commonly classified in Boletaceae but has been classified in some of the lamellose families.

Several smaller genera have been omitted from the list partly for the sake of brevity and partly because, in some instances, in my opinion some of them are not sufficiently distinct from some of the larger, well-established genera to be recognized as a separate taxon. Included in this latter category are such genera as *Porphyrellus*, *Xanthoconium*, *Xerocomus*, and *Phaeogyroporus*.

If an evaluation of the major taxonomic or morphologic characteristics of these fungi is attempted, one is confronted with the major difficulty of determining those characters which might be of potential significance in indicating phylogenetic relationships. In the following paragraphs an attempt will be made to discuss some of these characters which, in my opinion, possibly possess some such significance. The first among such features which might be considered is the apparent mycorrhizal association, since there appears to be rather close correlation, in many instances, between generic limitations and such an association. This correlation is perhaps best exemplified in the genera *Suillus* and *Fuscoboletinus* in which the mycorrhizal host for most species appears to be restricted to conifers. Does this close relationship assume some significance when contrasted to the situation in such genera as *Boletus*, *Leccinum*, etc. in which associations are apparently formed with both gymnosperm and angiosperm trees and shrubs? It is also interesting to speculate on the significance of an alpine habitat and its role in the evolution of the boletes. This is mentioned because in many instances possible intermediate species appear to be confined, or at least more common, in such areas. This is particularly apparent in the Sierra Nevada and Cascade mountain ranges of the western United States.

Turning to a consideration of the basidiocarp proper, the first character that should be mentioned as possibly having some significance is the size of the fruit body. It has often been intimated that a large, massive basidiocarp is more complex and, presumably, more advanced than the smaller form. If this be the case there is little difficulty in demonstrating that the genus *Boletus s. str.* includes large fruit bodies as do closely related genera such as *Tylopilus* and *Leccinum*. Smaller basidiocarps are more characteristic of such genera as *Gyroporus*, and, to a lesser extent, of *Suillus* and *Fuscoboletinus*.

The nature of the surface of the pileus appears to be of relatively little value except at the species level. The usual distinctly viscid pileus of many species of *Suillus* is a possible exception, but this same character appears in numerous other genera and seems to have arisen independently in various genera. It is not peculiar to any single higher taxon. Such characters as the color of the pileus, color changes either with age, upon bruising, or exposure, and the nature of the pileus epicutis are of apparently little special significance above the species level.

The hymenophore, in my opinion, exhibits some characters of possible significance. The most important of these is undoubtedly its configuration. In some genera, for example, the hymenophore may vary from an almost truly lamellose condition through various intergrading conditions to a truly poroid condition. In other genera only the truly poroid condition is known to occur. In addition, the color of the dissepiments and the color changes occurring during the maturation process are worthy of consideration. In most species of *Suillus*, for example, the tubes are yellow throughout all ontogenetic stages, but in such genera as *Boletus*, *Leccinum*, and *Tylopilus* the color of the hymenophore when young is likely to be quite different from that in age. In addition, the color of the pores, if different from the dissepiments, and the color changes which occur on bruising are of some significance, but more so at the species level. It should be pointed out that the rather common blueing reaction has often been shown to be erratic and is not always reliable as a taxonomic character; this is particularly true at ranks above the species. It is possibly worthwhile to mention the type of attachment of the "tubes" to the stipe. The most common attachment is adnate to adnexed to shallowly depressed. Less common are decurrent attachments, but this is often encountered in species of *Suillus*. In *Gyroporus* the tubes are characteristically noticeably depressed around the stipe.

The characters of the stipe are most significant in delimiting *Leccinum* from other genera. The presence of groups of caulocystidia, commonly referred to as glandular dots, on the surface of the stipe is, as far as is known, restricted to species of *Suillus*. There are certain structural similarities between these cystidia and the so-called scales or scabrosities in *Leccinum*. The reticulate surface on the stipe is of some significance since it is rather characteristic of the *Boletus edulis* series, but it does, however, rather commonly occur in other groups. The rather deeply reticulated to alveolate condition is seen in a few apparently closely related species of the genus *Boletellus*. Also of some importance is the presence of an annulus as well as the veil. Annulate boletes are most consistently encountered in species of *Suillus* and related genera. Annulate species, however, also occur in other genera as exemplified by *Pulveroboletus ravenelii*. Other characters of the stipe such as color, color changes, and nature of the surface appear to be of little significance at delimitations above the species level.

Of microscopic characters, the most obvious features for discussion are the characters seen in the spores. It has already been indicated that the color of the spores, at least in mass, as well as the nature of the epispore are used for major generic delimitations. Recent reports by Smith and others of amyloidy in bolete spores may prove to be of value as more investigations are carried out. Furthermore, there appears to be a consistency of spore size in the genus *Suillus* where the spores are generally smaller than in most other taxa. The opposite is true of spore size in *Leccinum*, where the spores are generally larger than in other genera. There also appears to be some slight constancy in spore shape in *Suillus*, and similar correlations, of course, can be seen in such genera as *Strobilomyces*. Features such as germ pores, truncate apices, and spore-wall thickness are of some importance but mostly at the species level.

Various kinds of cystidia are almost always found in the boletes. They are, therefore, of little help in formulating phylogenetic sequences or affinities in the group. The one major exception to this observation is found, once again, in species of *Suillus* and possibly *Leccinum*. The presence of clustered or fascicled cystidia on the stipe and in the hymenium is, in the case of *Suillus*, perhaps the most important single character. As is well known, these same cystidia give a characteristic dark-brown color reaction when mounted in dilute solutions of potassium hydroxide.

The hymenophoral trama is highly distinctive in all the boletes. Unfortunately it is the same or almost the same for all species. It is generally considered divergent except perhaps for the xerocomoid group in which the configuration is more-or-less parallel when young. Other microscopic characters including the nature of the epicutis, pileus trama, stipe trama, and even clamp connections appear to be of no great value at other than the species level. It is in the area of microscopic anatomy and such related areas as biochemistry and cytology that we can expect new and perhaps major breakthroughs in the field of morphology and phylogeny. The work of the biochemists at the University of Washington and Purdue University has given us good indications of the contributions that can be made in these areas.

In order to further lay the foundation for a phylogenetic discussion of this group of fungi, it is necessary to devote some space to a discussion of some of the related or satellite taxa of fungi. It is immediately obvious that there are two such groups of fungi to be considered. One constitutes the agarics or lamellate fungi and the other the puffballs or gastromycetes. If we direct our attention first to the agarics we find that several different groups have at one time or another been indicated as having some degree of affinity with the fleshy pore fungi. The most obvious agaric genus is *Phylloporus*, which is considered so closely related to the boletes that it is presently typically classified in the same family with them. It is, however, usually lamellate and, in the past as well as at the present, has sometimes been classified in the genus *Paxillus*. Much of the internal microscopic anatomy is similar to that of the boletes, including the features of the spores and hymenophoral trama. Frequently the lamellae become intervenose, resulting in the formation of large, highly irregular pores. Furthermore, anyone who has seen this fungus in the woods is immediately impressed with the close similarity of its appearance to the boletes.

Many workers have emphasized the similarity and possible affinity of members of the Paxillaceae and Gomphidiaceae to the boletes. The gomphidii are highly distinctive by virtue of their thick, waxy, strongly decurrent, distant lamellae. The basis for the possibility of close relationships of this group with the boletes is found in the similarity of the shape, size, and color of the spores, the conspicuous, usually common cystidia and the typically divergent gill trama. The Paxillaceae are also credited with having characters in common with the boletes. These characters, like those of the Gomphidiaceae, are largely microscopic and include features seen in the spores, trama, etc. There has been some suggestion that there may be closer affinity between these gilled fungi and the rough-spored boletes than with the smooth-spored.

So far this discussion has centered on the possible affinities with lamellose groups. Let us devote our attention now to the second group mentioned above, i.e., the gastromycetes. The recent publications of Smith and Singer have called to our attention several groups of these fungi which in some way show similarities to the boletes. Certainly the most obvious taxon worthy of consideration is the group of curious "puffballs" belonging to the genus *Gastroboletus*. In many ways these fungi very closely resemble the boletes, with the only major difference found in the absence of the ability to forcibly discharge basidiospores. Appearing related to this feature is the irregular orientation of the tubulose gleba in such a manner as to mechanically prevent the escape of the spores. In those few species in which the gleba is more-or-less permanently covered by a peridial layer there is no evidence of any spore deposit on the inner surface of the layer. Until recently there was only one species of *Gastroboletus* well known in the United States. An additional species has been recently reported from Michigan by Mazzer and Smith, and Trappe and I have a paper in press[1] in which we describe four or five additional species that we found primarily in the alpine to subalpine areas of the Pacific Coast. Other species have been reported from various regions of the world indicating that it might be a relatively large taxon of puffballs.

In addition to the gastroboletes other genera of gastromycetes which have been commonly credited as having possible affinities with the boletes are *Truncocolumella*, *Chamonixia*, and *Rhizopogon*. *Truncocolumella* is a small genus of hypogaeous fungi which are also relatively common in the western United States. The basidiocarps are often partially buried in moss banks but typically become exposed at maturity. These fungi appear quite distinct from the gastroboletes because of the lack of a well-developed stipe and more nearly lacunose or alveolate than tubulose gleba. The basis for considering *Truncocolumella* related to the boletes is the similarity of the trama of the glebal plates, a similar blueing reaction of the flesh when exposed or bruised, the alveolate to obscurely tubulose gleba, and the size and shape of the basidiospores. It should also be noted that in *Truncocolumella* the peridium completely surrounds the gleba, and thus it never becomes exposed as it does in most of the gastroboletes.

*Chamonixia* is distinguished from *Truncocolumella* most readily by the roughened nature of the epispore of the former. According to Smith and Singer there is some difficulty in distinguishing some species from both genera. I must emphasize that I have not seen material of *Chamonixia* and I rely on the studies of other workers. *Chamonixia* seems to show essentially the same similarities to the boletes as *Truncocolumella*, and, as indicated earlier, the roughened wall might possibly indicate some relationship to such genera as *Strobilomyces* and/or *Boletellus*.

The third genus mentioned above, *Rhizopogon*, is the farthest removed from the boletes. As is well known, *Rhizopogon* is a very large, widely distributed group of truly hypogaeous fungi which, in most

[1] Thiers, Harry D. and James M. Trappe. 1969. Studies in the genus *Gastroboletus*. Brittonia 21: 244–254.

instances, appear to form mycorrhizal associations with conifers. As in *Truncocolumella* the gleba is completely enclosed and the stipe-columella is lacking. The structural similarities to the boletes are microscopic and are found chiefly in the nature of the spores, poroid to lacunose gleba, and possibly the trama of the glebal plates.

It has been the purpose of this discussion thus far to superficially review the nature and availability of the raw materials we have to work with in formulating ideas or theories concerning the relationships of the boletes. In addition, a brief attempt has been made to familiarize the reader with the present-day species which have to be accommodated in our conjectures and discussions. With some trepidation I turn, therefore, to a discussion of the phylogeny and possible evolutionary pattern of the boletes. Perhaps the discussion should be initiated by considering the possible origin of these organisms. In so doing we are faced with the formidable obstacle of presenting a solution to a problem upon which there has never been general agreement in the past. We are also faced with the immediate decision of whether to search for the ancestral type in the agaric line or in the gastromycete series. As is well known, Smith and Singer in their studies on the gastromycetes emphasized their belief in the gastromycete ancestry of these fungi. It was their opinion that *Gastroboletus* was the probable immediate ancestor, and that closely related to this were *Truncocolumella* and *Chamonixia* and close to these latter groups was *Rhizopogon*. In the case of Smith, at least, more recent publications have indicated a change of outlook in this regard.

I confess to finding difficulty in supporting the gastromycete theory of the origin of the boletes. It appears to me that the gastroboletes represent the first probable step in a degeneration or simplification of an evolutionary series arising from the boletes. As mentioned above, the gastroboletes show two major trends diverging from the boletes. One is the already emphasized absence of forcible spore discharge, and the other is a pronounced tendency toward a hypogaeous habit. Along with the loss of forcible discharge of spores is the often strong disorientation of the tubulose gleba. In defense of my concept that simplification has occurred here, it seems to me that a simple change, perhaps a single gene mutation as observed by McKnight in his studies on a species of *Psilocybe*, could accomplish the initial alteration. It also appears credible that after having lost the ability to discharge spores and consequently losing the advantage of any selective value in the vertical orientation of the gleba, the "tubes" eventually could have become highly disorganized and irregular in orientation. Apparently associated with the glebal disorganization was the tendency for the peridium to enclose the gleba. There are some species of *Gastroboletus* in which the gleba seems strongly exposed but others in which it is often almost completely covered either with a continuation of the pileus peridium or by a highly differentiated membrane reminiscent of the type often seen covering the hymenophore of *Boletus edulis*.

As indicated above there is a definite trend toward a hypogaeous habitat. Both Trappe and I have observed that several of these fungi found at higher altitudes are distinctly hypogaeous and often are only slightly erumpent and never become fully elevated from the soil at maturity. Perhaps an indication of an independent trend toward enclosure by the peridium and a hypogaeous habitat can be detected in a species of *Tylopilus* found on the western coast of the United States. *T. humilis* is a very rare species and the only representative of that genus known from this area. The fruit bodies are definitely gastroid in appearance, exemplified by a poorly developed stipe and an irregular disposition of the epicutis. Furthermore, this species shows strong tendencies to become hypogaeous. It has not yet, however, lost the ability to forcibly discharge basidiospores.

An evaluation of the characters of some of the new species of *Gastroboletus* seems to lend further support to the ideas stated above. Trappe and I have found species of gastroboletes in the mountains of the western United States closely paralleling the microscopic structure of *Suillus*; others which show features strongly reminiscent of the *Boletus edulis* stirps; an additional species appearing close to the xerocomoid species, and, of course, *Gastroboletus turbinatus*, which is apparently closely related to the *Boletus luridus* complex. In addition, Mazzer and Smith have found *G. scabroides*, which is obviously closely related to *Leccinum*. It seems almost too much to believe that each of the groups of boletes mentioned above had a separate origin. On the other hand, it is conceivable to me that each of these groups could have given rise to the similar-appearing gastrobolete through a simple set of mutations or some other genetic phenomenon.

The connection between *Gastroboletus* and other gastromycetes does not seem as clear-cut or obvious to me. Whether any of the other taxa that were mentioned above might have been derived from

*Gastroboletus* is an open question. There are, however, sufficient similarities to warrant careful consideration. As has been previously indicated, the next genus in the possible evolutionary sequence after the gastroboletes is *Truncocolumella.* It appears to me that if this small group of fungi is closely related to *Gastroboletus* then it has suffered considerable change in its morphology. It is no longer at all suggestive of a bolete in gross appearance and it is necessary to pay close attention to anatomy to detect any remaining similarities. There is only the suggestion of a stipe, the gleba is distinctly alveolate to tubulose, and the trama of the glebal plates is distinctly boletoid. Further similarities are seen in the spores and color changes of the flesh. On analysis, the characters of this group seem to show little evidence to support the concept that this genus might have given rise to *Gastroboletus* or that it was derived from it. Similar observations can be offered for *Chamonixia*, another genus of more-or-less hypogaeous gastromycetes which possess certain characters perhaps interpretable as possibly indicating affinities with the boletes. As was the case with *Truncocolumella*, these features are once again found only in the microscopic anatomy of the fruit bodies. Smith and Singer seemed to believe that this genus was approximately at the same level of evolution as *Truncocolumella*, and they did not suggest that one was derived from another. It was inferred by them that because of the roughened spores of *Chamonixia* it might be involved in the ancestry of some of the rough-spored bodies.

Finally, the group of puffballs that appear to be farthest removed, yet perhaps still related to the boletes, is *Rhizopogon.* Any familiarity with this group of fungi shows that there is little or no evidence of a stipe, that the gleba is finely lacunose, that the gasterocarps are almost totally hypogaeous, and that the major microscopic features indicating possible affinities with the boletes are the basidiospores and the arrangement of the trama of the glebal plates, which is also reminiscent of the boletes. From the foregoing discussion I hope that it is apparent that I do not question the affinities between the three or four different genera of puffballs. Nor do I question the affinities of these groups to the boletes. I do, however, question the proposed direction of evolutionary stages and prefer to believe that they have been derived from rather than precedent to the boletes.

It seems reasonably safe to me to assume that if the boletes were not derived from a gastromycete ancestral form, then the ancestor should be searched for in some other group of fleshy fungi. I do not feel competent, based upon my personal observations, to do more than suggest a possible ancestor. It even seems highly probable to me that the ancestral type which gave rise to the boletes may have long ago become extinct or, perhaps, the magnitude of difference between the two groups as known at present has become so great that it is impossible to detect the relationships. Realizing the vast number of differences apparent at the present time and without suggesting any close affinity among present-day species, it seems possible that the ancestral type might have been similar to that we presently see in the cantharelloid fungi. It seems more satisfactory to me to search for an ancestral type in a group in which there are some indications of similar hymenophoral types, however primitive. Why isn't it possible that the early ancestral types might have had an hymenophore with similar configuration to that of the present-day cantharelloid group which, as development continued, produced the fleshy-poroid fungi on the one hand, and the lamellate fungi on the other? Even though there are certain readily apparent similarities between the boletes and such mushrooms as those in the Gomphidiaceae and Paxillaceae, I do not feel that they represent any ancestral line in conjunction with the boletes.

If we assume that the ancestral type of fungus giving rise to the boletes had a poroid-lamellate hymenophore, then it appears to me that the suilloid species of boletes are the most primitive. I believe this because several species in this group show a distinct tendency to develop very large, irregular pores. In fact, there are several species on the west coast of North America in which the pores are so large as to appear distinctly lamellose. In my opinion, further support in this contention is found in *Suillus* subg. *Boletinus* where, particularly in *S. decipiens*, there is a definite tendency to become lamellose or irregularly poroid. There are numerous other features found in *Suillus* which seem to lend weight to their primitive nature. There are, for example, no large, truly massive basidiocarps as found in most other genera; the spores are typically smaller and often of a different shape than in most other boletes. The hymenophore is frequently decurrent, a characteristic not commonly found in other species. In fact, this group of fungi, namely the genera *Suillus* and *Fuscoboletinus*, appears to be sufficiently distinct that I am more than a little tempted to suggest that they be placed in their own family.

It is not at all clear to me how the other groups of boletes might have arisen. Perhaps the answer is found in such species as *Boletus piperatus* or related species in which there are definite similarities to *Suillus* on the one hand and *Boletus* on the other. If, however, as most recently suggested by Singer, the gyroporoid boletes represent the most primitive line of these fungi, then the ancestor may not have been a suilloid type as known at present at all, but possibly an ancestral type similar to that which gave rise to *Suillus*. Regardless of the ancestor, those fleshy poroid fungi representing the genus *Boletus* and related genera seem to me to constitute a rather homogeneous group and, in my opinion, belong to one family. This family contains the largest number of genera and species and in all probability includes the most advanced of the boletes. Some attention will be given below to the possible affinities within the group.

So far nothing has been said concerning the rough-spored group of boletes. Presently two genera, *Strobilomyces* and *Boletellus*, are recognized and constitute the Strobilomycetaceae. There seems little doubt in my mind that the genus *Strobilomyces* is highly distinct and unique. The numerous characteristics which attest to this concept have already been elaborated. I strongly agree that this genus belongs in its own family, but I am unable to suggest any family tree for its evolution. It does not appear to me to be at all closely related to any of the other boletes in which the spore wall is roughened.

The genus *Boletellus*, on the other hand, appears to me to be a rather heterogeneous assemblage of species. As originally conceived, this genus was to include those boletes, other than *Strobilomyces*, which had roughened spores. It is because of the variety of types of ornamentation of the epispore that the group appears to be heterogeneous. It appears to me that there might be some basis for abandoning the genus and distributing the species among genera in the Boletaceae. Evidence for close relationship between these rough-spored species and members of the Boletaceae is seen in *Boletus mirabilis*, a species which rather clearly shows characters often noted in species of *Boletellus*. These characters include such features as the stipe which is often alveolate at least at the apex, the pileus which is characteristically almost globose and small in relation to the size of the stipe, and spores which, according to Singer, are often obscurely roughened.

Finally, a few comments regarding the possible relationships of the different taxa within the family Boletaceae—other than *Suillus* and related genera. As has been noted, *Gyroporus* has been generally considered the most primitive member of this group. The basis for such a supposition is found in characteristics of the hymenophore, the relatively small, pale colored spores, the mycorrhizal associates, and the relatively small size of the carpophores.

If we assume that *Gyroporus* represents the ancestral type for the family Boletaceae, then we must assume that other members have been derived either directly or indirectly from it. There do not appear to exist any strongly intergrading forms which connect the groups, at least in my opinion. This statement is not intended to refute the above concept but rather to imply that the possible intergrading forms have either become extinct or are so different that they can't be recognized at present. It is generally felt that the genus *Boletus* constitutes the central group or core of the family. It is the group showing the greatest diversity in form and other characters. It is also by far the largest genus in the group, and undoubtedly there are still many undescribed species. There appears to be very active evolution occurring in *Boletus*, resulting in many distinct stirpes or related units. Obviously, all groups have not maintained the same rate of change and we find some appearing considerably more distinct than others. It seems apparent to me that such taxa as *Tylopilus* and *Leccinum* have evolved to the point where they are amply distinct and are readily recognized, although other lines or groups have not yet reached this stage in their divergence from the parent stock. Included in this group would be such taxa as *Porphyrellus* and *Xanthoconium*.

In summary it seems possible, first of all, that the boletes arose from an ancestral type in which the hymenium was of the poroid-lamellose type as seen in the present-day cantharelloid fungi. I feel that a gastromycete series was derived from the boletes rather than ancestral to them. I believe that there are three major groups of boletes today—the *Suillus* and *Fuscoboletinus* group which possibly represent a distinct family; the Strobilomycetaceae with a single genus, *Strobilomyces*; and the large family Boletaceae which has numerous genera and species and seems to form the core of the fleshy poroid fungi. Finally, I believe that the *Suillus* line of boletes represents the most primitive group, but that within the Boletaceae possibly *Gyroporus* represents the prototype.

## Discussion

Donk: You mentioned the Cantharellaceae as possible ancestors of the Boletaceae. Was it by design that in this connection you showed pictures of the *Gomphus* group? Would you be prepared to replace this group by *Cantharellus cibarius* or some other element related to this latter species? There is an appreciable difference between *Gomphus* and *C. cibarius*.

Thiers: No, it was not by design. I have not seen enough to really say. I cannot point a finger to any certain species and say it is the ancestor.

Donk: The hymenium in *Gomphus* becomes lamellar (or rather folded) not because of a special structure that acts as hymenophore, but because the surface of the hymenium increases by the rapid addition of new basidia. This is an important difference from the Boletaceae because in the latter the dissepiments are formed as definite structures preceding the formation of the hymenium. There is perhaps only a superficial resemblance between the lamellate boletes and the more-or-less lamellate Cantharellaceae.

Thiers: As I pointed out, I was not saying that any of the present-day species represent a possible ancestral type: perhaps we should look for an ancestral type in that *type* of fungus.

Olexia: Exactly where do you put *Phylloporus* in your scheme?

Thiers: Frankly, I do not know where to put it.

Smith: On this question of *Phylloporus*, one of my collectors in northern Michigan collected a *Boletus piperatus* that is also a *Phylloporus*. It has perfect lamellae. The evidence accumulating would seem to indicate that *Phylloporus* can be somewhat safely regarded as a freak—a *Xerocomus* that has developed lamellae. This type of modification of the hymenophore does not necessarily involve much of a change in the genome.

Petersen: Has not the old *Boletinus merulioides* or *B. porosus* also been put in *Phylloporus* by some people? Is this valid?

Smith: It was put in *Paxillus* in the old days but I think Watling is accepting it as *Phylloporus* on the basis of an extremely sublamellate habit.[1] Of course, this will introduce another extraneous type into *Phylloporus*, which will soon become just a form-genus.

DONK: You mentioned *Gyroporus* as a possibly simple bolete. In this connection I would like to draw attention to some of the polypores that are terrestrial and perhaps not mycorrhizal: to *Polyporus* [*Albatrellus*] *confluens*, *P.* [*A.*] *ovinus*, and similar species occurring in North America that show a blueing of the flesh. Have you ever compared them with *Gyroporus*?

THIERS: No, I have not.

SINGER: I have compared them, and I could not find any affinity.

HEIM[2]: I was very interested by this communication. Some years ago I was convinced that *Boletus* was a monophyletic group but at the moment I am not so sure. May I use a diagram? Here you see the Agaricales, *Rhizopogon*, the Gasterales, and the Polyporales. There was a day when all mycologists assumed that there was no relation between the polypores and boletes. Then Dr. Singer said *Gyrodon* had certain affinities with the Polyporales and I still believe that. About two years ago, in the rain forests of New Caledonia, I discovered an extraordinary mushroom and published it. It is the genus *Meiorganum*. This is a *Boletus* which is a polypore. It has no mycorrhizal origin, no clamp connections, and the spores are very similar to those of *Gyrodon*. It is impossible to obtain in culture. I think that a genus from Australia, *Campbellia*, and another from South Africa, *Gilbertiella*, are probably nonmycorrhizal, and thus in the direction of *Gyrodon*. We know that several species of *Gyrodon* are mycorrhizal, but that some forms are saprophytic. If I am not sure about this, nevertheless *Gyrodon* is very close to the polypores and *Meiorganum*, which have exactly the same system of tubes as *Gyrodon*. I suppose there is a relation in this series linking the Polyporales with *Gyrodon*, but then what about *Boletinus*, *Porphyrellus*, and so on? And which is primitive? There is always the same question: what is the direction of evolution? Is it from the Gasterales to the Boletales,

[1] The reader is referred to Watling (1969. The genus *Paragyrodon*. Notes Roy. Bot. Gard. Edinburg 29: 67–73), where this species is referred to the genus *Gyrodon* [Ed.].
[2] The reader is referred to fig. 4 in Heim's paper [Ed.].

or the other way round, or from the Gasterales to the Agaricales, as Drs. Singer and Smith think, or inversely? It would be arbitrary to decide!

There is another problem. It is the series of *La Ratia* of Patouillard and *Paxillus sensu* Fayod. Dr. Singer discovered a very interesting mushroom, *Austerogaster*, in the same complex as *Phylloporus* and *Xerocomus*. The problem is the direction of evolution to *Paxillus*. Was it from the Agaricales, from *Xerocomus*, or indirectly from the Polyporales? And what is the relationship between *Xerocomus* and the polypores?

There is a relation between *Boletellus* and *Leccinum*, but *Boletellus* is large and probably heterogeneous. I suppose there are three sections of *Boletellus*, all quite different. *Heimiella* is another closely related genus of boletes from the Pacific, and with several species. It shows an extraordinary thing; a cuticle very near *Leccinum*, a long stipe, and reticulate spores, and some other characters of *Boletellus*. It is a very extraordinary genus and I suppose it is possible to consider *Heimiella* between *Leccinum* and *Boletellus s. str.* There are some other series but I haven't found out the beginning or the end but it is possible to define every series in one sense or another. I think *Porphyrellus* is not a good genus. *Suillus* is primitive, *Tylopilus* is a progressive genus, and *Phlebopus* is highly controversial. The position of *Gyroporus* (which Dr. Singer said is near *Gyrodon*), I suppose, is also progressive. These observations are all in favor of the great complexity of these series and actually I am not sure any longer that the Boletales is a monophyletic group. It is very complicated because the group includes many organisms as does the Agaricales.

And the last point is a reflection on your very interesting remark about certain relationships between the Clavariaceae and Boletaceae. I am very interested in the fact that the ornamentation of the spores of these two groups is about the same. There is also a great similarity of integumentation, as seen with the electron microscope. Certainly this rapport is a theoretical idea but I am sure that there is something to think about.

THIERS: Thank you, sir, I appreciate this. I agree with you concerning *Porphyrellus*.

SMITH: For Dr. Heim's information, I might say that there will be a paper in Mycologia in which two more mycologists dump the genus *Porphyrellus* in the waste basket.[3]

HACSKAYLO: You have been working with Jim Trappe and he is quite interested in the Tuberales and other truffle fungi noted in the ascomycetes to be the only groups that are probably mycorrhizal. You have used the mycorrhizal habit as a criterion. In the Basidiomycetes, *Rhizopogon* is not only mycorrhizal, but hypogaeous. Is there any connection or do you think this is just coincidence?

THIERS: I see no significance in the fact that they are both mycorrhizal. I am not sure that I really accept the idea that the hypogaeous Ascomycetes are the only mycorrhizal groups. Aren't there some of the helvellas that are possibly mycorrhizal?

BURDSALL: From what little knowledge I have, I would say that there are many of the Pezizales which will turn out to be mycorrhizal.

ROGERS: One of our graduate students has just finished a study on *Elaphomyces*, which he does not consider a truffle fungus. It is constantly mycorrhizal with conifers.

HACSKAYLO: But at this point at least we don't know of many of them being proven definitely. This is just guilt by association.

SMITH: I think in view of the work on *Rhizopogon* I might just make a few comments about the genus. At the present time I consider it a rather primitive genus because of the tremendous diversity of spore types, in amyloidity and other characters. This genus must have been evolving for a very long time, at least in the Pacific northwest, because the hymenium basically is what I call a trichodermial type—made up of filaments rather than just end-cells, as I commented briefly the other day. I am skeptical about any real relationship between *Rhizopogon* and the boletes at the present time; I think we can draw a line between that genus and the boletes. There may be a few chemical similarities such as the emulsification of some of the pigments in chloral hydrate, but this needs to be examined further in other groups to see how extensive the character is.

[3] The reader is referred to Smith and Thiers (1968. Notes on boletes—I. 1. The generic position of *Boletus subglabripes* and *Boletus chromapes*. 2. A comparison of four species of *Tylopilus*. Mycologia 60: 943–954) [Ed.].

ROLF SINGER

*Research Associate*
*Field Museum of Natural History, Chicago, Illinois*

# A REVISION OF THE GENUS MELANOMPHALIA AS A BASIS OF THE PHYLOGENY OF THE CREPIDOTACEAE

The genus *Melanomphalia* was first published by Christiansen (1936), who thought it to be related to *Gomphidius*. The topotype was revised by Singer (1955) after examination of authentic material of the type species of *Melanomphalia* which was shown to have no affinity to the Gomphidiaceae. In the same paper Singer (1955) gave a more comprehensive analysis of the anatomy and the spore characters of *M. nigrescens* and concluded that *Melanomphalia* was a good autonomous genus similar to *Cuphocybe* Heim, and therefore believed to belong to the Cortinariaceae rather than the Gomphidiaceae.

Somewhat earlier, Singer (1948a, 1948b) had separated a section *Thermophilae*, based on the Floridian species *T. thermophila*, from the typical *Tubariae* (section *Tubaria*). Section *Thermophilae* differed from section *Tubaria* in bearing spore ornamentation of type XI (Singer, 1951), which was similar to that of many *Crepidoti* such as *C. nephrodes*. Later, Singer and Digilio (1952) described a second species of section *Thermophilae*.

Since then, I have had an opportunity to observe a number of species from New Guinea and South and North America which seemed to belong either in *Melanomphalia* (like *Inocybe platensis* Speg.) or in *Tubaria* sect. *Thermophilae*. Observations of these specimens gradually narrowed the gap between these two taxa to the point where it appeared necessary to combine them into a single generic taxon, the emended genus *Melanomphalia*.

## Position

I am now convinced that *Melanomphalia* does not belong in the Cortinariaceae. There are certain similarities between *M. nigrescens* and the genera *Descolea* and *Cuphocybe*, although the latter may actually be identical with *Descolea*. *Descolea* has been recollected on numerous occasions and has been shown basically to have the characters of the Bolbitiaceae. The genus is not ectotrophically mycorrhizal and is strongly related to *Pholiotina*. On the other hand, *Melanomphalia* has a completely different anatomy of the epicutis. Several species of *Melanomphalia* have pleurocystidia while others do not have well-defined cheilocystidia. The ornamentation of their spores is strongly reminiscent of the spore ornamentation of certain *Crepidoti*, a type of ornamentation not known in either Bolbitiaceae or Cortinariaceae, only in the Crepidotaceae, few Tricholomataceae (*Fayodia*), few Strobilomycetaceae, and some Hymenogastrineae, especially "Secotiaceae." Since a relationship with both *Fayodia* and especially *Boletellus* is out of the question, considering the hymenophoral structure and the spore pigment, I must now consider *Melanomphalia* as belonging in the Crepidotaceae. This makes the distinction between Cortinariaceae and Crepidotaceae much more clear-cut. Wherever spore ornamentation is present in the Cortinariaceae, it is always an exosporial ornamentation. Spore ornamentation in the Crepidotaceae, where present, is perisporial and episporial, or only episporial in the sense that the ornamentation results from the insertion of heterogeneous matter, in the form of short columns, dots, or short ridges, in the episporial layer of the spores from where this heterogeneous material may or may not project, but it is not a case of a heterogeneous layer *superimposed* (although fractured), as in the case of the exosporial ornamentation.

The heterogeneous spore wall of the Crepidotaceae, especially of *Melanomphalia*, often gives the impression of punctate spores when these are seen not in optical section and not with a good immersion lens. Under these circumstances, they can easily be confused with spores of cortinariaceous genera such as *Cortinarius* and *Galerina*. The fruit bodies of these three genera often appear very similar macroscopically, so that they might be and probably have been confused. How can the *Melanomphaliae* be separated from the similar cortinariaceous genera? The species of *Melanomphalia* are not ectotrophically mycorrhizal as are those of *Cortinarius*. This circumstance makes it possible to suspect in the field that some collections, otherwise much like *Cortinarius*, *Inocybe*, or *Alnicola*, are actually none of these. Moreover, they could not be *Tubaria* or *Simocybe* which they also resemble because of the heterogeneous episporium of their spores. A number of species of *Melanomphalia* are even constantly lignicolous, or grow in typical anectotrophic stands of trees. Furthermore, the range of colors of spore prints in the genus, although often individually comparable with those of *Cortinarius*, is much wider. This is exemplified by *M. nigrescens* (where the spores are olive-fuliginous) and *M. omphaliopsis* (where they are milk-coffee colored—"Mosul" M&P). Finally, those species of *Melanomphalia* that for any other reason come close to *Cortinarius* in their appearance possess cheilocystidia quite different from those observed in some *Cortinarii*. No species of *Melanomphalia* has any affinity with *Cortinarius*.

If *Melanomphalia* is compared with *Galerina*, with two notable exceptions (*M. columbiana* and *M. emarginata*), the fruit bodies are quite different in appearance. Microscopically the genera are easily distinguished: in the species of *Galerina*, either the spores have a distinct plage area in the suprahilar region, or the hyphae of the carpophores have no clamp connections, the only exceptions being species of *Galerina* whose spores are devoid of ornamentation. On the other hand, the hyphae of all species of *Melanomphalia* have clamp connections, and no spores have a true "plage." Only on the spores of one species, *M. vernifera*, have I observed areas with homogeneous walls which may or may not be located suprahilarly but certainly are not restricted to this area. These nonornamented regions of the spore wall are not marked by a semicircle of ornamentation material and cannot be compared with or homologized to the plage of the *Galerina*. A similar situation has been observed in *Chromocyphella*, where the spores have ornamentation like that of *Melanomphalia* and also show wall areas without ornamentation which have been mistaken for plage. Incidentally, *Chromocyphella* is a typical crepidotaceous genus, but cyphelloid and without a hymenophore.

The habitus of *Melanomphalia* varies considerably. The stipe is almost exclusively central, however, never strongly eccentric, lateral, or reduced. The separation from *Crepidotus* is therefore quite easy, and the morphological hiatus between these two genera is so strong that no discussion of the limits is necessary in spite of their phylogenetic affinity. The same is true for the separation of *Melanomphalia* from *Tubaria*, since the latter genus always has homogeneous spore wall, even where this, as in *T. dispersa* (Pers.) Sing. [= *T. autoch-*

*thona* (Berk. & Br.) Sacc.] is wrinkled-uneven, and perisporial ornamentation is unknown in *Tubaria.*

The following is the revised generic description of *Melanomphalia.*

## MELANOMPHALIA CHRISTIANSEN
FRIESIA 1: 288. 1936.

Habit often omphalioid, or of a small *Paxillus, Cortinarius, Galerina, Inocybe,* with at first straight or incurved margin, generally rather thin-fleshy to moderately fleshy, lamellae at first ascendant or not, decurrent or adnate to narrowly adnexed. Stipe central, with basal mycelium, with or without a veil. Spores with a faintly heterogeneous to strongly heterogeneous episporium (ornamentation type XI) and often also a perisporial ornamentation, short ellipsoid to fusoid, without a true "plage," with or without an apical discontinuity but never with a strongly truncate germ pore, inamyloid. Cheilocystidia present, but sometimes scattered and basidiomorphous, more frequently dense and numerous, and of different shapes, clavate, fusoid, ampullaceous, ventricose-constricted, even capitate; pleurocystidia more frequently absent than present, but at times very prominent and even metuloid in one species. Clamp connections present. Epicutis an ixocutis, a cutis, or with bunched together dermatocystidioid terminal members of a trichodermium or else with ascendant terminal members of a cutis. Hymenophoral trama regular. Spore print varying from pale ochraceous to ochraceous-brown to olive fuliginous (for example "Mosul," "Yucatan," "ginger" of Maerz & Paul but also much paler and much darker). On the ground, on earth and on debris, but also frequently on dead woody matter, trunks, culms, or among mosses, in and outside the forest. North and South America, Europe (here probably adventitious), New Guinea, probably widely distributed, the largest number of species known from temperate and tropical South America.

Development of the carpophores and nuclear cytology not studied.

### KEY TO THE SPECIES

A. Epicutis gelatinized; pleurocystidia absent or inconspicuous, lamellae generally sinuate to adnexed, never broadly adnate and never strongly or distinctly decurrent. On leafmold, on earth or volcanic ashes, not lignicolous. Temperate species, of the southern hemisphere.

B. Cheilocystidia well differentiated and numerous, vesiculose; spores 10–12.5 x 5–6.5 μ; fall-fruiting. . . . 1. *M. viscosa* Sing.

B. Cheilocystidia rather scarce; generally spring-fruiting. . . C

C. Spores 10–11–(11.5) x 6–6.5–(7) μ; ixocutis poorly developed; pileus papillate. . . . . . . . . . . 2. *M. vernifera* Sing.

C. Spores 6.5–7 x 3.5–4.5 μ; ixocutis well developed underneath a velar layer; pileus obtuse. . . . . 3. *M. cortinarioides* Sing.

A. Epicutis not or scarcely gelatinized, or else lamellae decurrent or pleurocystidia very conspicuous. . . . . . . . . . . . D

D. Pleurocystidia conspicuous, ampullaceous, yellow, thick-walled (metuloid). Spores 10.5–12 x (5)–7.8–8 μ.
New Guinea. . . . . . . . . . . . . 4. *M. dwyeri* Sing.

D. Pleurocystidia different or absent. . . . . . . . . . E

E. Pileus lilac, then ochraceous; lamellae lilac, later rusty. Tropical South American species. . . . . . 5. *M. universitaria* Sing.

E. Pileus, even in youth, without any lilac tones. . . . . . F

F. Lamellae sinuate-decurrent, adnato-decurrent, broadly adnate, or plainly decurrent. . . . . . . . . . . . . . . . G

G. Spore print olive-fuliginous. . 6. *M. nigrescens* Christiansen

G. Spore print pale ochraceous to rusty-brown. . . . . H

H. Spores ellipsoid to short ellipsoid, 7–8 x 4.7–6.2 μ. Tropical North Argentina. . . . . . 7. *M. crocea* (Speg.) Sing.

H. Spores larger than 8 μ, and generally more elongated, often subfusiform. . . . . . . . . . . . . . . . . I

I. Lamellae deep gray, extremely broad (as broad as they are long); epicutis of interwoven chains of broad, short, voluminous cells 30–44 x 17–25 μ. Temperate species, growing on the earth in South Chile. . . . . . 8. *M. platyphylla* Sing.

I. Lamellae some other color, narrow to broad; structure of the epicutis as indicated above, or different. . . . . . . J

J. Stipe white or whitish, more rarely with a citrinous shade; spore ornamentation easily observed, of the *Crepidotus nephrodes* type, perisporium little developed. In tropical and subtropical forests and hammocks, in sandy soil, often attached to rotting leaves, fruits, etc. Basidia 1–2–3–4-spored; veil none. . . . 9. *M. thermophila* (Sing.) Sing.

J. Stipe more-or-less concolorous with the pileus. . . . K

K. Pileus characteristically umbonate. Veil present. South American species. . . . . . . . . . . . . . . L

L. Carpophores with the habit of *Galerina*; pileus 2–5 mm; stipe 27–40 x 1–1.5 mm.
Colombian species. . . . . . 10. *M. columbiana* Sing.

L. Habit of a small *Paxillus*; carpophores larger. Argentine species. . . . . . . . 11. *M. platensis* (Speg.) Sing.

K. Pileus not umbonate or papillate.
Veil present or absent. . . . . . . . . . . . . M

M. Pileus convex, not umbilicate. North American species with or without veil. . . . . . . . . . . N

N. Veil absent; lamellae narrow to moderately broad, dried "rose stone," "Kermanshaw," "fawn," "coffee," and close to crowded. . 12. *M. alpina* (A. H. Smith) Sing.

N. Veil present; lamellae rather broad to broad, dried "sepia," and close to subclose. . 13. *M. smithii* ad int.

M. Pileus when mature deeply and rather acutely umbilicate, habit of *Omphalina*. Veil none. Species of Northern and Central Argentina and Chile, perhaps also in Florida, U. S. A., growing in exposed habitats in the subtropical and in wooded areas of the semi-arid zones, on the ground in parks, on lawns, in semi-arid woods, generally gregarious. . . . . . . . 14. *M. omphaliopsis* (Sing.) Sing.

F. Lamellae adnexed to deeply sinuate or emarginate and narrowly adnexed. South American species. . . . . . . . O

O. Stipe not broader than 1.5 mm; pileus less than 7 mm broad, umbonate or papillate; veil present but somewhat fugacious; appearance of *Galerina*. . . . . . . . . . . . . P

P. Temperate species from South Chile; spores 8–8.5 x 6–6.5 μ; cheilocystidia clavate, basidiole-like, some with a slight constriction; lignicolous. . . . . . 15. *M. emarginata* Sing.

P. Species of the tropical belt but growing in the montane zone; spores 8.5–10.5 x 4.5–6 μ; cheilocystidia ampullaceous, some with a claviculate to capitate apex, some with attenuated or cylindric thin apex; bryophilous. . 10. *M. columbiana* Sing.

O. Stipe broader than 1.5 mm; pileus broader than 7 mm, umbonate or not; veil present or absent; appearance of *Inocybe* or *Gymnopilus*, or *Psilocybe*. Growing in temperate South America (if in North America, lamellae not deeply sinuate, see N above). Lignicolous. . . . . . . . . . . . . . . . . . Q

Q. Veil present; appearance of *Inocybe*; pileus umbonate; spores 8.5–9 x 4.5–5 μ; cheilocystidia scattered and basidioliform. . . . . . . . . . . . 16. *M. inocyboides* Sing.

Q. Veil extremely scanty or absent; appearance of a thin-fleshy *Gymnopilus*; pileus convex, then applanate or very slightly depressed in the center, without an umbo;
spores 7–8.5 x 5–6μ; cheilocystidia numerous
although not strongly projecting. . . 17. *M. pacifica* Sing.

(Veil scanty; appearance of *Psilocybe*; pileus papillate; spores 10–11.5 x 6–7 μ; terricolous. . . . . . . (see C above.)

## Description of the Species[1]

### 1. *Melanomphalia viscosa* Sing. *spec. nov.*

Pileo castaneo, ad marginem leniter dilutiore, in siccis "caldera" M&P in centro, "Saratoga" M&P ad marginem, per quintam partem radii pellucide striato, glabro, levi, viscoso, convexo, interdum umbone exiguo praedito, 24–30 mm lato. Lamellis fusco-griseis, ad aciem

[1] The colors given in M&P terms refer to the first edition of Maerz and Paul (1930). Some species are not illustrated in the present revision of the genus since they are illustrated in two forthcoming papers (Acta I.M.U.N.P, Recife, in press; Micoflora Australis, Ed. Univers. Chile, in press) or were illustrated in previous papers (Singer, 1955). The key is not thought to be fully representative of the natural affinities of the species, but rather stresses the outstanding characters and those characteristics that make it easy to identify the species. A subdivision of the genus in sections has not yet been attempted.

flavis, aut colore coffeaceo (cum lacte) et ad aciem pallide flavis, latis, ventricosis, confertis, sinuato-adnexis. Sporis in cumulo ferrugineis. Stipite ad apicem grisello-fusco vel grisello, basin versus brunneolo-pallido, in parte centrali et inferiore brunneo-armillato, haud viscido, apicem versus attenuato, 37–56 x 5–6 (in parte inferiore), x ± 3 mm (ad apicem). Contexto inodoro. Sporis 10–12.5 x 5–6.5 μ, interdum usque ad 8 μ latis in positione frontali, ellipsoideis, interdum apicem versus attenuatis, sed semper obtusis et rotundatis, interdum mucronatis, punctatis ex ornamentatione episporiali typi XI, perisporio vix vel bene evoluto et interdum laxe involucrante pallide brunneolo saepe fragmentato praeditis, applanatione vel depressione suprahilari praeditis, sed disco suprahilari non ornamentato destitutis, ferrugineo-ochraceo-brunneo-punctatis in fundo subpallidiore, membrana pro ratione tenui instructis, poro germinativo destitutis,

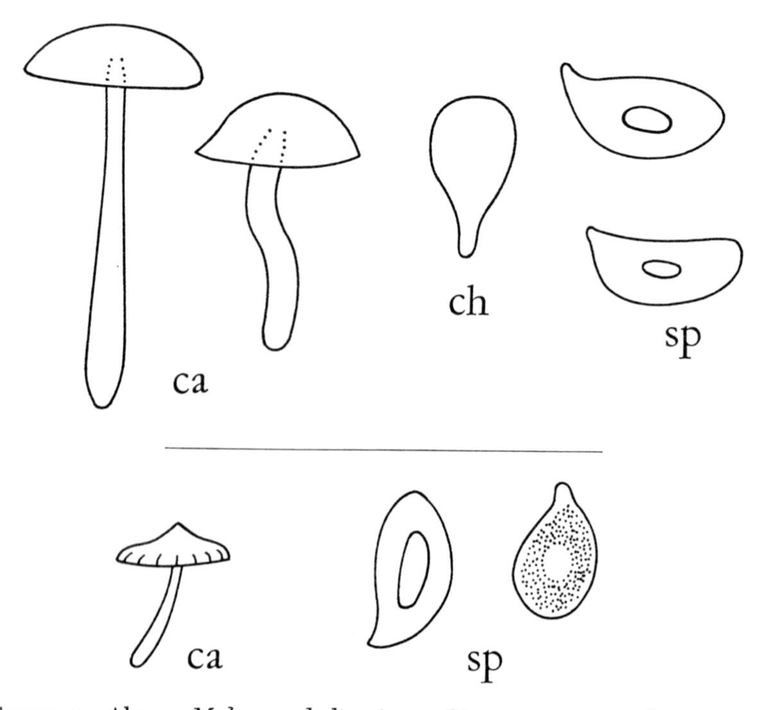


FIGURE 1. Above: *Melanomphalia viscosa* Sing., ca = carpophores × 1; – ch = cheilocystidium, × 1000; – sp = spores in outline × 2000. Below: *Melanomphalia vernifera* Sing., ca = carpophore, × 1; – sp = spores in optical section, × 2000.

membrana (sine perisporio) 0.3–0.5 μ crassa. Hymenio e basidiis efformato et cheilocystidiis marginato. Basidiis 28 x 8–8.5 μ, (2–3–)4–sporis. Cheilocystidiis 20–26 x 9.5–13 μ, vesiculosis et saepe pedicellatis vel clavatis, late rotundatis ad apicem, integris et levibus, tenuitunicatis, inclusionibus visibilibus destitutis, numerosis. Hyphis fibulatis. Epicute gelatinosa, ex hyphis pro ratione latiusculis (2–8μ) elongatis, repentibus, subparallelis vel intertextis, ferrugineo incrustatis pigmento, paucis vix incrustatis efformata; dermatocystidiis nullis; supra epicutem strato intermittente, probabiliter velari, visibili, ubi hyphae minus gelatinisatae et fortius incrustatae (2–12 μ diametro) sunt. Ad terram cinerosam et arenosam sub Nothofagis dombeyi humilibus extra silvam gregatim, autumnalis. Typus a R. Singer (M 6857) in Chile australi lectus et in herbario SGO conservatus est.

Material studied: CHILE: Volcán Osorno, west slope, at 1,000 m altitude, under the highest scrub trees (*Nothofagus dombeyi*), on soil mixed with volcanic ashes, 12–V–1967, Singer M 6857 (SGO), typus.

### 2. *Melanomphalia vernifera* Sing. *spec. nov.*

Pileo castaneo, glabro, subviscido pellucide striato, convexo, subacute papillato, 17–18 mm lato. Lamellis argillaceo-brunneis, ventricosis, sat latis, subconfertis, adnexis. Stipite pallido e strato sericeo tenui, probabiliter velari supra trama brunneum, aequali vel subaequali, cc. 22 x 2 mm. Contexto brunneo, inodoro. Sporis 10–11–(11.5) x 6–6.5–(7) μ, ellipsoideis vel ellipsoideo-subamygdaliformibus, raro subovatis, applanatione suprahilari praeditis vel destitutis, haud mucronatis vel minutissime mucronatulis e perisporio leniter angusteque protracto, poro germinativo destitutis vel rarius membrana in apice distali coroniformiter anguste perforata vel truncata, sed epi- et endosporio vix discontinuis, episporio materia heterogenea perforatis et punctatim vel lineariter ornamentato ferruginee in episporio pallidiore ita ut sporae in circumferentia asperulae videantur, perforationibus (ornamentatione) 0.7–1 μ altis, membrana tota 0.8–1.3 μ crassa, endosporio pallidiore, guttula elongata interna ornatis. Hymenio e basidiis efformatis, eis 24–34–(37) x 9–11.5–(12.5) μ tetrasporis, plerumque hyalinis, clavatis vel leniter constrictis. Cheilocystidiis subnullis vel paucis. versiformibus, 20–29 x 7–9 μ, plerumque ventricosis, interdum mucronatis vel constrictis. Pleurocystidiis nullis. Hyphis inamyloideis, fibulatis, in tramate hymenophorali regulari subhyalino vel ochraceo-hyalino elongatis, usque ad 11 μ latis. Hymenopodio ex hyphis axialibus hyalinis vel subhyalinis haud gelatinosis, 2–2.5 μ crassis efformato. Subhymenio subcellulari, ex elementis exiguis (usque ad 5 μ diametro) hyalinis efformato. Epicute pilei leniter gelatinosa, interdum intermittente, hyalina, hyphis glabris filamentosis 2–4.5 μ crassis, repentibus vel rarius ad apicem subascendentibus efformata. Hypodermio haud gelatinoso, ex hyphis latis cutem efformantibus, saepe breviusculis, brunneolis (ita ut in tramate pilei) consistente. Ad terram in silva anectotrophica, vernalis. Typus a R. Singer (M 7523) in Chile lectus et in SGO conservatus.

Material studied: CHILE: Osorno: Lago Puyehue, 14–IX–1967, under *Aextoxicum, Persea, Drimys, Gevuina*, Myrtaceae, R. Singer M 7523 (SGO), typus.

### 3. ***Melanomphalia cortinarioides*** Sing. *spec. nov.*

Pileo coriaceo-brunneo ("Arabian br" M&P), hygrophano, dehydratatione pallidiore, per medium radium pellucide striato, e velo subtili pallido in margine fibrilloso-pruinoso, convexo, obtuso, 6 mm cc. lato. Lamellis brunneolis ("Pekinese" M&P), mediocriter latis vel latiusculis, confertis, adnexis. Stipite dilute brunneolo, e velo ex integro sericeo-subpruinato, subaequali, cc 17 x 2 mm. Contexto brunneolo, inodoro. Sporis 6.5–7 x 3.5–4.5 μ, ellipsoideis, depressione suprahilari destitutis, punctatis, episporio heterogeneo typi XI, ochraceo-brunneis, guttula interna olei elongatula repletis, perisporio paullum manifesto. Hymenio e basidiis et perpaucis cystidiis inconspicuis sparsis, filamentosis, hyalinis efformato. Basidiis 20–21 x 6–7 μ, hyalinis, clavatis, fibulatis, paucis bisporis, plerumque tetrasporis. Acie lamellarum homomorpha. Hyphis fibulatis, in tramate pilei fortiter pigmento ferrugineo-ochraceo-brunneo incrustatis, incrustationibus granularibus vel zebraeformiter armillatis, haud gelatinosis. Epicute pilei pigmento minus incrustata, ex elementis longe filamentosis, gelatinosis, multis spiraliter curvatis, 1–3 μ latis efformata. Supra epicutem nec non in epicute partim immersa elementa velaria haud gelatinosa visa quae 8–10 μ crassa, multi-septata, hyalina, membrana ± 0.7 μ crassa praedita et levia sunt. Ad folia putrescentia acervata sub Nothofago dombeyi, Berberide darwinii, Chusquea couleu, Laurelia serrata, Saxegothaea, locis humidis et umbrosis, solitario.

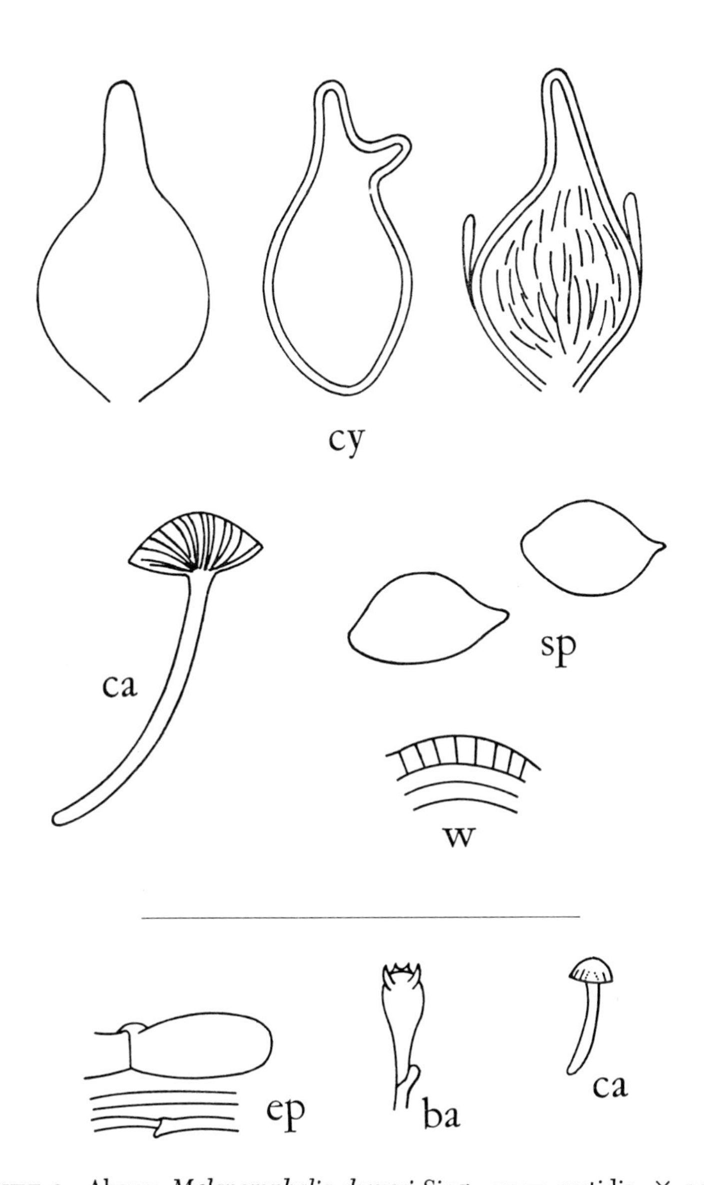


FIGURE 2. Above: *Melanomphalia dwyeri* Sing., cy = cystidia, × 1000; – ca = carpophore × 1, (from dried specimen); – sp = spores, outline, × 2000, w = detail of wall structure of spore.
Below: *Melanomphalia cortinarioides* Sing., ep = fragment of epicutis, × 1000; – ba = basidium, × 1000; ca = carpophore, × 1.

Typus a R. Singer (M 6054) in Argentina lectus et in BAFC conservatus est.

Material studied: ARGENTINA: Neuquén: Los Cántaros, 10–XI–1966 (verne), R. Singer M 6054 (BAFC), typus.

### 4. *Melanomphalia dwyeri* Sing. *spec. nov.*

Pileo brunneo, apparenter glabro, ± 20 mm lato. Lamellis fuscis, angustis vel moderatim latis, confertis, sub lente flavo-maculatis. Stipite brunneo (e velo?), subfloccoso, aequali, cc. 50 x 2 mm. Contexto tenuiusculo. Sporis 10.5–12 x (5)–7.7–8 μ, depressione et applanatione suprahilari nullis, poro germinativo nullo, sed ad apicem distalem membrana absentia ornamentationis discontinua instructis, ornamentatione typi XI e spinulis tenuissimis insertis in episporio pallide brunneolo atro-porphyrio-brunneis ornatis ex ea re punctatis, ovoideis vel fusoideo-ellipsoideis. Hymenio e basidiis et metuloidibus efformato. Basidiis hyalinis, tetrasporis. Cystidiis metuloideis ad latera aciem que lamellarum praesetibus, numerosis, 42–50 x 12–30 μ, subtus ventricosis, apice angustatis, interdum bifurcatis vel subcapitatis, flavido-incrustatis, apice 20–25 μ longo et 4–6.5 μ lato, obtuso praeditis. Epicute ex elementis appressis refringentibus, elongatis, membrana e flavido lutea instructis, paullum, ut videtur, gelatinisatis efformata. Ad terram sub truncis. Typus a R. E. P. Dwyer (no. 1726) in Nova Guinea lectus et in K conservatus est.

Material studied: NEW GUINEA: Wahgi Valley, Dwyer 1726 (K), typus.

This material was sent to Kew in dried condition and was studied by me at the Herbarium of the University of Michigan in 1958. The generic position was then not immediately clear but was first mentioned (as "Agaricales sp.") by Singer (1958b:277). The material was almost entirely blackened when received, but the blackening may be due to the manner of conservation. Later, it was suspected that the species belonged in *Melanomphalia* where it was outstanding because of its cystidia. The communication of 1958 was justified because this material was sent as one of the species causing the hysterical state known as the "Wahgi Valley frenzies."

### 5. *Melanomphalia universitaria* Sing. *spec. nov.*

Pileo lilaceo dein ochraceo-brunneo, squamuloso-granulari, levi, convexo, dein centro depresso, 30–33 mm lato. Lamellis lilaceis ("old lilac" M&P), demum ferruginascentibus, latis, mediocriter confertis, sinuatis et sat profunde decurrentibus. Stipite lamellis concoloribus, subtiliter subtomentoso, cavitie parva excavato, in parte centrali et inferiore latiore, 45 x 6.5 (ad basin) x 4.5 mm (ad apicem). Contexto sapore miti et odore nullo gaudente. Sporis 5.5 –6.5–(9) x 4–4.8–(6) μ, ellipsoideis vel breviter ellipsoideis, juvenilibus interdum subangularibus, inamyloidies, membrana firma sed tenuiuscula, interdum collabente, pallide ochraceo-brunneola vel subhyalina heterogenea, ornamentatione typi XI (spinulis tenuissimis brevissimis vel lieis cretatis, interdum reticulatis immersis episporio haud obscurioribus et hac e re vix contrastantibus punctatulis) instructa praeditis, membrana ± 0.5 μ crassa, perisporio vix visibili. Hymenio e basidiis et cystidiis efformato et cheilocystidiis marginato. Basidiis 24–32.5 x 6–7 μ, clavatis, tetrasporis. Cystidiis et cheilocystidiis aequalibus, ad acies dense aggregatis, 35–76 x 6–16–(19) μ plerumque ventricosis in parte inferiore, apice tenuiore, interdum bifurcato vel leniter constricto subaequali praeditis et sic ampullaceis, membrana firma usque ad 0.5 μ crassa et contentu colloideo nonnihil refringentibus, hyalinis vel pallide brunneolis, fortiter projicientibus. Hyphis fibulatis, inamyloideis, haud gelatinosis, in tramate hymenophorali regulari hyalinis vel pallide brunneolis, leniter subintertextis, diametro admodum variabilibus. Epicute pilei ex hyphis repentibus haud gelatinosis efformata sed hic inde ex ea fascicula hypharum ascendent quarum membra terminalia dermatocystidiformia sunt, 40–60 x 7.5–15 μ, subulatis vel cheilocystidiis similibus. Praeparationibus in KOH pigmentum flavidum exudantibus quod in medio dissolutum est. Superficie pilei KOH ope obscurescente : castanea. Ad frustula lignea putrida. Typus a R. Singer in Brasilia lectus (B 3387) et in BAFC conservatus est.

Material studied: Brazil: Pernambuco: Dois Irmãos, 13–VII–1960, R. Singer B 3387 (BAFC), typus.

The bunches of hyphae which end in dermatocystidia form the granular vestiment of the pileus. The colors as well as the pleurocystidia are characteristic for this species.

### 6. *Melanomphalia nigrescens* Christiansen

Friesia 1: 288. 1936.

This species has been macroscopically described fully and illustrated by Christiansen (1936) and by Lange (1940). The microscopical characters were described by Singer (1955).

### 7. *Melanomphalia crocea* (Speg.) Sing. *comb. nov.*

*Cantharellus croceus* Speg. An. Mus. Nac. Bs. As. 19: 263. 1909.

As for the macroscopical description, we refer to Spegazzini loc. cit., or Saccardo 23: 104.1912. The lamellae are not cantharelloid nor even venose, but rather narrow, and appear to be decurrent in the dried material.

Spores 7–8 x 4.7–6.2 μ, short ellipsoid or ellipsoid, with faint spinules perforating the episporium and a firm nonseparating perisporium, the latter and the tips of the spinules perhaps very slightly amyloid (?), otherwise inamyloid, without suprahilar depression, finely punctulate when seen from above, subhyaline to a very pale rusty in KOH. Basidia 27–28 x 7.7 μ. Cystidia none seen. Cheilocystidia perhaps present, irregularly filamentous. Hyphae with clamp connections, often with orange rusty internal and membranal pigment, inamyloid.

Material studied: ARGENTINA: Salta: Orán, Rio Seco, Spegazzini (LPS), typus. This species has not been seen in fresh condition. The colors and the spore characters should be sufficient to characterize this species as an independent representative of the genus *Melanomphalia.*

### 8. *Melanomphalia platyphylla* Sing. *spec. nov.*

Pileo atro-rubro-brunneo ("Mandalay" M&P), subtiliter fibrilloso vel fibrilloso-squarruloso, neque hygrophano nec viscido, convexo, obtuso, cc. 7 mm lato. Lamellis atro-griseis, horizontalibus, latissimis, didymis, admodum distantibus, late adnatis. Stipite brunneo ("kis kilim" M&P), fibrilloso-tomentoso-piloso, apparenter fortiter squarruloso in siccis sub lente, aequali, cc. 23 x 1 mm; velo haud viso. Contexto in parte exteriore atro-castaneo, parte centrali sub centro pilei subdilutiore pileo (eodem, colore quo tegumentum pilei gaudet), inodoro. Sporis 9–10.3 x 5–6.3 μ, ellipsoideis vel ellipsoideo-amygdaliformibus, applanatione suprahilari praeditis, ornamentatione VII–XI punctulatis, episporio in parte externa ut minime leniter heterogeneo, ferrugineo-brunneo (ornamentatione vix obscuriore) sed perisporio stramineo-subhyalino, appresso, endosporio stramineo, membrana tota 0.6–0.7 μ crassa, punctationibus ornamentationis sparsis vel densis, subirregulariter dispersis, interdum lateraliter protractis, mediocriter manisfestis, raro duabus vel quattuor sporis in uno sacco perisporiali, rarissime una spora in perisporio relaxato involucrata, perisporio continuo haud diffracto, disco sine ornamentatione in regione suprahilari nullo, poro germinativo et callo nullo, rarius ad apicem distalem membrana discontinua angustissime et haud manifeste nec truncate poriformiter insigni. Hymenio e basidiis efformato et cheilocystidiis marginato. Basidiis 30–35 x 6–9 μ, cylindraceis vel subclavatis, plerumque brunneolis e pigmento incrustante, tetrasporis. Cheilocystidiis e tramate tenuissimo acierum lamellarum excrescentibus, hoc tramate subtrichodermiali et ex elementis catenulatis efformato, membris terminalibus (cheilocystidiis) 25–66 x 11–28 μ, interdum leniter crasse tunicatis, levibus, stramineis vel ochraceis, haud incrustatis, vesiculosis vel ventricosis vel subcylindraceis. Pleurocystidiis nullis. Hyphis fibuligeris, in tramate hymenophorali regulari ex hyphis parallelis pigmento ferrugineo incrustatis, haud

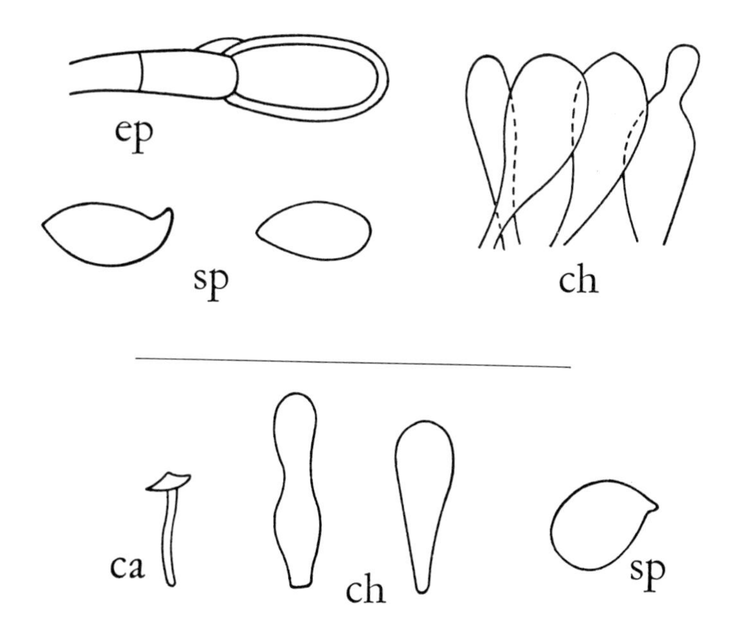


FIGURE 3. Above: *Melanomphalia thermophila* (Sing.) Sing., ep = fragment of epicutis hyphae; – sp = spores, outline, × 2000, – ch = cheilocystidia, × 1000.
Below: *Melanomphalia emarginata* Sing., ca = carpophore, × 1; – ch = cheilocystidia, × 1000; – sp. = spore, in outline, × 2000.

gelatinosis. Epicute pilei eadem structura qua acies lamellarum gaudente, cellulis terminalibus conglobatis ubi pileus squarruloso-asperulatus est, 30–44 x 17–25 μ. Hypodermio crasso ex elementis hyphalibus fortiter pigmento ferrugineo-incrustatis, subparallelis vel leniter intertextis, nonnullis cellulis perlongis aliis breviusculis; sub hypodermio carne pilei in parte marginali pilei paullum evoluto, sed in medio incrustationibus pigmenti minus fortibus differente. Ad terram in silva anectotrophica valdiviensi sub Eucryphia, Aextoxico, Saxegothaea, Podocarpo, Myrtaceis autumno, solitario. Typus a R. Singer (M 6727) in Chile lectus et in SGO conservatus est.

Material studied: CHILE: Valdivia: Hueycolla, 4–V–1967, Singer M 6727 (SGO), typus.

### 9. *Melanomphalia thermophila* (Sing.) Sing. *comb. nov.*

*Tubaria thermophila* Sing., Pap. Mich. Acad. Sc. Arts & Lett. 32: 145. 1948.

For description see the original publication. In additional material from Brazil the color of the pileus was found to correspond to "Agate" (M&P), the lamellae were "suntan" (M&P) and broad, the context was white, and there was no veil. These specimens were but slightly smaller than the type. Among very numerous 1–2–3-spored basidia, a few 4-spored individuals were observed, but the majority were 2-spored.

Material studied: U. S. A.: Florida: Highlands Co.: Highlands Hammock State Park, July–August 1942, R. Singer F 20 (FH), typus. F 20a, 20b, 624. BRAZIL: Pernambuco: Dois Irmãos, 13–III–1960, Singer M 3387a (BAFC).

### 10. *Melanomphalia columbiana* Sing. *spec. nov.*

Pileo pallide alutaceo vel ochraceo-brunneo, sicco "Arab" M&P, margine recto in juvenilibus, longe sulcato, conico vel rarius cylindraceo, dein conico, papillato, glabro, haud viscido, 2–6 mm lato, 3.5–6 mm alto; velo in juvenilibus minutissime albo-floccoso. Lamellis ochraceis, ascendentibus, angustis vel subangustis, subdistantibus, adnexis. Stipite pallide alutaceo vel brunneolo, pileo pallidiore, velo excepto glabro nudoque et haud viscoso, gradatim ad apicem attenuato, 21–40 x 0.5–1.5 mm; velo cortiniformi, laxo, tenui, dein sub lente fibrillis ascendentibus demum appressis, albis, subtilibus in superficie stipitis notabilibus. Contexto inodoro, sub umbone crassiusculo. Sporis 8.5–10.5 x 4.5–6 μ, characteristice amygdaliformibus, brunneolo-melleis vel partim aureo-melleis, membrana pro ratione subcrassiuscula (1–1.3 μ crassa) praeditis, inamyloideis, ornamentatione typi XI in episporio pro ratione tenui paullum contrastante et brevissima punctiformiter vel brevi-cristulate immersa, magis manifesta in latere exteriore sporarum, discontinua in apice distali nonnullarum sporarum qua ex re porum germinativum angustissimum haud truncatum simulare possit, sed poro germinativo vero nullo, ad apicem interdum callo mucronato praesente, discus haud ornamentatus in regione suprahilari de est. Hymenio e basidiis efformato et cheilocystidiis marginato. Basidiis 21.5–23 x 6–7 μ, tetrasporis vel perpaucis bisporis intermixtis. Cheilocystidiis 17–40 x 5–7 μ, ampullaceis, apice longo, haud incrassato vel capitato vel claviculato, 1.5–2 μ diametro praevisis, sed si capitata vel claviculata sunt, incrassatio diametrum usque ad 5–(6) μ habent, membrana hyalina instructis, tenuitunicatis. Pleurocystidiis nullis. Hyphis fibulatis. Epicute pilei haud gelatinosa ex hyphis repentibus hyalinis vel paulisper pigmento granulari-incrustatis et tunc ochraceis efformata. Hypodermio ex elementis cutem efformantibus, latioribus et magis pigmento incrustatis, brunneis haud gelatinosis efformato, nonnullis elementis breviusculis et latis, usque ad 26 x 20 μ e. gr. et membrana crassiuscula instructis. Ad et inter muscos, sed in Sphagno haud visa, solitario vel gregatim. Typus a R. Singer lectus (B 3587), in Columbia et in SGO conservatus.

Material studied: COLOMBIA: Boyacá: Arcabuco, 29–VII–1960, Singer B 3587 (SGO), typus. B 3570 (SGO), para-typus.

### 11. *Melanomphalia platensis* (Speg.) Sing.

Rev. Mycol. 20: 16. 1955; Sydowia 11: 321 1958.

*Inocybe platensis* Speg., An. Mus. Nac. Bs. As. 6: 124. 1899.

A type analysis has been published by Singer, Lilloa 25: 501, 1952. For macroscopic data, see Spegazzini, loc. cit.

### 12. *Melanomphalia alpina* (A. H. Smith) Sing.

Sydowia 11: 321. 1958.

*Kuehneromyces alpinus* A. H. Smith, Sydowia, Beiheft 1, Festschr. f. F.Petrak, 52: 1956.

For a description of this species see Smith, loc. cit.

Material studied: U. S. A.: Idaho: Papoose Creek, Seven Devils Mountain, 23–VIII–1954, Bigelow & Smith 46580 (MICH); typus.

### 13. *Melanomphalia smithii* Sing. ad int.

Pileus rich tawny when moist, fading somewhat but often reddish to reddish-tawny in age, unpolished to minutely squamulose, moist, convex with incurved margin, 10–30 mm broad. Lamellae ochraceous tawny becoming tawny or more reddish, close, depressed, broad, adnate to seceding; dried lamellae "sepia" (M&P) from the dehydrated spore dust. Stipe colored as the pileus but fibrillose from variously dispersed pallid brownish veil; fibrils darkened in age at base; stipe 30–50 x 2–3 mm; context cartilaginous, thin in the pileus, pallid buff to whitish. Fruit body without odor or significant taste.

Spores 8.2–11 x 6–8.8 μ, ellipsoid or rarely irregular in shape (e. gr. heart-shaped) or subamygdaliform, the wall moderately thick and smooth in outline, with very thin episporium which is indistinctly heterogeneous from a very fine and low ornamentation of type XI (Singer, 1951), under a strictly applicate, not loosening perisporium without plage or germ pore although a porelike discontinuity is sometimes visible at the apex of the spores, but this incomplete and never truncate; episporium deep rusty-mahogany in KOH; general color impression of spores light brownish melleous. Basidia 28–37 x 10–12.5–(14) μ, hyaline, at edges of lamellae some fulvous, (2)–4-spored. Cheilocystidia 19–29 x 12–18 μ, balloon-shaped, hyaline, smooth, scattered among basidia and basidioles. Pleurocystidia none. Hyphae with clamp connections, in the regular hymenophoral trama brownish melleous to ochraceous, slightly interwoven, not gelatinous. Hyphae of pileus epicutis generally repent and radially arranged, making up a cutis although some of the uppermost are slightly ascendant and the terminal members at times cystidioid, ventricose, and tapering to the tip, or on the contrary, clavate, always broadly rounded, e. gr. 32 x 9.5–11 μ, the repent hyphae usually very variable in length and diameter, some very short and broad, all or almost all incrusted by an ochraceous-ferruginous brown pigment. Neither this or the subjacent hypodermium gelatinized, the latter still more pigment-incrusted. Pileus trama below scarcely or poorly pigment-incrusted and subhyaline.

On debris of a beaver dam, consisting mainly of aspen remainders, gregarious.

Material studied: U. S. A.: Wyoming: Pole Mt. area, 9–VII–1950, A. H. Smith 34646 (MICH).

This material, at first inspection very similar to the preceding species, differs in the presence of a veil, in somewhat larger spores, and rather broad to broad and perhaps less crowded lamellae. Since I have seen no fresh material of either this or the preceding species, I am not fully certain that this is not a young stage of the latter. Veil and spore size are, however, important characters in the delimitation of the species of *Melanomphalia*, and I believe that this new species, here described *ad interim*, is rather specifically different from *M. alpina*. Most macroscopical data are taken from A. H. Smith's notes.

### 14. *Melanomphalia omphaliopsis* (Sing.) Sing. *comb. nov.*

*Tubaria omphaliopsis* Sing. in Sing. & Digilio, Lilloa 25: 397. 1952.
Description of this species see Lilloa, l.c.

Material studied: ARGENTINA: Tucumán: Ciudad, 7–I–1949, R. Singer T 89 (LIL), typus. 20–III–1949, T 236 (LIL), para-typus. San Pablo, 29–III–1949, Singer T 306 (LIL), para-typus. Anta Muerta, 19–XII–1949, Singer T 789 (LIL), para-typus. Capital, 28–III–1951, Singer M 1448 (LIL), para-typus. Catamarca: Andalgalá, in an irrigation ditch, 22–I–1952, Singer T 1797 (LIL). Mendoza: Capital, open places on clay soil, 2–I–1959, A. Ruiz Leal 20.357 (BAFC, Herb. Ruiz Leal). CHILE: Santiago: Quebrada at 2 km west of Ranque, 6–VII–1967 Singer M 7142 (SGO).—also some collections by W. Lazo from Central Chile. U. S. A.: An apparently inedited species (*Omphalina subfulviceps* Murr., FLAS) from Gainesville, Florida, may possibly turn out to be the same species.

### 15. *Melanomphalia emarginata* Sing. *spec. nov.*

Pileo melleo (12–F–6, M&P), glabro, levi, nudo, convexo, dein ad marginem applanato ita ut subumbonatus appareat, margine anguste sterile: projiciente, diametro ± 6 mm. Lamellis griseis, dein ferrugineo-ochrascentibus e sporis, latissimis, distantibus, profunde et anguste sinuato-emarginatis et subadnexis. Stipite stramineo-flavidulo, glabro velo excepto, aequali, cc. 20 x 1 mm; velo subtili et fugaci, e fibrillis appressis brunneolis efformato. Contexto inodoro.

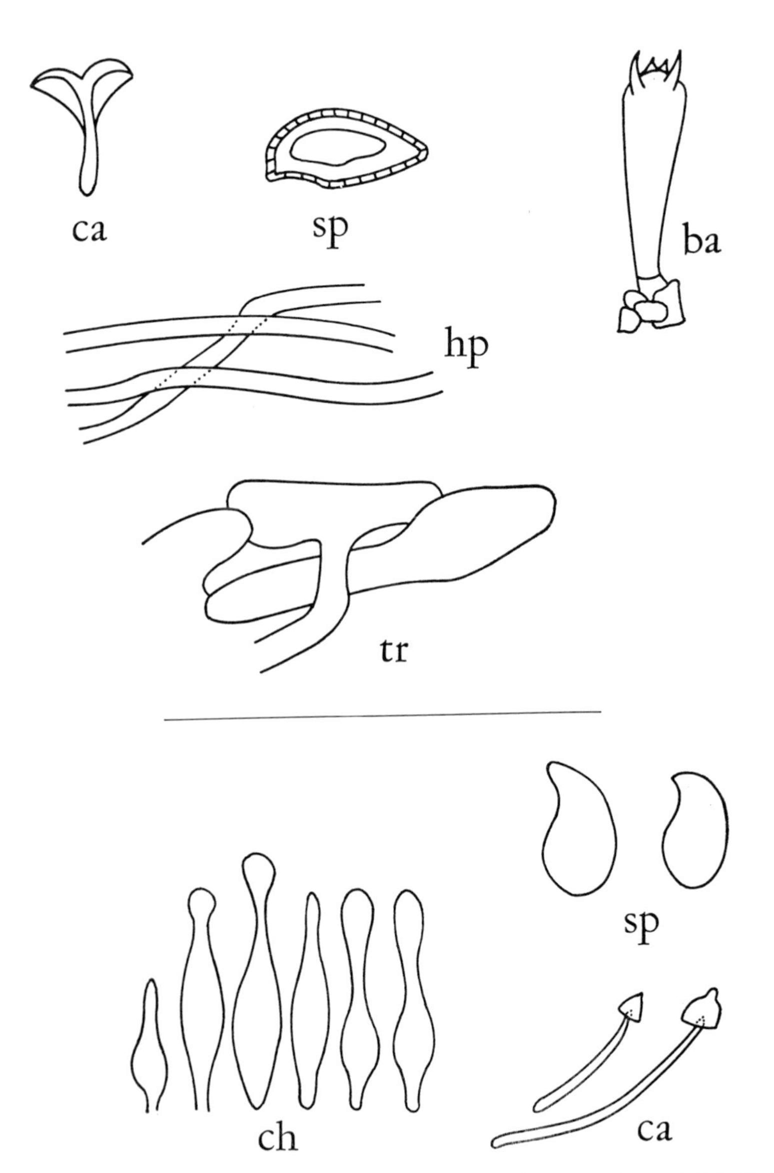


Figure 4. Above: *Melanomphalia omphaliopsis* (Sing. in Sing. & Digilio) Sing., ca = carpophore, × 1; — sp = spore in optical section, × 2000; — ba = basidium with subhymenial elements, × 1000; — hp = hymenopodium, × 1000; tr = hymenophoral trama, × 1000.
Below: *Melanomphalia columbiana* Sing., ch = cheilocystidia, × 1000; — sp. = spores, in outline, × 2000; — ca = carpophores, × 1.

Sporis 8–8.5 x 6–6.5 μ, ellipsoideis vel breviter ellipsoideis vel breviter subamygdaliformibus, raro depressione suprahilari gaudentibus, endosporio pallido, episporio ochraceo, perisporio hyalino haud relaxante, ornamentatione typi XI moderatim conspicua sed supra episporium projiciente et interdum perisporium nonnihil perforante, e spinulis subtilibus efformata instructis, punctatulis, sine disco haud ornamenta suprahilari, ochraceo-brunneolis, bene pigmentatis ut minime in spinulis ornamentationis, poro germinative vero destitutis, sed interdum in apice distali membrana discontinua anguste sed paullum manifeste nec truncate poriformi praeditis. Hymenio e basidiis efformato et cheilocystidiis marginato. Basidiis 26–33 x 7–9 μ, hyalinis vel flavidis, tetrasporis. Cheilocystidiis 25–30 x 6.5–8.5 μ, basidiolis similibus, clavatis et saepe constrictis, tenuitunicatis, numerosis. Pleurocystidiis nullis. Hyphis fibuligeris, in nulla parte gelatinosis. Epicute pilei cutem efformante, ex hyphis haud incrustatis, haud gelatinisatis, filamentosis efformata. Hypodermio ex elementis latioribus, saepe constrictis ad septa, saepe breviusculis constante. In tegumentis et tramate pigmento flavo praesente quod facile in medium dissolvitur. Dermatocystidiis in pileo interdum praesentibus, e membris terminalibus hypharum efformatis et cheilocystidiis simillimis. Ad lignum in Sphagneto putrescente, autumnalis, solitario. Typus a R. Singer in Chile lectus (M 6746) et in SGO conservatus.

Material studied: CHILE: Valdivia: Cordillera Pelada, bog near Mirador ("Turbera Hernandez"), at 900 m altitude, 6–V–1967, Singer M 6746 (SGO), typus.

### 16. *Melanomphalia inocyboides* Sing. *spec. nov.*

Pileo margine ochraceo vel sordide alutaceo ("honeysweet" vel "maple" M&P), centro brunneo ("Mohawk" vel "Alamo" M&P), glabro sed haud manifeste radiatim innate fibrilloso et in juventute in zona marginali subtiliter lanoso-fibrilloso, levi, convexo, dein applanato circum umbonem, demum saepe concavo, constanter umbonato, 10–13 mm lato. Lamellis argillaceis ("honeysuckle" M&P), ventricosis, sat latis, subconfertis, profunde sinuatis et adnexis. Stipite juvenili flavo ("buttercup" M&P), dein ochraceo-luteo (11–I–6, M&P) e fibrillis ("spruce y" M&P) flocculosis veli, solido, subaequali vel leniter gradatimque basin versus incrassato, basi plus minusve subbulbosa praedito, 28–37 x 2–3 mm, in parte inferiore 3·5–4 mm diametro. Contexto flavobrunneo ("Yucatan" M&P), inodoro. Sporis 8.5–9 x 7.5–8.5 μ, subamygdaliformibus (in polis ambobus

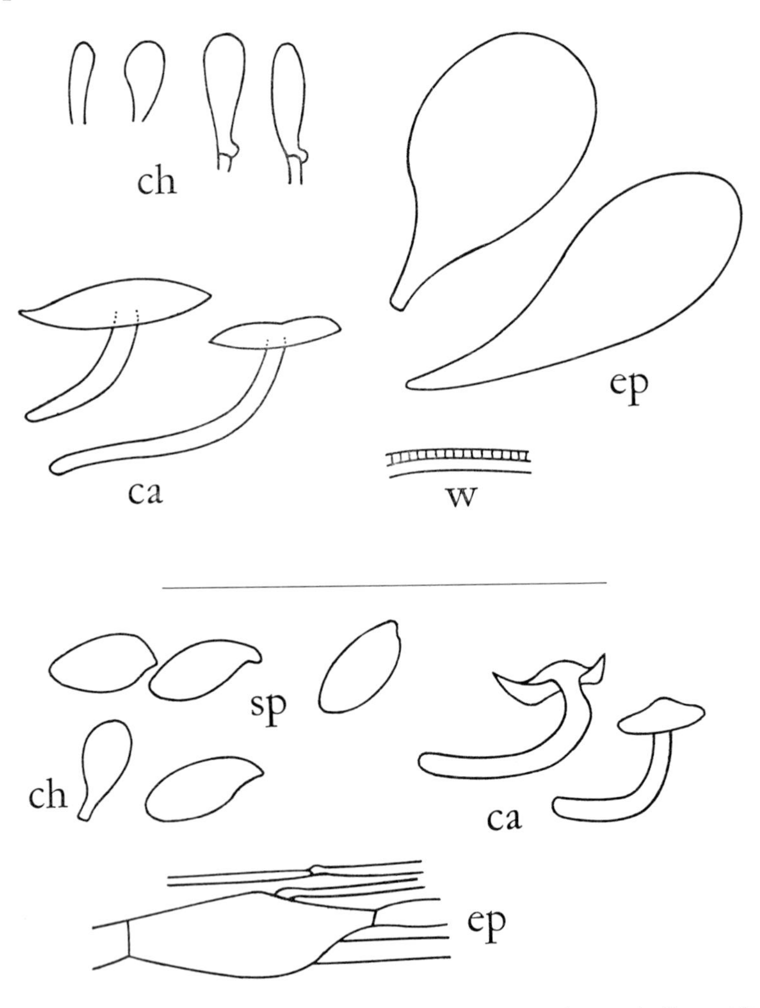


FIGURE 5. Above: *Melanomphalia pacifica* Sing., ch = cheilocystidia, × 1000; – ep = cells of the epicutis, × 1000; – ca = carpophores, × 1; – w = detail of spore wall structure.
Below: *Melanomphalia inocyboides* Sing., sp = spores, outline, × 2000; – ch = cheilocystidium, × 1000; – ca = carpophores, × 1; – ep = epicutis of pileus × 1000.

subacutis), ornamentatione typi XI punctulatis, e columnis exiguis interdum cristulate dilatatis admodum subtilibus et brevibus in episporio immersis consistente ornatis, nonnullis paullum ornamentatis, juvenilibus sine ornamentatione, hyalinis, dein (in ammoniaco) ochraceis vel (in KOH) ferruginascente-ochraceo-brunneis disco ornamentatione destituto suprahilari nullo, poro germinativo nullo, applanatione vel depressione suprahilari praesente. Hymenio e basidiis efformato et cheilocystidiis marginato. Basidiis 14–29.5 x 7.5–8.5 μ, tetrasporis. Cheilocystidiis sparsis et forma sua basidiolas in mentem revocantibus, 15–16 x 6.5–8.5 μ, haud conspicuis; pleurocystidiis nullis. Hyphis fibuligeris, in tramate hymenophorali haud gelatinosis, regulariter dispositis. Epicute pilei ex hyphis non gelatinosis cutem efformantibus, 3–5 μ latis, pigmento ferruginascente-ochraceo incrustatis, tenuitunicatis et interdum membrana sinuato-flexuosa instructis efformata. Hypodermio ex elementis partim tumescentibus cutem efformantibus, haud gelatinosis, usque ad (in tumescentiis) 33 x 15 μ. Supra epicutem et ad superficiem stipitis strato velari fragmentario visibili. Hyphis tramatis sub hypodermio etiam pigmento incrustatis. Ad lignum dicotyledoneum putridum inter muscos in silva sparsa nothofaginea (*Nothofagus obliqua*) mixta, vernalis. Typus a R. Singer in Chile lectus (M 7768) et in SGO conservatus.

Material studied: CHILE: Valdivia: 25 km south of the city along the road to Unión, 12–IX–1967, Singer M 7768 (SGO), typus.

### 17. *Melanomphalia pacifica* Sing. *spec. nov.*

Pileo intense cinnamomeo vel cinnamomeo-fulvo, macroscopice glabro in humidis, sed in siccis manifeste subtomentoso, levi, evelato, convexo, dein applanato vel centro leniter depresso, 17–34 mm lato. Lamellis ferrugineo-brunneis, mediocriter latis, interdum latis, confertis, profunde sinuatis et anguste adnexis vel adnexis. Sporis in cumulo brunneis ("Yucatan" M&P). Stipite intensius brunneo quam pileus, evelato, glabro, nudo, aequali vel attenuato basin versus, 21–56 x 2.5–5 mm. Contexto flavido-fulvidulo; sapore miti; odore nullo. Sporis 7–8.5 x 5–6 μ, ellipoideis vel ovoideis vel seminibus mali similibus, applanatione, rarius depressione leni suprahilari praeditis, intense ferruginescente-ochraceo-brunneis e spinulis ex episporio emergentibus et inter eum et perisporium projicientibus tenuissimis brevissimis columnaribus, ornamentationem typi XI efficientibus, endosporio hyalino praeditis, spinulis colore intensiore cum episporio contrastantibus, punctulatis, episporio interdum subruguloso, ornamentatione excepta ferrugineo-ochraceo-hyalino, guttula centrali olei impletis, disco ornamentatione destituto suprahilari destitutis, poro germinativo destitutis. Hymenio e basidiis efformato et cheilocystidiis marginato. Basidiis 19–27 x 6.5–7.6 μ, plerumque hyalinis, clavatis vel ventricosis et constrictis, (2)–4-sporis. Cheilocystidiis numerosissimis sed paullum projicientibus, 10.3–18.5 x 4.5–6.5 μ, ventricosis vel multum frequentius clavatis et pedicellatis, raro in parte superiore constrictis, late rotundatis, hyalinis vel pigmento intracellulari dissoluto flavo impletis. Pleurocystidiis nullis. Hyphis fibulatis, plerumque flavis. Epicute pilei ex hyphis repentibus filamentosis flavis et flavo-incrustatis non gelatinosis efformata; ex eis ascendent fascicula elementorum cystidiformium, his vesiculosis vel ventricosis vel late clavatis, flavis, interdum aurantiaco-ferrugineo incrustatis vel flavido-incrustatis, membrana ferruginascente vel pallide ochrascentis et frequenter pedicello longo praeditis, rarius his dermatocystidiis solitariis, 28–63 x 16–26 μ. Ad lignum emortuum in silva, gregatim, autumnalis. Typus a R. Singer in Chile lectus (M 6807) et in herbario SGO conservatus est.

Material studied: CHILE: Llanquihue: 10 km north of Colaco, 9–V–1967, Singer M 6807 (SGO), typus.

## PHYLOGENETIC RELATIONS OF THE GENUS MELANOMPHALIA

As mentioned before, I include the genus *Melanomphalia* in the Crepidotaceae. I do so for the reasons indicated above and believe, as I did with relation to the section *Thermophilae*, that the genus is intermediate between *Tubaria* and *Crepidotus*. Since the key difference between *Tubaria* and *Melanomphalia* consists of the structure of the spore wall, it may correctly be asked whether it would not be wiser to treat the complex *Tubaria-Melanomphalia* in the same way as the genus *Crepidotus* has been treated, viz., maintaining the sections *Tubaria* and *Thermophilae* (=*Melanomphalia s.l.*) in the same manner as the sections *Echinosporae* and *Crepidotus*, *Crepidotus* being parallel to *Tubaria* and *Echinosporae* to sect. *Thermophilae*. I hesitate doing so for several reasons. In the first place, the differentiation of species in *Tubaria* is not comparable to that in *Melanomphalia*, remaining rather restricted in the former genus and showing altogether new and different tendencies in *Melanomphalia*, such as presence of pleurocystidia, different pigments, a wide variety of lamellar attachment, habit, and spore-print color, and a distinctly greater variation in the range of epicuticular structures. In no case have I had difficulty in separating species of *Tubaria* from species of *Melanomphalia* nor have I been able to establish definite lines of affinity between them.

The stipe in *Melanomphalia* is constantly central even in mature specimens, an important character in distinguishing these species from *Crepidotus.* The relationship between *Melanomphalia* and *Crepidotus* is therefore similar to that between *Pholiota* and *Pleuroflammula, Gymnopilus* and *Pyrrhoglossum*, or *Psilocybe* and *Melanotus*, and is not comparable to the relationship between centrally stipitate and eccentric forms in *Pleurotus, Polyporus*, or *Marasmiellus* where no hiatus exists between the forms of pleurotoid and nonpleurotoid habit. But this is not the only difference between the two genera. Species with small round spores, species without clamp connections, and mainly white or whitish (pigmentless) species are very important and numerous in *Crepidotus* and absent in *Melanomphalia.* The spores tend to be somewhat larger in *Melanomphalia* than in *Crepidotus.* The lignicolous habitat, as the basis of the pleurotoid development, is much more widespread in *Crepidotus* than in *Melanomphalia* and is actually normal in the former. Thus here, too, one may say that the evolution in these two genera has followed different paths.

The frequent presence of a veil in *Melanomphalia*, as well as the central stipe and the constant presence of clamp connections, would make it appear that *Melanomphalia* is the more primitive of the two genera. It is only reasonable, then, to look principally in the genus *Melanomphalia* for direct connecting lines, bridges, that lead from the Crepidotaceae to other groups. Now that we have a new definition and a better knowledge of the melanomphalias, we may attempt to look for the ancestors of the whole group.

But before doing so, it should be stated that the absence of data on the cytological characters, particularly on the number of nuclei in the young spores, is a considerable handicap not only in the discussion of the relations of *Tubaria* with *Melanomphalia* and *Crepidotus*, and, for that matter, with *Pleurotellus*, but also in the discussion of the derivation of the present-day melanomphalias. We have only insufficient data on *Tubaria* and *Crepidotus*, and none on *Melanom-*

*phalia.* For the time being, we assume that both uninucleate and binucleate spores exist in the family. It is hoped that some more thorough work on the cytology of the Crepidotaceae will eventually be forthcoming.

It is of course possible to derive the Crepidotaceae from some Tricholomataceae such as *Fayodia.* However, in this case we would have to assume a gradual intensification of the pigmentation of the spores and an abrupt change in pigment of the carpophore hyphae. We would be at a loss to explain the gelatinization of some strata in the carpophores observed in so many *Crepidoti,* and of the epicutis as observed in so many Crepidotaceae. Furthermore, it would appear that in this case the species with the least-pigmented spores, such as *Melanomphalia crocea* and *M. omphaliopsis,* would be the most primitive, but this is obviously not the case; on the contrary, these species seem to be farther removed from *Fayodia* than some others, and since they are without veil, veiled species would have to be derived from them.

It would appear, therefore, that the coincidence in spore ornamentation is not always of phylogenetic importance as far as we can see now. It would also seem unnecessary to prove that the same type of ornamentation has originated independently at several points and levels of the Basidiomycetes, remembering that it exists equally, as mentioned before, in *Boletellus,* but also in the Aphyllophorales, in certain species of *Grifola,* and *Scutiger* (Scutigeraceae) (Singer, Snell, & White, 1945).

This suggests immediately a comparison of those genera of Gastromycetes, Hymenogastrinae, where spore ornamentation of type XI has been discovered (Singer, 1951; Singer & Smith, 1959; Smith & Reid, 1962), viz., in *Setchelliogaster, Cribbea,* and *Hymenogaster.* Apparently, species are known in *Hymenogaster* which have a similar spore ornamentation to *Cribbea* and *Setchelliogaster* so that it is "rather tempting to try and connect *Cribbea* up to *Hymenogaster*" (Smith & Reid, 1962) and *Setchelliogaster* to *Cribbea,* on the "secotiaceous" level.

Indeed, *Cribbea* has a number of characters in common with *Setchelliogaster*: the spore-wall structure, the structure of the exoperidial epicutis, the relatively broad and large spores, the stipecolumella, the nonpulverulent, fleshy gleba. Both these genera are separated from each other principally by the lack of gelatinous layers in *Setchelliogaster*, the lack of true leptocystidia in the same genus, and the absence of pseudobasidia there. However, the gelatinous layers are not present in *Cribbea andina* (Speg.) Sing., (Singer, Wright, & Horak, 1963); on the other hand, there is a gelatinous outer peridial layer in a species mentioned by Singer and Smith (1960) in connection with *Setchelliogaster*, namely, *Secotium eburneum* Zeller, which has the same essential characters as *Setchelliogaster* otherwise. Although there are no leptocystidia of the *Cribbea*-type in the known species of *Setchelliogaster*, there are, nevertheless, frequently pseudoparaphysoid or basidiole-like cystidia in both *S. tenuipes* and *S. tetrasporum*, such as we find in several species of *Melanomphalia*. The gap between *Cribbea* and *Setchelliogaster* is thus not sufficient to insert these genera in different families. Consequently, I consider *Setchelliogaster* as another genus of the family Cribbeaceae (Singer, Wright, & Horak, 1963). This is done in spite of the fact that we are essentially in agreement with Smith and Reid (1962), who regard *Cribbea* as a side branch, "an endline of development," but the two genera are certainly related enough to have originated from a common ancestor. We assume that this common ancestor, as in so many comparable cases, gave rise to an Australian branch of the genus—with gelatinized trama—and a South American branch, *C. andina*, with nongelatinous trama. From this, as has to be expected then, arose the American genus *Setchelliogaster*.

One of the differences between *Cribbea* and *Setchelliogaster* appeared to be the mainly four-spored basidia in the former and the mainly two-spored basidia in the latter. This, however, is a differentiation based on insufficient knowledge. The American *Cribbea* has two-spored basidia, and a new species of *Setchelliogaster*, *S. tetrasporum*, has 4-spored basidia. In view of the importance of this new species, I shall here describe it:

### ***Setchelliogaster tetrasporum*** Sing. *spec. nov.*

Gastrocarpio magnitudine Thaxterogastero magellanico comparando, forma irregulari, gleba haud exposita, late rotundato aut flexuoso-depresso in parte superiore, gradatim in stipitem transeunte. Peridio tenui, haud gelatinoso, nec viscido, subtiliter venosulo sub tegumento tenuissimo arachneoideo, venis aut vix manifestis aut elevatis et avellaneobrunneis, colore basico grisello-pallido. Gleba ochraceo-brunnea, structura eam. Se tenuipedis in mentem revocante. Stipite submultiplici-connato, irregulariter curvato et furcato, in siccis cavitiebus excavato, albo, in columellam continuato. Columella simplici vel ramosa, alba, plus minusve percurrente. Sporis 11.5–13.5 x 8.5–9 μ, plerumque 13–13.5 x 9 μ, ellispoideo-mucronatis, appendice hilari hyalino axialiter symmetrice disposito, mucrone apicali brunneo (concolori), obtuso, 2–2.5 x 1.7–2 μ, ornamentatione type XI lineis punctationes hic inde connecticantibus (typo IV, IIIb) e spinulis subtilibus brevibus nec nonnonullis cristulis interdum catenulatis ex episporio natis et spatium inter id et perisporium hyalinum vel subhyalinum occupantibus, ornamentatione brunnea, in fundo pallido contrastante, spinulis 1.5–2 μ altis, tenuissimis vel crassiusculis, mucrone subporiformiter perforato materia heterogenea qua de causa membrana hic discontinua est, sed poro vero germinativo absente nec apice umquam truncato, endosporio subhyalino tenui praesente. Hymenio e basidiis efformato, cystidiis nullis visis. Basidiis 21–31 x 12.5–13.5 μ, vesiculosis vel clavatis, tetrasporis; sterigmatibus rectis et symmetricis, acutis, 4–6 μ longis. Hyphis fibuligeris in tramate hymenophorali regulari dense dispositis, haud gelatinosis, subintertextis, e pigmento membranali sed haud incrustante stramineis. Subhymenio ex elementis irregularibus, nonnullis subisodiametricis et subglobosis usque ad 12 x 9 μ, hyalinis efformato. Hyphis tramatis haud gelatinosis. Epicute peridii bistratosa, strato externo arachnoideo ex hyphis omnino repentibus 2–10 μ latis; strato interno ex elementis horizontaliter catenulatis, catenulis tangentialiter dispositis, sed multiseptatis, multis latis (10 μ et multum magis latis) et perbrevibus e. gr. 13–15 x 10 μ vel 40 x 20 μ, pigmento incrustante brunneolo-ochraceo vel ferrugineo-brunneo tinctis, haud gelatinosis efformato. Semihypogaem sub Nothofago obliqua in humo et terra crescens, autumnalis. Typus a R. Singer in Chile lectus (M 6923) et in herbario SGO conservatus est.

Material studied: CHILE: Malleco: Cordillera Nahuelbuta, San Carlos, 16–V–1967, Singer M 6923 (SGO), typus.

This species as well as the other native South American species, ***Setchelliogaster brunneum*** (Horak) Sing. comb. nov. (bas. *Hypogaea brunnea* Horak, Sydowia 17:299. 1964[1]), has the same covering of the peridium as I find in some of the species of *Melanomphalia* as far

[1] Horak himself corrected the name to *Setchelliogaster* when sending out the reprints of his paper (1964). His species differs from *S. tetrasporum* in 2-, rarely 1- or 3-spored, basidia.

as their epicutis or hypodermium of the pileus is concerned. It is less epithelium-like than in S. *tenuipes.* So if we take the Cribbeaceae and the Crepidotaceae as whole, in the way we have defined them, and compare their characters, there is not a single character as found in any member of the Crepidotaceae that has not a forerunner in some member of the Cribbeaceae, other than the hemiangiocarpous development and the spore-discharge mechanism. Such characters are:

1. Spore ornamentation type: The nonornamented spores are found in S. *aurantiacum.*

2. The tendency to lose clamp connections (in *Crepidotus*) is also found in *Cribbea,* especially *C. andina.*

3. Gelatinous strata (in *Crepidotus*) are also found in three species of *Cribbea.*

4. Gelatinous epicuticular layers (in three species of *Melanomphalia,* rarely in *Crepidotus*) are also found in "*Secotium*" *eburneum* and *Cribbea lamellata.*

5. The absence of pigments, as well as presence of the orange, brown, yellow, red-brown, and fuscous pigment range, all are found likewise in *Cribbea,* "*Secotium*" *eburneum,* and *Setchelliogaster. Cribbea lamellata* is also described as purple when young. This is probably a pigment comparable to that in *Melanomphalia universitaria.* These pigments are frequently found to incrust the hyphal walls.

6. Inamyloid spores and hyphae are constant in both families.

7. The pleurocystidia of *Melanomphalia* and apparently some species of *Crepidotus* are also present as pseudoparaphysoid bodies in *Setchelliogaster tenuipes* and as leptocystidia in *Cribbea,* some of which have thick-walled metuloid cystidia mixed in (as seems to be the case in *C. reticulata*) apparently corresponding to the metuloids of *M. dwyeri.* In *C. gloriosa* the cystidia show somewhat refractive contents as do those of *M. universitaria.* Species of *Melanomphalia* without distinctive pleuro- and cheilocystidia, in this regard, are comparable to *S. brunneum* and *S. tetrasporum.*

8. The specific type of apical discontinuity of the spore wall as observed in some species of *Melanomphalia,* which does not correspond with the typical germ spore of other agarics and gastromycetes, has its counterpart in the spore structure of three *Setchelliogaster* species.

9. There is absence of a characteristic odor and the absence of characteristic discolorations of the trama when exposed to the oxygen of the air, another characteristic of most Cribbeaceae and most Crepidotaceae.

10. The habitat is subhypogaeous to terricolous, with a tendency toward lignicolous habitat, the characteristic habitat of most *Crepidoti* and many melanomphalias. None are ectotrophically mycorrhizal.

11. All Cribbeaceae and Crepidotaceae share the regular hymenophoral trama.

12. The lamellar configuration of the hymenophore of the Crepidotaceae is anticipated by a lamellate to sublamellate gleba in *Cribbea lamellata*; a reminiscence of a chambered gleba may be seen in the forked lamellae of *Melanomphalia crocea.*

Although it is always relatively easy to construct bridges between related groups and thus show their affinity, it is always considered difficult to demonstrate a definite direction of evolution.

After Singer and Smith (1960) showed the probability of a direction from the Hydnangiaceae to the Elasmomycetaceae and eventually the Russulaceae, Romagnesi (1967) objected (*a*) that the step from the Hydnangia to the Elasmomycetaceae was insufficiently documented; (*b*) that the pseudoamyloid reaction of the membranes as found in *Octavianina* could not be considered as related to or a forerunner of the amyloid reaction because one could not quote a single case where both reactions were found in a single group; and (*c*) that the primitive species in the Basidiomycetes had colorless hyaline, thin-walled spores, whereas the spore color was rather deeper in those "Astrogastraceae" we linked with the Russulaceae. I may say, however, that I do not share the first opinion for reasons best documented by Malençon and other French mycologists; as to the second objection, we have pseudo-amyloid spores in *Lepiota acutesquamosa* and all other species of section *Echinatae*, whereas the section *Amyloideae*, with *L. amyloidea*, differs exclusively by strongly amyloid spores (Smith, 1965). A similar case exists between *Clavicorona pyxidata* and *C. coronata* (Leathers & Smith, 1967). Furthermore, I may say that white-spored agarics may well be derived from groups with pigmented spores. What makes the reverse more acceptable for Romagnesi is the fact that Singer (1929) as well as others has stated that the most primitive Russulas appear to be white or whitish spored, and therefore any direct ancestor of them should be expected also to have pallid spores. There are, however, two facts against this reasoning. First, species with hyaline spore walls and pallid spore masses in the gleba do occur in *Hydnangium* as well as in other representatives of the astrogastraceous series; second, the pigment of the spores being concentrated in the walls, it is logical to expect the species with particularly thick-walled spores to be also those with the darkest color in spore mass. Since it is obvious that the gasteromycetous spore has the same tendency as that of the Tricholomataceae to become thick-walled when "overmature," with the additional advantage that thick walls have for spores which are not immediately released, we must expect that some gasteromycetous spores will reach darker colors than do their agaricoid derivates.

Returning, then, to the series under discussion (Cribbeaceae-Crepidotaceae), we have four indications for the direction their evolution has taken:

1. Once it is established that, in all probability, the astrogastraceous series leads from gasteroid forms to agaricoid forms, it would seem more acceptable that in any case of demonstrated affinity among agarics and gastromycetes, the same sense of progression is followed. This thought has also been expressed by D. P. Rogers (although in the reverse sense) in his communication to the International Botanical Congress in Stockholm, 1950. It is certainly not forceful, but it is an argument as good as any other. Indeed, it does not seem very probable that the same forces of morphological determination reacting to the climatic changes to trigger a change in carpophore development and spore-discharge apparatus should also, and apparently at the same time and level, have been active to produce the reverse effect. The example of the relation between Tuberales and Pezizales is not relevant here, because under completely different conditions a reverse development could well be imagined, although the Malençon-thesis on their phylogeny—even if accepted in many mycological texts—is not at all convincing to me. On the contrary, I believe very strongly that an evolution from the cleistocarp toward hemiangiocarpous, pseudoangiocarpous, and gymnocarpous development is, in several series, repeated all over the system of evolutionary lines throughout the fungi. It cannot be denied that in the lowest stages of development in the fungi, whenever carpophore development is

initiated, a closed carpophore results (the sporocarpium of the Zygomycetes: Endogonaceae; the cleistothecia of the Eurotiales; the phleogenaceous Heterobasidiomycetes–Auriculariales–of the type of *Hoehneliomyces javanicus*, etc.).

2. A progression from larger to small, lighter-colored to darker spores, if accepted as indication of every evolutionary trend in the higher Basidiomycetes as Romagnesi wishes it, is evident in the direction *Cribbea* to *Setchelliomyces* to *Melanomphalia*, and even more convincingly so if we take into consideration the white-spored genus *Mycolevis* (see below).

3. If the number of genera is any indication, it grows in the direction toward the Agaricales, as does the number of species in each genus: Cribbeaceae: 2 (–3) genera; *Cribbea*: 4 species; *Setchelliogaster*: 4–5 species; and Crepidotaceae: 4 genera; each with more than a dozen, some with several dozens of species, with the exception of *Pleurotellus* (if this is accepted as generically different from *Crepidotus*).

4. Clamp connections present in the majority of the species on the gasteromycetous side, in a smaller majority on the agaricoid side.

If this discussion may be extended to material I have not seen, I would draw attention to the genus *Protoglossum* Mass. The type species, *P. luteum* Mass., has been studied recently by A. H. Smith (1966). The spore characters have been compared with those of *Cribbea* and evidently the structure of the peridium would also fit into this genus. The absence of clamp connections would make Massee's genus comparable to *C. argentina*. The genus would be distinguishable from *Cribbea* by the absence of cystidia and of a columella (at least this is not mentioned in the original description) and the hypogaeous development. It would differ from *Setchelliogaster* also in the absence of a columella (?) and in the absence of clamp connections and a different structure of the epicutis of the peridium. Could this be the ancestral genus of both *Cribbea* and *Setchelliogaster* as postulated above? And which, if any, is the relationship with a genus we have thus far (and continue to have) included in the gasteroboletaceous series: *Austrogaster*, which has spores like *Protoglossum*?

Without studying better material of *Protoglossum* I cannot answer these questions. However, it becomes quite evident that the more we descend in the Hymenogastrineae, the less clear-cut become the family differences as recognized in the Agaricales, and the more evident it becomes that the higher (secotiaceous and agaricoid) forms can be derived from common gasteromycetous ancestors. I would like to recall the case of *Maccagnia* from which both agaricaceous and tricholomataceous fungi may be derived, and of *Hydnangium* which appears to be the starting point of both hygrophoraceous, tricholomataceous, and at the same time astrogastraceous and russulaceous forms. We may also recall that *Galeropsis* is somehow intermediate between the Bolbitiaceae and the Strophariaceae. It may well be expected that the opposite situation would prevail, viz., a stronger differentiation as we descend to the less agaricoid and more gasteroid forms, if the latter were to be derived from the former. Thus, continuing research on the hymenogasteraceous forms, with modern anatomical approach, tends to make it increasingly more probable that the Agaricales, by a progressive evolution, have evolved from the Hymenogasterineae.

A further genus recently published and obviously significant for the subject of the present contribution is *Mycolevis* A. H. Smith. The spore ornamentation is sufficiently similar to that of the Cribbeaceae in order to be inserted in that family as proposed by Smith. *Mycolevis* is obviously a primitive form, lacking a stipe, lacking cystidia, and lacking a germ pore, with little pigmentation in peridium and spore wall, and with no gelatinization anywhere in the trama. Thus, to a certain degree it is intermediate between *Cribbea* and *Setchelliogaster*. It differs from *Protoglossum* in the presence of a columella, an amylaceous crust on the spore wall, and hyaline spores.

Resuming our survey of the genera with ornamentation type XI in the spores, we have now at our disposal a series of genera with different character combinations but evident affinities to each other which, taken as a whole, represent another bridge between true Agaricales and true Gastromycetes.

## References

CHRISTIANSEN, M. P. 1936. *Melanomphalia* n. gen. Friesia 1: 287–289.

HORAK, E. 1964. Fungi austroamericani. VII. *Hypogaea* gen. nov.—aus dem *Nothofagus*-Wald der patagonischen Anden. Sydowia 17: 297–301.

LANGE, J. E. 1940. Flora agaricina danica. vol. V. Copenhagen 105 p.

LEATHERS, C. H., AND A. H. SMITH. 1967. Two new species of clavarioid fungi. Mycologia 59: 456–462.

MAERZ, A., AND M. R. PAUL. 1930. A dictionary of color. McGraw Hill, New York. 207 p.

ROMAGNESI, H. 1967. Les russules. Bordas, Paris. 998 p.

SINGER, R. 1929. Eine neue *Russula*-Art, *Russula mairei* nov. spec. Arch. Prot. 65: 306–314.

———. 1948a. New and interesting basidiomycetes. II. Pap. Mich. Acad. Sci. 32: 103–150.

———. 1948b. Diagnoses fungorum novorum Agaricalium. Sydowia 2: 25–42.

———. 1951. The Agaricales (mushrooms) in modern taxonomy. Lilloa 22: 1–832.

———. 1955. Le genre *Melanomphalia* Christiansen. Rev. Mycol. 20: 12–17.

———. 1958a. New genera of fungi. X. *Pachylepyrium*. Sydowia 11: 320–322.

———. 1958b. Observations on agarics causing cerebral mycetisms. III. A russula provoking hysteria in New Guinea. Mycopath. & Mycol. Appl. 9: 275–278.

SINGER, R., AND A. P. L. DIGILIO. 1952. Prodromo de la flora agaricina argentina. Lilloa 25: 5–462.

SINGER, R., AND A. H. SMITH. 1959. Studies on secotiaceous fungi. VI. *Setchelliogaster*. *Madroño* 15: 73–79.

———. 1960. Studies on secotiaceous fungi. IX. The astrogastraceous series. Mem. Torrey Bot. Club 21: 1–112.

SINGER, R., W. H. SNELL, AND W. L. WHITE. 1945. The taxonomic position of *Polyporoletus sublividus*. Mycologia 37: 124–128.

SINGER, R., J. E. WRIGHT, AND E. HORAK. 1963. Mesophelliaceae and Cribbeaceae of Argentina and Brazil. Monographs of South American Basidiomycetes, especially those of the east slope of the Andes and Brazil. VI. Darwiniana 12: 598–611.

SMITH, A. H. 1965. New and unusual Basidiomycetes with comments on hyphal and spore wall reactions with Melzer's solution. Mycopath. & Mycol. Appl. 26: 385–402.

———. 1966. Notes on *Dendrogaster*, *Gymnoglossum*, *Protoglossum* and species of *Hymenogaster*. Mycologia 58: 100–124.

SMITH, A. H., AND D. A. REID. 1962. A new genus of the Secotiaceae. Mycologia 54: 98–104.

## Discussion

Petersen: I am impressed by the bridges between the gastromycetes and agaricoid forms, but as an aphyllophorologist, I would want to ask this question. If the agarics came from the gastromycetes, then where did the gastromycetes come from?

Singer: If it were true that the Agaricales did evolve from the Gastromycetes, then it would be much less probable that the genera which showed definite affinities both with Aphyllophorales and primitive Gastromycetes would still exist. It is possible that these forms would have disappeared to a much larger degree than the forms bridging the Agaricales to the Gastromycetes. As I conjectured in one of my papers on phylogeny, I would suggest that you look for the gastromycete origin very low in the classification of the Basidiomycetes, and I would consider that this is just as difficult a problem as finding the link between the Basidiomycetes and Ascomycetes.

Petersen: So it would seem that either we have direct links between aphyllophoraceous forms and gastromycetes, or that the aphyllophoraceous forms are degenerate from the agaricaceous forms. Is this true?

Singer: This is possible. The discussions here suggest that the Aphyllophorales may not be a natural order. Apparently, some Aphyllophorales have originated from forms very close to the Heterobasidiomycetes. On the other hand, it is quite possible to imagine that some of the forms of Heterobasidiomycetes with internal spore production may actually be close to certain gastromycetes. But these are mere conjectures. I would rather dwell on the bridges between the gastromycetes and agaricales, because these are facts.

Both: In your study of the bolete flora of Florida, and further in your experience in South America, you have indicated quite clearly that the boletes become scarcer going south across the equator. Now you have presented the same trend in *Melanomphalia* going north. Is this simply because you have worked in South America?

Singer: This may well be so, because these melanomphalias, since I have come to a definition of the genus, have been looked for and found mostly in southern South America. It may very well be that there are quite a few more species in other regions, like the Gulf (of

Mexico) region and in the southern (United) States. On the other hand, it is quite possible, perhaps even likely, that there is a petering out of species going north. This would tend to be parallel to what happens in the genus *Paxillus* and the Paxillaceae, which in general are represented by many more species in southern South America than they are in any other place in the world, and they tend to peter out as they go north. We must assume, therefore, that there was a southern origin of the Paxillaceae. In the series that Moser and Horak have characterized as the Paxillogastraceae, composed of three genera, it is no coincidence that they were discovered in southern South America. In the boletaceae, just the opposite happens. We have just two species of the genus *Boletus* in southern South America, and these are strictly linked with *Nothofagus*, which makes one think naturally that they have migrated with *Nothofagus* coming from Australia.

BURDSALL: I cannot envision what pressures could force the establishment of a forcible discharge mechanism of spores in underground forms. To have developed the spore discharge mechanism in so many gastromycete lines, and to have them all come out the same, is improbable. I cannot see that there would be enough selection to develop a system like this.

SINGER: You do not mean to ask how this type of basidium could originate, but how this form of violent discharge of basidiospores could originate. I want to make these two concepts clear. They are very important and very often confused.

A gastromycete is not necessarily hypogaeous. Neither are the hymenogasteraceous fungi to which I referred as hypogaeous. At best they are semihypogaeous and many of them are strictly epigaeous. We have quite a few like *Weraroa* which, for instance, is undeniably very strongly related to the Strophariaceae. The species are lignicolous, and not one member of that group is hypogaeous. Moreover, your question probably implied the origin of the violent discharge mechanism in gastromycetes with internal spore production, eliminating the Phallales for instance.

Now I must state that several kinds of gastromycetes evolved functional systems that are capable of propelling the spores into the air.

BURDSALL: Why do these groups arrive at exactly the same spore-discharge mechanism? The Ascomycetes have developed several different spore-discharge mechanisms which accomplish the same purpose.

SINGER: By the same token, one could ask why the Discomycetes, the agarics, and the Aphyllophorales have arrived at the same carotenoid pigmentation at different levels. I don't see why they couldn't or why they should be different, because this may have been the only mechanism possible at the time. But if you say that only one type originated, then I want to remind you of a paper recently published by Greenling and me on the agarics of the Congo, in which we described a different type of eubasidium or functional basidium in the genus *Pseudohiatula*, in which we have perfectly symmetrical spores as you find in the Ascomycetes. In *Pseudohiatula* the sterigmata are exactly as you see them in other groups of agarics or aphyllophorales, but the spores come out at right angles where the proximal end is attached to the tip of the sterigma. At the tip there is a perfectly symmetrical spore which is round. If you saw it isolated, you would think it was a gastromycete. There are at least two types of eubasidium, then, that seem to have originated in different parts of the gastromycetes.

PETERSEN: I suspect that this is only a partial answer. This second type of spore production was described recently, and in one genus. Many, many species have been described and studied in detail which do have one type. So even if we were to say that two types originated, then this would not nearly account for the number of bridges which have been described between the agarics and the gastromycetes.

SINGER: I cannot quite see why it should, because this mechanism is an answer to the same necessity which arises at one particular moment. When the gastromycetaceous fruiting structure comes to a certain level of evolutionary development, it is in need of some spore-discharge mechanism, because this will make its spore distribution more effective. There aren't many ways open to it. You will find curved sterigmata in many gastromycetes and now in one group these curved sterigmata bear naturally symmetrical spores. The necessity arises as a consequence of the development of the gastromycete organism.

BOTH: It seems to me that your statement is of fundamental importance, but will have to do ultimately with the environment at the time when these fruiting structures developed. A forced discharge will give a fairly wide spread of spores. It also may mean that the wind conditions in the atmosphere were not as they are today. I think the question of spore discharge may become very important in the entire question of the evaluation of these things.

DONK: When you analyze the mechanism of forcible discharge of the ballistospore you will recognize that it is coupled with an enormously complicated structure, including, for instance, the curved sterigmata tipped by spicula of a very narrow diameter, the production of the "water drop" on which Buller insisted, and the asymmetrical spores. It is difficult to imagine that this complicated mechanism could have arisen spontaneously more than once as a whole, like Athene from the head of Zeus, complete with helmet, spear, and so on. This is my principle objection to accepting the agarics as being derived from the gastromycetes.

The basidia of the gastromycetes are very variable from genus to genus. The spores may be sessile or formed on sterigmata (often of different lengths) which are straight and not tipped by spicula; the spores may also arise from the sides of the basidium, and so on. In every instance of assuming the origin of a particular group of agarics from gastromycetes, not only must it be accepted that the complete mechanism of spore discharge comes into being, but also that the transforming basidium must be reorganized into a neat clavate structure with the curved sterigmata on top and all of the same length. This transformation is so complicated that I cannot imagine that it could have arisen more than once.

This is not all: the gasteromycetous fruiting body will have to be adapted not only for supporting favorable spore discharge but also each time result in an agaric. But let me stop here.

SINGER: Supposedly, we had a very highly evolved basidium already, which accrued step-by-step. It would be more difficult to imagine that this arose in a very primitive basidiomycete as you have postulated in the Aphyllophorales.

In those forms where this evolution has actually happened, as in *Macowanites*, and other genera that are really already at the agaric level, this kind of basidial irregularity is not found. It is found only at a much lower level. I interpret this, therefore, as the flexibility of an organ which is primitive, not as degenerative.

DONK: I think that *Macowanites* is a case in which the mechanism of ballistospore discharge may have disappeared "recently." If I remember correctly, Dr. Smith has once told me that it occurred that specimens collected as species of *Russula* turned out not to produce a spore print. They lacked the ability to discharge their spores and turned out to be species of *Macowanites*.

SINGER: Every mycologist has had it happen that some specimen of agaric did not give a spore print. We are always bothered by that. This does not prove that all these specimens are inclined to become gastromycetes. There are many factors that will influence the living specimen of agaric not to shed a spore print, even leaving it in the refrigerator.

PETERSEN: Smith said also that when he looked at the basidia of these fruiting bodies, they were gastroid and the spores symmetrical on the sterigmata.

SINGER: In many species of *Macowanites* the configuration of the basidia is quite like that of an agaric. If we assume the opposite theory, of the loss of violent discharge, I cannot see why it would be lost. Where is it lost? In hypogaeous forms, where it is not needed. But here we must insist that it is not hypogaeous fungi alone, but mostly epigaeous and semihypogaeous forms which have lost this property.

BURDSALL: I am not so sure that violent discharge is the most efficient spore-dispersal mechanism because selection must be heavily against this system for survival. Certainly it is efficient, but I could see how it could be easily lost. With such a refined mechanism, mutation of very few genes, I think, could easily produce a gastromycete-type basidium.

SINGER: If, by accident, you lost an efficient apparatus for spore dispersal, why is a long line of gastroid agarics formed which ends in ever more gastroid conditions? You can start with an agaric and produce a gastroid state by modifying certain conditions. But I cannot see why this inefficient organism should start a long evolutionary line of gastromycetes which are not grouped together. In the Agaricales, you have only *Russula* and *Lactarius* in the whole astrogastraceous series.

PETERSEN: Would you care to comment on species numbers in all of the Agaricales?

SINGER: Species numbers are sometimes taken as an argument in favor of this theory of degeneration. Species number, however, is always very large in the final ramifications of the system, because the taxa are actively evolving.

DONK: I wish to remind you that the loss of forcible discharge of spores (but I am not committing myself for the moment) seems to have occurred in many places among the Hymenomycetes. I remind you of *Xenolachne* described by Dr. Rogers, one of the resupinate Tremellaceae, where you will find some kind of gastromycetous state, viz., the loss of active spore discharge. Something similar is found in *Phleogena*, an auriculariaceous fungus, without forcible spore discharge; it produces so-called dry spores apparently requiring mechanical disturbance to achieve distribution by air currents.

LIBERTA: I might respond to Dr. Singer's thought that he could not see how the highly specialized efficient spore-discharge mechanism could degenerate. I think it is a fairly well-accepted principle that as an organism becomes more highly specialized it faces a greater chance of extinction. A basidium with a highly specialized spore mechanism could very well begin degenerating, perhaps many times and in many different ways by very small genetic changes.

SINGER: I would accept what you say, and I think I have said it before. We have forms, as in *Psilocybe*, in which we know that this has taken place. We can control this phenomenon. What I cannot imagine is that from this degraded *Psilocybe* you can construct a further long line of evolution. Wouldn't you rather expect this to happen from the normal *Psilocybe* carpophore?

ALEXANDER H. SMITH

*Professor of Botany and Director of the Herbarium*
*University of Michigan, Ann Arbor, Michigan*

# THE ORIGIN AND EVOLUTION OF THE AGARICALES

Much has been said on the subject of the origin of the fungi in general and of the hymenomycetes in particular—so much so that it is difficult to devise a new approach. As I interpret my assignment today, it is to try to present this subject with a different slant or emphasis than one finds in the literature in general. To do this, I am of course drawing largely from my own experience over a forty-year period of studying the organisms in question. Presumably in this time I have developed a few ideas: it is you who are to judge their value.

My plan of presentation is to extrapolate backward in time from certain existing models to see if they have any meaning in speculation on the phylogeny of the group. My hunting ground has been the continent of North America north of Mexico. My coverage (an adequate sample of the group from each vegetationally significant region) of the area is hopelessly incomplete, and I must apologize for having only one lifetime in which to do the work—but if the study of over 80,000 collections has any meaning, perhaps there is some justification for my discoursing on this subject. I know of no continent better suited to the study of species evolution in the fleshy fungi than North America. There are the rain forests of the Pacific northwest with pockets of arid or semiarid land in the rain shadows: 10–15 inches per year of precipitation at Sequim, Washington, on the Olympic peninsula and 70–100+ inches per year sixty miles west. There is the arid southwest with its alpine patches of conifer forests where, in the summer, the rainfall is more than adequate for fleshy fungi. You will see the Smoky Mountains here in the southeast for yourself. New England presents its own pattern of differences and certainly Laurentide Provincial Park in Quebec must be mentioned as among the best for fleshy fungi on the continent—if the collector can stand the black flies. The central Great Lakes region, the Ozarks, the Gulf

Coast, and the Great Plains all add to a diversity of habitats ideal for particular groups of fleshy fungi, but the important areas are those where one type of habitat merges into another. The number and pattern of transition areas, at least to me, seem to be legion, and they all need to be studied.

It should also be pointed out that a single season in an area is hopelessly inadequate to study a mushroom flora. It takes many seasons of concentrated effort. I know this first-hand just from working the area in southeastern Michigan. In fact, it is this necessity to work many seasons in an area on a full-time basis that has kept the development of knowledge of our group from increasing as rapidly as it has in other groups where the same type of field problem is not encountered. But one season is always better than none.

After World War II there were extensive virgin forests of hardwood—beech-maple, birch-hemlock, and also jack pine—on the upper peninsula of Michigan. To a large extent these were butchered according to the traditional pattern of the American lumber industry and the debris (consisting of cull logs and stumps as well as the tree tops) was left to rot. Fires were kept out of the hardwood areas for the most part, young growth came up as sprouts and was so dense in beech-maple areas that essentially a giant damp chamber soon existed on the ground where the slash lay. This was ideal for the development of wood-rotting fungi and therefore for the study of them.

From 1950 to 1965 I made as intensive a study of one of these areas (in Chippewa County, Michigan) as time and energy permitted. The incidence of different species was noted in a general way, but a particular study was made of *Pluteus.* In the course of about six summers' work it seemed to me that between 25 and 50 per cent of the pieces of wood the diameter of your wrist or larger were invaded by some member of the *Pluteus cervinus* group and a considerable number of variants around *P. cervinus* and *P. salicinus* were noted.

Then late in August, 1957, I walked into one of my plots and every piece of wood, almost literally, larger than the diameter of one's thumb had a *P. cervinus*-type basidiocarp on it. I brought home 200 samples, of which 80 were finally studied and numbered. Five days later there was hardly a *Pluteus* basidiocarp to be found in the area, and marked basidiocarps had decayed and disappeared in that length of time. The quantitative observations made the first day and again two days later clearly showed that *P. cervinus*-like fungi had invaded at least 99 per cent of the available substrate in the area and, obviously, my previous estimates of the saturation of the substrate by *Pluteus* were off by 50 to 70%. Two days of collecting in a three-day period had refuted the observations of at least six seasons of sampling.

This should be of interest to all mushroom collectors, but of greater interest to me was the diversity in the characters found in the basidiocarps and the difficulty in classifying them to species. Variations in the shape and size of cheilocystidia, thick- and thin-walled "metuloids," the degree of apical ornamentation of the "metuloids," and macroscopic features such as odor, taste, color changes, pigmentation of pileus, and ornamentation of the stipe were all cataloged. Variants were present representing all stages of diversity from the presence or absence of a single character to various combinations of characters usually used for the recognition of species in the group. I have not described these variants as yet, and for the moment, detailed descriptions would appear superfluous. Let us focus instead on the diversity of this gene pool as represented by the collections studied. It was seen very soon in the course of the collecting that certain obvious types were repeated frequently in the plot. Others were proportionately more rare, and finally there were some represented by a single basidiocarp. Microscopic study showed that in some cases groups could be subdivided further and at one time in a working manuscript I described about 60 "variants" in the group.

This experience with *Pluteus* was one of the clearest demonstrations I have ever seen of genetic diversity in the population of one relatively small group of wood-inhabiting fungi. To me it seemed to be evidence, if we assume that the characters concerned were gene-controlled, of introgression. From a basic population of (my estimate) less than a half-dozen parent species, about all possible combinations of characters turned up in the basidiocarps in this slashing.[1] I think these observations, crude as they were, have a bearing on our subject today: I could (and did at one time) take my long list of variants and arrange them in a "phylogenetic" series starting with basidiocarps showing a difference in one character and

[1] Slashing: the debris left on the forest floor after usable timber has been removed—lumberman's term [Ed.].

working up in sequence to the greatest number of differences as represented by the combinations in individual basidiocarps. Thus, populations with a difference in one character in this scheme were considered to be more closely related than those differing by two, etc.

But finally, all this seemed meaningless to me as far as a "true" evolutionary sequence was concerned. My impression of the situation in retrospect is that in these slashings there obtained an ideal situation for explosive development of a group of organisms and that obviously it occurred, since I found the evidence. Moreover, in the "melting pot" of gene exchange on a wholesale basis, and the opportunity for the progeny of the exchanges to survive, these fungi apparently behaved like other plants in similar situations, except in vascular plants each variant in a hybrid swarm is no longer recognized as a species. In the fungi—fleshy fungi in particular—we are not sure the variants are hybrids in the usual sense of the term, and we have a tendency at present to describe species and genera, and worst of all to propose elaborate systems of classification based on these combinations of features and to speak of evolutionary lines leading from one group to another. More pertinently, I think that probability favors the conclusion that many of the species in such lines originated independently from the same gene pool more-or-less simultaneously and that variants of lesser to greater complexity did not originate in an orderly sequence. Rather, I prefer to think in terms of clusters of variants all originating from a basic gene pool.

But how does all this relate to the origin of the higher fungi? Let us first consider the basic ingredients of the problem: the abundance of substrate allowed for survival of a larger progeny of *Pluteus* species than usual from the spore rain of this group over a period of years (not more than 15 in this instance), and the amount of inoculum was tremendous. In addition, the time involved was short by all estimates, as these pertain to the evolution of species. But it was perfectly clear that gene exchange of some sort had occurred at a rapid pace and on a large scale. If this were a single case I would hesitate to generalize from it, but in forty years of field experience I have found the pattern repeated with variations, depending on the genus and the ecological niche it occupies, and have come to the conclusion that periodic explosions of the *Pluteus*-type represent a major pattern for the evolution of (or at least the initiation of) variants which, if they can successfully compete, eventually become recognized as species.

In 1947 I found a similar situation with *Cortinarius* in the Mt. Hood National Forest of Oregon. My two previous seasons in the area had indicated 80–100 species of *Cortinarius* present in the area but in 1947 I found 192—recognizing species at the same level as used currently by European mycologists. In 1952, it occurred in the genus *Galerina* on the warmest and driest year known for Mt. Rainier National Park. The moss-covered run-off areas near the mountain glaciers had an exceptionally diverse flora, and a few years later some of these same areas were again under ice throughout the year. In *Leccinum*, it is evident to me that when Michigan was logged for its pine and burned over during the first quarter of this century, the extensive development of aspen-birch forests allowed the survival of many variants from a residual gene pool which escaped the holocaust, creating a situation similar to that reported in *Pluteus*, but here dealing with mycorrhiza-formers. As you can see, personally, the evidence for this type of evolution appears convincing on the basis of my experience, and I prefer to project it backward into geologic time and state that in all probability the basic pattern was the same then as it is today. In one form or another this principle is well known and recognized in the studies of the evolution of other plant groups. My own observations on the higher fungi merely indicate that it can also be applied here.

Lacking any significant fossil record, we do not know what the ancestors of the fleshy fungi were like, but on the basis of their present life habits, ecology, and the structure of the sporocarp, we might assume there was some similarity to present-day types. We might also extrapolate backward in time to postulate what these primitive types were like, for this is what we are here to talk about. I assume that complicated structures arose from simpler types, but that evolution of types progressed both in the direction of increasing complexity as well as back toward simpler types (regression), depending on the opportunities for survival.

If we consider the conditions that prevailed on the earth at the time (or period) in which plants were adjusting to a terrestrial life, it would seem likely that organic matter accumulated on land in great quantity, lacking any agent to reduce it. We could even call this accumulated organic matter "sludge" since we know that the algae soon developed the ability to live on wet earth, so clearly there was an evolution of green plants from a truly aquatic habitat to one above the water line. But there is *no reason in the world* to assume that only one species of alga made this move. Moreover, we know from present-day evidence that vascular plants at various levels of evolution have lost their chlorophyll and have become adjusted to highly specialized modes of existence. We can reasonably extrapolate back in time to the various algae which evolved to a terrestrial existence and assume that, given contact with organic materials, any of them could have adjusted to a saprobic manner of life, or even to a mildly parasitic type of existence. Such organisms would be classed as fungi, but obviously would have retained any of the reproductive patterns which favored survival, even to motile reproductive cells.

In this situation there would have been an excellent opportunity for explosive evolution, with opportunities for many species to give rise to variants which in return would survive and evolve further at a rapid pace. Clusters of variants could be expected, between which there would be all degrees of genetic isolation. I would defy anyone to outline a tree of evolution given these circumstances. This pattern of evolution would repeat itself each time a mutation allowed exploitation of a new source of food for the variant, or each time ecological changes occurred to suddenly produce a new nutrient source, such as the charred remains of material following a fire (caused by lightning) and the release of assorted minerals in relatively favorable proportions. Thus, the pyrenophilous fungi could have originated and survived—even to our time—because of the frequency of lightning fires, but they are not thought to be closely related as a group as judged by similarities in their reproductive apparatus. As in *Pholiota*, however, there is evidence that in at least some instances there has been development of diversity after adjustment to the habitat had been made by a particular taxon. This is far more obvious in the coprophilous fungi, which must have evolved at a later time. For the most part, these fungi consist of an assortment of species of various relationships whose ancestors at one time or another made the same ecological adjustment, undoubtedly triggered by similar conditions prevailing in the habitat.

On this basis it is easy to see the difficulties in regard to the concept that the fungi represent a single "line" with branches from it.

In the evolution of the fungi, gene pools featuring various patterns of sexual reproduction certainly existed at various times. On the basis of populations now present, it is entirely reasonable to assume that some types of sexual reproduction in these pools featured exogenous and others endogenous formation of meiospores, but it would be a mistake to assume that the latter were Ascomycetes and the former Basidiomycetes. Both the ascus and the basidium as we know them now are highly evolved structures both functionally and morphologically and must have reached their present diversity by evolving from less specialized types.

On the basis that most ascomycetes have 8-spored asci and basically the production of 8 nuclei follows subsequent to meiosis, and the fact that a number of extant basidiomycetes exhibiting this pattern still persist, it is reasonable to assume that the production of 8 meiospores was common to many of the prototypes. Hence, a derivation of at least a large segment of Ascomycetes and some of the Basidiomycetes can be assumed from this common mixed prototype. In one direction there was a trend (evolution) toward exogenous production of meiospores and in the other a refinement in the production of such spores endogenously.

Thus far I have said nothing about types of ascocarps or basidiocarps. If one starts with a filamentous thallus (and this assumption seems irrefutable), what type of fruiting structure would be most likely to evolve under the circumstances, assuming that the first type of spore was air-borne? In both spore production types, simple club-like sporocarps are still present; obviously a "parallel" type of development, since the reproductive structures are totally different. But is the production of an upright cylindric to clavate sporocarp a primitive structure? For all these fungi, basically saprobic in their nutrition, I think it is. Consider two things: (*a*) There has been a tendency in the primitive gene pool to produce air-borne spores. (*b*) The simplest and relatively most efficient type of sporocarp for aerial dispersal of spores would be a column of hyphae projecting up from the substratum as far as the relative humidity of the habitat would permit. The hyphal end-cells would (or could easily) bend toward the surface of the sporocarp and very soon in the course of evolution a layer of such cells would be produced and a hymenium established. The simplicity of the model appeals to me because it would appear to involve no major change in the genome of the species to produce it.

How many times it originated independently in the parental prototype-polygenic pool no one can say, but it would be naïve to assume that it originated only once. Such a primitive fungus, if it were a basidiomycete, should show certain basic features: (*a*) eight nuclei produced in the "meiotic package"; (*b*) basidia of a primitive type—the hyphal end-cell should not be morphologically greatly differentiated from those end-cells remaining sterile. I would not expect all the spores to be produced apically.

If we search in the Cantharellales we can find species with basidiocarps which fulfill the requirements of a primitive species quite admirably. True, in some the apex of the hyphal column has flattened and spread out with the "hymenium" forming on the under surface. The surface varies from smooth to wrinkled depending on the species. In the genus *Cantharellus* we find the following primitive features: (*a*) relatively long, narrow basidia; (*b*) a rudimentary "hymenophore" (the surface smooth to wrinkled); (*c*) frequently more than 4 spores per basidium, and in the *C. infundibuliformis* group occasionally lateral sterigmata bearing spores, usually only one lateral on a basidium; (*d*) basidiocarp hyphal structure relatively simple; (*e*) sterigmata typically large and bowed (the "*Clavulina*-type"); (*f*) gymnocarpic development of the basidiocarp—this feature is imperative if the previously made rationale is correct.

It appears to me, on the basis of the possible introgression which forms the rationale for my "rosette" pattern of evolution, that each time a variant finds an opportunity for sudden rapid evolution, approximately the same type of hyphal modifications will sort into various combinations to produce what we now call "species." Thus, a dimitic type of basidiocarp could arise independently at several stages in the evolution of the basidiocarp in the Cantharellales—at the cylindric, the clavate, the pileate, and finally at the hemiangiocarpic stage, if such a stage actually exists in this order. In other words, there is no better rationale to support the idea that the dimitic species all evolved from a single parent with a dimitic hyphal system than there is for assuming that all basidiocarps bearing an hymenophore of teeth are closely related. Thick-walled hyphae could also arise *de novo* in most any stirps of any genus. In order to plot the evolution of a group not only must the anatomical features of the hyphae be studied but also these should be correlated with other sets of characters: chemical reactions, stages in evolution indicating a distinct increase in efficiency of spore dispersal, or levels showing more refined adaptations to insure the success of the spores produced. The latter are largely macroscopic features such as presence or absence of a veil, a poroid versus a lamellate or smooth hymenium, gelatinous layers which reduce evaporation from the basidiocarp, etc.

If one traces possible relationships through the fleshy fungi using such correlations it is apparent (at least to me) that an overall correlation of characters can be shown, probably indicating relationship in the progression from the simple clavarioid fungi through *Clavariadelphus*, *Craterellus*-like fungi (with more-or-less smooth hymenium) and *Cantharellus* to *Hygrophorus* on one hand and *Clitocybe* on another.

These need not be visualized as "lines," I suppose. We have all used that term and probably will continue to do so, but they should be visualized rather as intermediate gene-pools from which constellations of variants arise. These variants eventually result in other gene pools at some later time from which more variants may arise. Whether progressive or regressive evolution is evident will be very difficult to plot from the surviving progeny of the interaction. It is this type of major intergradation that Singer and I used to indicate major areas of relationship in our studies on secotioid fungi.

If this procedure is projected from the *Clavaria-Clavariadelphus* group through *Craterellus-Cantharellus* and *Hygrophorus-Clitocybe* into the gill fungi, *Clitocybe* is seen as the key genus in the evolution of the latter group. One major character, as judged by the overall diversity in the gill fungi, is color of spore deposit. In *Clitocybe* this color may be white, yellowish, pinkish, or pale tan. In this gene pool, therefore, one finds the beginnings or "roots" for the elaboration of the chromospored groups—it is but a step from *Clitocybe* to yellow-spored and brown-spored species, which reach their maximum diversity in *Cortinarius* with approximately 800 species in North America. As the brown-spored gene pool built up, there was a bifurcation of rough- and smooth-spored species with the smooth finding a center in *Pholiota* in the sense of the Smith and Hesler monograph. In the smooth-spored group starting in *Galerina*, a discontinuity in the apex of the spore began to differentiate, presumably as the wall of the spore thickened (?). The genes for this structure, however, apparently were absent in the *Cortinarius* pool. In the *Pholiota* gene pool of about 200 species in North America we find so many degrees of development of the apical germ pore of the spore wall that it was found impossible to maintain the genus *Kuehneromyces* as described by Singer and Smith. Also in this pool we found the development of chrysocystidia of so many different morphological types, and with such variable content from one species to another, that any attempt to use these organs as a diagnostic generic character did not appear advisable. It is in this gene pool apparently that we find the beginnings and "mature" development of chrysocystidia on a large scale, but it is also obvious that chrysocystidia originated in many other groups of hymenomycetes as well. Hence, they appear to be a response to a general need encountered by various groups in the course of their metabolic processes.

The significant conclusion about chrysocystidia and the development of a broad germ pore on the basidiospores of some species is that there has been brought into focus a combination of features characteristic of most of the species in the subfamily Stropharioideae of Singer (Strophariaceae). Singer's subfamily as described covers a new gene pool derived from *Pholiota* by what appears to many as a rather insignificant change in the color of the spore deposit in highly developed species from rusty-brown to purplish-brown or actually violaceous. Please remember that this difference must be regarded as a color spectrum—the species show many levels in the evolution of the pigment(s), and there are very confusing intermediates. I regard Singer's subfamily as basically a single gene pool and would now apply one name to the genera *Stropharia*, *Naematoloma*, and *Psilocybe*. The noted French mycologist Lucien Quélet came to the same conclusion long ago, and this concept was also followed by Kühner and Romagnesi. I suspect that none of the gene changes necessary to accomplish the modifications in the characters discussed, or in any others pertinent to the subject, need be great at any single stage, and it is this possibility above others which convinces me that there is a strong probability that evolution actually progressed in the direction outlined.

As in any truly natural classification one finds groups which do not connect readily to other known groups of fungi. The Amanitaceae constitute such a group, as do the Coprinaceae, to a lesser extent. The angular-spored species with a reddish spore deposit probably originated from *Lyophyllum* or an ancestral *Lyophyllum* gene pool in which the genes for spore color were retained.

One of the most unique trends of evolution is the one which has received the most attention early in the study of relationships of secotioid fungi to the Agaricales, namely the astrogastraceous series (of some authors). This series includes species which connect to *Lactarius* and *Russula* but in which the basidia do not forcefully discharge spores. Since the International Botanical Congress in Montreal I have continued to study these fungi, most recently from the standpoint of a gene pool from which these fungi could have evolved. The following are my current thoughts on this problem: *Hydnangium s. str.* does not belong in the series—at least some species are related to *Laccaria. Octavianina nigricans*, *O. asterosperma*, *O. tuberculata*, *O. lutea*, *O. lonigera*, and *O. laevis* are certainly peripheral to the series on the basis of the peculiar, often dextrinoid spore ornamentation. The presence of sphaerocysts in the fruit body peridium of this group does not necessarily mean a relationship to the true "astrogasters" since the feature is scattered throughout the Hymenogastraceae, even in *Rhizopogon* where it obviously is not phylogenetically significant. The important fact to me is that it is only in the Hymenogastraceae that one finds a large gene pool of species showing various degrees of development of heteromerous tissue, and to my knowledge this is the only group in the higher fungi for which this can be said to be true. In the pattern of development of the known species and in the large number of taxa involving *Rhizopogon*, *Hymenogaster*, and the astrogasters including *Martellia* and more advanced forms, it is a sizable gene pool. This situation reminds me strongly of the chrysocystidial types in *Pholiota*. It is clear to me (although perhaps not to some of you) that the heteromerous type of tissue as found in *Russula*, for instance, had its origin in the pockets of inflated cells adjacent to the inner surface of the peridium (or actually involving its tissue), which normally (under conditions of continued development) would have developed into spore-producing cavities. As development slowed down a zone of these immature cavities became arrested in development and appear in sections of the peridium as an inner zone of pockets of inflated cells, for localized cell inflation is the first stage of differentiation in locule formation.

The "astrogasters" at the *Martellia* level appear to be related to the group near *Hymenogaster*, which bears colored spores with wrinkled, nonamyloid ornamentation. The reason for venturing this evolutionary opinion is, again, the probably slight changes in the genome necessary to pass from an *Hymenogaster s.l.* to a *Martellia.* Other features, however, support this view. The basidia in the astrogasters and in the hymenogasters are of a relatively advanced type. Many "species" with 2-spored and some 1-spored basidia are known, the basidia typically large and in a distinct palisade. This is in decided contrast to what is found in *Rhizopogon,* for instance. To me the basidial characters of the fungi in the *Hymenogaster* gene pool indicate reduction from agaric ancestry, also perhaps documented by the gross morphology of the basidiocarp. The studies of Singer and Smith on *Thaxterogaster* actually lend themselves to this interpretation rather than to the conclusion that the thaxterogasters have evolved from other Gastromycetes. Moreover, Thiers and Smith[1] described some hypogaeous taxa which were true *Cortinarii* because a copious spore-deposit was found on the inner surface of the veil even though the basidiocarps were up to 3 inches below ground level. These discoveries more-or-less terminate the argument about the direction of evolution in this group, as I find it very difficult to maintain that spore discharge reached full development *before* it could possibly be of any use in matters pertaining to survival of the species.

At the Chicago meeting of the American Association for the Advancement of Science a number of years ago I suggested, with my tongue in my cheek so to speak, that the relationship of *Cortinarius* to the astrogasters might be the direction in which evolution actually occurred. But the case for that conclusion is stronger now than it was before. That such a line as this might eventually revert to an agaric state is not improbable at all. The genes necessary to effect this change may have been rendered inoperative under certain conditions but were capable of functioning again when for some internal or external reason the inhibition was removed. By that time, in all likelihood the species reacting were no longer either typical hymenogasters and not yet typical astrogasters. Obviously, linkage between the controls for heterometerous tissue and amyloid spore ornamentation occurred at some time and a totally new trend of evolution emerged in its own direction. This idea certainly explains why the stipe in

[1] Thiers, H. D., and A. H. Smith. 1969. Hypogaeous Cortinarii. Mycologia 61: 526–536.

*Lactarius* and *Russula* is such an ineffective organ as compared to that of more specialized agarics.

In short, I still hold that the astrogastraceous fungi evolved into the genera *Lactarius* and *Russula*, but that they arose from a specialized gene pool, one in which at least one direction of evolution proceeded from the brown-spored agarics. The amyloid feature of spore ornamentation could have originated *de novo*, as it has in so many other groups. Although controversial, the idea is worth further study. In any event, this rather circuitous path of supposed evolution adequately accounts for all the unique features of the Russulaceae and does it from a very rational point of view.

There is not time today to apply my gene pool approach to all groups of hymenomycetes even if I had the knowledge to do so, but I should like to leave you with a few observations as follows:

1. The more one collects and studies fleshy fungi, the more one comes to realize that sooner or later almost all combinations of known characters will be found in some segment of the population of a genus. To me this argues for caution in proposing new genera.

2. An adequate study of diversity in fleshy fungi has yet to be made in any area (including Europe). Smith, Thiers, and Watling have become painfully aware of this in *Leccinum*, for instance. Such a study must be made for several areas before we can draw any final conclusions as to directions of evolution and "connections" from one group to another.

3. The major areas of rapid evolution can be accurately ascertained by thinking in terms of gene pools and introgression among the members. When we learn enough about a group, the relationships of it and within it will be found obvious, not limited to one or two rare taxa, and that age-old model of an extinct ancestral type which perhaps has been overworked will be laid to rest. We must think in terms of ancestral gene pools.

4. In arriving at ideas about relationships we should emphasize the taxa which are intermediate between given groups. The more intermediates we find the better is our evidence for what Singer and I have called a phylogenetic connection.

5. We should be wary of placing much weight on the meaning of a single odd species, perhaps already overemphasized in the past.

6. Above all, we need an accelerated program of studying these

fungi with proper refinements in both field and study procedures in order that the characters present in a collection be accurately observed and recorded. Too much of the collector's time is spent in searching fruitlessly through the literature, trying to compare species concepts extant on one continent with those bearing the same name from another continent.

In closing, I should like to make a plea to the European mycologists that they cooperate in a program to establish neotype specimens for all European species for which no holotype exists. We should legislate special rules for these particular neotypes: (*a*) that they be collections accurately and adequately studied in the fresh state; (*b*) that they favorably compare with the characters given in the original description; (*c*) that this concept be published in detail (accompanied, it is hoped, with an adequate photograph or painting); (*d*) that the neotype be collected in the country from which the species was originally described. In the recent work by Corner on the Cantharellales we have good examples in *Craterellus cornucopioides* and *Cantharellus cibarius* of what the lack of such neotypes leads to in the way of confused concepts, and eventually to confused ideas on evolution and phytogeography.

## Discussion

LUTTRELL: I have heard repeated references to the lack of a fungus fossil record. From one point of view I am not certain that the presence of some fossil record would not be even a greater handicap. It has been said that nothing inhibits speculation so much as a few facts. I think if we are going to use the lack of a fossil record as some sort of general excuse for possible failure in working up the evolution of the fungi we are first under some obligation to examine the fossil record and know just how inadequate it is. In short, I am not sure that we can say there is a lack of a fossil record.

SMITH: As far as the anatomy of the fruiting body is concerned, we do not have structures comparable to those of the ferns or other groups. We do have records of mycelium with clamp connections going way back to the Pennsylvanian Period.

LUTTRELL: I think we also have to state just what we think this fossil record would contribute to the particular argument. We might be under some obligation to point out the points in our argument for which the fossil record would be critical. This is not especially pertinent to this particular group, however, but to the fungi in general.

SMITH: Our studies are based to a great extent on very perishable material. We draw conclusions or speculate in regard to the evolution of structures that simply cannot be demonstrated as having been present in the Eocene or Miocene or anything like that. If we knew that polypores actually occurred in the Pennsylvanian this would be of great help. We could at least state that there were polypores back that early, and we would know that the fungi had reached considerable diversity at that time. There would be no reason why some other groups could not have evolved to the same level by the same time, but we do not know this.

One of my points was that I thought evolution in the higher fungi was going on at a rapid pace, and it is our group that is perhaps one of the latest groups of plants to undergo this extremely rapid evolution. In my class I use the example of the influence of Henry Ford on the evolution of the fungi. Before Ford invented the tractor every farm in the midwest had a manure pile and there were coprophilous fungi all over the place. Now the substrate is getting very rare. It is spread out very early, the fungi do not have a chance to grow on it to the same extent, and the evolution of coprophilous fungi may have slowed down materially.

LUTTRELL: One of those things that bears some weight is that there were highly developed fungi, perhaps polypores, or perhaps certain highly developed Ascomycetes, back in the Devonian.

SMITH: Before you can say it was an ascomycete, you have to have a fossil ascus. You might have something that looks like a *Xylaria*, but it wouldn't mean anything. It might just as well be a basidiomycete, unless you actually have the reproductive structures preserved.

SINGER: In a paper in the American Journal of Botany,[1] I described *Phellinites*. This discovery indicates that polypores of the *Fomes* type go pretty far back. I would not apply this alibi of lack of fossil record generally. I would certainly restrict it to some higher fungi.

SMITH: But I think you have to admit, though, that the number of examples that we have of the fungi in the fossil record is exceedingly meager compared to any other group.

[1] Singer, Rolf, and Sergio Archangelsky. 1958. A petrified basidiomycete from Patagonia. Amer. J. Bot. 45: 194–198.

SINGER: When you first told me this story about the "*Pluteus* explosion," I had the nightmare that mutations and gene exchanges take place before our eyes at the present time. I call this a nightmare because it would be something rather catastrophic to taxonomic work. In relation to this, what evidence do you have that the other forms of *Pluteus* that appeared later on these trunks were not brought in by spores from other places, so that the parent group from the first population that you observed would be genetically stable?

SMITH: I cannot exclude the possibility of introduction from outside, for this is all speculation. But I would answer you this way. Previous to this work, we had been convinced of rather few species of the *Pluteus cervinus* group in the Great Lakes region. On this basis I have not found any new species in all the time I have worked there. I assume that this localized "explosion" was due to the natural group that happened to be occupying the area at the time. The "explosion" resulted from an interaction, because of a very generous supply of nutrient material. At that point, the whole past project I had done on *Pluteus* suffered a sudden shock. I could not make the spores of these entities germinate. I was going to contact Dr. Raper on this. I had hoped to do all the mating experiments, and catalog the mating reactions of all these plutei. I failed in this because I could not germinate a spore. Because they are supposed to be wood-rotting fungi, *Pluteus* species should culture very easily, but they seem to be very difficult. I think Fergus had one collection of *P. cervinus* that germinated. We know very little about the ability to germinate the spores of this genus.

RAPER: I know of no case where there has been bona fide interspecific crossing between Basidiomycetes. *Sistotrema* is about as close as anything I know of. I am understandably a little puzzled about your speaking of gene pools, the interaction of these organisms, and the tremendous number of species, because in all probability they are genetically isolated.

SMITH: This is the reason why I hesitated. Why do you get so many variations, though?

RAPER: The only means I can see for getting an "explosion" like that is that in *Pluteus* the various species exploit a particular niche. That is, they get in there and overwhelm all the competition. This would seem to me a more likely explanation.

SINGER: From what we know about spore dispersal, we can certainly assume that they might have come from the outside.

SMITH: They have never been described from anywhere, so this proposition begins to get weaker. This was my point.

RAPER: In your previous examination of this area would you not want to include the presence of one or a few of these many, many types over the years?

SMITH: As I recall, I had most of the parent species, at least the ones that I consider to be parents, and probably ten or fifteen of these so-called variants that differed in one character from the parent species. So I would say that, without deep thought on it, there is plasticity in the species.

MCLAUGHLIN: What is your evidence that all of the variations you observed were genetically determined? If the fungi were in nature, all you knew was that environmental conditions were not controlled. It seems to me that you couldn't know for sure the genetic constituency of the variations.

SMITH: I can collect, say, ten fruiting bodies from different mycelia, and if the composition of characters is similar, I assume that these are genetic characters. A violet reproduces a violet, whether parthenogenetic or not. Some of these fungi, like *Mycena*, are parthenogenetic. What you are introducing is an assumption that we make for all things that we find in nature—that they are constant. That would also apply to *Pluteus*, but I haven't proven this. I failed in the culture work, which is where I hoped to really concentrate my efforts, and I must find some way of germinating the spores before the proof that you desire can be presented.

NOBLES: I wonder whether these could have fruited in the haploid condition. When I assayed a whole series of monospores from one fruit body there was far more variation among the monocaryotic mycelia than there was when I had an equal number of dicaryotic mycelia of the same species. Is it possible that you might have fruiting in the monocaryotic condition and so get this wide variation?

SMITH: We worked on this aspect in this material. Of course, one of the big factors in this study was the presence or absence of clamp connections, because this is a character by which the species of the *P. cervinus* group have been distinguished. We have examined close to 200,000 septa in the *P. cervinus* group to establish ratios, presence, and number of clamps present per hundred septa for many of these specimens. We are finding that we have some species that are regularly characterized by almost 90% clamps, some by about 50% clamped septa or less than that, and some down at the margin of 5% where you can find a very low incidence of clamps. One of the underlying problems in this study was the behavior of clamp connections, but we haven't gotten anywhere with that, except for the statistical part that I summarized.

JESSOP: You have discussed a lot of differences in the *Pluteus* group. You counted up as many as 60 different combinations. What is the "group" to which you refer?

SMITH: It is the section *Pluteus*. The parent species would be *P. cervinus*, *P. salicinus*, *P. brunneodiscus*, *P. magnus*, *P. washingtonensis*, and perhaps one or two others. No more than that, as far as I can tell. This is all based on circumstantial evidence, but how else can one explain this variation, even though one admits that they came in from somewhere else? It seems to me that there must be some gene exchange to produce them. We have another situation in Petersen's *Ramaria*. How can one explain a large genus like *Ramaria* where all the species look alike and differ in a relatively few characters? Is this gene exchange?

PETERSEN: One must bear in mind that this is Smith saying that they all look alike. I don't subscribe to that idea.

RAPER: I feel that Smith's theory is fine sexually. But I am worried about the gene exchange at this particular time if these are good species. I think there is very little evidence of any interspecific hybridization and there is only a little evidence in the ascomycetes.

Even one fruit body per year of each of the 60 species in the whole area of the upper peninsula could account for this when the conditions were exactly right. I do not mean to cast aspersions on your competence as a collector, but you might have missed a few of them.

SMITH: Suppose we stretch the time limit and accept these 60 as species. How did they originate?

RAPER: They may have come from the same species originally, but become isolated in the meantime by various processes. I am not talking about immediate isolation, either, but about isolation back about the ice age or something of that sort. Small pockets and time to fix genetic characters.

SMITH: But I think the probability of that is smaller than the probability of gene exchange.

RAPER: The trouble is that we have no reported precedence in this whole group for interspecific gene exchange. When and if you learn how nature germinated all those spores—because obviously they germinated—the problem can be solved.

ROGERS: I think the problem here is whether we have genetic variation at all. I have seen two extremes of *Hericium*, that with massive base of mycelium and long, unbranched spines, and that with a highly branched system coming out of the selfsame mass of mycelium. Here was the widest variation that the genus was apparently capable of, coming out of the same knot of mycelium. I think that we have yet to establish that this is actually a detectable gene pool, that genetic variation and crosses have taken place, and that combinations and mutations have occurred, which we must do before we can discuss it profitably.

BURDSALL: I was wondering if you could make a little clearer the origin of *Russula* and *Lactarius* in the astrogastraceous series.

SMITH: *Macowanites* and *Arcangeliella* resemble unexpanded basidiocarps of *Russula* and *Lactarius* particularly. At the level of *Cystangium* and *Martillea* you have *Rhizopogon*-like fruiting bodies with various degrees of a columella, but a closed basidiocarp. The anatomical features are still the same, however. Globose amyloid spores and heteromerous trama are found in the whole series. It is a very clean-cut series—one of the best established I think, in mycology, as far as the number of taxa is concerned. And in this particular series, the true agaric forms are the most recently evolved.

BURDSALL: Do you assume the development of the forcible spore-discharge mechanism a second time in this group?

SMITH: Yes, but it is not coming *de novo* because it was probably in the original agaric ancestors. You have got to think in terms of a circle. In another series one can start with *Cortinarius*, go to *Thaxterogaster* and *Hymenogaster*, and return to the agarics again. But the last agarics have a different type of fruiting body than the original parent agaric stock.

BURDSALL: I find it difficult to see how you would get selection, especially in a series, for this mechanism.

SMITH: The whole problem, of course, is speculative, but there is apparent selection in the hymenogasters. The problem you are pointing up is what there is in the environment that selects for survival in the Hymenogastrales. There are a lot of *Rhizopogon* and *Cortinarius* species that fruit underground. How do they disseminate? I have collected one of these hypogaeous cortinarii from Priest Lake, Idaho, to the Salmon River country, and they occur on the mountain slopes in southern California. They do not shed their spores where the wind can get at them, but they still move around. They must be eaten by something or get carried around that way by rains, by spores leaching through the duff, or something like that. Somehow, there is survival value and frankly we don't know what it is.

PETERSEN: Are we assuming two selective processes—first a selection for nonviolent discharge of spores to get into the hypogaeous astrogastraceous series, and then another selection for violent discharge to get back into *Russula* and *Lactarius*? Your primitive agaric type is apparently gymnocarpic. But you have conjectured that *Lactarius* and *Russula* came from a hypogaeous group. Somewhere in between there must have been a loss of violent spore discharge and then a later gaining of the same thing. So, after a selection in one direction, there seems to be a selection in the opposite direction.

SMITH: Yes, but the forces of the environment could do this. In fact, you would have the same selective forces at work on the *Thaxterogaster* end of that line and on the *Macowanites* end of the astrogastraceous line. In these there is no violent discharge. Violent spore discharge only begins again when the fruiting body begins to open up.

SINGER: If you refer to this first loss of functionality in the basidium-sterigma complex as an inhibition, then could this inhibition be caused by the closed fruiting body? If so, why doesn't the inhibition appear immediately? Why is there still violent discharge in those cases of agarics with gastromycetous stages, like the *Psilocybe* of McKnight, or even in some of the agarics closely related to genera of gastromycetes, such as *Weraroa*, where you still have these asymmetric sterigmata?

SMITH: Evolution in nature is always expressed in ever-spreading patterns. All lines possible develop. There could arise precisely what you are suggesting. Certainly the evidence is there, but I would suggest that in one place the opposite thing occurred. At sometime or other there was a back-mutation which led again to spore discharge.

SINGER: If this is true, then you are presenting a whole series of genera and families within the gastromycetoid sphere which through it all maintained the genes for violent spore discharge.

SMITH: They also maintained the genes for large basidia and large sterigmata which are practically straight but very easily become bowed. In many species of *Thaxterogaster* one can find straight and bowed sterigmata in the same fruiting body. So all this can't be reduced to just a simple procedure. My answer in a sense would be, how else can you explain the Russulaceae? We have this line. What is the most rational explanation for how it developed?

SINGER: If you admit once that this is the way from the gastromycetes to the agaricales, wouldn't it be more logical to admit it in other cases, too?

SMITH: I am basing my argument on forms that are still extant. We do not have any more gastroid forms that connect up to *Brauniellula*, so we propose that *Brauniellula* arose from the Gomphidiaceae. If we look to *Gastroboletus* as Dr. Thiers indicates, we go possibly to *Truncocolumella*, but that has only two species. We do not have a really sizable connection on which to base an argument.

PETERSEN: Of course, your statement must be based in the context of time. A few years ago you would not have had *Thaxterogaster* to talk about. You could not have made these statements. When you say that there is no connection to a certain taxon, this must be tempered by the possibility that the connection may be uncovered in two days or twenty years.

One is reminded of the postulation of the constitution of the solar system. At a certain point, someone theorized that another planet must be out there, because of the other forces and orbits which could be measured. When they looked for it, Uranus was indeed there.

SMITH: If someone finds the intermediates then we can change the map. Our process here essentially is one of mapping the possibilities. This is all theoretical, believe me.

SINDEN: Would you remove the Russulaceae from the Agaricales?

SMITH: I have debated this at length with Drs. Shaffer and Singer, and it depends on the philosophy of the investigator. If one were to draw a horizontal line and say that the gill fungi are at a certain level, then one must maintain a taxon for those with a certain type of hymenophore. In this instance, one would put the Russulaceae in the gill fungi. If one were to arrange the taxa on a vertical line according to relationships, then this family would be placed as a part of the Russulales. The vertical arrangement would be the more natural, in my opinion. Dr. Shaffer, in the recent edition of the keys which he uses for his class, has each of these secotioid groups at the end of its respective agaric family. He is trying to put them in a vertical arrangement.

SINDEN: On that basis the Russulales would come away from the Agaricales.

SMITH: Yes. There would be the Russulales, Agaricales and other orders.

HEIM:[2] For me this is the series Asterosporée. One can proceed from a primitive agaric like "lactario-russula," which is at the base of the primitive group I think, which also contains *Russula archaea. R. archaea* is from Madagascar and is a primitive species of these mushrooms. Its fruit bodies are very heavy because, although there are sphaerocysts, they are not very significant. From here there is an evolution of the mixed group *Lactarius-Russula*, as well as section *Compactae* of *Russula* (or *Compacti* of *Lactarius*) with spores which are very pale, and a further evolution and progression in the direction of subgenus *Russularia*, the many species of which are very friable, have little endurance, and bear very yellow spores. But at this stage there is an appearance of a pseudoangiocarpic form in the tropics and this form I shall speak about in my paper. Over a long time, I have concluded that angiocarpy evolved from pseudoangiocarpy. Pseudoangiocarpy persists in the direction of the subgenus *Pellicularia* the fruit bodies of which have many sphaerocysts, and a great increase in the lightness of the fruit body. This has an ecological explanation, namely, adaptation to the conditions of life that dictate the presence of sphaerocysts. After this there is an overevolution, degra-

[2] The reader is referred to fig. 2 of Heim's paper [Ed.].

dation or decadence, I suppose, in the direction of *Elasmomyces*, *Macowanites*, *Martellia*, and *Asterogaster*, and so on.

*Lactariopsis* is separated from the "lactario-russula" group (*Russularia*, *Russula*, *Lactarius*). Pseudoangiocarpy in *Lactarius* is the same as that in *Russula*. *Lactariopsis* goes in the direction of angiocarpy and the subterranean groups of mushrooms like *Elasmomyces*; and *Russula* does exactly the same through pseudoangiocarpy to *Macowanites*, *Asterogaster*, etc. Finally, there is the *Gastrolactarius* and *Gastrorussula* group. Before the general line of "lactario-russula," which perhaps came from the primitive lactarii at the start of this evolution, there was a double line in the direction of *Russularia* and in the direction of *Gastrorussula*.

SMITH: There are two points that have been left out of the discussion. One concerns the Pacific northwest, where most of these fungi occur. In that area there is an extremely diverse population of conifers as compared to most other areas of the world. There also seems to be the largest population of rhizopogons in the world, and I am of the opinion that *Rhizopogon* is a relatively primitive genus. The presence of a large number of species of *Rhizopogon*, the presence of a large number of species of conifers, and the presence of the large number of species of the lower Asterosporeé seem to indicate that this is a region in which primary evolution has occurred, and apparently in which there has been survival of a large number of primitive as well as highly evolved forms. Since the Asterosporeé are so abundant in this area, and since this is where we find most of the species of *Hymenogaster*, all of the ecology and the quantitative aspects of the whole problem seem to fit this theory.

Now I have left out hemiangiocarpy because I feel that it is advanced in most agarics. In *Macowanites* the origin of the enclosed gleba does not require a veiled peridium, the latter merely remains attached to the stem. You do not have to have a veil, and I do not think a veil would figure prominently in this. You (Heim) know more about this than I do, because I haven't seen any of these hemiangiocarpic forms.

HEIM: There are no sphaerocysts or very few sphaerocysts in the angiocarpic forms because it is necessary for the angiocarpic form to swell in size under the ground. The astrogastraceous part of the line is heavy because of the buried environment, whereas it is normal that the russulas and others are light because they are outside. The fruit bodies with sphaerocysts are light and the filamentous species are heavy because of response to their ecological situation.

SMITH: I described a genus called *Mycolevis*, the fruit bodies of which were about the specific gravity of elder pith and exceedingly light, but they occur in this same area where rainfall is so heavy. If it comes to a matter of the water content I don't think your theory is necessarily true.

I would suggest a theory on this, depending on the strength of the plasma membrane. If a fruit body has small cells and these enlarge into globose structures, they can exert much more effective pressure than the growing of hyphae and intussusception in, let us say, a peridium. The enlargement of these pockets of globose cells might actually exert more pressure in pushing away dirt than the growth of filamentous hyphae.

PILÁT: I believe that the Russulaceae is a very modern family that includes evolutionary development, mycorrhizae, and everything else you might want.

SMITH: This approaches the idea that they were the last to evolve in this chain of asterosporous fungi. *Cortinarius* is probably of recent origin, too, and where the environmental pressures are severe, *Cortinarius* tends to fruit underground.

SINGER: I agree with Dr. Pilát. *Agaricus* also tends to fruit underground under severe conditions .

SMITH: So does *Russula*. So the russulas really aren't out of the ground too long in this kind of habitat.

HEIM: All of the russulas are not mycorrhizal. I have found a parasitic *Russula* with an annulus, growing on living trees and living leaves. There is no relation with the normal growth form.

SINGER: I agree completely. I have collected lactarii which are absolutely nonmycorrhizal. They occur in the tropical rain forests in Amazoña in primitive virgin areas which have remained undisturbed.

ROGER HEIM

*Professeur*
*Muséum National d'Histoire Naturelle, Paris, France*

# THE INTERRELATIONSHIPS BETWEEN THE AGARICALES AND GASTEROMYCETES[1]

The conquest of contemporary natural history by what may be called transformist theories was bound, even as early as the middle of the nineteenth century, to extend to the fungi and especially to the Macromycetes, in spite of the extreme paucity of paleobotanical source materials relating to this immense domain. As is true for living categories of the animal and vegetables kingdoms, our ever more extensive knowledge of forms has inevitably concentrated attention more and more on the problems of the relationships among the great cryptogamic groupings: algae and fungi, Ascomycetes and Basidiomycetes, and even more sharply, in the course of the last thirty years, among the two fields that together make up the Hymenomycetes in the classical sense of the term: the Agaricales, Boletales, Asterosporales, on the one hand, and the Aphyllophorales and the Gasteromycetes on the other. The descriptive and embryological studies of one after another of these areas have led to advances that have aroused ever increasing interest, which in turn has been reinforced and enriched by the discovery of new genera, new groupings, made both in the tropics and in other regions theretofore hardly penetrated.

If the Gasteromycetes and the agarics have found themselves in the forefront of this topical interest, this has been because they offered by their physiognomies, by the diversity of their characters, and at the same time also by the resemblances among them, by homologies, by points of agreement that marked them—they offered, I say, a spectrum of silhouettes and arrangements among which one could make comparisons, perceive resemblances, note differences, all with particular ease. Little by little embryogeny helped to deepen these

[1] I am particularly indebted to Mr. R. Gordon Wasson, my friend, for his priceless suggestions and help in the writing of this paper.

researches, while the importance of biotic relationships, and of ecological factors, made itself felt in these discussions. Anatomical studies also cut deep into them assisted by phytochemistry and sporogenesis, and finally leading to a better knowledge of the tegumentary structure of the spores. The study of the fungi, which with the older authors was confined to the adult forms and often even to dried specimens, rapidly expanded, thanks to the expeditions carried out in the field, notably in the tropics and in America and by mycologists rather than mere collectors. Techniques of culture allowed us to obtain living carpophores in the laboratory, and the embryologists, profiting thereby, sought to discover more precise stages in the ontogeny of the plant. So successful were these efforts that after the early findings to which they led, on the one hand on the part of Brefeld, and on the other by de Bary, Fayod (1889), and Patouillard (1900), setting forth and delimiting already the fundamental theories regarding the relationships of the Agaricales and the Gasterales, there sprang up a whole series of monographs. Other authors such as Fischer (1925), Gäumann (1928), Holm (1954), Kühner (1926), 1945), and Savile (1955) added their thoughts and data. An astonishing fact: the progress that resulted from the confrontation of the two theories, the one deriving the Agaricales from the Gasterales, and the other the reverse, was in both cases substantial, supported in each case by positive evidence, and in each case making use of the Aphyllophorales, which were always present and under discussion with the other two. It seems to me, however, that the arguments favoring one of the paths are more solidly based than those favoring the other. In this paper I will try to establish that conclusion.

In focusing our attention on this matter I shall do my best to define the successive achievements about the problems discussed, of those mycologists who have interested themselves in these questions of phylogenesis: the interrelations between the Agaricales and the agaricoid Gasterales. But I shall only touch from afar the position of the Phallales and the Aphyllophorales, for this would lead us too far afield. I shall not even deal with the Plectobasidiae (Endogastrinae of Malençon), whose "ascent reaches its terminus at the exact limits of this group" and of which "the agaricimorphic bearing that they sometimes adopt is only a superficial resemblance."[2]

[2] This and all other quotations were translated into English by the author, as an aid to the reader [Ed.].

No one can fail to be struck by the part played in these discussions by the strong leaning of each participant in favor of concepts, the objectivity of which has not always been above reproach, the sequence of ideas revealing sometimes an *a priori* intuition or conviction inspired by a rather personal evolutionary view. Nothing allows us to transpose the hypothetical deductions arrived at in phylogenesis into a chronological sequence in evolution. Naturally, we should like to compare the synchronous species observed today, species distinct in their probable origin, with what they may have been as they developed over a period of time. But this is only a vain pretension.

We shall have more to say on this, but now we conclude: the interpretation of the facts is a domain apart from the facts themselves. Phylogenesis in mycology is not yet a science. It is an intellectual game based on arguments that often escape rigorous control and that almost always escape the test of experiment. And experiment can be brought to us through cultures—when feasible—and through morphogenesis studied in the laboratory.

## The Series of the Asterosporae (fig. 1)

Malençon's (1931) treatise on the Asterosporae constitutes the basic work that all later writers on the relations between Agaricales and Gasteromycetes either have or should have taken into account. It stands as the first overall contribution to this theme, according to which the "barrier erected between Plectobasidiae and Hymenobasidiae may well prove to have been a mere convention." True, certain allusions had already pointed to plausible relationships between Lactario-Russulae and hypogaeous asterosporae, notably in the report by Bucholtz (1903). But the credit goes to Malençon for having assembled the genera scattered throughout the various systematic groupings of which the kinship to the Russulaceae seemed to him beyond dispute. Separating out the characteristics that make up the personality of the russulae and the lactarii, he showed that the clearest, the most reliable indices, were to be found among some of the hypogaeous or gasteroid forms, which sometimes presented extremely important alterations all the more marked in the simpler forms that seemed better suited to conditions of life underground. Confronted with the dilemma whether to consider them as primitive genera or as derived from agaricoid forms, he had already adopted the hypothesis that the Lactario-Russula group constituted the starting point and he saw in the others inferior and regressive states. At the same time he explained how by development of the ventral region, asymmetrical spores of the Lactario-Russulae had been able

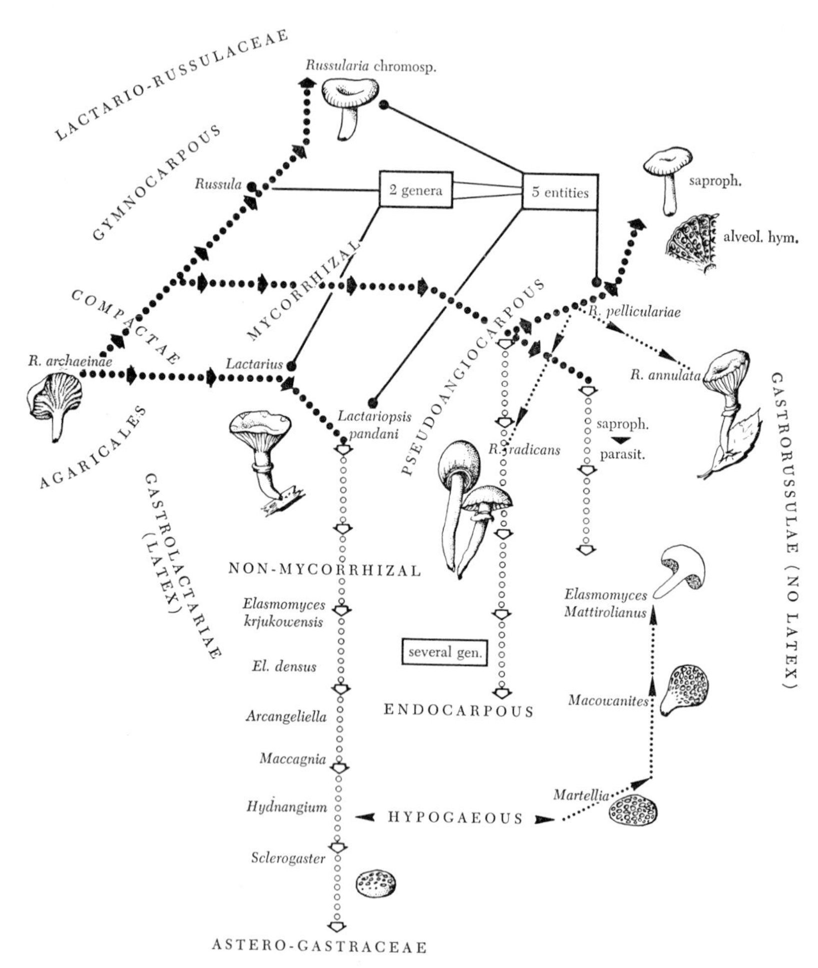


FIGURE 1. Phylogenetic pathways of the series Asterosporae.

by a process of standing up to show the way to the classical aspect of the spores of russulose agarics, whereas the spores of the hypogaeous forms little-by-little tended toward a symmetrical-ovoid form with respect to an axis. By assigning to each organ a "coefficient of dynamism," he was able to evaluate the progressive decadence of each characteristic, leading even to their complete disappearance, "a morphological regression that accompanies the progressive attentuation of the peridium of the *Russula* into *Elasmomyces*, *MacOwanites*, *Arcangeliella*, *Hydnangium*, *Clathrogaster*, *Martellia*, *Gymnomyces*, and *Sclerogaster*": reduction of the stipe, maintenance then abandonment of the columella, the hymenial base becoming restricted to a small veined basal platform, a gradual disappearance of the sphaerocysts, cystidia becoming more and more rudimentary, a continuous regression of the sporal asymmetry toward the spherical form reached and maintained in *Sclerogaster*.

The clarity of the exposition led to compelling conclusions: "The Asterosporae series constitute a natural grouping of genera with exceedingly close affinities in spite of apparent dissimilarities in their extreme forms. This series does not seem to show a connection with any other similar series (Coprinaceae, Phallaceae, Battarraeaceae, Sclerodermataceae, etc. . . .)."

A second series of investigations in the course of our expedition to Madagascar lent its support to the essential conclusions of Malençon by the discovery and study of annulate lactarii and russulae (R. Heim, 1937–1938).

The discovery of *Lactariopsis pandani* Heim and the disclosure of pseudoangiocarpy leading to the formation of an annulus of secondary origin invited a comparison with *Elasmomyces mattirolianus* Cav. (Heim, 1936). There also followed a juxtaposition of the development of *Lactariopsis* with that of *E. kijukovensis* Buch., primitively subgymnocarpic then angiocarpic, the type of evolution therefore being the same. In other words, *Lactariopsis* constituted the bridge between *Lactarius* and *Elasmomyces*. The characteristics of the *Lactariopsis* entity revealed the three-phase development with precision: initially gymnocarpic, then fusion and centripedal, finally rupture and centrifugal, the mode of development being closely related to that which Kühner (1926, 1948) had pointed out in certain annulate boletes (*Ixocomus flavus*) which were pseudoangiocarps. On the other hand, the spore ornamentation of *Lactarius adhaerens* Heim, which was apparently annulate, was reminiscent of *Clathrogaster vulvarius* Petri.

As to the annulate russulae or those inseparable from them we divided the ones in Madagascar into two stirpes, *Radicantes* and *Discopodae*, to which there used to be attached two gymnocarpous sections, the *Aureotactinae* for the first, the *Heliochrominae* for the second. But even within a given *annulate* species, an *exannulata* variety has proved how fleeting and inconstant the marginal veil is (Heim, 1937, 1938). From the sum total of these considerations we drew a conclusion favorable to the evolution of the Lactario-Russulae in the direction of the pseudoangiocarpic forms and perhaps of hemiangiocarpic forms with double ring. For a long time they kept closed, and finally came the degraded, asterogastraeus Russulaceae adapted to a subterranean life.

Since these publications our documentation (Heim, 1943; etc.) on the pseudoangiocarpous russulae has been considerably enriched by the collection of numerous African species, notably of the subgenus *Pellicularïae*, especially during these latest years in Cameroun in central Africa, and in Gabon. It is especially noteworthy to observe that certain of these annulate forms are parasites on living plants, therefore freed of all mycorrhizal ties, and that another (*R. alveolata* Heim) has an alveolate hymenium. Later, on the occasion of a study of *E. densus* Heim (1959), which I had collected in Thailand, I took up again the study of several species of Asterogastraceae, revealing the tangle of characteristics of the genera *Elasmomyces*, *Arcangeliella*, and *Macowanites*, the Siamese mushroom constituting perhaps "the principal link both pedicellate and epigeal that separates *Lactarius* from *Arcangeliella* and from *Hydnangium*, this species, being *Arcangeliella* by its latex, *Elasmomyces* by its ontogeny and morphology, and *Lactarius* by its anatomy and milk, deserves to constitute a new genus."

Thus we see how a chapter dealing with the Asterogastraceae offers us one of the richest examples of the close ties that bring together asterosporous forms—gymnocarpic, agaricimorphic, pseudoangiocarpic, and gasteroid-angiocarpic. If our own collections in the tropics have permitted us to contribute some new examples to these phyla, we must give full recognition to the contributions made by R. Singer, A. H. Smith, and H. Horak, studying other genera and new species, either hypogeous or subhypogeous, apodal or with columella, all of them more-or-less characterized by axial symmetry. They have added to our knowledge of the Asterogastraceae as many valued items as they have presented us with species, most of them South American, with carpophores either closed (*Martellia* Matt.), or barely pedicellate (*Cystangium* Singer & Smith, 1960), or with columella completely preformed (*E. nothofagi* Horak, 1964b).

One of the arguments Singer uses to defend his thesis lies in the difficulty of attributing to the Lactario-Russulae a plausible origin among the Agaricales. But the presence of the spherocysts in the Asterosporaceae can be explained very well by the hypothesis of Reijnders: the mass of spherocysts constitute "a system that confers a larger volume to the carpophore." This character is gradually lost as one perceives the presence of regressive or preferably decadent characters, since it is linked to the lightness of the flesh in the epigeal carpophores of these two genera of Agarics: *Russula* and *Lactarius*. As for the axis around which are grouped the spherocysts, and which is constituted by a laticifer, this arrangement is also found here and there, especially in the medullary tissue of the stipe of *Gyroporus*. This character must be thought of as a function of the conditions of life. It has lost its utility, attached as this was to an aerial habitat, when, having become hypogaeous or gasteroid, the mushroom had to resist the pressures of its milieu. If you find this explanation anthropomorphic, we can arrive at the same destination by invoking only the influence of the environment: the epigeal life of the evolved forms of the Lactario-Russulae in the direction of the pseudoangiocarpic forms that—as I have already shown—would constitute a stage toward the gastrorussulae. (Besides, the spherocysts do not belong exclusively to the russulae, as one finds them in the derms belonging to diverse genera of agarics and boletes.) I would add that the marginal veil of the pseudoangiocarpic russulae in Singer's concept would be suitable for the forms affiliated with primitive species, which seems in the event unbelievable, especially when given the other characteristics: spores either aculeolate or alveolate, flesh pellicular, and—we have observed this recently in Gabon—adaptation of the annulate russulae to a parasitic life on the leaves and stems of living trees, reaching appreciable height.

Reinjders concluded his remarkable study with these words: "The discovery with more certainty of these agaricoid remains among the gasteromycetes would argue in favor of their descent from the Agaricales." This is the thesis that I have defended for a long time and to which I have brought various kinds of supporting evidence in these last few years that I report in the present study.

## The Polyphyletic Complex of the Boletales: Boletaceae and Boletogastraceae

### A. *Mono- or polyphyletism of the Boletales?*

I shall not linger on the position of the Boletaceae among the Hymenomycetes, considering simply that it is hardly possible to see in this immense group the signs of a monophyletism, or at least of a general proximity, for this is contradicted by too much evidence. If the thesis of Karsten maintained their place among the polypores, that of Patouillard (1900) linked them with the agarics because of their fleshy consistency, the separability of the hymenium (which is not natural), and their pores "clearly derived from anastomosed and radiating gills"; for Patouillard, *Phylloporus* and *Paxillus* clearly became intermediary links in the direction of *Xerocomus*, an hypothesis unanimously admitted today, confirmed apart from other reasons by the lamellate trama. Later, some went even further, admitting that the Boletales entered into the aggregate of the Agaricales, but without bringing to such a conception any further solid criteria.

More recently certain reservations have been expressed about this latter grouping. E. J. Gilbert (1931) has contributed to this skepticism, Singer has perceived affinities between the gyrodons and the polypores, and I in turn have reinforced this point of view by the discovery in New Caledonia of a new genus *Meiorganum* Heim (1966), close to the gyrodons and neighboring genera (*Campbellia* Cke. & Mass., *Gilbertiella* Heim, *Gyrodon* Opat.), these genera seeming to constitute a link between the boletes and the polypores, contrary to the generally accepted belief held today.

The researches carried out in these recent years on the tropical boletes by Jacqueline Perreau (1961, 1967) and myself (Heim, 1957, 1966), as well as those of many other authors, will permit us to draw up shortly a phylogenetic chart of this vast aggregate to which the taxonomic term Boletales is applied.

Such a chart (fig. 2) might well be drafted with the following remarks kept in mind: the Meiorganaceae lead the lignicolous Polypores in the direction of the mycorrhizal boletes by way of the gyrodons, a channel that would perhaps explain the evolution of the polypores (and therefore the Aphyllophorales) in the direction of the boletes;

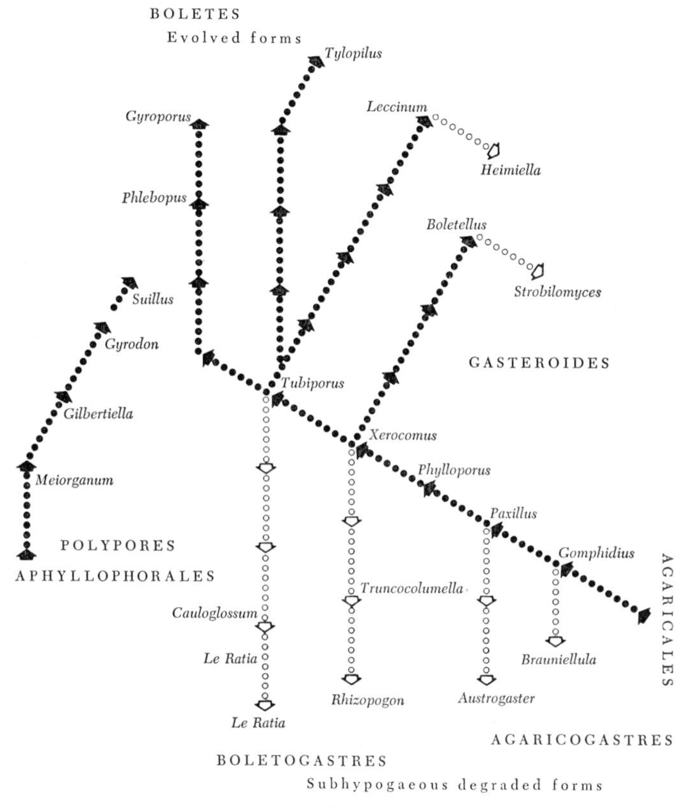


FIGURE 2. Phylogenetic pathways in the Boletales.

the Boletogastraceae, having emerged via a decadent route, embrace the rhizopogons, derived from *Truncocolumella*, and *Le Ratia* by way of *Cauloglossum* as we shall see later; from *Xerocomus*, a vast group, one reaches *Phlebopus* and *Gyroporus*, in accordance with the indicated conditions of a progressive evolution; the Strobilomycetaceae manifest gasteroid affinities, as does *Boletellus*, a complex genus not homogeneous, and perhaps *Heimiella*, which even recalls *Krombholzia* (= *Leccinum*).

These above-mentioned genera, like *Gyroporus* in a separate phylum and also *Tylopilus* manifest a progressive evolution that is very plausible.

In any case, it seems that the hymenial tube is an organic element of convergence, just as the gill is another. The agarics with either tubulate or alveolate hymenium are altogether different, and we know that such a disposition—the alveolate or tubulate hymenium—is found by homology in various tropical fleshy agarics (Heim, 1945) that are exceedingly different: *Mycena, Clitocybe, Marasmius, Omphalia, Lentinus,* and *Russula,* where they may sometimes constitute subgeneric sections, or even genera or subgenera that are distinct: *Poromycena, Phlebomarasmius, Mycomedusa,* etc.

We think therefore that no peremptory argument favorable to the unity of the world of boletes among the Agaricales can be invoked against the thesis of the polyphyletism of the Boletales, where the decadent tendencies in the direction of the hypogaeous forms are to be found in precisely the same manner as among the Russulaceae, the Agaricaceae and the Aphyllophorales.

Therefore, two groups with hypogaeous forms seem to us to be in close relations with the agaricimorphic boletes: *Le Ratia* and *Rhizopogon.* I shall speak more specifically about them.

### B. *The* Le Ratia *series*

On several occasions we have dealt with the endemic New Caledonian genus *Le Ratia* Pat. (Plate XIII) of which two species have been described by Patouillard: *L. similis* and *L. smaragdina,* a third, *L. coccinea,* by Massee and Wakefield, and the last, *L. atrovirens,* by me. In a memoir I (Heim, 1968b) have re-examined the aggregate of problems raised by the existence of these mushrooms, in which the variability of the carpophores within a single species—*L. similis*—extends from a form entirely closed and apodal to a type of open agaricoid.

I have drawn attention (Heim, 1951b) to the presence on the spores of a rudimentary germinative pore corresponding to a thinning of the episporic partition, and I insisted on the remarkable variability and the instability of the basidia. I argued in favor of the relationship of this genus with *Cauloglossum,* at the same time mentioning a possible relationship with the boletes. Having succeeded in cultivating them on an artificial medium, I observed the presence of clamps[3] and of peculiar organs, sporoid ocelli. Furthermore, my pupil Mlle. Mélendez-Howell, in her thesis on the ultrastructure of germinative pores, defined the tegumentary composition of the spores and of the pore of *Le Ratia* (Mélendez-Howell, 1967).

In 1966 and 1967 the collection of numerous specimens of *Le Ratia similis* in New Caledonia illuminated the extreme polymorphism of this species, in which the details of the gleba, the columella, and the stipe were brought to light: a true stipe, thin or thick, or scarcely outlined, or the columella extending it, sometimes in continuity with the cortex, simple or multiple, sometimes complete (reaching the apex of the carpophore), this latter being either locellate or not, with extremely variable alveoles, small or large, either completely disordered or indicating a lamellate orientation, certain specimens deprived of columella showing such a disposition of the elements of the gleba. These profound differences brought us a new element to confirm the thesis of degradation, all the more telling because in culture the primordia disclosed in general an agaricoid physiognomy, with predominant stipe, but a bulging, closed pileus of angiocarpic origin.

From *L. similis* we pass to *L. smaragdina*, which in nature is sessile, deprived of any hint of a columella, thinning briefly and for the most part at the base. But the primordia that appear in artificial cultivation reveal the existence of a stipe rather thin for all its length. Thus, the differentiation of the mushroom is precocious, and the primordium, quickly differentiated judging by the independent stipe, remains partly angiocarpic; in other words, exhibits an incomplete hemiangiocarpy: the stipe has persisted in its elongation while the sporophore fails to reach the point of opening.

It is in the remarkable *Le Ratia atrovirens* Heim (1968a) that the sterigmata of the agaricoid origin stand revealed in all their plenitude. Thus, in this case the pileus remains conspicuously open around the foot, which is attached on the inside to the inferior face of the

[3] I remain extremely skeptical concerning the advantage that one can derive from using the presence or absence of clamps to support an argument favoring an evolved or a primitive form, as the case may be. Thus, *Le Ratia* only shows clamps in culture. In *Secotium s. l.* where the relationship is close one finds either the absence of clamps (*S. arizonicum*), or variability as to clamps (*S. macrosporum*), or the presence of clamps (*S. gueinzii*), according to the species.

summit of the pileus. Not a true columella, it is a *complete and free pedicel.* As for the gleba, it is at the same time alveolate and lamellate. The spore gives no indication of a germinative pore or a tegumentary thinning. In fact, the hilar system seems to play the role of a germinative pore.

Thus, the extreme variability of the receptacles of diverse species of *Le Ratia*, reminiscent of adult states of superior forms from which they seem to be derived, is especially suggestive. The first stages reveal to us their delayed agaricoid and hemiangiocarpic character—so delayed that they will not succeed in reaching their adult stage. It is the juvenile state that belongs to the agaricimorphic form: an individualized foot, an open and lamellate hymenium (*L. atrovirens*). In other words, the polymorphism that stands revealed in the diverse carpophores, their gleba, their pedicel, their columella, presented to us in a single species, through irregular morphological aspects, almost teratological, leads me to conclude that in them we may become aware of a degradation for which we will find the proof elsewhere, in the anatomical details. The embryogenesis certainly seems to support the thesis of degradation rather than the opposite.

The study of the ultrastructure of the germinative pore permits us to assemble still more closely the five species of *Le Ratia* and *Cauloglossum* that we have examined (Heim, 1968a). This study renders apparent a profound similarity in the tegumentary sporal composition of these mushrooms and especially in the details of the apex: complete absence of pore in *Le Ratia atrovirens*, corresponding to a slight thinning hardly perceived in *L. coccinea*, and much clearer in the three other species, *L. similis*, *L. smaragdina*, and *C. transversarium.* The sporal dimensions, for their part, run from *Cauloglossum*, where they are extremely small (about 6.2 x 4 μ) up to *L. atrovirens* (about 17 x 9 μ). There can be no doubt therefore that the affinities of *Cauloglossum* (= *Ropalogaster*) bring it nearer to *L. similis* and *L. smaragdina* than these last to *L. coccinea* and *L. atrovirens.* We still have to discover from which Agaricales or which Boletales, or toward which group among these, the phyletic movement is directed.[4]

[4] Smith and Singer (1959) have dealt with the problem of the gasteroid forms of the Boletales and notably with *Le Ratia* and *Truncocolumella* at the same time as *Gasteroboletus* and *Chamonixia.* Our latest inquiries into *Le Ratia* (Heim, 1967, 1968a) make it unnecessary for us to revert here to the considerations of these two authors, which in part link up with ours on this genus. As for those that

### C. Rhizopogon *series*

In 1934, in conformity with an earlier allusion of Malençon (1931) to such a phylogenetic connection, I already suspected a relationship between the rhizopogons and the boletes, and we proposed the establishment of a tribe Rhizopogoneae, calling attention to the discovery of *Hypomyces chrysospermus* on *Rhizopogon à propos* this thesis. Subsequently, Malençon (1938) described his genus *Dodgea* (*occidentalis*), a synonym of *Truncocolumella* Zeller with its columella not percurrent and a relictual hint of a stipe. He interpreted it as close to the subterranean rhizopogons, which, in their turn, offer no longer any pedicular or axial trace.

After having stated the essential anatomical features of this entity, and notably the paraphyses, as extremely variable, Malençon concluded that the new American genus represented the first of the known intermediate taxa "between the evolved forms of complex organization such as the boletes and the angiocarpic forms represented by *Rhizopogon.*"

Our first observations on *Le Ratia* from New Caledonia led me to consider this genus "as one of the first links in the degradation of the Boletales in the direction of an underground life and the angiocarpic state, of which the final stage seems to be identified with the rhizopogons," and of which two other intermediate stages might appear, on the one hand in certain *Cauloglossum,* nearer still to the boletes, and on the other hand, and much more surely, *Truncocolumella* (Heim, 1948, 1951b).

If it seems difficult to trace such a path and a meaning in the phylogeny that brings together these diverse genera, it appears just as sure that ties with the Boletales are virtually certain insofar as *Rhizopogon* and *Truncocolumella* are concerned. So far as *Le Ratia* and *Cauloglossum* are concerned, the affinities with the boletes remain still hypothetical although plausible. From this our position is derived, which maintains, at least provisionally, a distinction between these two phyla.

---

relate to *Truncocolumella,* which Smith and Singer have studied in detail in North American species, their researches only confirm the opinion of Malençon and Heim. I lack space here to extend my remarks on this subject. [I would say the same with respect to the remarkable Australian genus *Cribbea* Smith and Reid (1962). I shall take up this discussion again when we examine the Gasteroboletaceae.]

## Polyphyletic Complex of the Secotiae (Agaricales) Cyttarophyllaceae series (= Galeropsidaceae Sing.)

### A. *Positions and Heterogeneity of the Secotiae*

The problem of the Secotiaceae is without doubt one of the richest of the domain dealt with here, and in recent years this is in large part due to the publications of R. Singer and A. H. Smith on this group, above all for the South American forms. It is certain that in the light of their works the heterogeneous character of this aggregate becomes even more emphatic than it was before. These basidiomycetes offer us a remarkable juxtaposition of gasteroid and agaricoid characters that justify numerous observations and hypotheses as a result of profound analyses. As one approaches the adult states in the growth of the carpophores, several of the species which were studied first by Bucholtz (1903) manifest an increasing separation between the resemblances in the direction of the agarics and those that remind one of the gasteromycetes. In other words, it is above all in the young states that one finds suggestions of a relationship with agaricoid types. But these agaricoid types would seem to argue in favor of a gymnocarpic origin for the Gasteromycetes, which has moved Singer to perceive an argument here in favor of an evolution in the direction of angiocarpy, an argument favorable to the thesis that he defends concerning the derivation of the agarics from the Gasteromycetes. But this implies a correlation between the peridium of the latter and the veil of the Agaricales, which according to some results from a confusion. Thus, Reijnders (1963, p. 363) found that there was an indefensible comparison here: "The veils of the Agaricales and of many peridia of the Gasteromycetes are not at all homologous and whoever has studied the development of the Agaricales is inevitably led to the conclusion that these structures cannot be considered as the result of a process of degradation, as was suggested by Brefeld, von Höhnel, and Lohwag." And by a comparison of the growth phases of *Secotium agaricoides* and *S. novae-zelandiae*, the primordia of which are identical before the growth of the gleba with those of the Agaricales, he came to the conclusion that the term "peridium" was ambiguous and that we have applied this word quite simply to "all the envelopes of the Gasteromycetes, whether homologous or not." For him, "the veils of the Agaricales do not play a role in a process of degradation." If it were otherwise, we should have to admit that the feebly angiocarpic agaricoid forms like the clitocybes and gymnopili were the most evolved, whereas there are many arguments in favor of the opposite hypothesis. Gymnocarpy is not secondary; it is the end of the road of a close relationship with the Aphyllophorales. Reijnders dwelt at length on this argument drawn from ontogenesis in an analysis that was clairvoyant. The veil that "emanates" from the Agaricales is "an adjunction, an excrescence, that shows up where it is needed . . . , independently of the place occupied by the species in the system." It cannot be compared with the peridium. And secondary pseudoangiocarpy or angiocarpy corresponds to efforts to "return" to fundamental characteristics. According to him, Singer's thesis deriving the Agaricales from the Gasteromycetes would lead to an admission that the order of succession in the evolution of the agarics "had already been placed at a level inferior to the Gasteromycetes" in the course of the general evolution of the Basidiomycetes. How are we to explain that the primordial forms of these agaricoid gasteromycetes reveal types of growth that one finds also in the Agaricales, except by the fact that being extremely close to these latter they derive them therefrom? In other words, they are decadent or degraded forms of Agaricales, registering a morphological decline, as Malençon (1955) and R. Heim (1951a, etc.) have agreed, at least for a category of Gasteromycetes, a term that covers at least two distinct domains, one progressive and the other regressive or, better, decadent.

Reijnders brought an argument of weight to bear on the conception of the decadent evolution of the phylums that have arrived at "the end of the road." According to Singer, the veil of the Agaricales has no meaning, whereas it is invested with a profound meaning in the case of the Gasterales. In fact, this reasoning, criticized by a certain number of mycologists, could be extended to the formation of the stipe and the pileus. How are we to explain that the hypogeous gasteromycetes form these two organs in their primordia *at the beginning*, whereas they have as yet *no functional role*? Better yet (Reijnders), "How are we to explain, on the basis of the derivation of the Agaricales from the Gasteromycetes, that certain ones among these last, the Secotiaceae for example, are stipitocarpic, others pileocarpic, yet others hymenocarpic?"

The problem of the spherocysts of the Lactario-Russulae deserves also to be analyzed. We all know that Singer explained this appearance as a derivation from the cellular subhymenium of certain gasteromycetes by the compression of the upper loculi of the gleba where the spherocysts would find their refuge in their intimate association in a mass. Here again, Reijnders demonstrated that the mass of spherocysts was formed by a rising channel (*Russula olivacea* and *R. emetica*), which contradicted Singer's scheme leading to the adoption of a descending route.

With much reason Singer lays emphasis on abnormal carpophores as a means to elucidate the problem of the relations between the Agaricales and Gasteromycetes, especially between *Cortinarius* and *Thaxterogaster*. He would explain these phenomena as a kind of atavism linked to an hypogaeous life. These instances of teratology are at least as well explained by a decadent adaptation that provokes a morphogenetic instability.

Clearly, one of Singer's key arguments has had to do with the Lactario-Russulae, which would find their point of departure in the Gasteraceae, whereas one has difficulty discovering among the agarics an acceptable explanation of their original parentage. Reijnders pointed out that this was as much as to say that all the Agaricales descend from the Gasteromycetes, and independently of each other. For Singer (1962b) the instability of the method of sporification in the Gasteraceae constitutes an argument in favor of the descent of the Agaricales from the Gasteromycetes. But this would mean that evolution always represents progress, a statement that many examples contradict in the animal and vegetable kingdoms. We see nothing in this variability of the method of sporification that argues in favor of the primitive origin of the gasteroids.

Among the Exogastrinae we may distinguish two general groups of gasteroid Agaricales that lend themselves to comparison with the Secotiaceae *s.l.*; one, made up of the Cyttarophyllaceae, that groups together the agaricimorphic and pedicellate forms, leading us to think of *Podaxon* but linked with *Conocybe* and *Bolbitius*. The other group assembles several subgroups in close relations with the closed forms on the one hand and the agarics on the other, the aggregate of these latter corresponding to diverse phyla closely associated with the hemiangiocarpic Agaricales: *Cortinarius*, *Gomphidius*, etc., perhaps the lepiotas, and *Mycena*, which suggest relations comparable to those that link the Asterogastraceae to the Russulaceae.

### B. *The* Galeropsis *or Cyttarophyllaceae series*

When I (Heim, 1931) resumed the study of *Galera besseyi* Peck, which Patouillard had reported from Madagascar, I noted that this remarkable species with globular pileus and the margin embracing the stipe offered lamellae anastomosed in compartments after the fashion of a bee-hive, but a bee-hive with irregular and unequal compartments. For this reason I was led to inquire whether some relationship with *Podaxon* could not be suspected. The study of *G. paradoxa* Matt., from Ethiopia, brought a new example of this peculiarity. These observations led me to characterize such forms in a new generic entity that I called *Cyttarophyllum* (Heim, 1931), "a specimen that deserved perhaps to be inscribed among those that supported the thesis concerning the close relations between Agaricaceae and Gasteromycetes," a position developed in my work on the ochre-spored agarics. Already at this time I advanced the hypothesis, based on the study of certain "*groupes de passage*" that these were "not primitive forms, but rather degraded forms" (Heim, 1931, 1948, 1950), a position that linked up with that which Malençon had formulated in so far as the Asterosporales were concerned. Singer (1936) pointed out that J. Velenovsky had announced a mushroom from Moravia under the name *Galeropsis desertorum* Vel. & Dvor. that seemed to be extremely close to *Cyttarophyllum.* I pointed out (Heim, 1937) that such mushrooms had been already known by Kalchbrenner on the Cape of Good Hope (*Bolbitius liberatus*). Zeller (1939, 1943) described two "*Secotium*" that could be compared to *Cyttarophyllum.* Finally, in a memoir devoted to the genus *Galeropsis* Velen. (= *Cyttarophyllum* Heim) I (Heim, 1950) published a survey of the diverse species already described as belonging to this genus, which had wrongly been multiplied by various authors, and offer an "exceptional variability of characters: spores, height and silhouette of the carpophores, interlamellary anastomoses." I set forth the arguments based on physiognomy, hymenium, anatomy, spores, ontogenesis, and bibliography, which led me to support the "thesis that we had always defended since 1931, namely, that the Gasterales (gasteroid forms) had their origin in the Agaricales, from which point degraded groups could be defined," the variability of these characteristics supporting the position that we had taken in this regard.

Later, in his remarkable study on *Torrendia*, Malençon (1955) characterized *Cyttarophyllum* [of which *Galeropsis* may be considered a synonym although Pilát (1948; Pilát *et al.*, 1958) separates the two entities] in a series Cyttarophyllaceae.

Quite recently, on the occasion of my visit in Bihar (Eastern India, in 1967), I gathered a remarkable mushroom, at first fleshy, of which the physiognomic features were those of the other *Cyttarophyllum*, but whose spores are entirely different (Heim, 1968b). This species, which constituted a new genus *Cyttarophyllopsis* (*C. cordispora* Heim), was hygrophilous in contrast to *Cyttarophyllum*, growing in fields with sparse herbal cover after extremely abundant rains. These indications raise a double problem of which interest the reader may judge: we may perceive a close resemblance between certain aspects of two morphological types that are decidedly different in spore details. We may imagine that another origin may be attributed to *Cyttarophyllopsis*, the spores of which are small (4–5 x 3–4 μ), triangular, exceedingly pale (approaching cream), and are deprived of any indication of a germinative spore. Therefore they differ profoundly from those of *Galeropsis* with basidiospores of the *Conocybe* or *Bolbitius* type. These latter spores appear gross, extremely variable, largely amygdaliform, with a large overflowing germinative pore of which optical examination and ultrastructure study reveal as similar to those of the Strophariaceae, *Montagnites*, and certain *Secotium* (Heim, 1951a; Meléndez-Howell, 1967—pore type $\beta_1$). If the position of *Cyttarophyllopsis* remains a question, its existence is clothed with a general significance of high interest.

We must add that *Secotium erythrocephalum* (Tul.) (*Cytophyllopsis* Heim, 1951a) offers a close similarity to the *Galeropsis* "symmetrical" spores and complete envelope, "that is to say, that the base of the envelope is joined to the summit of the stipe in the adult stage, but it manifests a more accentuated degree of angiocarpic degradation, its sporal characteristics alienating it in any case from all the other forms of *Secotium* that have been described." Singer and Smith (1958c) attached this species to the genus *Weraroa* Sing. obedient to the rules of nomenclature, my Latin diagnosis of the genus *Cytophyllopsis* having been published in December, 1958, seven years after my description in French.

In truth nothing authorizes us to think that *Cytophyllopsis ery-*

*throcephalum* belongs to the ochre-spored *Galeropsis*. Here there may well be a case of convergence. Nevertheless, the spores of this species sometimes reveal a somewhat octagonal profile, and Singer's hypothesis concerning a relationship with the Strophariaceae is quite plausible, as well as the relationship that he assigns to his entity *Thaxterogaster* with the Cortinariaceae.

I will mention (Heim, 1968b) one more North American species that Dr. D. P. Rogers entrusted to me and which corresponds to a hygrophilous *Bolbitius* with pileus almost completely closed, with spores similar to those of other representatives of this genus, and with extremely long stipe. I judged it to be not a fleshy and hygrophilous *Galeropsis*, but rather a *Bolbitius* adapted to climatic conditions of great humidity, as is the *Cyttarophyllopsis* of Bihar. But it is of interest to note here again the angiocarpic convergence between the aspect of the pileus and its "galeropsisioid" profile characterizing the *Bolbitius*, with the form of *C. cordispora*, which differs from the *Bolbitius* by the sublignose character of the stipe and the spores that are entirely different, but which presents the same sudden growth in very wet soil.

Thus, we find that entirely different climatic conditions—dry on the one hand, extreme rains on the other—can bring about the same physiognomic results. Here is an hypothesis that might well be extended to other cases.

### C. *Other diverse series with agaricoid and gasteroid forms*

*Rhodophyllaceae series:* Rhodophyllus *and* Richoniella

Romagnesi (1933) insisted on the indisputable rhodogoniosporous character of the subterranean genus *Richoniella*, an extreme link prolonging *Rhodophyllus* agaricimorphs whose spores are similarly "crystallographic."

*Coprinaceae series:* Coprinus, Gyrophragmium, Montagnites

The genus *Montagnites* (= *Montagnea*), "dry-"gasteroid, adapted to the dunes of the littoral zone and to sandy earths, has been considered as an agaricoid gasteromycete since the days of Patouillard, and *Gyrophragmium* and even more certainly *Xerocoprinus arenarius* may be joined to it, these three genera being related to *Coprinus* (see Zeller, 1943).

*Gomphidiaceae series:* Gomphidius *and* Brauniellula

It was with another genus of the Agaricales, *Gomphidius*, that A. H. Smith and R. Singer (1958) suspected connections when they treated a remarkable underground fungus of western America, of which they made *Brauniellula*, belonging to the Secotiaceae *s.l.* The columella strong and percurrent, adnate or free, continuous with the stipe; the gleba lacunose; the spores and the long cystidia are sometimes incrusted; the biotic relationships with *Pinus contorta*—all militate in favor of this thesis. The photographs of *B. nancyae* (*op. cit.*, p. 935, fig. 9) make evident the "contorted" structure of the gasterocarp, a sign found here again of the teratological state of "suffering" of such fungi in the subterranean milieu. The authors went further: they brought together two known species belonging to this hypogeous branch of *Gomphidius* (*G. leucosarx* Sm. & Sing. from *G. septentrionalis*, *B. nancyae* Sm. from *G. helveticus*) and placed *Secotium albipes* Zeller in this same new genus, having in mind the pigments—red—and the amyloid reaction of the trama and the peridial hypha. This example supports in its turn the heterogeneity of the artificial grouping of the Secotiae.

*Paxillaceae series:* Paxillus *and* Austrogaster

Let me note that the genus *Austrogaster* (*A. marthae* Sing.), discovered by Singer in Patagonia (1962b), may be considered according to him as a paxillaceous *Secotium*. Here again there appears the link between *Paxillus*, a genus considered one of the Agaricaceae, and secotiums. There remains the question: which descended from the other? The origin of *Paxillus* among the Agaricales certainly remains enigmatic, but nothing prevents our thinking that the genus *Austrogaster* is also a decadent entity derived from the Paxillaceae, both being derived from a branch of the Boletales. We are here, of course, in the wide open field for hypotheses.

*Thaxterogaster series*

I refer the reader to the studies of Singer and A. H. Smith (1958a; Singer, 1962b) and to their conclusions insofar as the Secotiaceae *s. str.* are concerned, where the genera *Thaxterogaster* and *Endoptychum* constitute two most important aggregates for systematic studies and deductions of a phyletic order. The intimate relationship of the secotiums of the genus *Thaxterogaster* Sing. having brown, ornamented spores, with the Cortinariaceae and more especially with *Myxacium*, was placed in evidence well by Singer and Smith. It is necessary to clarify the precise position of *Secotium conei* Heim (1951a) of New Zealand, equipped with a cupulate volva, extremely close to the *Thaxterogaster* affiliated with the Cortinariaceae, but probably different, as Singer and Smith think. Horak has created a subgenus *Volvigerum* to accommodate this species, and Heim (1966) has proposed to elevate this entity to the rank of genus. Thus, the conception of the two American authors in the end separates *Thaxterogaster* with ornamented spores of the *Cortinarius* type from *Secotium s. str.* with smooth spores, and to this is added *Volvigerum*, at the same time volvate and with spores having more numerous ornaments. I might mention that the pigments bring *Volvigerum* close to *T. porphyreum*. As for the delicate problem of whether clamps are present or absent, we shall return to this elsewhere.

*Strophariaceae series*

The genus *Weraroa* Sing. (= *Cytophyllopsis* Heim) might well constitute the secotioid form (*W. erythrocephalum*) of a phylum of the Strophariaceae in the opinion of Singer and Smith (1958c).

*Endoptychum series*

Following the discovery in the western United States of a species of the Secotiaceae, *Endoptychum depressum*, Singer and Smith (1958b) assembled and defined the four species of *Endoptychum*, gasteroid forms with probable affinities with *Psalliota* in their opinion. Thus, the discussions concerning *Thaxterogaster* and *Endoptychum* that Singer and Smith have conducted lead plausibly to the introduction of the Cortinariaceae into the phylogenetic tree, bringing together the Agaricales and the Gasteromycetes and, on the other hand more hypothetically, *Psalliota* (or *Agaricus*). This taxonomic contribution to our knowledge strengthens greatly the theses that affirm the general character of the kinship between agarics and gasteromycetes, but the problem of the very meaning of evolution and partial evolution remains wide open.

*Neosecotiaceae series: (*Secotium guenzii = Neosecotium*)*

Among the Secotiaceae we find leucosporous species with closed pilei, with asymmetrical spores and incompleted pores (*S. guenzii*)

(Heim, 1951a; Singer & Smith, 1960), or inconstant (*N. macrosporum* Sing. & Sm.). Such characters have led these two authors to link these species to the lepiotas of the genus *Leucocoprinus*. This problem needs reexamination.

*The case of* Cryptomphalina

I have described (Heim, 1962) this genus in the form of the species *Cryptomphalina sulcata* of Thailand, in which the edges of the pileus hug closely the top of the stipe and of which the hymenium offers an hypophyllic area carpeted with incomplete pleats and with oscillating lines oriented in the radial sense. This leucosporous agaric, close to the Omphaliaceae, does not lend itself to assimilation with any gasteroid form in spite of its physiognomic characteristics, but it carries the stamp of a tendency that is confirmed by the criterion of hymenial disorders.

## THE PRIME EXAMPLE OF *Torrendia*

The important work of G. Malençon (1955) on the embryology and the phyletic position of *Torrendia* seems to us to have contributed strongly to sharpening our argument in favor of a decadent derivation of certain phyla in accordance with a most suggestive table (fig. 3). This basic memoir seems unfortunately to have escaped the attention of most authors (but not A. F. M. Reijnders) who in these last years have interested themselves in the domain of the phylogeny of the Gasteromycetes.

As to the structure of *Torrendia*, Maleçon demonstrated that it rode astride two types of construction believed until then to be incompatible. In this remarkable genus, the plectobasidial structure became partly hymenobasidial in the course of its development, yet the similarity of its constitution affects all parts of the mushroom,— "volva, stipe, endoperidium, pseudopileus are made of similar materials." *Torrendia* thus seemed as though it were constituted of an "undifferentiated mycelial mass disseminated throughout its cellular formations with manifest basidial potentiality." To summarize, *Torrendia* offers an example of a kind of confusion of the two states, hymenobasidial corresponding to the fertile territory and plectobasidial corresponding to the nutritive mass.

For Malençon, the decadence was translated by "the passage of the free hymenium into the imprisoned hymenium" and was begun by pseudoangiocarpy and hemiangiocarpy and finally consumed in some measure by having arrived at the angiocarpic stage. It is what I have included in the phyletic position of the Lactario-Russulae, still agaricimorphic but subangiocarpic.

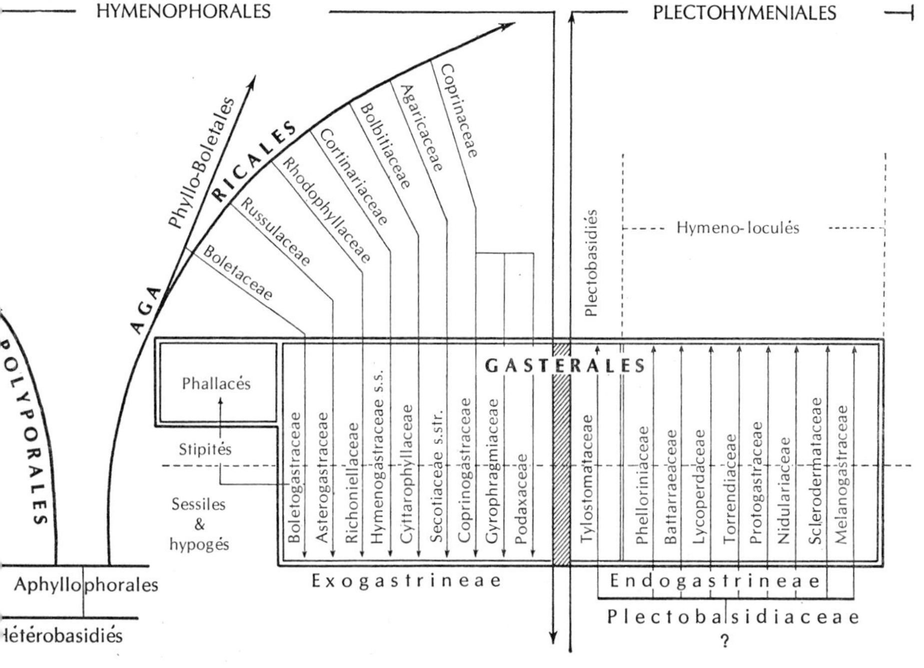


FIGURE 3. General phylogeny of the higher Basidiomycetes (after Malençon).

Little by little, this degradation of *Torrendia* has left its organic evidence: loss of regularity, friability, rumpling, winding sinuosities, division by the mechanism of anastomoses, even to the extreme simplification of the orbicular cavity. "The fruits, completely degraded, no longer have regularity, nor base, nor apparent summit."

## CONCLUSION

As for the Agaricales, Rolf Singer (1962a) has examined successively and carefully the arguments for and against three theories that explain the derivation of the Agaricales as coming from the Gasterales, or alternatively from the Aphyllophorales, or from both. We refer the reader to that summary for further details. While defending the thesis of a derivation of the Agaricales from the Gasterales (which in the opinion of H. Lohwag [1924, 1941] would supply the primitive arsenal whence the ensemble of Macromycetes may well have emerged), Singer has not thought it necessary to take into account the conception that argues in favor of decadent adaptations of the Agaricales in the direction of the Exogastrinae-Gasterales. Yet the situation is that this fourth conception has been elaborated in sundry papers and critical writings formulated notably by G. Malençon, A. F. M. Reijnders, H. Romagnesi, and me. By leaving our arguments unanswered the other side might possibly run the risk of giving the impression that it avoids a confrontation.

We have already called attention to certain replies of Reijnders to the thesis of H. Lohwag, developed as it has been by R. Singer and A. H. Smith. For my own part I have already brought into the discussion certain new observations on *Le Ratia* and *Galeropsis* which assuredly do nothing to weaken the thesis of degraded phyla: the embryology of agaricoid forms, the nature and the variations of the veils, the decadent reactions (if indeed not teratological) of many organs belonging to the gasteroid forms—all of which support our thesis. G. Malençon in his remarkable study of *Torrendia* has summed up perfectly, in my opinion, the objections to Lohwag's thesis: "When one examines the simplest Exogastrinae—*Gastrosporium*, *Hymenogaster*—one is struck by the highly marked degree of evolution that their hymenophore presents. How then should we explain the existence of such a differentiation, so far below the scale proper to the forms, otherwise than by an inheritance received from elsewhere? In short the persistence of the superior organization has survived side by side with the collapse of the vegetative apparatus" (*loc. cit.*, 1955, p. 125).

Of course my paper does not try to cover all the Gasteromycetes, and by the same token it has left in the shadow the Aphyllophorales and the details of the Agaricales, the Boletales, and above all, the Phallales, these last representing in our opinion the *summum* of evolution in the fungi: the Plectohymeniales and Phallales can be considered as overevolved and they have nothing in common with the Hymenophorales.

As seems indicated by this focusing of attention on the filiation of

Hymenomycetes and Agaricales, the discussions of gasteroid forms and their development in recent years reveal a weakness that is indisputable as far as their extension to phylogenetic considerations is concerned. This is the result in part of the complexity of the organisms under discussion, in part of the multiplicity of influences that may have been brought to bear on the mushrooms beginning with the nature of the supporting agencies, and in part and most importantly, of the almost total absence of material evidence derived from paleontology. These difficulties affect a domain, namely that of phylogenesis, that strictly speaking is not a true philosophy, much less a science. We are facing a body of arguments that would like to be proofs but that are in part shaped by intuition. If the merit of a discussion is to force us to arrive at an explanation both logical and impartial, it can lead to only one conclusion: we are dealing with a close-packed intellectual game based on facts that neither experiment nor evidence from the past permit us to control. In truth, the merit of each author lies in his effort to explain, in the play of his mind exercising his power of synthesis, demonstrating at the same time and more surely a certain prescience and premonition. It is less a question of the facts than of their interplay and confrontation.

I have recalled successively the essential premises on which the theses facing us are supported. Each has its strengths, each its weaknesses. But if the plausibilities carry most mycologists toward one theory rather than another, no certainty, no essential and sufficient peremptory indication underlies such a verdict. In the end this rests on an evolutionary postulate inspired, according to one or another, by a Darwinian preference, or Lamarckian, or mutationist, or neo-Darwinian, etc. Thus, one admits or does not admit the adaptive importance of the environment and the potentialities of regressivity or decadence. The Macromycetes—both the Ascomycetes and the Basidiomycetes—are struck by weighty influences leading to *convergence*, accidental resemblances, nature having at its disposal only a limited arsenal of forms, and by the perfidious reflection of *homology*: but a resemblance is *a priori* meaningless. Nevertheless, without infringing on our skepticism, we believe that new arguments will spring out of pure cultures and of the experimental study in the laboratory of carpophores, of mutations and teratological cases, of refined anatomy and embryology, as well as of a deeper knowledge of the chemistry, of the biotic ties, and finally of the more precise structure of the teguments, of the ornaments, and of the germinal pores of the sporal elements. In short, the future is far from being closed, and this indeed is one of the justifications of the remarkable symposium that we are completing. But let us not forget that in each homogeneous group of genera or species, the principle of the subordination of characters plays a different role. Whereas in one case those of a chemical or embryological or histological or even sporal order will be of prime importance, elsewhere everything will be opposed to such a statement of the relationships.

But humility cautions us all to consider first and foremost the facts as they are brought to us, not the theories that try to mobilize or dragoon the facts for or against. Today for most of the families of Agaricales there have been discovered living examples under the headings of angiocarps, gasteroids, either hypogaeous or epigaeous, with pileus either closed or slightly open. Here is the rich harvest that repeated observations during the last twenty years have brought to our doorstep.

And so, in conclusion, we render homage to the mycologists whose command of a large spectrum of mycological groupings is the essential reason for the interest with which they have imbued the subject of phylogenesis. To this explanation we must add the manifest curiosity, the incisive desire to explain, the need to bring together, to read in some measure the past, to reconstruct the modalities of evolution in the world of the mushrooms, a need all the greater perhaps because the mycologists know that they are not fully utilizing what is needed to realize their ambition. Whatever may be the value, the truth, the future of each of these theories that stand confronting each other, we must render justice to those whom the desire to know, to build, takes even beyond the reasonable or imperious limits of science. As for the rest, the reconstruction of the past—on this only the future may perhaps pronounce judgment.

## References

Bucholtz, F. 1903. Zur morphologie und systematik der Fungi hypogaei. Ann. Mycol. 1: 152–174.

Fayod, V. 1889. Prodrome d'une histoire naturelle des Agaricinés. Ann. Sci. Nat., Bot. VII. 9: 181–411.

Fischer, E. 1925. Zur Entwicklungsgeschichte der Fruchkörper der

Secotiaceae. Festschr. Carl Schroeter Geob. Inst. Rübel, Zurich 3: 571–582.

GÄUMANN, E. 1928. Comparative morphology of fungi. New York. 541 p.

GILBERT, E. J. 1931. Les Bolets, in les livres du mycologue, vol. 3. Paris. 254 p.

HEIM, ROGER. 1931. Les liens phylétiques entre les Agarics ochrosporés et certains Gastéromycètes. C. R. Ac. Sci. 192: 291–293.

———. 1936. Sur la parenté entre Lactaires et certains Gastéromycètes. C.R. Ac. Sci. 202: 2101–2103.

———. 1937. Observations sur la flore mycologique malagache. V. Les Lactario-Russulés à anneau: Ontogénie et phylogénie. Rev. Mycol. 2: 1–42.

———. 1938. Prodrome à une flore mycologique de Madagascar et Dépendances. I. Les Lactario-Russulés du domaine oriental de Madagascar. Essai sur la classification et la phylogénie des Astérosporales. Paris. 196 p.

———. 1943. Remarques sur les formes primitives ou dégradées de Lactario-Russulés tropicaux. Boissiera 7: 266–280.

———. 1945. Les Agarics tropicaux à hyménium tubulé. Rev. Myc. 10: 3–61.

———. 1948. Phylogeny and natural classification of Macro-fungi. Trans. Brit. Mycol. Soc. 30: 161–178.

———. 1950. Le genre *Galeropsis* Velenovsky (*Cyttarophyllum* Heim), trait d'union entre Agarics et Gastérales. Rev. Mycol. 15: 3–28.

———. 1951a. Notes sur la flore mycologique des Terres du Pacifique Sud. III—Sur les *Secotium* de Nouvelle-Zélande et la phylogénie de ce genre. Rev. Mycol. 16: 29–153.

———. 1951b. Notes sur la flore mycologique des Terres du Pacifique Sud. IV—Le genre néo-calédonien *Le Ratia* Pat. Rev. Mycol. 16: 154–158.

———. 1952. Les voies de l'évolution chez les champignons. Colloque intern. du C.N.R.S. sur l'évolution et la phylogénie chez les végétaux. L'Année Biol. 28: 27–46.

———. 1957. Les Champignons d'Europe. Paris. 2 vol., 818 p.

———. 1959. Une nouvelle espèce de Gastrolactarié en Thaïlande. Rev. Mycol. 24: 93–102.

———. 1962. Contribution à la flore mycologique de la Thaïlande. Rev. Mycol. 27: 124–159.

———. 1966. Les Meiorganés, phylum reliant les Bolets aux Polypores. Rev. Mycol. 30: 307–329.

———. 1967. Notes sur la flore mycologique des Terres du Pacifique Sud. VII. Les énigmes des *Le Ratia.* Rev. Mycol. 32: 11–15.

———. 1968a. Notes sur la flore mycologique des Terres du Pacifique Sud. VIII. Nouvelles recherches sur les *Le Ratia.* Rev. Mycol. 33: 137–154.

———. 1968b. Deuxième mémoire sur les Cyttarophyllés. Bull. Soc. Mycol. France 84: 103–116.

HOLM, L. 1954. Classification et phylogénie des Gastéromycètes. Rapp. 8th Congrès Int. Bot. Paris, sect. 18, 19, 20: 54–60.

HORAK, E. 1964a. Fungi Austroamericani. V. Beitrag zur Kenntnis der Gattungen *Hysterangium* Vitt., *Hymenogaster* Vitt., *Hydnangium* Wallr. und *Melanogaster* Cda. in Südamerika (Argentinien, Uruguay). Sydowia 17: 197–205.

———. 1964b. Fungi Austroamericani. VI. *Martellia* Matt., *Elasmomyces* Cav., und *Cystangium* Sing. & Smith in Südamerika. Sydowia 16: 206–213.

KÜHNER, R. 1926. Contribution à l'étude des Hyménomycètes, et spécialement des Agaricacés. Le Bot. 17: 5–215.

———. 1945. Le problème de la filiation des Agaricales à la lumière de nouvelles observations d'ordre cytologique sur les Agaricales leucosporées. Bull. Soc. Linn. Lyon 14: 160–166.

———. 1948. Place des Bolets dans l'ensemble des Basidiomycètes et rapports des diverse espèces de Bolets entre elles. Bull. Soc. Nat. Oyonnax 2: 37–48.

LOHWAG, H. 1924. Entwicklungsgeschichte und systematische Stellung von *Secotium agaricoides.* Österr. bot. Z. 73: 161–174.

———. 1941. Anatomie der Asco- und Basidiomyceten. Handb. der Pflanzenanat. 6 (II, 3).

MALENÇON, G. 1931. La série des Astérosporés. Trav. crypto. déd. ā L. Mangin. 1: 337–396.

———. 1938. *Dodgea occidentalis* Malenc., nouveau genre et nouvelle espece de Rhizopogoneae. Bull. Soc. Mycol. France 54: 193–203.

———. 1955. Le développement du *Torrendia pulchella* Bres. et son importance morphogénétique. Rev. Mycol. 20: 81–130.

MÉLENDEZ-HOWELL, L. M. 1967. Recherchés sur le pore germinatif des basidiospores. Ann. Sci. Nat., Bot. 12e sér., 8: 487–638.

PATOUILLARD, N. 1900. Essai taxonomique sur les familles et les genres des Hyménomycètes. Lons-le-Saunier. 184 p.

PERREAU, J. 1961. Recherches sur les ornementations sporales et la sporogenèse chez quelques espèces des genres *Boletellus* et *Strobilomyces.* Ann. Sci. Nat., Bot. 12e sér., 2: 399–489.

———. 1967. Recherches sur la différenciation et la structure de la paroi sporale chez les Homobasidiomycètes à spores ornées. Ann. Sci. Nat., Bot. 12e sér., 8: 639–746.

PILÁT, A. 1948. On the genus *Galeropsis* Velenovsky. Stud. Bot. Cechosl., Prague 9: 177-185.

PILÁT, A., *et al.* 1958. Gasteromycetes. Flora C.S.S.R., Prague. 862 p.

REIJNDERS, A. F. M. 1963. Les problèmes du développement des carpophores des Agaricales et de quelques groupes voisins. La Haye. 412 p.

ROMAGNESI, H. 1933. Le genre *Richoniella,* chaînon angiocarpe de la série des Rhodogoniosporés. Bull. Soc. Mycol. France 49: 433–434.

SAVILE, D. B. O. 1955. A phylogeny of the Basidiomycetes. Canad. J. Bot. 33: 60–104.

SINGER, ROLF. 1936. Studien zur Systematik der Basidiomyceten: *Galeropsis,* ein gasteromycet. Beih. Bot. Cbl. 56: 147–150.

———. 1962a. The Agaricales in modern taxonomy. 2nd ed. Weinheim. 915 p.

———. 1962b. Monographs of South American Basidiomycetes, especially those of the east slope of the Andes and Brazil V.—Gasteromycetes with agaricoid affinities (Secotiaceous Hymenogastrineae and related forms). Bol. Soc. Argen. Bot. 10: 52–67.

SINGER, ROLF, AND A. H. SMITH. 1958a. Studies on Secotiaceous fungi. I. A monograph of the genus *Thaxterogaster.* Brittonia 10: 201–216.

———. 1958b. Studies on Secotiaceous fungi. II. *Endoptychum depressum.* Brittonia 10: 216–221.

———. 1958c. Studies on Secotiaceous fungi. III. The genus *Weraroa.* Bull. Torrey Bot. Club 85: 324–334.

———. 1960. Studies on Secotiaceous fungi. VII. *Secotium* and *Neosecotium.* Madroño 15: 152–158.

SMITH, A. H., AND D. REID. 1962. A new genus of the Secotiaceae. Mycologia 54: 98–104.

SMITH, A. H., AND R. SINGER. 1958. Studies on Secotiaceous fungi.

VIII. A new genus in the Secotiaceae related to *Gomphidius*. Mycologia 50: 927–938.

———. 1959. Studies on Secotiaceous fungi. IV. *Gastroboletus*, *Truncocolumella* and *Chamonixia*. Brittonia 11: 205–223.

ZELLER, S. M. 1939. New and noteworthy Gasteromycetes. Mycologia 31: 1–32.

———. 1943. North American species of *Galeropsis*, *Gyrophragmium*, *Longia* and *Montagnea*. Mycologia 35: 409–421.

# LIST OF DISCUSSANTS

Mr. Terry L. Amburgey
Department of Plant Pathology
North Carolina State University
Raleigh, North Carolina

Mr. Gerald M. Baker
Department of Botany
Utah State University
Logan, Utah

Mr. Ernst E. Both
Buffalo Museum of Science
Humboldt Park
Buffalo, New York

Dr. Lynn R. Brady
College of Pharmacy
University of Washington
Seattle, Washington

Dr. Harold H. Burdsall, Jr.
USDA Forest Service–
Forest Disease Laboratory
Laurel, Maryland

Dr. Stanley Dick
Department of Botany
Indiana University
Bloomington, Indiana

Dr. Virginia Dublin
1410 Garden Lane
Champaign, Illinois

Dr. Charles Leonard Fergus
Botany Department
Pennsylvania State University
University Park, Pennsylvania

Dr. Larry F. Grand
Department of Plant Pathology
and Forestry
North Carolina State University
Raleigh, North Carolina

Mr. John H. Haines
Department of Botany
Oregon State University
Corvallis, Oregon

Dr. David L. Hanks
Botany Department
Brigham Young University
Provo, Utah

Dr. Richard T. Hanlin
Department of Plant Pathology
University of Georgia
Athens, Georgia

Dr. James D. Haynes
Department of Biology
State University College at Buffalo
Buffalo, New York

Mr. Charles R. Jessop
2132 Keith Road
Abington, Pennsylvania

Dr. James W. Kimbrough
Department of Botany
University of Florida
Gainesville, Florida

Dr. Michael J. Larsen
Canada Department of Forestry and Rural Development
Sault Sainte Marie, Ontario, Canada

Dr. Anthony E. Liberta
Department of Biological Sciences
Illinois State University
Normal, Illinois

Dr. E. S. Luttrell
Department of Plant Pathology
University of Georgia
Athens, Georgia

Mr. David J. McLaughlin
Botany Department
University of California
Berkeley, California

Dr. Royall T. Moore
Department of Botany
North Carolina State University
Raleigh, North Carolina

Dr. Paul D. Olexia
Department of Biology
Kalamazoo College
Kalamazoo, Michigan

Dr. René Pomerleau
Forest Research Laboratory
Sillery, Quebec, Canada

Miss Celia Dubuvoy Rudoy
Instituto de Biologia
Laboratorio de Criptogamia
Universidad Nacional Autonoma de Mexico
Mexico, D.F.

Dr. James W. Sinden
Gossae-Zuerich
Switzerland

Dr. Jack Vozzo
Manned Space Center
NASA Headquarters
Houston, Texas

Dr. A. L. Welden
Department of Biology
Tulane University
New Orleans, Louisiana

# INDEX

Italicized numbers indicate synonyms, illustrative, or tabular material.

# PLATES

Plate I (Lentz)

Fig. 1–Tomental generative hypha with clamp connection (*Panus rudis*); fig. 2–Clamped generative hypha (*Cortinarius californicus* A. H. Smith); fig. 3–Simple-septate generative hypha (*Lopharia crassa*); figs. 4, 5–Skeletal hyphae (*Microporus xanthopus* [Fr.] O. Kuntze); fig. 6–Arboriform skeletal hypha (*Ganoderma lucidum* [Fr.] Karst.); fig. 7–Binding hyphae (*Ganoderma curtisii* [Berk.] Murr.); fig. 8–Binding hyphae (*Polyporus squamosus* Huds. ex Fr.); fig. 9–Binding hypha (*Microporus xanthopus*).

Plate I (Lentz)

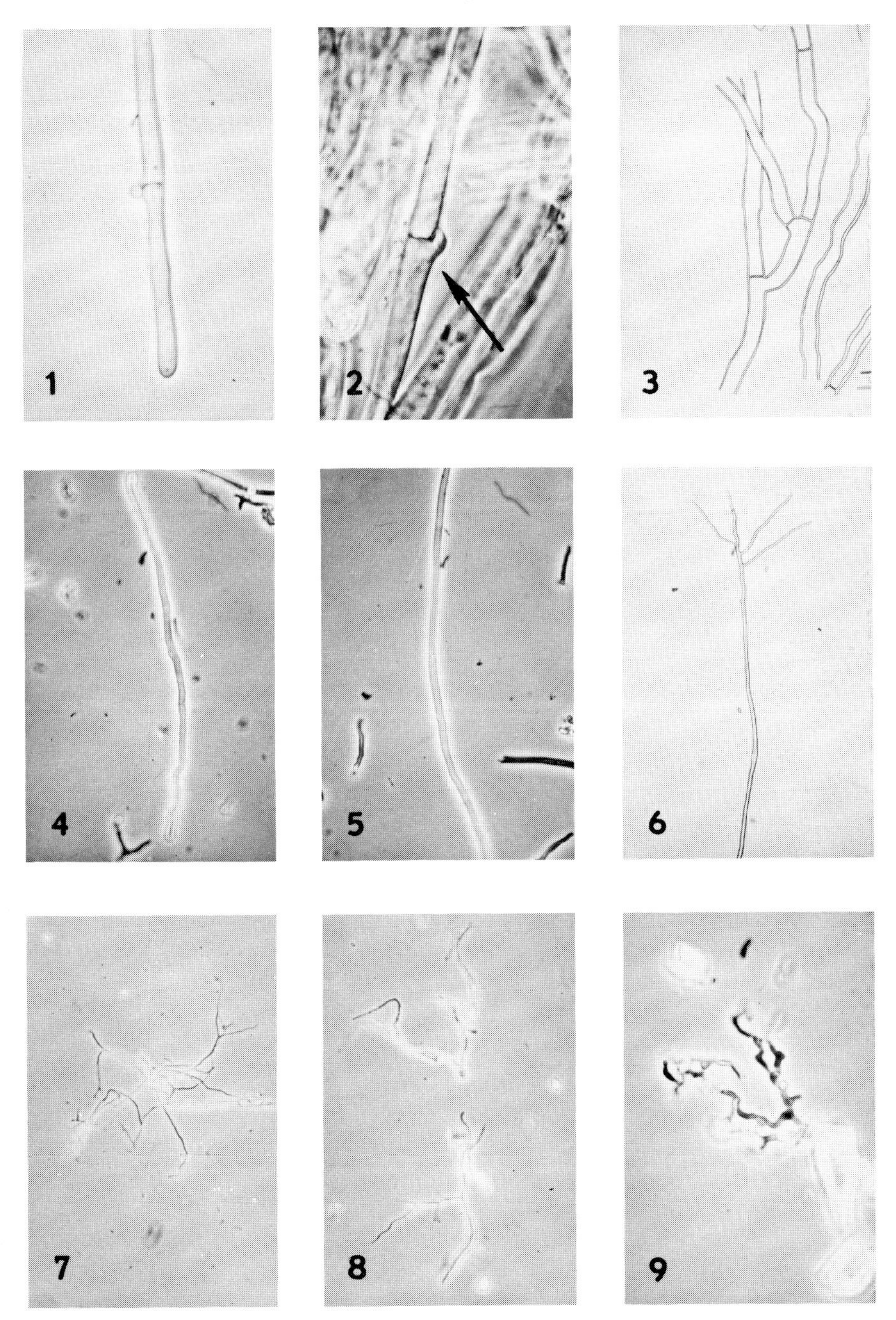

PLATE II (Lentz)

Fig. 10—Interweaving lateral hypha branching from generative hypha (*Donkia pulcherrima*); fig. 11—Basidiocarp habit (*Amphinema byssoides*); fig. 12—Basidiocarp section with cystidia (*Amphinema byssoides*); fig. 13—Generative hyphae in culture, showing simple septa, a single clamp, and multiple clamps (*Stereum ochraceoflavum*); fig. 14—Globose basidiospores with amyloid warts (*Vararia peniophoroides*); fig. 15—Subfusoid basidiospores with amyloid plaques (*Vararia investiens*); fig. 16—Dichohyphidia and broadly ellipsoidal basidiospores (*Vararia tropica*); fig. 17—Generative hypha with multiple clamps (*Donkia pulcherrima*); fig. 18—Generative hyphae in culture, with single clamp and multiple clamps (*"Peniophora" sanguinea* [Fr.] Höhn. & Litsch.—*Phanerochaete*).

PLATE II (Lentz)

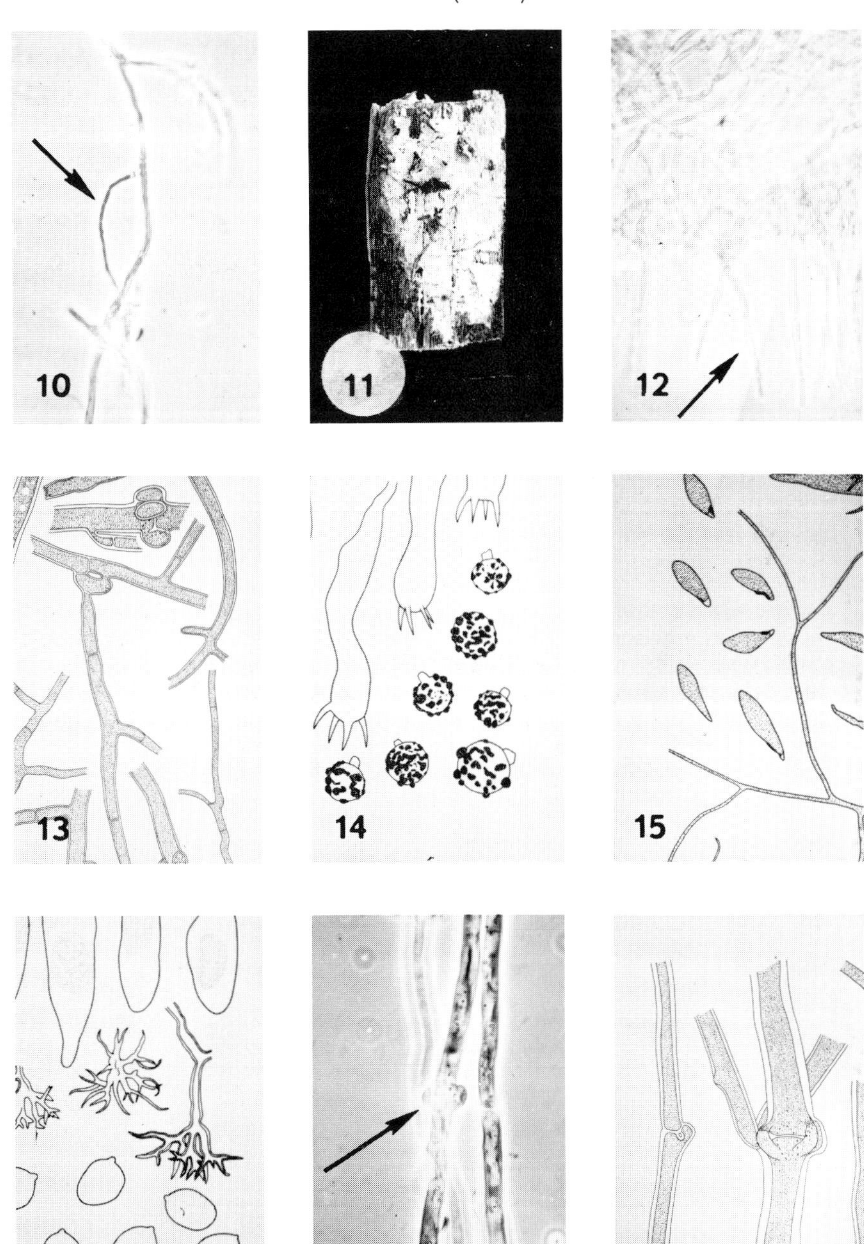

PLATE III (Lentz)

Fig. 19–Hyphae with H-connection ("*Peniophora*" *cremea–Phanerochaete*); fig. 20–Contextual hypha with young branch developing at right angle, and broad, thick-walled, colored basal hypha (*Botryobasidium pruinatum* [Bres.] Donk–*Dimorphonema*); fig. 21–Oil-filled hypha ("*Corticium*" *ravum* Burt); fig. 22–Vascular hyphae with waxy contents (*Laetiporus sulphureus* [Fr.] Murr.); fig. 23–Waxy vascular hypha from the stipe (*Collybia velutipes* [Fr.] Quél.); fig. 24–Branched tip of waxy vascular hypha (*Laetiporus sulphureus*); fig. 25–Skeletocystidium developing from a generative hypha (*Echinodontium tinctorium* [Ell. & Ev.] Ell. & Ev.); fig. 26–Basidiocarp section (hymenium on left) with sanguinolentous skeletal hyphae terminating as pseudo-cystidial skeletocystidia (*Stereum sanguinolentum* [Fr.] Fr.); fig. 27–Gloeocystidia (*Peniophora rufa* [Fr.] Boidin).

Plate III (Lentz)

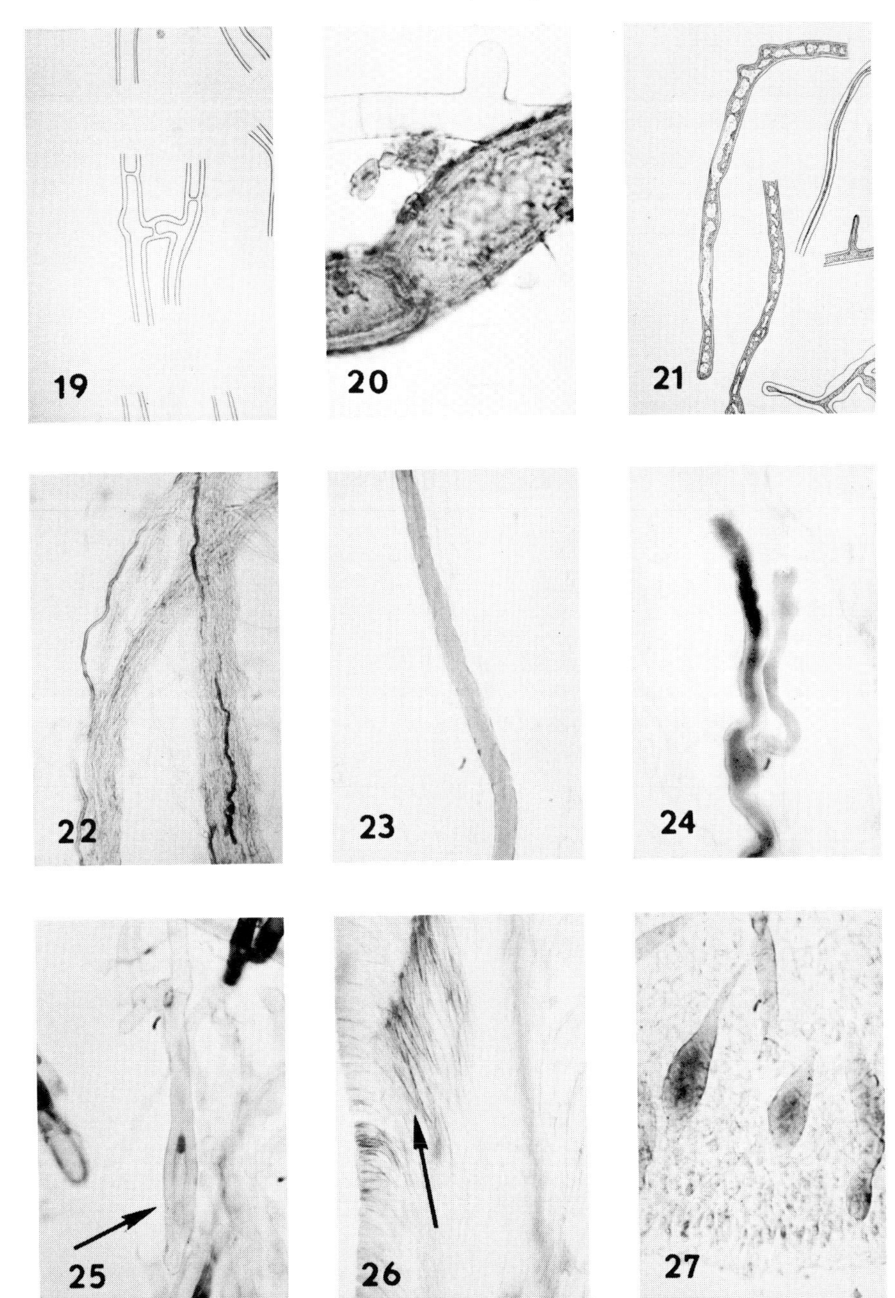

Plate IV (Lentz)

Fig. 28–Spherocysts as seen in a cross-section of a stipe (*Russula* sp.); fig. 29–Vesicular bodies among hyphae of the context adjacent to the sub-hymenium (*Chondrostereum purpureum*); fig. 30–Stephanocysts on generative hyphae in culture (*Hyphoderma guttuliferum* [Karst.] Donk); fig. 31–Simple setae protruding from the hymenium (*Hymenochaete sallei* Berk. & Curt.); figs. 32, 33–Embedded setae of context (32) and dissepiment (33) (*Inonotus glomeratus*); fig. 34–Section of spiny basidiocarp with stellate setae (*Asterodon ferruginosus* Pat.); fig. 35–Stellate seta (asteroseta) (*Asterostroma andinum* Pat.); fig. 36–Cystidia (pleurocystidia), showing origin from trama of a lamella (*Marasmius coherens* Groves).

PLATE IV (Lentz)

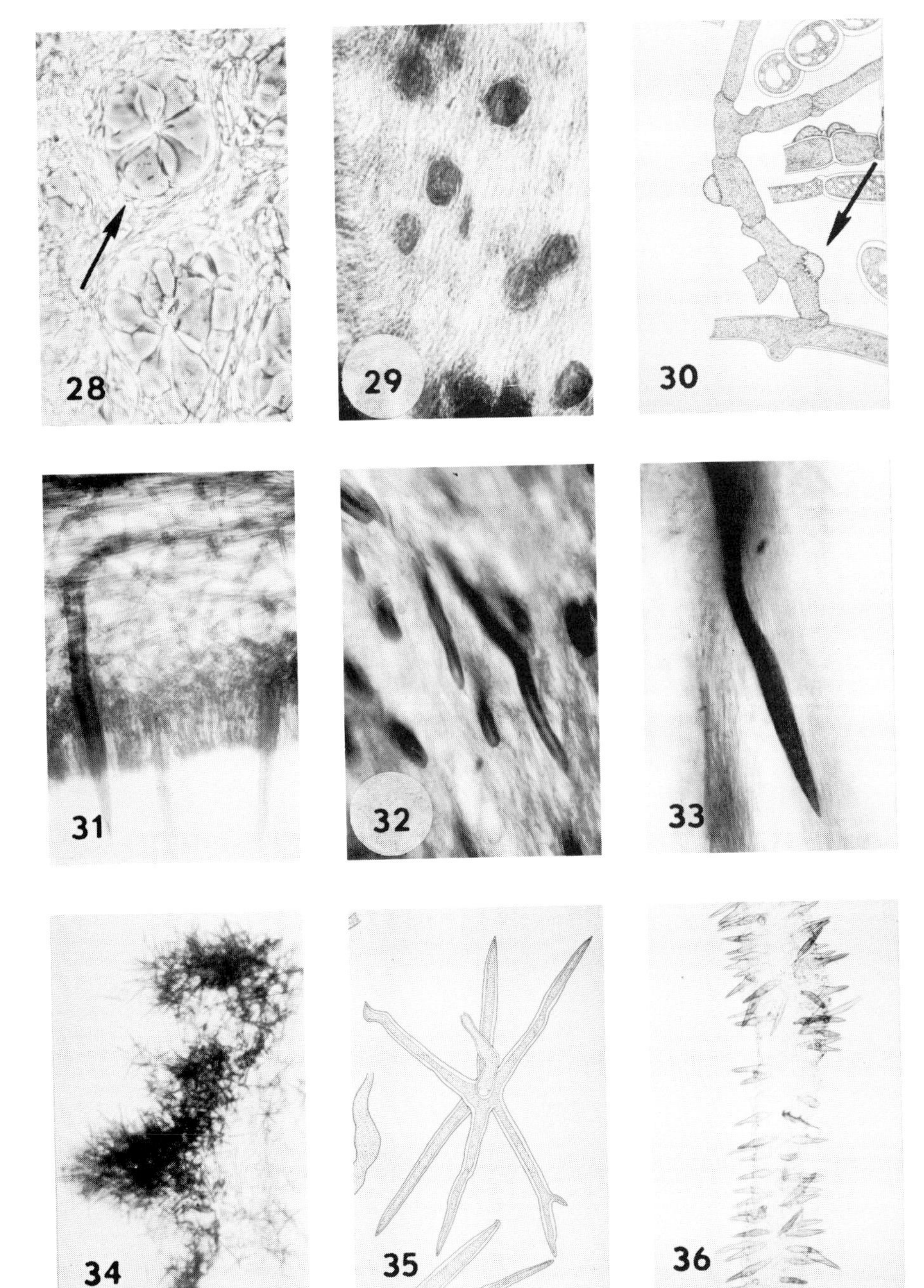

PLATE V (Lentz)

Fig. 37–Incrusted cystidia, showing origin from clamped generative hyphae (*Echinodontium taxodii* [Lentz & McKay] Gross); figs. 38, 39–Digitate cells, intergrading structures, and pileocystidia of the pileus surface (*Marasmius coherens*); fig. 40–Tomental generative hyphae appearing as "hairs" of the pileus surface (*Panus rudis*); fig. 41–Hairlike generative hyphae appearing as caulocystidia on the stipe (*Collybia velutipes*); fig. 42–Peglike hyphal fascicle originating from the hymenial region of a basidiocarp (*Heterochaete tenuicola* [Lév.] Pat.); fig. 43–Peglike hyphal fascicles, showing origin from the region in and near the hymenium (*Mycobonia flava*); fig. 44–Dendrohyphidium (*Laeticorticium minnsiae* [H. S. Jacks.] Donk); fig. 45–Dichohyphidium (*Vararia investiens*).

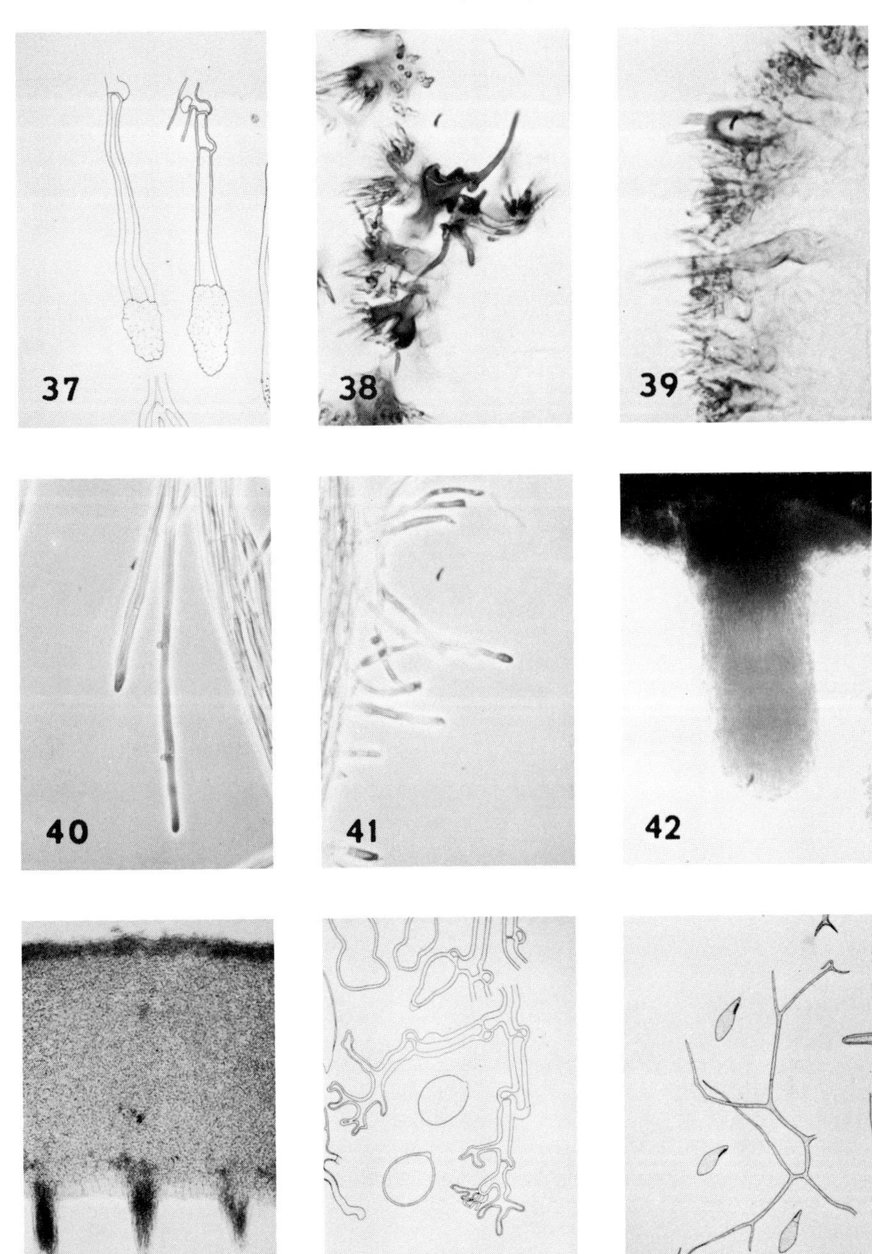
37
38
39
40
41
42
43
44
45

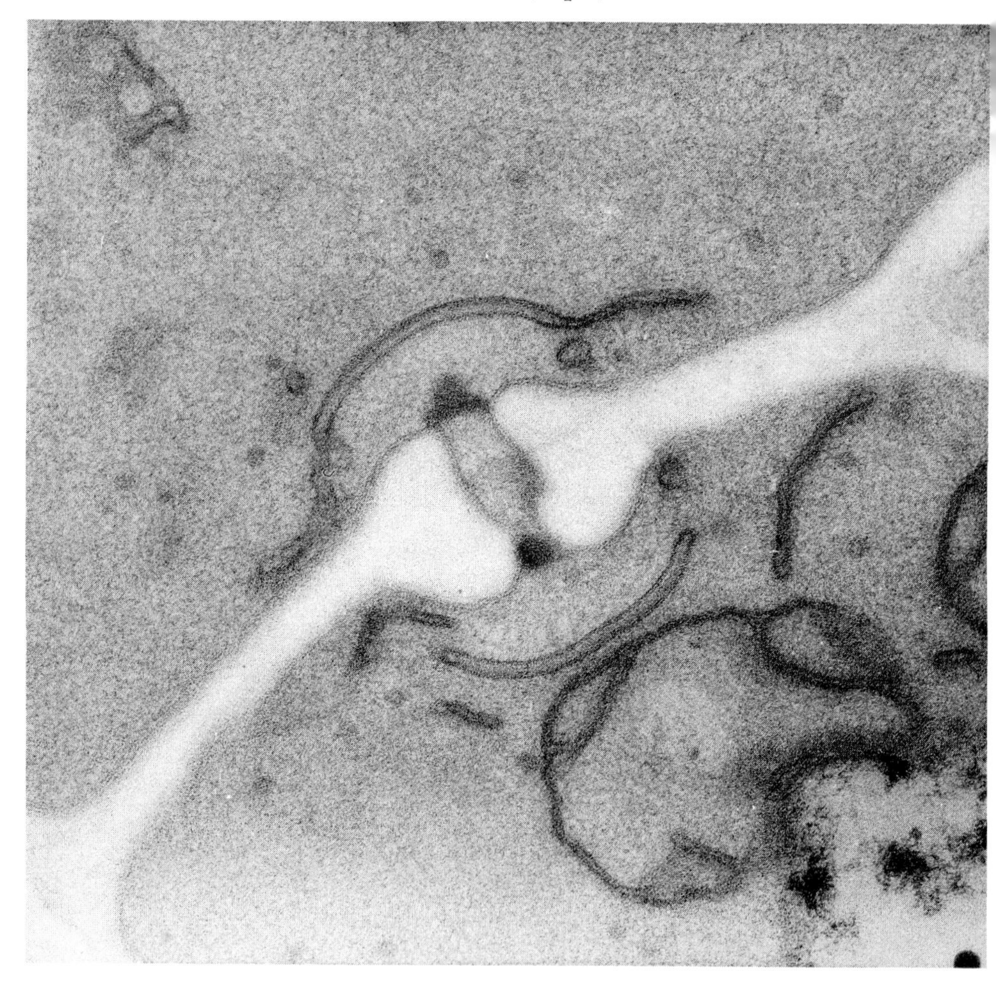

PLATE VI (Raper and Flexer Discussion)
Septum of *Exidia glandulosa*, X15,000. This type of septum is characterized by the dolipore formed by the centrally flared apertural margins which are blocked by electron dense plugs that may be homologous to those observed in ascogenous hyphae (see Moore, The Fungi, vol. I, 1964) and by the overarching, membranous parenthesomes. Such septal configurations are typical of Basidiomycete hyphae, including the clamp connections (see Jersild *et al.*, Arch. Mikrobiol. 57: 20. 1967). (Photo courtesy of Royall T. Moore.)

PLATE VII (Miller)

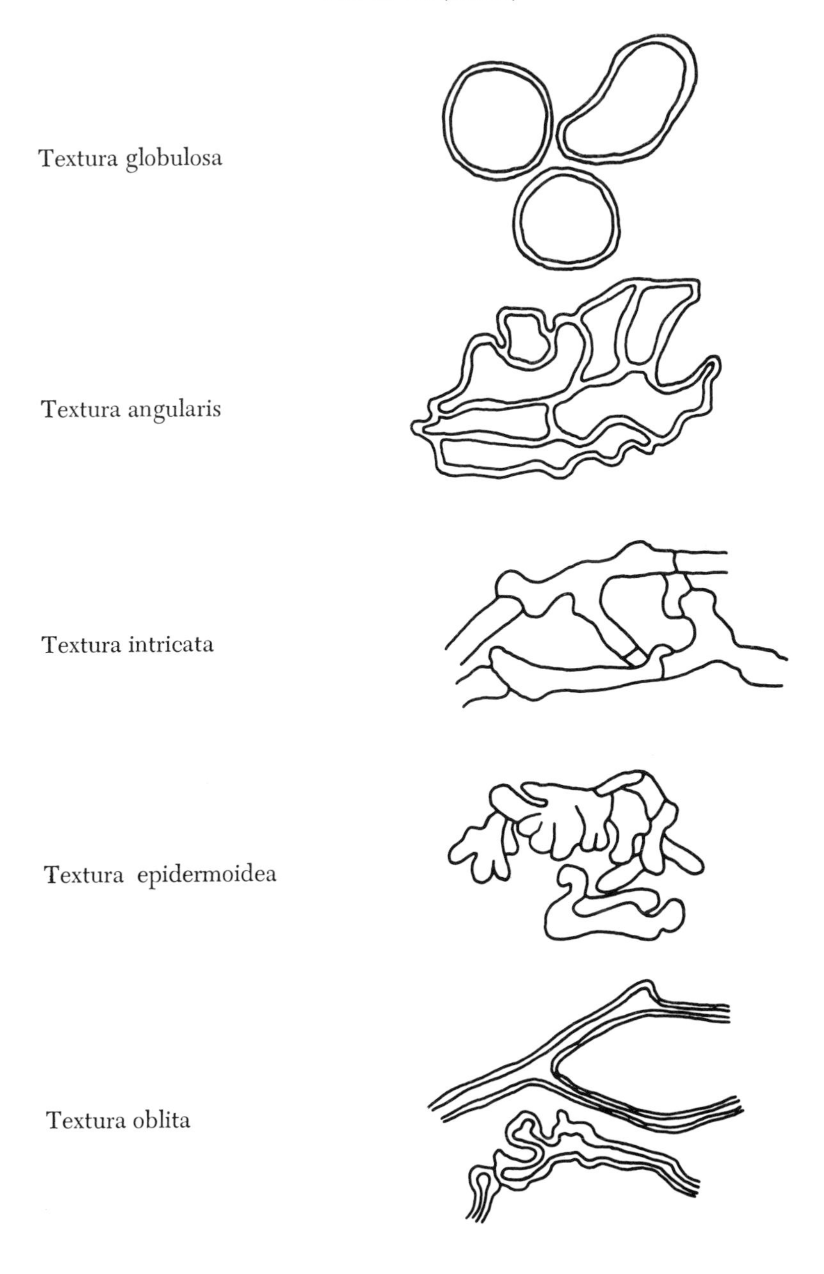


Tissue types found in culture (X700).

Plate VIII (Miller)

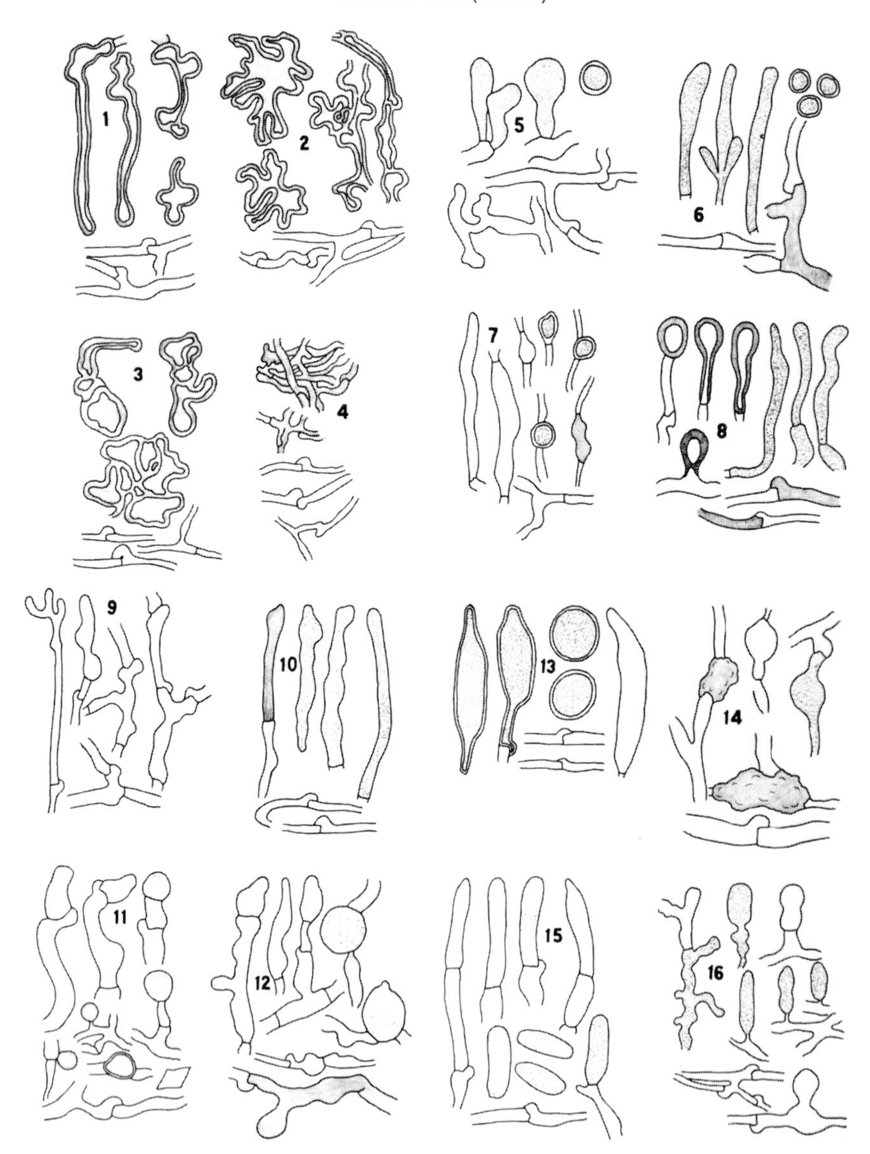


Fig. 1–*Xeromphalina brunneola*; fig. 2–*X. kauffmanii*; fig. 3–*X. campanella*; fig. 4–*X. cauticinalis*; fig. 5–*Lentinellus pilatii*; fig. 6–*L. ursinus*; fig. 7–*L. pilatii*; fig. 8–*L. cochleatus*; fig. 9–*Pleurotus ostreatus*; fig. 10 –*P. ulmarius*; fig. 11–*P. sapidus*; fig. 12–*P. candidissimus*; fig. 13–*P. dryinus*; fig. 14–*Phyllotopsis nidulans*; fig. 15–*Pleurotus cystidiosus*; fig. 16–*Hohenbuehelia petaloides*. (X700)

PLATE IX (Miller)

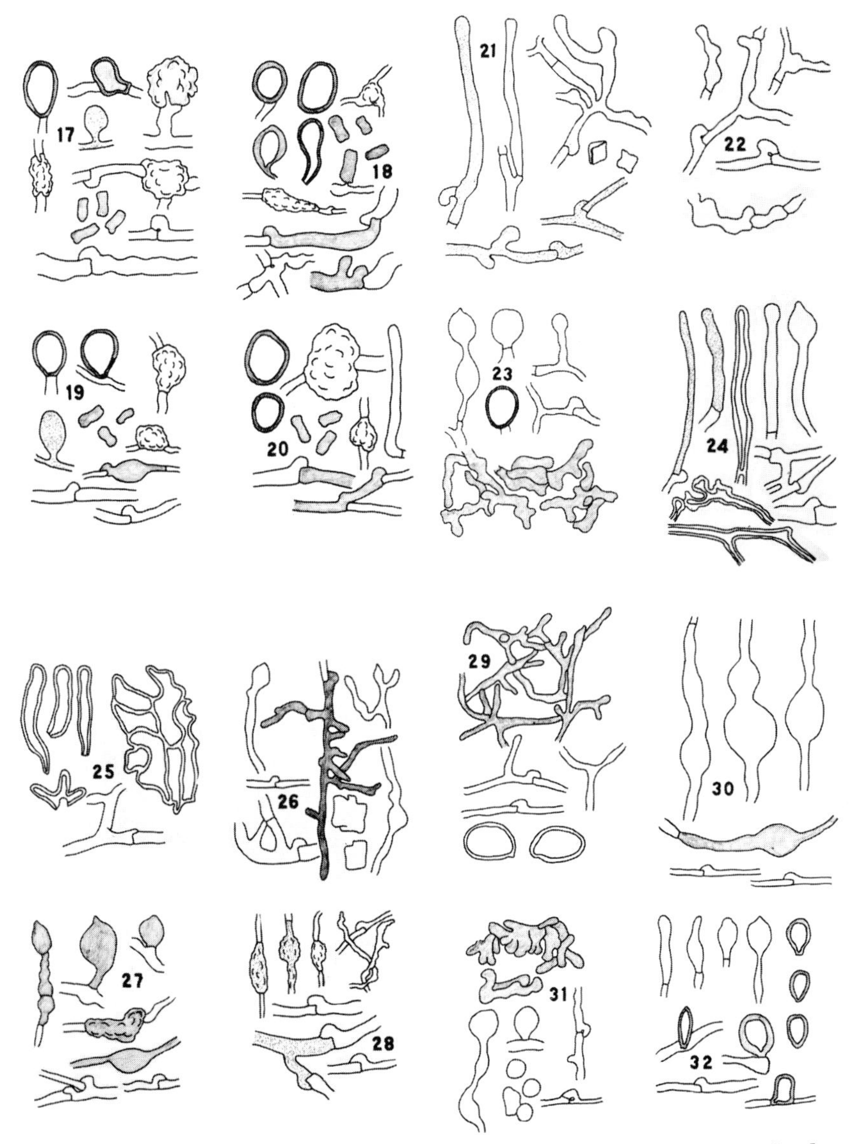


Figs. 17–20—*Pleurotus elongatipes*; figs. 17–18—Idaho; fig. 19—Maryland; fig. 20—New York; fig. 21—*Panus rudis*; fig. 22—*P. conchatus*; fig. 23—*P. tigrinus*; fig. 24—*P. fragilis*; fig. 25—*Panellus ringens*; fig. 26—*P. mitis*; fig. 27—*P. stipticus*; fig. 28—*P. patellaris*; fig. 29—*Collybia radicata*; fig. 30—*Tricholomopsis platyphylla*; fig. 31—*Oudemansiella mucida*; fig. 32—*Rhodotus palmatus*. (X700)

Plate X (Miller)

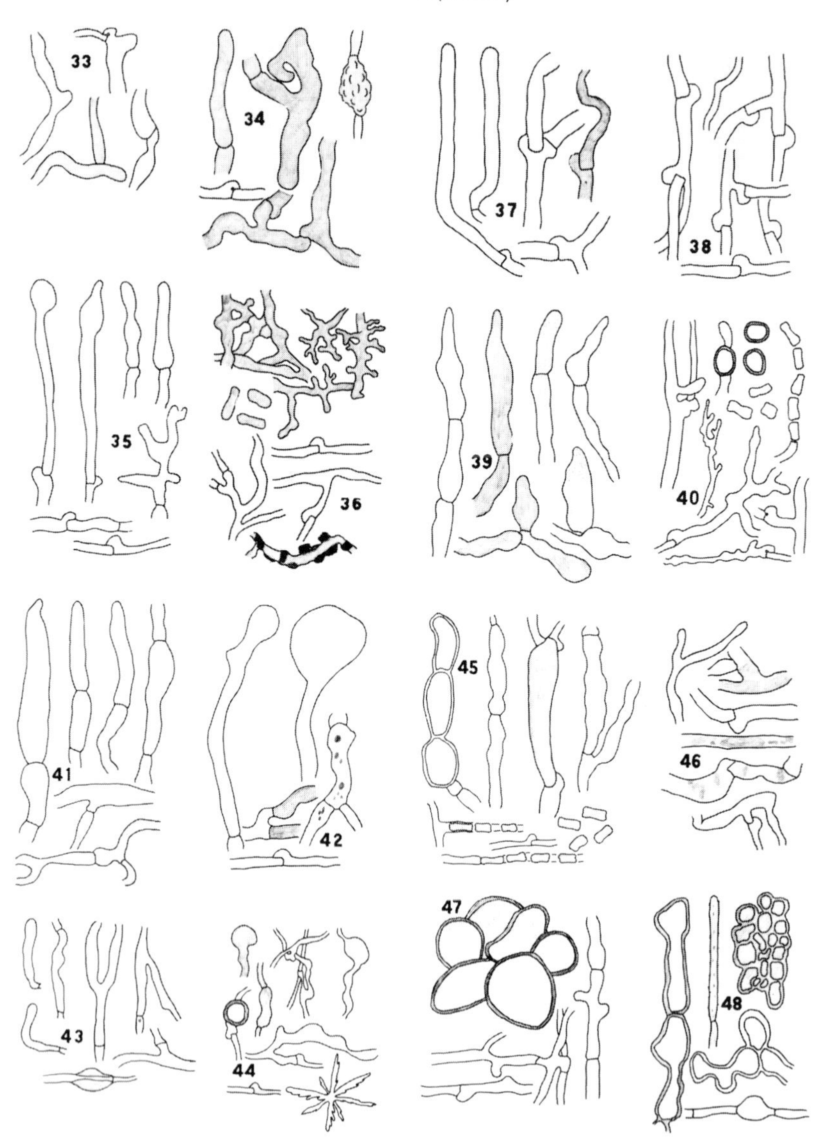


Fig. 33–*Clitocybe infundibuliformis*; fig. 34–*C. maxima*; fig. 35–*C. albirhiza*; fig. 36–*C. illudens*; fig. 37–*Collybia dryophila*; fig. 38–*C. conigenoides*; fig. 39–*C. albipilata*; fig. 40–*Flammulina velutipes*; fig. 41–*Tricholoma flavobrunneum*; fig. 42–*T. album*; fig. 43–*T. albobrunneum*; fig. 44–*T. portentosum*; fig. 45–*Armillaria albolinaripes*; fig. 46–*Panellus serotinus*; fig. 47–*Armillariella tabescens*; fig. 48–*A. mellea*. (X700)

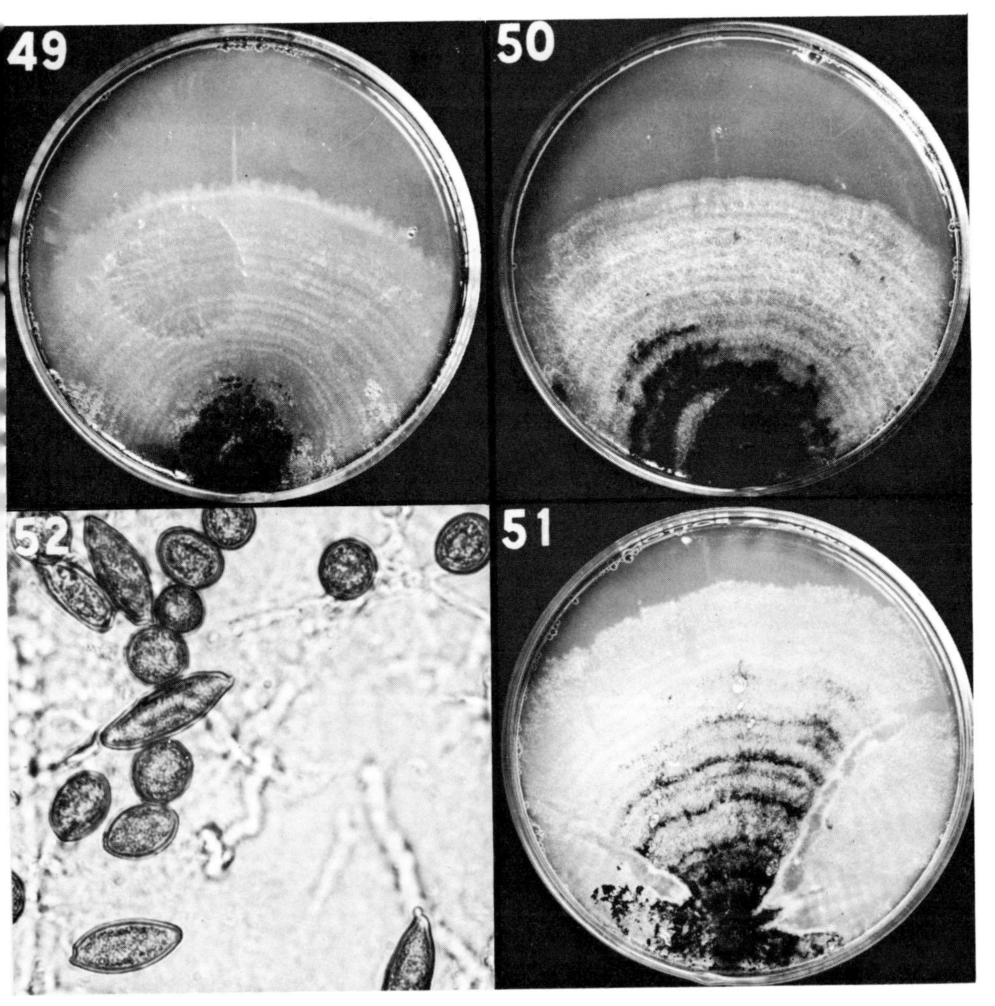


Figs. 49–52–*Pleurotus dryinus*; figs. 49–51–Growth on Petri plate after 49 days at 25° C.; fig. 49–OKM 4889, Idaho; fig. 50–RLG. 5107, New York; fig. 51–Lowe 10009, Arizona (X1); fig. 52–Thick-walled dark brown chlamydospores which are typically found in all cultures (X700).

Figs. 53–56—*Lentinellus cochleatus*; fig. 53—Fruiting bodies at 18° C. develop in Petri plate (X2); figs. 54, 56—Typical fruiting bodies which developed at 23° C. (X4); fig. 55—Typical fruiting body which developed at 18° C. (X4); fig. 57—*L. ursinus* fruiting body which developed at 25° C. (X5); fig. 58—*L. pilatii* coraloid fruiting body which developed under variable temperature conditions (X6).

PLATE XIII (Heim)

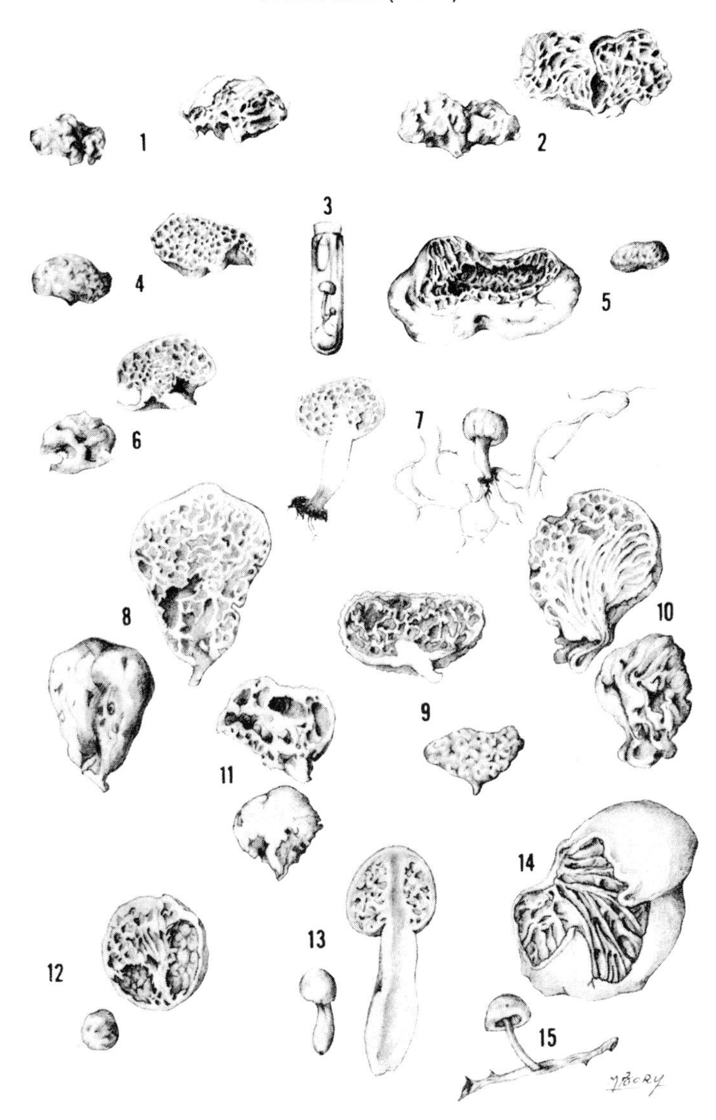


Variation in the fruiting bodies of *Le Ratia.* Figs. 1–3, *L. smaragdina*; figs. 4–13, *L. similis*; figs. 14, 15, *L. atrovirens.* Figs. 4, 6, 13 represent fruiting bodies preserved in liquid; figs. 1, 2, 5, 14 represent dried specimens (all reduced to 1/10). Figs. 1, 2, 4, 6, 8, 11 are enlarged 6:5; figs. 7, 12, 13 X2.7; and 5, 15 X3.6 (all but 7 and 15 are longitudinal sections). Fig. 3 represents two primordia produced in artificial culture (reduced to 1/10).